Veterinary Genetics

Veterinary Genetics

F. W. Nicholas

Associate Professor, School of Animal Husbandry,
University of Sydney

CLARENDON PRESS · OXFORD

Oxford University Press, Walton Street, Oxford OX2 6DP

Oxford New York Toronto
Delhi Bombay Calcutta Madras Karachi
Petaling Jaya Singapore Hong Kong Tokyo
Nairobi Dar es Salaam Cape Town
Melbourne Auckland

and associated companies in
Berlin Ibadan

Oxford is a trade mark of Oxford University Press

Published in the United States
by Oxford University Press, New York

First published 1987
Reprinted (with corrections) 1988, 1991

British Library Cataloguing in Publication Data
Nicholas, F. W.
Veterinary genetics.
1. Animal genetics
I. Title
575.1'024636   QH432
ISBN 0-19-857569-6

Library of Congress Cataloging in Publication Data
Nicholas, F. W.
Veterinary genetics.
Bibliography: p.
Includes index.
1. Veterinary genetics.   I. Title.
SF756.5.N53   1987   636.08'96042   85-8928
ISBN 0-19-857569-6

Printed in Great Britain
  by Bookcraft Ltd.
Midsomer Norton, Avon

Preface

Veterinary genetics encompasses those aspects of genetics that are relevant to animal diseases and to animal production. In writing this book, my intention has been to give equal coverage to both areas.

It is hoped that this book will be useful as a textbook for veterinary students and as a reference book for veterinarians engaged in practice and/or research. In addition, many chapters are directly relevant to non-veterinarians involved with animal production. All chapters in Parts I and III, for example, are certainly relevant to such people. Furthermore, topics such as segregation analysis, immunogenetics, and the evolution of resistance to chemicals in parasites and in pathogens are becoming increasingly important to non-veterinarians involved in animal production. Consequently, many of the chapters in Part II should also be of interest to a wider audience.

A significant portion of veterinary genetics involves population and quantitative genetics, neither of which can be explained satisfactorily without the aid of simple mathematics and statistics. So as to render the actual text less alarming to those who do not feel at home with either of these disciplines, most of the mathematics and statistics have been presented in appendices at the end of the relevant chapters. Besides clearing the text of much algebra, this practice has the added virtue of assembling important derivations and examples in compact, independent and readily accessible units.

A few features common to all chapters should be mentioned. When first encountered in the text, a new term appears in italics. This indicates that a definition or at least an explanation should be found somewhere nearby. From time to time, whole clauses or sentences are in boxes; these represent a significant statement or conclusion. The most important of these are assembled together at the end of each chapter, in the form of a chapter summary. Following the summary, an annotated list of key references is given, with the aim of providing readers with some idea as to where they should first look for further information. Readers wishing to gain access to the latest information on particular topics would be well advised to search through the relevant abstracting journals, which in the case of veterinary genetics are *Animal Breeding Abstracts*, *Index Veterinarius*, and *Veterinary Bulletin*.

No attempt has been made in this book to review thoroughly the literature on inherited diseases; the choice of one particular example to illustrate a point has often meant the omission of several other equally relevant examples. Consequently, readers requiring specific information about a particular defect or disease may find that it is not mentioned in the text. As an aid to such readers, a list of references that contain details on particular defects and diseases is given in the Introduction to Part II. Also, a list of monographs dealing with the genetics of particular species is presented just prior to the complete list of references at the end of this book.

Sydney   F. W. N.
May 1986

Acknowledgements

This book was commenced during a fruitful period of study leave spent in the Genetics Laboratory, University of Oxford. I wish to thank Professor Walter Bodmer for his substantial assistance at that time.

Friends and colleagues have been very helpful and tolerant at all stages of this project. For reading and criticizing drafts of various chapters, I wish to thank Kevin Bell, Walter Bodmer, Keith Brown, Rex Butterfield, Henry Collins, Brian Farrow, Peter Healy, Ian Hughes, Bob Jolly, Daria Love, Peter Outteridge, Laurie Piper, Grant Poolman, Mike Stear, Adam Torkamanzehi, and Greg Willis. Particular thanks are due to Keith Hammond, who read the whole of Part III, and to John James and Chris Moran, who read the whole book. John James also deserves thanks for providing me with the definition of additive relationship that appears in Chapter 13.

In addition, I wish to thank Merrilee Baglin and Jan Graham for help in many different ways; Jan Rowe for working beyond the normal call of duty in typing almost all of the manuscript, and in helping with the proofs; Stephen Brown for performing the calculations required for a number of illustrations, and for developing and running the software for the handling of references; Edna Federer, Peta Madgwick, and Melinda Shorten for help with references and permissions; John Roberts for the artwork; and Mike Stear and Rob Wilson for tolerating the delays in other jobs.

The passage of this book through its final stages has been greatly eased by the excellent cooperation and assistance received from the staff of OUP in Oxford.

I wish to pay a special tribute to Chris Moran, who has willingly and happily involved himself in the development of my ideas about the teaching of genetics to veterinarians, and who will have to live with this book in the future.

Finally, I thank my wife Jan, not only for her considerable help with proof-reading and indexing, but also for persevering during the whole of this undertaking.

To my parents

Contents

Part I
Genetics

1
Basic genetics

1.1 Introduction

Some readers of this book will already have a sound knowledge of basic genetics obtained from biology courses at school and/or university. Some will have attended biochemistry courses that included a good genetics component. Other readers, however, will have attended university sufficiently long ago to have been taught very little of the genetics that their children are now learning at school. Most importantly, all readers are likely to need an easily available review of basic genetics when reading the rest of this book. This chapter provides such a review. It concentrates on the general principles of genetics that apply to normal, healthy animals. The exceptions to these principles are often the basis of genetic diseases, which will be discussed in subsequent chapters.

Some of the important and basic concepts of genetics will be introduced by considering sex determination in higher animals. The advantages of using sex determination for this purpose are several: it provides a logical introduction to chromosomes, to meiosis and to simple or Mendelian inheritance, and it clearly illustrates the phenomenon of chance variation to which we will return many times throughout this book.

1.2 Sex determination

For many thousands of years, humans have observed two phenomena in relation to sex determination in higher animals: first, that there is considerable variation in the numbers of each sex among the offspring of pairs of parents; and second, that despite this variation, the overall numbers of males and females across families are approximately equal. While these two observations have been common knowledge for a long time, and despite many attempts to explain them down through the ages, it is only recently that the biological basis of sex determination has been reasonably understood. The major breakthrough occurred as soon as chromosomes could be examined under a microscope.

1.2.1 Chromosomes

When cells are treated in the manner outlined in Table 1.1 and are then viewed under a light microscope, structures called *chromosomes* become clearly visible. They are scattered randomly within clusters, and each

Table 1.1. The steps involved in preparation of cells for the viewing of chromosomes

---

1. Obtain rapidly dividing cells from tissues such as bone marrow, or stimulate other cells such as peripheral blood leucocytes or skin cells to divide by culturing them in an appropriate medium for several days.

2. Stop cell division by adding colchicine.

3. Treat cells with hypotonic solution so as to swell them.

4. Fix the cells with a mixture of methanol and glacial acetic acid.

5. Drop a sample of cell suspension on a microscope slide.

6. Stain.

For detailed descriptions of techniques for each domestic species, see Hare and Singh (1979).

cluster contains all the chromosomes from just one cell. In order to study chromosomes more closely, a suitable cluster is chosen and photographed, as shown in Fig. 1.1. From the photographic print, all chromosomes are cut out individually with scissors, and arranged in order of size on a sheet of paper. An arrangement such as this provides a picture of the complete chromosome complement or *karyotype* of a cell (Fig. 1.2). If many such arrangements are examined from normal, healthy individuals of both sexes of any species of mammal or bird, then two facts become evident.

> *Each species has a characteristic karyotype, and within any species, each sex has a characteristic karyotype.*

Karyotypes of different species differ in the shape, size and number of their chromosomes. Within any species, all the chromosomes occur in pairs. In individuals of one particular sex, both members of each chromosome pair have the same size and shape. In the other sex, all but two chromosomes occur in such pairs, with the remaining pair consisting of two chromosomes of different size and shape. In this unequal pair, one chromosome has the same shape and size as members of one of the pairs in the opposite sex.

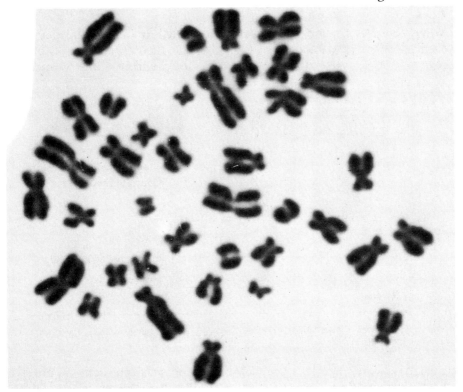

Fig. 1.1. The chromosomes of a cat, as seen through a light microscope. The treatment outlined in Table 1.1 has caught cells midway through the process of duplicating themselves. This process is called mitosis and is described in Section 1.2.4. Each unit in this figure consists of two rod-like structures joined together at a constricted point. Each rod-like structure is a *chromatid*, and the constriction is a *centromere*. The two chromatids have just been formed from one original *chromosome*. If the cell division were allowed to proceed, the centromere would split and each separate chromatid would then be called a new chromosome. For convenience, we talk of each pair of chromatids joined at the centromere as being just one chromosome, referring in fact to the chromosome that has just given rise to them. (Courtesy P. Muir.)

The difference in karyotype between the two sexes is the key to sex determination. In mammals, the two chromosomes that form the unequal pair occur in males and are called the X and the Y chromosomes. One of the pairs of chromosomes in the cells of female mammals consists of two X chromosomes. Thus in mammals, males are XY and females are XX. The X and the Y chromosomes are known as *sex chromosomes* and are illustrated in Fig. 1.3. In birds, the sex chromosomes are given different names, and their relationship to sex is the opposite to that in mammals; male birds are ZZ and female birds are

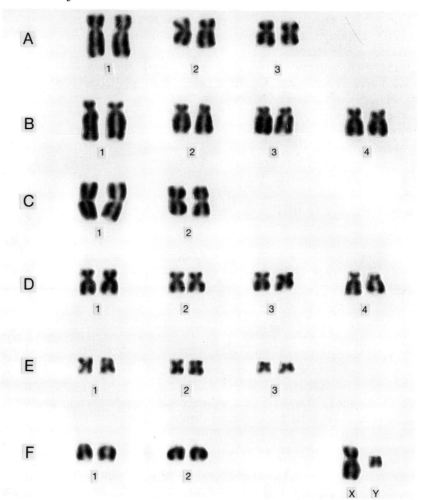

Fig. 1.2. The karyotype of a cat as obtained by rearranging individual chromosomes cut out from a photographic print of Fig. 1.1. The pairs of autosomes are arranged and numbered in groups, according to an internationally-agreed convention described by Ford *et al.* (1980). (Courtesy P. Muir.)

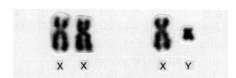

Fig. 1.3. The sex chromosomes of a female (XX) cat and a male (XY) cat. (Courtesy P. Muir.)

ZW. For convenience we shall refer only to mammals in the following discussion, although all statements apply equally to birds if the names of the sexes are reversed.

All the chromosomes in a cell apart from the sex chromosomes are called *autosomes*. Within any species, males and females have the same set of autosomes, occurring in pairs. The sex chromosomes plus the autosomes constitute a *genome*, which is the total set of chromosomes in a cell. Genomes in which chromosomes occur in pairs are said to be *diploid*, and the two members of a pair are called *homologues*. In order to emphasize that chromosomes occur in pairs, the total number of chromosomes is called the 2n number, where n is the number of pairs. For example, the number of chromosomes in the karyotype illustrated in Fig. 1.2 is 2n = 38. To enable identification of each pair of chromosomes in a karyotype, the autosome pairs are labelled according to an internationally-agreed convention, as can be seen in Fig. 1.2. The two sex chromosomes are usually placed at the end, and are labelled individually with the appropriate symbol. More sophisticated methods of identifying individual chromosomes are described in Chapter 4.

As noted above, the difference in sex chromosomes between the two sexes is the key to sex determination. The reason why XX individuals are females and XY individuals are males is explained in Chapter 4. For the present, we shall ask simply why is there so much variation in the numbers of XX and XY individuals in the offspring of pairs of parents, and yet at the same time approximately equal numbers of each sex overall? The answer lies in an understanding of the process of gamete formation.

1.2.2 Meiosis

Meiosis is the process of *gamete* formation in which sperm cells are formed in testes of males and egg cells are formed in ovaries of females. The main result of meiosis is that each sperm and each egg contains one member of each pair of chromosomes. Containing exactly one half of the usual diploid number of chromosomes, gametes are said to be *haploid*. The union of a sperm with an egg at fertilization produces a fertilized egg with the usual diploid number of chromosomes.

The process of meiosis commences with a normal cell containing the usual diploid set of chromosomes. To make the explanation easier, we shall consider what happens to just one pair of chromosomes (the sex chromosomes) in one sex (females), as illustrated in Fig. 1.4. In order to distinguish the two X chromosomes in females, we shall refer to them as X_p (paternal: originating from the father) and X_m (maternal: originating from the mother).

Meiosis occurs in two stages. Meiosis I begins with each chromosome duplicating itself, giving rise to two replicate *chromatids* joined at a

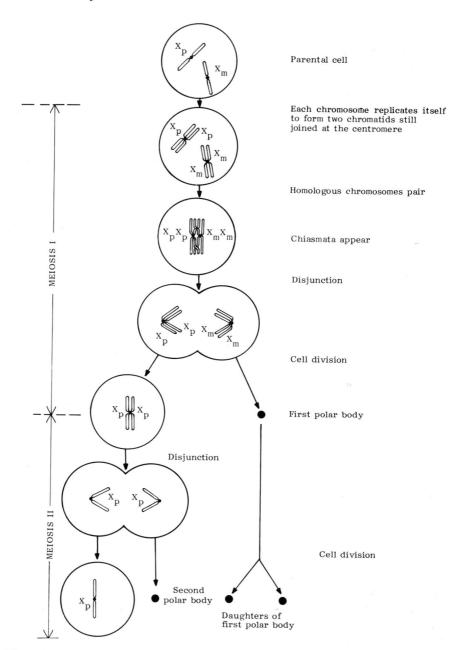

Parental cell

Each chromosome replicates itself
to form two chromatids still
joined at the centromere

Homologous chromosomes pair

Chiasmata appear

Disjunction

Cell division

First polar body

Disjunction

Cell division

Second
polar body

Daughters of
first polar body

Fig. 1.4. A summary of meiosis in a female, in which one diploid parental cell gives
rise to a haploid gamete, a first polar body and a second polar body. Sometimes the
first polar body divides into two. The processes that are shown to occur here for
the sex chromosomes also occur for all pairs of autosomes. Subscripts p and m indicate paternal and maternal origin respectively.

constricted point called the *centromere*. Homologous chromosomes, in our case X_p and X_m, line up next to each other in the centre of the cell, in a process known as *pairing* or *synapsis*. Because each chromosome has already replicated itself into two chromatids, there are now four chromatids side by side in the cell; two X_p chromatids and two X_m chromatids. The two X_p chromatids are still joined at their centromere, as are the two X_m chromatids. At this stage, cross-like structures joining homologous chromatids become visible. Known as *chiasmata*, these structures reflect a process called *recombination* or *crossing-over*, in which homologous chromatids each break at the same site and, in the process of re-uniting, exchange segments. In order to simplify the present discussion, we shall continue to refer to the chromatids simply as X_p or X_m, realizing that, as a result of crossing-over, any one chromatid may in fact consist of parts of both X_p and X_m. (For a full discussion of the genetic implications of crossing-over, see Section 1.7.) In the next stage of meiosis I, the two centromeres are pulled to opposite ends or *poles* of the cell, with the result that the two X_p chromatids move to one pole of the cell and the two X_m chromatids move to the other pole. Since this process involves the two pairs of chromatids disjoining from their previous paired arrangement, it is known as *disjunction*. In the final stage of meiosis I, the original cell divides into two daughter cells, one containing the two X_p chromatids still joined at the centromere, and the other containing the two X_m chromatids, also still joined together at their centromere. Following disjunction in females, only one daughter cell continues to function normally; the other degenerates into an inactive dark-staining structure known as the *first polar body*.

It is entirely a matter of chance as to which of the daughter cells remains functional. Consequently, there is an equal chance of either the two X_p chromatids or the two X_m chromatids ending up in the functional daughter cell. (In Fig. 1.4, it happens to be the X_p chromatids that have survived.)

In meiosis II, the two chromatids in the functional cell move apart (disjoin) and the cell divides into two daughter cells, with each containing one chromatid which is now called a chromosome. Once again, only one of the two daughter cells remains functional; the other degenerates into the *second polar body*. And once again, it is entirely a matter of chance as to which of these two daughter cells becomes the second polar body.

It is evident that in females, only one functional gamete results from each cell that originally underwent meiosis. It is also obvious that, irrespective of which daughter cell ultimately remains functional, all gametes produced by females are the same in the sense that each contains one X chromosome. For this reason, females are known as the *homogametic* sex.

In males, meiosis is basically the same as described above: a disjunction followed by a cell division in meiosis I and in meiosis II. There are, however, two important differences. The first difference is that the X and the Y chromosomes have only a small homologous region at the end of one arm, and so do not synapse as other chromosome pairs do. Instead, they usually join together end-to-end, as shown in Fig. 1.5. Their subsequent disjunction is normal, however, and gives rise to two functional daughter cells at the end of meiosis I, one containing two X chromatids still joined at their centromere, and the other containing two Y chromatids still joined at their centromere. The second difference between meiosis in females and in males is that polar bodies are not formed in males. Instead, both of the daughter cells formed at the end of meiosis I undergo a cell division in meiosis II, as shown in Fig. 1.5, giving rise to four functional gametes (sperms), two of which contain an X chromosome and two of which contain a Y chromosome. Since males produce two different types of gametes, they are known as the *heterogametic* sex.

The synapsis or pairing of chromosomes that occurs during meiosis is facilitated by a protein structure called the *synaptonemal complex*, which 'zips' two homologous chromosomes together. Two synaptonemal complexes are shown in Fig. 1.6.

Having now produced the gametes, the next stage is fertilization which, genetically speaking, is largely a matter of chance.

1.2.3 *Chance and variation*

> *Fertilization involves the random choice of a male and a female gamete.*

Since all female gametes contain an X chromosome, the chance or probability of a female gamete containing an X chromosome is one. Males, on the other hand, produce equal numbers of X-bearing gametes and Y-bearing gametes. There is, therefore, a chance of $\frac{1}{2}$ that a particular sperm will contain an X and the same chance that it will contain a Y. It follows that the chance of obtaining an XY zygote is $1 \times \frac{1}{2}$ which equals $\frac{1}{2}$. Similarly, the chance of obtaining an XX zygote is $1 \times \frac{1}{2}$, or $\frac{1}{2}$. We can represent this situation using a common genetic device called a checkerboard, as shown in Fig. 1.7. In using a checkerboard, the proportion at the head of each column is multiplied by the proportion at the head of each row to give the expected proportions of offspring in the centre of the checkerboard.

We have now seen how meiosis enables the production of an expected equal proportion of each sex, which accounts for one of our original observations. How can we account for the second observation, concern-

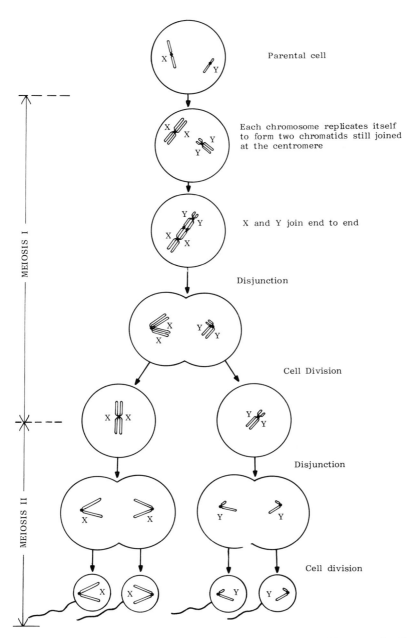

Parental cell

Each chromosome replicates itself to form two chromatids still joined at the centromere

X and Y join end to end

Disjunction

Cell Division

Disjunction

Cell division

Fig. 1.5. A summary of meiosis in a male, in which one diploid parental cell gives rise to four haploid sperms. With the exception of the end-to-end pairing, the processes that are shown to occur here for the X and Y chromosomes also occur for all pairs of autosomes.

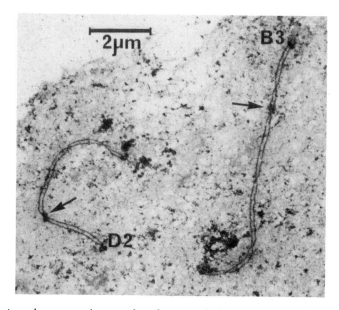

Fig. 1.6. An electron micrograph of two of the 19 synaptonemal complexes that occur in a meiotic cell of a cat, corresponding to two pairs of autosomes (B3 and D2; see Fig. 1.2). Each complex consists of two parallel *lateral elements* separated by a *central element*. In this micrograph, lateral elements are visible as thin lines, but chromosomes themselves are not visible. However, homologous chromosomes, which at this stage each consist of two tightly-interwoven replicate chromatids, are known to be located one on each side of the relevant complex, with each chromosome being associated with one lateral element. The numerous points at which a pair of lateral elements cross each other do not represent chiasmata, but are simply the result of the synaptonemal complex being twisted. Arrows indicate centromeres. (Courtesy of C. Gillies.)

		Male gametes	
		$\frac{1}{2}$ X	$\frac{1}{2}$ Y
Female gametes	all X	$\frac{1}{2}$ XX	$\frac{1}{2}$ XY

Fig. 1.7. A checkerboard illustrating the determination of sex in mammals.

ing the considerable variation in numbers of each sex among the offspring of different pairs of parents? There is just one fact that enables us to explain this variation.

> *Each fertilization is an independent event.*

By this we mean that irrespective of whether an X or Y bearing sperm is successful with a particular unfertilized egg, the result of that fertilization has no bearing on subsequent fertilizations, even if they occur at around the same time. For example, in a sow that ovulates 10 eggs, the chance that the last egg fertilized will give rise to a male is exactly $\frac{1}{2}$ irrespective of whether the 9 other eggs gave rise to 9 males, or to 9 females, or to any combination of males and females.

> *Because each fertilization is an independent event with just two possible outcomes, the expected number of either sex among any number of offspring follows the binomial distribution.*

This fact enables us to calculate the chance of obtaining a particular number of males and females in one or more families. Some examples of such calculations are given in Appendix 1.1. The answers are somewhat surprising to those who have not given much thought to probabilities. For example, the chance of obtaining an equal number of males and females in the offspring of four matings in a herd of cattle is 6/16 which is less than one half. Thus, the chance of observing exactly the numbers of each sex that you expect is surprisingly low. The reason for this is that there are quite a few other possible outcomes, each of which has a certain probability of occurring, which can be calculated using the formula in Appendix 1.1. We conclude that because each fertilization is an independent event, all possible outcomes in relation to the distribution of sex within a family can occur, from the one extreme of all males to the other extreme of all females; and the probability or chance of each outcome occurring can be calculated.

We have now provided adequate explanations for each of the observations described at the beginning of Section 1.2. In so doing, we have discussed chromosomes, meiosis, simple inheritance and chance, each of which is basic to an understanding of genetics. In order to complete the cycle of reproduction on which we embarked when discussing meiosis, we need to pass, by a process known as mitosis, from the zygote to an adult capable of producing its own gametes.

1.2.4 Mitosis

The growth of a single-celled zygote into a multicellular adult involves a mechanism whereby the number of cells can be expanded rapidly, while at the same time ensuring that each cell produced has exactly the same set of chromosomes as the original single-celled zygote. Mitosis is such a mechanism. For convenience we shall consider just two chromosomes (the sex chromosomes) in a male; the process is exactly the same for all chromosomes in both sexes. As shown in Fig. 1.8, mitosis begins when each chromosome duplicates itself to

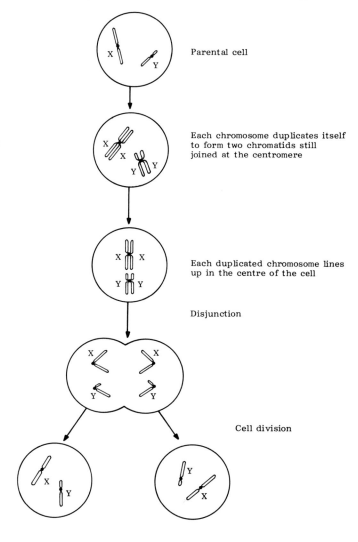

Fig. 1.8. Mitosis in a male. The process is exactly the same for all chromosomes, and in all cells of either sex.

form two chromatids still joined at the centromere. Each duplicated chromosome moves to the centre of the cell but does not, as in meiosis, synapse with its homologue. This stage, which is known as *metaphase*, is the one at which chromosomes are viewed. Karyotypes, therefore, are of metaphase chromosomes. After metaphase, the centromere splits and the chromatids separate (disjoin), one going to each pole of the cell. A constriction forms in the centre of the cell and two daughter cells are formed, each containing both an X and a Y. In this way, the two daughter cells have exactly the same set of chromosomes as did the original cell.

In both meiosis and mitosis, we have seen that chromosomes are capable of duplicating themselves. How do they achieve this? Fortunately, we now have sufficient knowledge of the biochemical nature of chromosomes to be able to answer this question and to explain a few other processes as well.

1.3   The biochemistry of inheritance

Chemically, chromosomes consist of mostly deoxyribonucleic acid (DNA) with a small amount of protein called histone. The latter has a binding and structural function while the former constitutes the genetic information that is passed from parent cell to daughter cell, and from one generation to the next.

1.3.1   DNA

While the chemical composition of DNA had been known for quite some time, it was not until 1953 that its structure was first discovered, by Watson and Crick. In 1962, the Nobel Prize was awarded to Watson and Crick, in recognition of their discovery, and also to Wilkins, who provided much of the X-ray diffraction data that led to the discovery.

A molecule of DNA consists of two strands joined very specifically together. Each strand consists of a linear arrangement of *nucleotides*. All nucleotides of DNA contain an identical sugar molecule (deoxyribose) and an identical phosphate group, but their third component, a nitrogenous base, exists in four different forms (Adenine: A; Guanine: G; Thymine: T; Cytosine: C) giving rise to four different nucleotides, as illustrated in Fig. 1.9. A strand of nucleotides is held together by covalent bonding between the $5'$ phosphate of one nucleotide and the $3'$ OH of the adjacent nucleotide, as shown in Fig. 1.10. The two strands that constitute a DNA molecule are held together by very specific hydrogen bonding of A with T, and G with C (Fig. 1.11). The result of this is that one strand of a DNA molecule is complementary to the other; the sequence of bases in one strand can be predicted from the base sequence in the other strand. The base sequence of a section of two complementary strands of DNA is shown in Fig. 1.12.

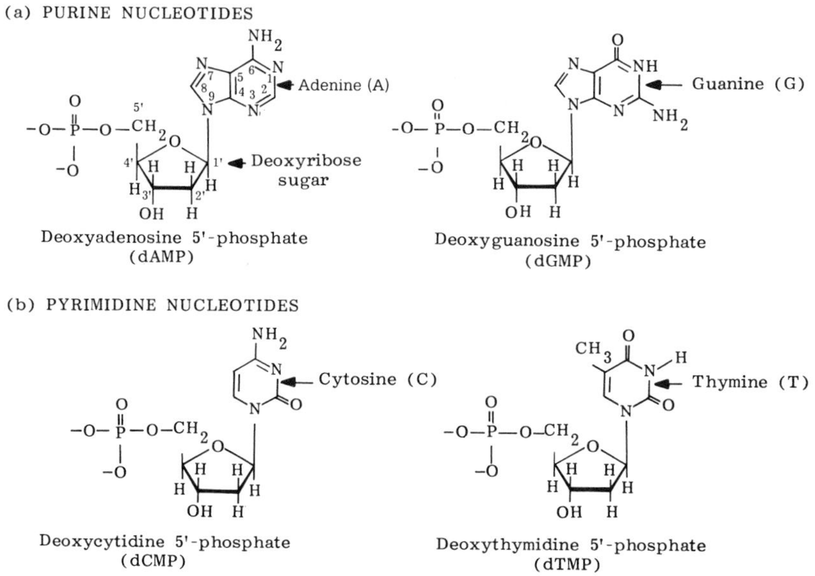

(a) PURINE NUCLEOTIDES

Deoxyadenosine 5'-phosphate
(dAMP)

Deoxyguanosine 5'-phosphate
(dGMP)

(b) PYRIMIDINE NUCLEOTIDES

Deoxycytidine 5'-phosphate
(dCMP)

Deoxythymidine 5'-phosphate
(dTMP)

Fig. 1.9. The chemical structure of the four different nucleotides that are the building blocks of DNA. (a) The bases adenine and guanine have a similar structure and are called *purines*. (b) Cytosine and thymine are called *pyrimidines*. Notice that the phosophate group is attached to the 5' carbon atom of the sugar molecule, and that an OH group is attached to the 3' carbon atom of the same molecule; these are the structures by which nucleotides bond together covalently, forming chains, as shown in Fig. 1.10. (From *An introduction to genetic analysis* (2nd edn) by Suzuki, D. T., Griffiths, A. J. F., and Lewontin, R. C. Copyright © 1981 by W. H. Freeman and Company. All rights reserved.)

A further consequence of these pairing arrangements is that the two strands occur together in a spiral or helix. Because two strands are involved, it is known as a *double helix* (Fig. 1.13).

The most important aspect of the Watson–Crick model of DNA is that it immediately suggests a mechanism for self-replication. If the double helix begins to unwind from one end with the two strands separating, then appropriate nucleotides already present in the cell will be attracted to the now unpaired bases of each strand, forming a new and complementary strand for each of the original strands. Thus, as the unwinding proceeds (Fig. 1.14), two double helixes are produced from one original double helix; a DNA molecule has replicated itself. Several enzymes are involved in DNA replication, but two of the main ones are *DNA polymerase* and *DNA ligase* (Fig. 1.14).

It remains now to relate our knowledge of the structure of a DNA molecule to the structure of a chromosome as seen through a microscope. Unfortunately, our understanding of this relationship is very incomplete. We do know that if the total DNA in a mammalian cell were to be formed into one double-stranded molecule, it would be

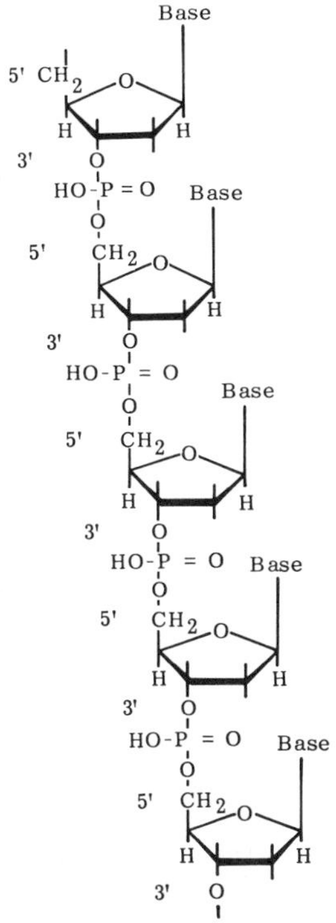

Fig. 1.10. The basic structure of a strand or chain of DNA, showing how the *sugar–phosphate backbone* is formed by phosphate linkages between the 3′ and 5′ carbons of adjacent sugar molecules. Note that any DNA chain has a 3′ carbon free at one end (called the *3′ end*) and a 5′ carbon free at the other (called the *5′ end*). (After Symons 1981.)

174 cm long, which is more than 7000 times longer than the total length of metaphase chromosomes viewed through a microscope! Obviously, therefore, a chromosome is composed of DNA that is very tightly folded or coiled. This raises the question as to how such a tight coil is unwound each time a chromosome replicates itself. The histone proteins are certainly involved in chromosome replication, but the actual mechanism is not yet known. Despite this gap in our knowledge, it seems certain that the Watson–Crick model of DNA structure is the key to chromosome replication.

The Watson–Crick DNA structure is also the key to an understanding of the way in which genetic information is stored in the chromosomes, and is transmitted to the cell in such a way as to produce a particular effect. In fact, the sequence of bases in a DNA molecule has a very specific meaning recorded in the form of a code.

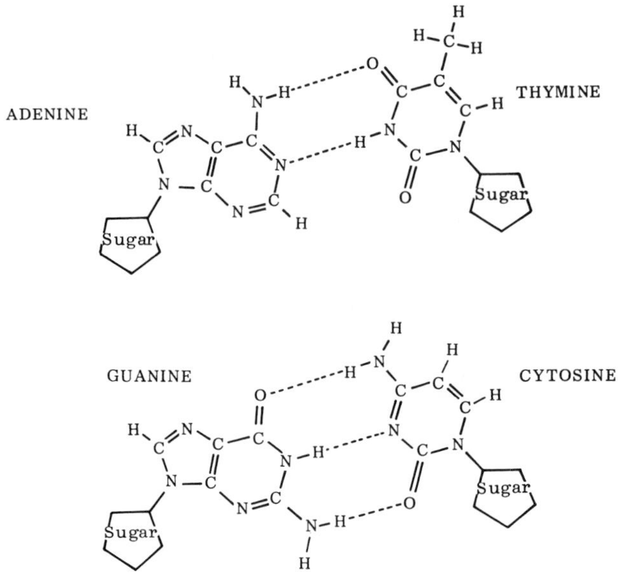

Fig. 1.11. The two types of *base pairs* (bp) that hold two strands of DNA together. Dotted lines represent hydrogen bonds. The lengths of double-stranded DNA is usually measured in terms of the number of base pairs or the number of *kilobase pairs* (kbp) where kilo indicates 1000. Similarly, the length of single-stranded DNA is measured in bases (b) or in kilobases (kb). These latter terms are sometimes used in relation to double-stranded DNA; the meaning is the same with either set of terms. (After Symons 1981.)

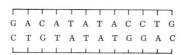

Fig. 1.12. The base sequence of two complementary strands of a segment of a molecule of DNA. Note that A (adenine) always pairs with T (thymine), and G (guanine) always pairs with C (cytosine).

1.3.2 The genetic code

Proteins are chemical compounds with a wide range of specific roles in various cells; they are involved in transport (e.g. haemoglobin), support (e.g. collagen), and in the immune response system (e.g. antibodies). In addition, there are many proteins that are enzymes which catalyse the innumerable biochemical reactions that occur in living cells. Finally, the commercial products commonly obtained from animals either consist almost solely of protein, e.g. meat and wool, or have protein as an important component, e.g. milk and eggs.

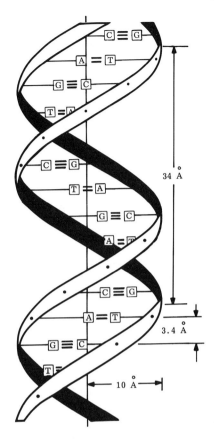

Fig. 1.13. A diagrammatic model of the Watson–Crick double helix. The two ribbons represent the sugar–phosphate backbones of the two DNA strands. The vertical line represents the central axis around which the two DNA strands wind. The structure repeats at intervals of 34 Å (one Å equals 10^{-10} metres), which corresponds to 10 base pairs. (After Symons 1981.)

Proteins consist of one or more polypeptides, each of which consists of a chain of amino acids. There are 20 different amino acids, and there are innumerable different combinations of some or all of the 20 amino acids in polypeptide chains of various lengths. Each particular polypeptide has a specific sequence of amino acids conferring upon it a specific set of physical and chemical properties.

The information necessary for the production of a specific sequence of amino acids is contained in code-form within the sequence of bases in a particular segment of a DNA molecule.

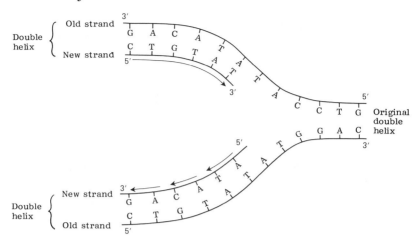

Fig. 1.14. Replication of DNA. Note that each new molecule of DNA consists of one old strand and one new strand. The formation of a new strand by the addition of nucleotides is accomplished with the aid of the enzyme DNA polymerase. However, this enzyme can add nucleotides only at the 3′ end of a chain, and hence replication can occur only in the 5′ → 3′ direction. Consequently, one new strand (top) is synthesized continually, while the other strand (bottom) is assembled in small segments which are each synthesized in the 5′ → 3′ direction. These segments in the second new strand are subsequently joined together by another enzyme called DNA ligase. The ability of the enzymes DNA polymerase and DNA ligase to perform the functions described above has now been put to good use in work with recombinant DNA (see Sections 2.3 and 2.4).

What type of code is required? In theory, the simplest code would involve letting one DNA base equal one amino acid. However, since there are 20 amino acids but only four bases, this would leave 16 amino acids unspecified. What if two adjacent bases coded for a particular amino acid? With four possible bases in each of two positions, this would provide 4 × 4 = 16 different combinations of bases, which is still insufficient to code for 20 amino acids. Not surprisingly, therefore, the genetic code is a *triplet* code, with most of the possible sequences of three adjacent bases coding for a particular amino acid. With 4 × 4 × 4 = 64 different possible triplets and only 20 amino acids, there is obviously some *redundancy*; in fact the first two bases of a triplet are often sufficient to specify a particular amino acid. Three triplets do not code for any amino acid and are known as *nonsense* triplets or *stop* triplets; they bring about the termination of a polypeptide chain. Another triplet (the one that codes for methionine) acts as a *start* signal for polypeptide synthesis. The complete genetic code is shown in Table 1.2. Equipped with the genetic code, we can now follow the processes involved in the synthesis of proteins.

Table 1.2. The genetic code

Codon†	Amino acid	Codon†	Amino acid
UUU	Phenylalanine	UAU	Tyrosine
UUC		UAC	
UUA		UAA	Stop
UUG		UAG	
CUU	Leucine	CAU	Histidine
CUC		CAC	
CUA		CAA	Glutamine
CUG		CAG	
AUU	Isoleucine	AAU	Asparagine
AUC		AAC	
AUA		AAA	Lysine
AUG	Methionine	AAG	
GUU	Valine	GAU	Aspartic acid
GUC		GAC	
GUA		GAA	Glutamic acid
GUG		GAG	
UCU	Serine	UGU	Cysteine
UCC		UGC	
UCA		UGA	Stop
UCG		UGG	Tryptophan
CCU	Proline	CGU	Arginine
CCC		CGC	
CCA		CGA	
CCG		CGG	
ACU	Threonine	AGU	Serine
ACC		AGC	
ACA		AGA	Arginine
ACG		AGG	
GCU	Alanine	GGU	Glycine
GCC		GGC	
GCA		GGA	
GCG		GGG	

† Following the usual convention, the triplets shown are actually mRNA codons, and are therefore described in terms of RNA bases, namely uracil (U), cytosine (C), adenine (A), and guanine (G).

1.3.3 Protein synthesis

The synthesis of polypeptides begins with the relevant segment of a DNA molecule unwinding and the two strands separating. The sequence of DNA bases in one of the strands (called the *template strand*) acts as a template for the synthesis of a different nucleic acid molecule. The DNA does this by attracting towards it appropriate nucleotides of a slightly different composition compared with those that are the components of DNA. Since these different nucleotides contain a ribose sugar rather than a deoxyribose sugar, the resultant chain of nucleotides is called ribonucleic acid or RNA. The enzyme that catalyses this process is called *RNA polymerase*. Three of the bases in RNA are the same as in DNA, and the fourth, uracil (U), occurs instead of thymine. The formation of a complementary strand of RNA on the DNA template completes the first stage of protein synthesis, known as *transcription* (Fig. 1.15). Before the next stage can commence, the RNA molecule has to move from the nucleus, where the chromosomes are, to structures called *ribosomes* in the cytoplasm, where polypeptide is synthesized. Because it is the above RNA molecule that carries the code between these two sites, it is called *messenger* RNA or mRNA. Its triplets are called *codons*.

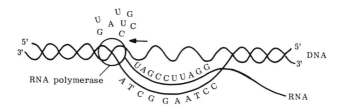

Fig. 1.15. Transcription, in which the enzyme RNA polymerase catalyses the synthesis of RNA from just one of the DNA strands (the template strand) in the double helix. The arrow indicates the direction of transcription, which is always from the 3′ to the 5′ end of the template strand of DNA. (From *An introduction to genetic analysis* (2nd edn) by Suzuki, D. T., Griffiths, A. J. F., and Lewontin, R. C. Copyright © 1981 by W. H. Freeman and Company. All rights reserved.)

The second stage of protein synthesis involves a second type of RNA known as *transfer* RNA or tRNA. For each of the 20 amino acids, there are one or more specific tRNA molecules which bind to the relevant amino acid and which have a nucleotide triplet (called an *anticodon*) that is complementary to the relevant mRNA codon. Being complementary, a tRNA anticodon is attracted towards the relevant mRNA codon, thus bringing the correct amino acid into position in the polypeptide chain. This second and final stage of protein synthesis is called *translation*. A summary of polypeptide synthesis is given in Fig. 1.16.

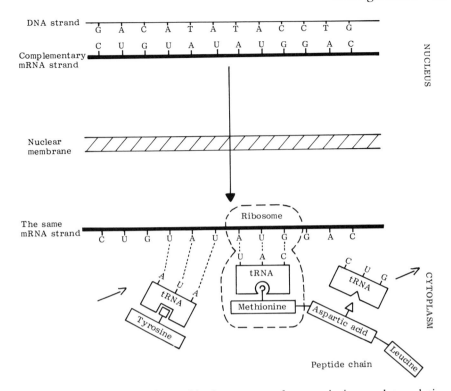

Fig. 1.16. Synthesis of polypeptide by means of transcription and translation. This diagram is strictly true for only a very limited range of polypeptides of *eukaryotes* (organisms whose cells have a separate nucleus). One such polypeptide is the protein histone. For most eukaryotic polypeptides, the mRNA strand undergoes a series of alterations (as shown in Fig. 2.14) before leaving the nucleus. If the nucleus were removed from the diagram, then it would be a true representation of polypeptide synthesis in *prokaryotes* (organisms whose cells lack a nucleus).

From the above account, it is evident that particular segments of DNA code for particular polypeptides.

> *A segment of DNA that includes all the nucleotides corresponding to all the amino acids in a particular polypeptide is called a structural gene.*

It is common for structural genes to be called simply 'genes', and we shall follow this practice. In Chapter 2, the meaning of 'gene' shall be expanded slightly to include nucleotide sequences on either side of the structural gene. Sometimes, the term 'gene' is used to describe segments of DNA that have as their sole function the production, via transcription, of either tRNA or ribosomal RNA (see Section 4.2.1). However, unless otherwise stated, the term 'gene' can be taken to mean 'structural gene'.

Until 1977, it was thought that each gene consisted of just sufficient nucleotides to code for its polypeptide. But it is now known that in higher organisms, most genes are split into sections, only some of which are represented in the mRNA that produces polypeptide. Sections that are represented in mRNA by the time it reaches the ribosomes are called *exons* (because they are the *ex*pressed regi*ons* of the gene). The remaining sections are called *introns* (because they were originally called *intr*agenic regi*ons*). Another name for intron is *intervening sequence*. The discovery of split genes is just one of the remarkable advances in knowledge that have resulted from the development of various techniques for the manipulation of DNA that are collectively referred to as *recombinant DNA technology* or *genetic engineering*. In Chapter 2 we shall review these techniques, and we shall also consider split genes in more detail.

1.3.4 Types of DNA

Despite the obvious importance of genes, not all DNA consists of genes. In fact, only a small proportion of total DNA consists of genes; probably less than 10 per cent and maybe as little as 1 per cent. What do we know about the rest of the DNA? This question is best answered by considering the different types of DNA that exist in a mammalian cell. Most of the examples in the following account will be drawn from cattle, which are the only domestic animals extensively studied in this regard.

The most common type of DNA consists of *unique* or *single-copy* sequences, which account for around 57 per cent of total DNA in cattle. These single-copy sequences are dispersed throughout the genome, occurring in segments with an average length of around 1000 bases. A small proportion of single-copy DNA accounts for most genes.

A second category of DNA is called *interspersed-repetitive DNA*. It consists of a small number of sequences having an average length of around 300 bases, with each sequence occurring up to 10^4 to 10^5 times in the genome. Sequences that are repeated are called *repeat units*. In the case of interspersed-repetitive DNA, these repeat units are interspersed among the segments of single-copy DNA, and are thus also dispersed throughout the genome. Interspersed-repetitive DNA accounts for about 20 per cent of bovine DNA.

The remaining DNA is called *satellite* DNA, which is DNA having a density sufficiently different from the major portion of DNA to enable it to form a separate peak or 'satellite' when the total DNA from various cells of an organism is subjected to *density gradient centrifugation* using caesium chloride or a similar compound. This process separates total DNA into portions according to the proportion of guanine (G) plus cytosine (C); because a G–C base-pair is heavier than an A–T pair, DNA that is particularly rich in G–C will travel further through a gradient of caesium chloride or a similar compound. Satellite DNA is *highly-repetitive*, by which we mean that it consists mostly of one or a small number of short repeat units, each occurring a very large number of times (10^6 or more). Unlike interspersed-repetitive DNA, satellite DNA is repeated tandemly, i.e. many copies of the same repeat unit are located side-by-side. Thus satellite DNA is not dispersed throughout the genome. Instead, it is located in regions of *constitutive heterochromatin*, which is the name given to dark-staining, condensed regions that occur at fixed sites in the genome. These sites include regions immediately adjacent to most centromeres, regions at the ends (*telomeres*) of some chromosomes, and certain other isolated regions.

Bovine DNA contains eight different satellites, which together constitute around 23 per cent of total bovine DNA. Three of these satellites have been studied in detail by Pech *et al.* (1979) and by Streeck (1981). Leaving aside certain complications, these three satellites can be said to contain approximately 7.4×10^6 copies of a 23-base repeat unit.

In summary, we conclude that:

Bovine DNA consists of approximately

1. *57 per cent single-copy DNA*
2. *20 per cent interspersed repetitive DNA*
3. *23 per cent satellite DNA.*

It is not possible to say how representative these figure are, because similar studies have not been conducted in other domestic mammals. We do know, however, that humans and mice have about the same proportion of single-copy DNA, but only 4 per cent and 10 per cent respectively of satellite DNA.

There is another category of DNA that should be mentioned. It is *fold-back* DNA, which occurs throughout the genome. The essential components of fold-back DNA are two short nucleotide sequences near to each other in the one DNA strand, with one sequence being complementary to the other, such that if the two strands of the DNA molecule become separated, then the two sequences on the one strand can pair with each other, resulting in the DNA strand folding back on itself. The two complementary sequences are called *inverted repeats*. The segment of DNA between the first base in the first sequence and the last base in its complementary sequence is called fold-back DNA. Using the inverted repeats ATCG and CGAT as an example, Fig. 1.17 illustrates the types of structures that can arise from fold-back DNA. It is thought that there are about 10^4 to 10^5 fold-back sequences in the mammalian genome, making up from 1 per cent to 5 per cent of total DNA. More than 60 per cent of inverted repeats are separated and

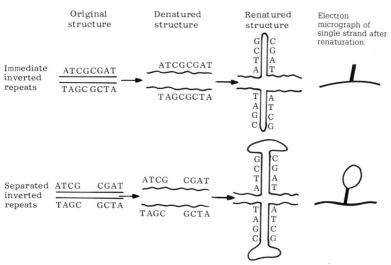

Fig. 1.17. Fold-back DNA. The original structure is a normal molecule of double-stranded DNA. The complementary sequences ATCG and CGAT are adjacent (top) or near to each other (bottom) in the same strand. Since the second sequence is really a repeat of the first sequence but in an inverted (complementary) form, the two complementary sequences are called *inverted repeats*. The two strands can be separated (*denatured*) by heating, which disrupts the hydrogen bonds between base pairs. If the solution of single-stranded DNA is slowly cooled, then bases pair again (*renaturation*). During renaturation, a sequence of bases will pair with the first complementary sequence it encounters. If DNA concentration is low, complementary sequences *within* a DNA strand are much more likely to pair, forming 'hairpins' (top right) or 'lollipops' (bottom right), than complementary sequences on different strands. Since A always pairs with T, and G with C, it follows that if there are two complementary sequences in one strand of DNA, then there must be the same two sequences in reverse order in the other strand, as shown above. Thus, in fold-back DNA, the base sequence of the two complementary sequences in the two strands considered together is the same whether read from left to right or from right to left. Such a sequence is said to show *dyadic symmetry*. By analogy with palindromes, which are words or sentences which read the same in either direction (ABLE WAS I ERE I SAW ELBA), the base sequence of the complementary sequences in fold-back DNA is called a *palindrome*, and fold-back DNA is sometimes called *palindromic* DNA. (After Lewin (1980). *Gene expression Vol. 2. Eukaryotic chromosomes* (2nd edn). Copyright 1980 John Wiley and Sons Ltd. Reproduced by kind permission of John Wiley and Sons Ltd.)

hence give rise to 'lollipops' rather than 'hairpins' (see Fig. 1.17). The average length of the loop of the lollipop is thought to be between 1000 base pairs (bp) and 3000 bp.

A segment of fold-back DNA in which the inverted repeats are separated is called an *insertion sequence* (IS). Specifically, an IS is a sequence of nucleotides that is not internally repetitive, flanked by inverted repeats. For example, Streeck (1981) discovered a bovine

IS which is 611 bases long, flanked at one end with the sequence GCCGGGGA, with its complement TCCCCGGC at the other end. This particular IS was discovered in one of the three bovine satellites discussed above, where it occurs 3.5×10^4 times. Little is known about insertion sequences in mammals, but it does appear as if a particular IS may occur many times throughout the genome, at more or less random positions, and often in different places in different individuals.

Insertion sequences have been known for some time in plants and in bacteria, where they have been shown to have the remarkable ability to move themselves from one part of the genome to another. Obviously, if an IS inserts itself in the middle of a functional gene, it could alter that gene's effect, or even inactivate it. Insertion sequences are one type of a group of DNA segments called *transposable genetic elements*, all of which have the ability to remove themselves from one particular site in a chromosome, and to insert themselves into some other site in the same or in a different chromosome. We shall discuss another type of transposable genetic element in Section 10.4.4, in relation to the rapid development of multiple antibiotic resistance in bacteria. In the meantime, we shall simply note that the discovery of insertion sequences in mammals is sufficiently recent for us to know very little at present about their nature and role in the mammalian genome.

1.4 Mutation

We have now seen how DNA replicates itself and how it gives rise to protein. Although the processes involved are remarkably elegant and usually operate faultlessly, mistakes do occur from time to time. Many mistakes have no effect at all, as they are corrected by the cell's own repair mechanisms. Uncorrected mistakes in DNA replication, however, result in an alteration in the DNA in at least one of the daughter cells. And because DNA replication is usually so faultless, the altered DNA will be passed on unchanged to all descendent cells; until the next mistake occurs.

> *Uncorrected mistakes in DNA replication are called gene or point mutations.*

They can involve the substitution of one nucleotide for another, or the addition or deletion of one or a few nucleotides.

In addition to gene or point mutations, there are also *chromosomal mutations*, which are changes in chromosome structure or number. Identifiable diseases resulting from gene mutations are discussed in Chapter 3, while defects associated with chromosomal mutations are the subject of Chapter 4.

If a mutation occurs in cells other than those that give rise to sex cells, then it is called a *somatic* mutation. The stage of development of the individual when a somatic mutation occurs will determine the total number of cells that contain the altered or mutant DNA; the earlier the mutation occurs, the larger the number of cells affected.

In contrast, a mutation that occurs in cells that will give rise to sex cells is known as a *germ-line* mutation, which may lead to the formation of a gamete that contains the altered DNA and/or chromosome. If this gamete is successful in fertilization, the mutation will be passed on to the resultant offspring, in every cell of whom it will be faithfully reproduced.

1.5   Genes, alleles, and loci

The different forms of a segment of DNA that can exist at a particular site in a chromosome are called *alleles*. The particular site or position of a gene in a chromosome is called the *locus* (plural loci). The word 'gene' is commonly used in the sense of either allele or locus. When used in this way, the appropriate meaning of the word is usually quite evident from its context.

If an offspring results from the union of a sperm with entirely normal DNA and an egg with an altered or mutant DNA segment in one of its chromosomes, then that offspring will have one normal chromosome and one mutant chromosome making up the pair of relevant homologues. More specifically, there will be one normal allele and one mutant allele at the relevant locus. We shall give these two alleles the symbols *B* and *b* respectively. Individuals with two different alleles at a particular locus are said to be *heterozygous* at that locus. In contrast, if an individual has two copies of the same allele then that individual is *homozygous*.

Although any one individual can have a maximum of only two different alleles at a locus, the number of different alleles in a population of individuals can be much greater than two. If more than two alleles exist in a population at a particular locus, then that locus is said to have *multiple alleles*. But irrespective of the number of alleles in a population, any diploid member of that population will either be heterozygous for any two of the alleles, or homozygous for any one.

1.6   Simple or Mendelian inheritance

> *The expected outcomes of particular matings involving homozygotes and/or heterozygotes can be easily predicted with the aid of a checkerboard.*

1.6.1 Single locus

Consider, for example, the mating of a heterozygote (*Bb*) with a homozygote (*bb*). This is exactly analogous to the situation with sex chromosomes, where the male is XY and the female is XX; the male produces two types of gametes in equal proportions and the female produces just one type of gamete. Consequently, the results of the mating *Bb* × *bb* can be represented in exactly the same way as that used for the inheritance of sex discussed earlier, with the aid of a checkerboard as shown in Fig. 1.18. It can be seen that the result of the mating *Bb* × *bb* is expected to be an equal proportion of *Bb* and *bb* offspring. The separation of alleles at a locus during meiosis is called *segregation*, and the ratio of different types of offspring resulting from the mating of particular parents is known as the *segregation ratio*. For the mating *Bb* × *bb*, the segregation ratio is 1*Bb* :1*bb*.

Gametes from
heterozygous parent

$\frac{1}{2}$ B $\frac{1}{2}$ b

Gametes
from   all *b* $\frac{1}{2}$ Bb $\frac{1}{2}$ bb
homozygous
parent

Fig. 1.18. A checkerboard illustrating the expected results of a mating between a heterozygote (*Bb*) and a homozygote (*bb*), for an autosomal locus.

A checkerboard can be used to predict the outcome of any particular mating involving a single locus. The segregation ratios expected from all possible types of mating with respect to a single locus are listed in Table 1.3.

Table 1.3. Segregation ratios expected from all possible types of mating in relation to a single autosomal locus, as obtained from a checkerboard

Type of mating	Segregation ratio among offspring				
	BB		*Bb*		*bb*
BB × *BB*	1	:	0	:	0
BB × *Bb*	1	:	1	:	0
BB × *bb*	0	:	1	:	0
Bb × *Bb*	1	:	2	:	1
Bb × *bb*	0	:	1	:	1
bb × *bb*	0	:	0	:	1

1.6.2 *More than one locus*

In the absence of any evidence to the contrary (see Section 1.7), it is assumed that segregation at one locus is independent of segregation at other loci. This was the assumption first made by Mendel, and it is true for many situations observed in higher animals today. If segregation at each locus is independent of segregation at other loci, then the combined chance of obtaining a gamete with a particular allele (say *B*) at the first locus and a particular allele (say *d*) at the second locus is simply the product of the probabilities associated with each allele independently. For example, if an individual is heterozygous at two loci (*BbDd*), then there are four possible types of gametes, *BD, Bd, bD,* and *bd,* which will be produced with equal frequency. The results of independent segregation at two loci are shown in Fig. 1.19.

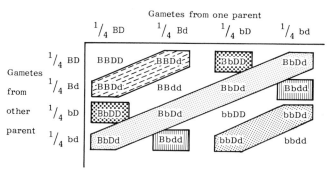

Fig. 1.19. A checkerboard illustrating the results of a mating between individuals that are both heterozygous at each of two loci. Each cell of the checkerboard, being the product of $\frac{1}{4}$ and $\frac{1}{4}$, occurs with a frequency of $\frac{1}{4} \times \frac{1}{4} = \frac{1}{16}$. Identical offspring are enclosed in the one shaded pattern.

Combining all cells of the checkerboard having identical offspring, it can be seen that the segregation ratio is:

1 *BBDD*:2 *BBDd*:1 *BBdd*
2 *BbDD*:4 *BbDd*:2 *Bbdd*
1 *bbDD*:2 *bbDd*:1 *bbdd*

Although checkerboards quickly become rather large, in principle they can be used to derive expected segregation ratios for any type of mating involving any number of independently segregating loci.

> *The passage of genes from one generation to the next is called inheritance, and the patterns of inheritance just described are called simple or Mendelian.*

More specifically, the above patterns of inheritance illustrate simple *autosomal* inheritance because they describe what happens in relation to loci on autosomes. Some loci, however, are on the X chromosome and consequently have different patterns of inheritance, as illustrated in Fig. 1.20 and as summarized in Table 1.4. Such loci are said to be *sex-linked* or *X-linked*, the latter being a better term because it is more specific. Very few loci have been identified on the Y chromosomes, but those that are located there are said to be *Y-linked*.

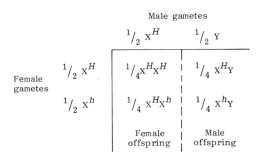

Fig. 1.20. A checkerboard illustrating the expected results of a mating between a heterozygous female (X^HX^h) and a particular male (X^HY), in relation to an X-linked locus. In the notation used for X-linkage, X and Y indicate chromosomes and the superscripts H and h refer to different alleles.

Table 1.4. Segregation ratios expected from all possible types of matings in relation to an X-linked locus, as obtained from a checkerboard.

Type of mating	Segregation ratio					
	Among females				Among males	
	X^HX^H	X^HX^b	X^bX^b		X^HY	X^bY
$X^HX^H \times X^HY$	1 :	0 :	0		1 :	0
$X^HX^b \times X^HY$	1 :	1 :	0		1 :	1
$X^bX^b \times X^HY$	0 :	1 :	0		0 :	1
$X^HX^H \times X^bY$	0 :	1 :	0		1 :	0
$X^HX^b \times X^bY$	0 :	1 :	1		1 :	1
$X^bX^b \times X^bY$	0 :	0 :	1		0 :	1

It is important that simple or Mendelian inheritance be clearly understood, as many important genetic diseases are simply inherited.

At the beginning of this section it was implied that sometimes good evidence is obtained to indicate that segregation at two or more loci is not completely independent. We shall now examine why this should be so.

1.7   Linkage

There are at least many thousands of different genes each coding for a different polypeptide. But there is only a relatively small number of chromosomes, being constant for a particular species and mostly ranging between 30 and 80 among different species of higher animals (see Chapter 4). Inevitably, therefore, each chromosome consists of many different genes, each of which has a specific position (locus) on that chromosome. If chromosomes were inherited as integral units, then for all the loci on a particular chromosome, the alleles present in that chromosome would always segregate together. Consider, for example, one chromosome containing allele B at one locus and allele D at another locus, and its homologue containing alleles b and d respectively. If chromosomes segregated as integral units, then only two types of gametes would result, namely BD and bd, with equal frequency. In contrast, four different gametes (BD, Bd, bD and bd) would be produced in equal frequency if the loci segregated independently.

In practice, chromosomes are not inherited as integral units. Instead, as described in Section 1.2.2, recombination or crossing-over occurs when homologous chromosomes are synapsed during the first stage of meiosis. During synapsis, breakages and rejoining of chromatids occur. If the two broken ends of a broken chromatid rejoin, then that chromatid will still be inherited as an integral unit. If, however, a break occurs in the same position in two adjacent chromatids, then sometimes the rejoining will bring together a segment from each chromatid into resultant cross-over or *recombinant* chromatids, as illustrated in Fig. 1.21. If the two chromatids are *sister chromatids* (from the one homologue, and therefore joined at the centromere), then the cross-over will have no effect, since sister chromatids are exact copies of each other. If, however, the two chromatids are *non-sister chromatids* (one from one homologue and one from the other), then the cross-over results in the reciprocal exchange of genes between homologous chromosomes. To the extent that breakages occur more or less randomly along the length of each chromosome, there is a direct relationship between the physical distance separating two loci on the one chromosome, and the number of cross-overs between them. Unfortunately, this number cannot be directly measured. What can be measured in certain matings is the *recombination fraction*, which is the proportion of gametes from one parent that can only have resulted from crossing-over during meiosis in that parent. The relationship between the recombination fraction and the distance between two loci enables the construction of *linkage maps*, in which loci are positioned according to the recombination fraction between them.

If two loci are very close together on the same chromosome, then the observed recombination fraction is quite low and the loci are said to be

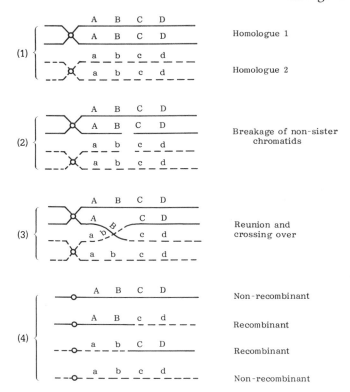

Fig. 1.21. The four stages involved in crossing-over between a pair of homologous chromosomes. The exchange of chromatid segments results in an exchange of genes at particular loci.

tightly linked. For example, in Chapter 8, we shall discuss some loci with recombination fractions between them of less than 1 per cent. Loci further apart on the same chromosome have more cross-overs between them and hence have a higher recombination fraction. For loci that are far apart on the same chromosome, there is a sufficient number of cross-overs for recombinant gametes to be just as frequent as non-recombinant ones, which gives the maximum value of recombination fraction of 50 per cent. Loci that are sufficiently far apart on the same chromosome to have a recombination fraction of 50 per cent are said to be effectively *unlinked* even though they actually belong to the one chromosome. They are said to be effectively unlinked because they segregate independently, as if they were on different chromosomes.

By observing the recombination fractions among many pairs of loci within a species, groups of linked loci become evident. Since the loci within each group are linked, they must be located on the same chromosome. It follows, therefore, that if enough loci have been identified in a species, the number of *linkage groups* will equal the number of pairs of chromosomes. In most species, however, an insufficient number of loci are known for this relationship to be observed.

The traditional method of constructing linkage maps involves observing the products of sexual reproduction. An illustration of the use of this method in calculating recombination fraction is given in Section 3.4. For a more sophisticated linkage study involving unplanned matings in cattle, see Hines *et al.* (1981). Identification of the chromosome corresponding to a particular linkage group gives rise to a *genetic map*. Such maps can be constructed with the aid of various chromosomal rearrangements, as described by Shoffner (1981).

Although the sexual method of constructing linkages and genetic maps is still widely used, a 'parasexual' method has been exploited recently to add considerably to our knowledge of linkage groups. The method involves the use of *somatic cell hybridization*, which is the fusion in the laboratory of two different populations of somatic cells. In linkage studies, the cells come from two different species, and at least one of the cell populations is a self-perpetuating *clone*, where the word 'clone' indicates a group of genetically identical cells derived by mitosis from a common ancestor. To illustrate the use of somatic cell hybridization in the construction of linkage and genetic maps, we shall consider the work of Heuertz and Hors-Cayla (1981), who fused a self-perpetuating clone of Chinese hamster cells with either cattle fibroblasts or cattle lymphocytes, and obtained clones of hybrid hamster–cattle cells. Cells in the hybrid clones contained a mixture of hamster chromosomes and cattle chromosomes. A common feature of such interspecific hybrid clones is that not all of the chromosomes in each hybrid cell are successfully duplicated during every mitosis; there is a gradual loss of chromosomes from the hybrid cells, with chromosomes from one of the parent lines being lost more rapidly than chromosomes from the other parent. In the case of the hamster × cattle hybrid clones described above, cattle chromosomes were preferentially lost.

So long as a particular cattle chromosome remains in a hybrid clone, the polypeptide products of all the functional loci on that chromosome should be detectable in a sample of cells taken from that hybrid clone. But as soon as that particular chromosome is lost from that hybrid clone, so too will all the gene products of that chromosome be lost. Now, a particular chromosome will disappear from different clones at different times. Thus, if several clones are established at the same time, and if a sample of cells is taken from each clone at a given time later, then any particular chromosome will be present in some clones but not in others. Consequently, all the gene products from that chromosome will be present together in some clones, and all will be absent from the remaining clones. Thus, groups of polypeptides that are always *concordant* (all present, or all absent) must be the products of loci on a single chromosome. Loci that are shown in this manner to be located on the same chromosome are said to be *syntenic*, and the group to which they belong is called a *syntenic group*, which is really the same thing as a linkage group.

By taking a sample of cells from each of many different hybrid clones and testing each sample for the presence of 17 different cattle enzymes, Heuertz and Hors-Cayla (1981) detected the three syntenic groups shown in Table 1.5. Each of the remaining 9 enzymes showed no concordance with any other enzyme; they are said to be *asyntenic*, and are most likely each located on a different chromosome. Similar studies in other species have shown similar results. For example, PGM3, SOD2 and ME1 have been shown to be syntenic in cats, chimpanzees, and humans as well as in cattle. And PGD and ENO1 are syntenic in mice, rats, hamsters, sheep and humans as well as in cattle.

Table 1.5. Syntenic groups discovered in a somatic cell hybridization study of 17 enzymes in cattle (from data provided by Heuertz and Hors-Cayla 1981)

Syntenic group number	Enzymes	
	Name	Symbol
1	Lactate dehydrogenase B	LDHB
	Peptidase B	PEPB
	Triose phosphate isomerase	TPI
2	Phosphoglucomutase 3	PGM3
	Malic enzyme 1	ME1
	Superoxidase dismutase 2	SOD2
3	Phosphogluconate dehydrogenase	PGD
	Enolase 1	ENO1

Studies like those described above indicate which loci are located on the same chromosome, but do not indicate the identity of the chromosome concerned. In most species, however, it is now possible to identify accurately each different chromosome, according to techniques described in Chapter 4. Obviously, if individual chromosomes are identified at the same time as polypeptides are detected, then it is possible to assign each syntenic group and each ungrouped polypeptide to a particular chromosome, by observing the concordance between polypeptides and chromosomes. An excellent account of the use of this technique in constructing a chromosome map of the domestic cat is given by O'Brien and Nash (1982).

A further refinement of the above technique can be used to indicate the location of loci within a chromosome. Suppose, for example, that there is an exchange between a cattle chromosome and a hamster chromosome in a hamster–cattle hybrid clone, such that only the tip of a particular cattle chromosome remains in the clone. It follows that if the loss of the remainder of that cattle chromosome coincides with the loss of all but one of the members of that chromosome's syntenic group of polypeptides, then the remaining polypeptide must be the product of a locus located at the tip of that particular chromosome. For a review of this and other refinements of the somatic cell hybridization technique, see Ruddle and Kucherlapti (1974); for a more general view of cell hybridization, see Irvin (1976).

The level of accuracy of the above methods of linkage analysis or genetic mapping in relation to the location of genes is quite sufficient for most purposes. At best, however, such studies can provide only an approximate indication of the location of a particular locus, with the level of accuracy never being better than ±5000 kb of DNA, even under ideal circumstances. In contrast, when recombinant DNA techniques are applied to gene mapping, much more accurate gene maps can be drawn. The increasing use of these latter techniques, as reviewed by Ruddle (1981), will greatly increase our knowledge of genetic maps in the future.

For a complete set of genetic maps of all organisms studied, see O'Brien (1984) or later volumes of the same publication. As an example, Fig. 1.22 shows the genetic map of the domestic chicken.

1.8   X-inactivation and dosage compensation

Among the many coat colours seen in cats, the mosaic of orange and non-orange, which is known as tortoiseshell (Fig. 1.23) is one of the most attractive. The non-orange hairs are due to an X-linked allele o, which is responsible for the production of black and brown melanic pigment, and the orange hairs result from the action at the same locus of the O allele, which prevents the production of melanic pigment. Since both alleles must obviously be present in order to produce the mosaic of orange and non-orange, tortoiseshell cats must be heterozygous, Oo, at this X-linked locus. But why do some parts of the body express the effect of the orange allele, while other parts express that of the non-orange allele? And why are the patches of non-orange and orange approximately equal in total area, and why are they scattered more or less randomly throughout the coat?

The answers to these questions lie partly in another observation first made in cats, by Barr and Bertram, who in 1949 reported that the nucleus of non-dividing nerve cells in females usually contains a

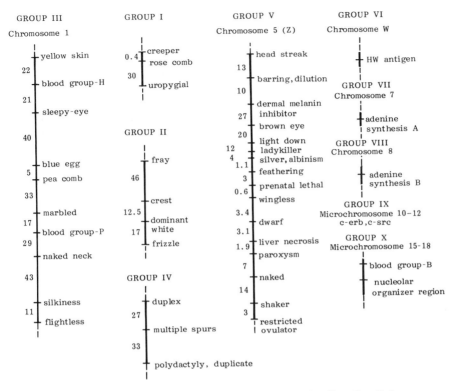

Fig. 1.22. The genetic map of the domestic chicken. The first five linkage groups were known as long ago as 1936. Since then, five more linkage groups have been discovered. Seven of the 10 linkage groups have been allocated to their respective chromosome (see Fig. 4.1 for a karyotype of the chicken). The numbers between adjacent loci indicate *map distance* (in units called *centimorgans*), which equals 100 times the recombination fraction (RF) for values of RF up to approximately 10%. Beyond that, map distance is an overestimate of 100 RF, for reasons that are clearly explained by Suzuki *et al.* (1981, Chapter 6). (Redrawn from Somes 1982.)

dark-staining body whereas that in males does not (Fig. 1.24). This dark-staining body, which is now called a *Barr body* or *sex chromatin*, is actually an X chromosome that was late in replicating during mitosis, and which subsequently became very highly condensed.

Drawing on similar observations in mice, Lyon (1961) suggested that the highly condensed X chromosome seen in female cells is the result of one of the X chromosomes (chosen at random) becoming inactive in each cell of all female embryos at an early stage of development. This is known as the *Lyon hypothesis*. All descendents of each cell in which inactivation first occurred will have the same inactive X chromosome. Furthermore, since the Lyon hypothesis postulates that the choice of X for inactivation is random, it follows that each of

Fig. 1.23. A tortoiseshell cat with white spotting. The white spotting is due to an allele at an autosomal locus. (Courtesy R. Green.)

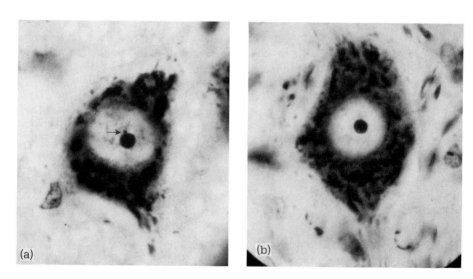

Fig. 1.24. Motor neurones from the hypoglossal nucleus of a mature female cat (a) and a mature male cat (b). The dark-staining body (arrowed) in the female cell is a Barr body. (From Barr and Bertram, 1949.)

the X chromosomes in normal females will be active in approximately one-half of all female cells.

The process of *random X-inactivation*, as postulated by Lyon, provides an adequate explanation for tortoiseshell coat colour; each patch of orange represents the cells that descended from a cell in which the non-orange allele was inactivated, and vice versa. In addition, the apparently random distribution of patches, and the approximately equal total area of orange and non-orange, are to be expected if the inactivated X is chosen at random.

In passing, it should be noted that since tortoiseshell cats are heterozygous at an X-linked locus, they must have two X chromosomes, in which case they should be female. Normal male cats, having only one X chromosome, can be either orange ($X^O Y$) or non-orange ($X^o Y$), but not tortoiseshell. Thus it is a fairly safe bet that any tortoiseshell cat will be female. Very occasionally male tortoiseshells are reported, but they usually turn out to be abnormal males having an extra X chromosome, as described in Chapter 4.

The result of random X-inactivation is that females are *mosaics*, consisting of two distinct populations of cells derived from a common source; in one population of cells the maternal X chromosome (i.e. the X that came from the mother) is inactive, and in the other cell population, the paternal X is inactive. In principle, the two populations of cells will be distinguishable in a female with respect to any X-linked locus at which that female is heterozygous.

In practice, the most informative illustration of the existence of two populations of cells in a female comes from the female mule.

Mules are the result of mating a horse (as dam) to a donkey (as sire). Since the X chromosomes of horses and donkeys can be distinguished in terms of length and the position of the centromere, and since all female mules must have one X chromosome from each parent, the two X chromosomes in female mules can be identified as being either maternal or paternal in origin. Furthermore, the enzyme glucose 6-phosphate dehydrogenase (G6PD), which is X-linked, occurs in different forms in horses and donkeys, and these different forms can be distinguished using electrophoresis, as illustrated in Fig. 1.25.

In order to demonstrate the existence of two populations of cells, Rattazzi and Cohen (1972) took skin biopsies and grew the resultant cells in culture medium. Single cells were then removed from the mixture of cells, and were cultured to produce a clone. When clones from skin biopsies of female mules were examined, it was found that there were just two types of clones: those that exhibited only the horse form of G6PD and those in which only the donkey form of

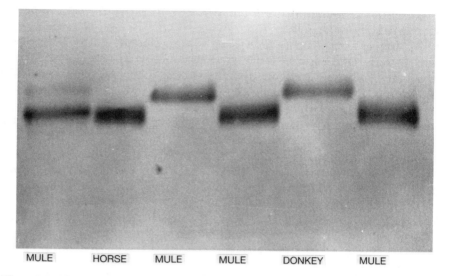

MULE   HORSE   MULE   MULE   DONKEY   MULE

Fig. 1.25. Electrophoresis of the X-linked enzyme G6PD in the horse, mule and donkey. The two different forms (horse and donkey) of the enzyme can be distinguished according to the distance travelled in a cellulose acetate gel through which an electric current is passing. Clones from mules exhibit either horse or donkey G6PD, but never both. (After Rattazzi and Cohen 1972.)

G6PD was present; in no clone were both horse and donkey G6PD present (Fig. 1.25). Furthermore, in each clone in which the horse form of G6PD was active, the donkey X chromosome was always late replicating, and vice versa (Table 1.6). As a result of this and other evidence, the Lyon hypothesis is now regarded as being essentially true for most mammals.

Table 1.6. Association between late replication of X chromosomes and G6PD type in 14 clones of female mule fibroblasts (D = donkey; H = horse)

		G6PD type	
		H	D
Late replicating X chromosome	H	0	5
	D	9	0

From data presented by Rattazzi and Cohen (1972).

The only well-documented exception to random X-inactivation so far recorded is in kangaroos, where it is the paternal X chromosome that is inactivated. The reason for this is not known, but it does give rise to some interesting phenomena. For example, female kangaroos with exactly the same genotype at an X-linked locus can have different phenotypes, depending on which of their X chromosomes came from their father. There is some evidence that paternal X-inactivation also occurs in other Australian marsupials, but further research is needed to confirm this.

Obviously the end result of X-inactivation in females is that each female cell has the same amount of gene product from X-linked genes as do male cells. Thus, X-inactivation is a mechanism that compensates for the difference in gene 'dosage' between males and females in relation to X-linked genes. This effect of X-inactivation is called *dosage compensation*.

> *While dosage compensation for X-linked genes as a result of X-inactivation appears to occur in all mammals, it does not occur for Z chromosomes in birds.*

There is a breed of chicken called Barred Plymouth Rock. The barred feather pattern is the result of bands that lack melanin, being distributed across feathers that would otherwise be entirely black. The bands are due to the action of an allele B at a locus on the Z chromosome. If dosage compensation occurred for Z-linked genes in chickens, then males (with two doses of B, namely $Z^B Z^B$) would have the same barred phenotype as females (with one dose of B, namely $Z^B W$). In fact, Barred Plymouth Rock males are lighter than Barred Plymouth Rock females, because birds with two doses of the B gene produce less melanin (narrower black bands and wider white bands) than birds with one dose of the B gene (Fig. 1.26).

In addition to these observations, sex chromatin has never been observed in birds. And, with the recent discovery by Baverstock *et al.* (1982) of an enzyme locus on the Z chromosome of domestic chickens and other species of birds, it has now been shown that males produce twice as much of this enzyme as females do. All this adds up to strong evidence that dosage compensation for Z-linked genes does not occur in birds.

Fig. 1.26. Feathers from a female (*left*) and a male (*right*) of the Barred Plymouth Rock breed of chicken. Note that the black bands are narrower and the white bands are wider in the male feather than in the female feather. This is because the male has two Z chromosomes, and hence two active Z-linked barring genes, while the female has only one Z chromosome, and hence has only one active barring gene. (Courtesy of K. W. Washburn.)

1.9   Summary

The process of sex determination illustrates many important aspects of basic genetics. Simple techniques are now available to enable structures called chromosomes to be viewed and photographed through a micro-scope. The total chromosome complement of a cell is known as a karyotype. Each species has a characteristic karyotype, and within each species, each sex has a characteristic karyotype, the latter being of vital importance in relation to sex determination.

Meiosis is the process of cell division by which haploid gametes are formed.

Fertilization involves the random choice of a male and a female gamete, and each fertilization is an independent event. It follows that the expected number of either sex among any number of offspring follows the binomial distribution.

After fertilization, a new animal develops as a result of mitosis, which is a process of somatic cell division enabling all cells in an animal to have exactly the same set of chromosomes.

The major chemical component of chromosomes is deoxyribo-nucleic acid (DNA). The Watson–Crick model for the structure of DNA immediately suggests a simple biochemical mechanism for replica-tion of DNA that occurs during mitosis and meiosis. The same model is also the key to an understanding of the way in which genetic informa-tion is stored in the chromosomes, and is transmitted to the cell. In fact, the sequence of bases in a segment of a DNA molecule acts as a template for the production of a specific sequence of amino acids which form a particular polypeptide. The information necessary for the production of an exact sequence of amino acids is stored in the DNA base sequence in the form of a genetic code. A segment of DNA that includes all the nucleotides corresponding to all the amino acids in a particular polypeptide is called a structural gene. In eukaryotes, most structural genes are split into exons and introns. Only a portion of the DNA in a eukaryotic cell consists of unique-sequence or single-copy DNA. The remainder is either interspersed-repetitive DNA (around 10^4 to 10^5 copies) or satellite DNA (usually more than 10^6 copies).

Although meiosis and mitosis are very efficient processes for accurately replicating DNA, mistakes (called gene or point mutations) in DNA repliction do sometimes occur. In addition to gene or point mutations, there are also chromosomal mutations, which are changes in chromosomal structure and/or number. Mutations occurring in cells other than those that give rise to sex cells are called somatic mutations. Mutations occurring in cells that give rise to sex cells are called germ-line mutations, which may lead to the formation of gametes that contain altered DNA.

The different forms of a segment of DNA that can exist at a particu-lar site in a chromosome are called alleles. The particular site or posi-tion of a gene in a chromosome is called the locus. Individuals with two different alleles at a locus are heterozygous, while those with two copies of the same allele are homozygous. The expected outcomes of particular matings involving homozygotes and/or heterozygotes can be easily predicted with the aid of a checkerboard.

Alleles at different loci usually segregate independently. If, however, two loci are close together on the same chromosome, then they are said to be linked, and their alleles do not segregate independently. The physical location of loci relative to one another can be depicted in linkage maps, which can be constructed from observing the results of particular crosses, or from the use of somatic cell hybridization.

The random distribution of orange and non-orange patches in tortoiseshell cats is a good example of random X-inactivation, in which one of the two X chromosomes (chosen at random) in each cell becomes inactivated at an early stage of embryonic development in most female mammals. Because it effectively compensates for the difference between males and females in dosage of X-linked genes, X-inactivation is said to result in dosage compensation for X-linked genes.

1.10 Further reading

Ayala, F. J. and Kiger, J. A. (1984). *Modern genetics* (2nd edn). Benjamin/ Cummings Publishing Company, Menlo Park, California. (One of several textbooks that cover all important areas in genetics.)

Davern, C. I. (Ed.) (1981). *Genetics: readings from Scientific American*. W. H. Freeman, San Francisco. (An excellent collection of 29 articles that review many areas of genetics, written by some of the people who have made major contributions in these areas. Also included is a translation of Mendel's original paper, and a list of all the 133 articles on genetics that have appeared in *Scientific American* up to July 1980.)

Fincham, J. R. S. (1983). *Genetics*. John Wright and Sons, Bristol. (Another general textbook.)

Gardner, E. J. and Snustad, D. P. (1984). *Principles of genetics* (7th edn). John Wiley and Sons, New York. (Another general textbook.)

Gartler, S. M. and Riggs, A. D. (1983). Mammalian X-chromosome inactivation. *Annual Review of Genetics* **17**, 155–90. (A detailed review.)

Green, E. L. (1981). *Genetics and probability in animal breeding experiments*. Oxford University Press, New York. (A very full account of probability, segregation and linkage analysis, using examples from laboratory animals.)

Nobel lectures, published in the journal *Science*. At the time of the Nobel Prize presentations, each recipient gives a lecture, which is subsequently published in *Science*. Traditionally, Nobel lectures are reviews of the research that led to the award, with particular emphasis on how the relevant breakthroughs occurred. In the case of the 1962 award to Watson, Crick, and Wilkins for the discovery of the structure of DNA, only Wilkins' lecture follows the traditional format; the lectures by Watson and by Crick focus on some of the remarkable advances in molecular biology that rapidly followed the discovery of the structure of DNA. The abbreviated references for these lectures are: Crick (1963) **139**, 461–4; Watson (1963) **140**, 17–26, and Wilkins (1963) **140**, 941–50.

O'Brien, S. J. (Ed.) (1984). *Genetic maps* 3. Cold Spring Harbor Laboratory, New York. (This book is published every two years, and contains an up-to-date account of genetic maps in all organisms studied.)

Orel, V. (1984). *Mendel*. Past Master Series. Oxford University Press, Oxford. (An account of Mendel's life and work, written by the director of the Mendel museum, which is located in the monastery where Mendel carried out his pioneering work.)

Peters, J. A. (Ed.) (1959). *Classic papers in genetics*. Prentice-Hall, Englewood Cliffs, New Jersey. (A collection of important original research papers. It is a valuable experience to read how an important discovery was first reported.)

Rieger, R., Michaelis, A., and Green, M. M. (1976). *A glossary of genetics and cytogenetics* (4th edn). Springer-Verlag, New York. (A very useful dictionary of genetical terms.)

Suzuki, D. T., Griffiths, A. J. F. and Lewontin, R. C. (1981). *An Introduction to genetic analysis* (2nd edn). W. H. Freeman, San Francisco. (A thorough introduction to genetics, with an emphasis on explaining how discoveries were made.)

Watson, J. D. (1976). *Molecular biology of the gene* (3rd edn). Benjamin, New York. (A detailed textbook of molecular genetics, written by one of the men who first determined the structure of DNA.)

Appendix 1.1
The chance of obtaining a particular number of males and females

In general, if p is the chance of a male and q the chance of a female, so that $p + q = 1$, then the chance or probability that among n offspring there will be r of one sex is

$$\text{prob }(r) = \binom{n}{r} p^r q^{n-r},$$

where

$$\binom{n}{r}$$

is the number of different ways of arranging r of one sex among a total of n offspring, and is equal to

$$n \times (n-1) \times \ldots \times (n-r+1)/\{r \times (r-1) \times \ldots \times 1\}.$$

For example, to obtain the chance of getting 4 males in a litter of 4, we set $r = n = 4$ and $p = q = \frac{1}{2}$, giving

$$\text{prob }(4) = \binom{4}{4} \tfrac{1}{2}^4 \tfrac{1}{2}^{4-4}$$

$$= \frac{1}{16}$$

As a second example, let us ask what is the chance of obtaining an equal number of males and females in the offspring of four matings in a herd of cattle? In this case, $r = 2$, $n = 4$ and $p = q = \frac{1}{2}$, which gives

$$\text{prob}\,(2) = \binom{4}{2} \tfrac{1}{2}^2 \tfrac{1}{2}^{4-2}$$

$$= 6 \cdot \tfrac{1}{2}^4$$

$$= \frac{6}{16}$$

2
Recombinant DNA

2.1 Introduction

Because it is so important in relation to basic biology, and in relation to several practical applications of direct relevance to animal health and production, it is essential that readers of this book have a basic understanding of recombinant DNA. In the following account, we shall review the important recombinant DNA techniques, and shall then illustrate their use in two relevant areas; the involvement of the domestic chicken in the discovery of split genes, and the production of a vaccine for foot-and-mouth disease.

2.2 Restriction enzymes

In 1970, it was discovered that certain strains of bacteria produce enzymes that are able to degrade foreign DNA that enters the bacterial cell. Because these enzymes play a key role in a phenomenon known as host restriction (whereby a particular bacterial strain causes a decrease in the infective ability of a certain bacterial virus), they are called *restriction enzymes*. Three of the people involved with the discovery of restriction enzymes, namely Arber, Smith, and Nathans, were awarded a Nobel Prize in 1978. More than 200 restriction enzymes are now known, and a standard method of naming them has been developed. For example, *Bam*HI is the first enzyme obtained from the H strain of *Bacillus amyloliquefaciens*, and *Hind*III is one of three enzymes obtained from the d strain of *Haemophilus influenzae*. Restriction enzymes bind to specific sequences of nucleotides called *recognition sequences*, and each enzyme cuts the DNA at a specific *cleavage site* that is normally located within the recognition sequence. In most cases, recognition sequences are palindromes, by which we mean that the sequence of bases in *both* strands of DNA considered together is the same when read from left to right and from right to left (see Fig. 1.17). A list of commonly used restriction enzymes together with their recognition sequences and cleavage sites, is given in Table 2.1.

Table 2.1. Some common restriction enzymes. For a complete list, see Roberts (1983). Notice that each recognition sequence is a palindrome

Source	Enzyme	Recognition sequence, and cleavage site (↑↓)
Arthrobacter luteus	*Alu*I	3′ –T–C–G–A– 5′ 5′ –A–G–C–T– 3′
Bacillus amyloliquefaciens H	*Bam*HI	3′ –C–C–T–A–G–G– 5′ 5′ –G–G–A–T–C–C– 3′
Escherichia coli	*Eco*RI	3′ –C–T–T–A–A–G– 5′ 5′ –G–A–A–T–T–C– 3′
Haemophilus influenzae R$_d$	*Hind*III	3′ –T–T–C–G–A–A– 5′ 5′ –A–A–G–C–T–T– 3′
Haemophilus aegyptius	*Hae*III	3′ –C–C–G–G– 5′ 5′ –G–G–C–C– 3′
Providencia stuartii	*Pst*I	3′ –G–A–C–G–C–T– 5′ 5′ –C–T–G–C–A–G– 3′
Serratia marcescens	*Sma*I	3′ –G–G–G–C–C–C– 5′ 5′ –C–C–C–G–G–G– 3′

It can be seen that

Each restriction enzyme has a specific recognition sequence and cleavage site.

Some restriction enzymes, such as *Sma*I, cut through both strands of the double helix at the same place, leaving two *blunt ends*. Others, such as *Eco*RI, cleave each strand at a different position, leaving some nucleotides unpaired. Because these unpaired nucleotides attract complementary nucleotides into the vacant positions, enzymes such as *Eco*RI are said to create *sticky ends*.

Now, when a particular restriction enzyme is added to DNA, the DNA is cleaved in as many positions as there are cleavage sites for that enzyme; the DNA is said to be *digested* into a number of fragments by the restriction enzyme. Since cleavage sites are not positioned at regular intervals along a DNA molecule, the fragments resulting from digestion will usually be of different lengths and hence different weights. If the mixture of fragments is placed at one end of an agarose gel and subjected to electrophoresis, then the fragments will migrate along the gel at a rate inversely proportional to the logarithm of their weights. Thus at any given time after the electric field was first applied, there will be as many different bands on the gel as there were different sized fragments in the original mixture. And the relative position of each band will indicate the weight of each fragment.

Consider, for example, a relatively small DNA molecule such as that of the bacteriophage λ, whose total DNA consists of a single molecule 49 kilobases (kb) long (Fig. 2.1). If λ DNA is digested with *Eco*RI, and if the resultant fragments are subjected to electrophoresis, then a picture like that in Fig. 2.2 emerges. Comparison of the results of digestions with several different enzymes enables the drawing up of a *restriction map* of the DNA, an example of which is illustrated in Fig. 2.3. For a complete set of restriction maps of all organisms studied to date, see O'Brien (1984), or more recent volumes of the same publication.

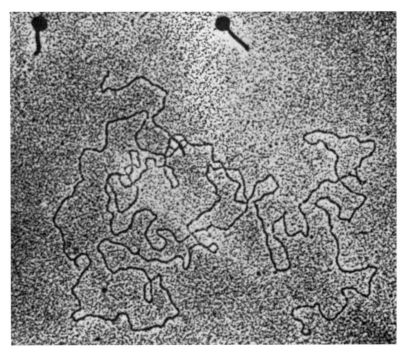

Fig. 2.1. Two intact λ bacteriophages (top) and the double-stranded DNA molecule from a third λ bacteriophage. The two ends of the DNA molecule are clearly visible. (From Grobstein 1977, courtesy of J. D. Griffith.)

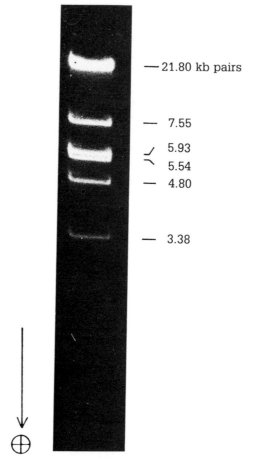

Fig. 2.2. Electrophoresis of λ DNA digested with *Eco*RI. The bands have been made visible under ultra-voilet (UV) light by the addition of the dye ethidium bromide, which stains DNA and fluoresces under UV light. Each band corresponds to a different DNA fragment, whose size in kilobase (kb) pairs is given beside the gel. Since λ DNA is linear, the existence of six fragments indicates that there must be five *Eco*RI cleavage sites in λ DNA. (From Old and Primrose 1985.)

2.3 Complementary DNA

There is a class of tumour-inducing viruses whose entire genome consists only of RNA. Because RNA is not capable of self-duplication, the only way for these viruses to propagate is for their RNA to be transcribed 'backwards' into DNA, which duplicates and then produces more RNA. The enzyme that catalyses the reverse transcription of RNA into DNA is called *reverse transcriptase*. This enzyme was discovered in the same year as restriction enzymes, and it also led to the award of a Nobel

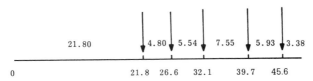

Fig. 2.3. An *Eco*RI restriction map of λ DNA. The horizontal line represents the double-stranded DNA molecule illustrated in Fig. 2.1. Cleavage sites are indicated by arrows, with the numbers under each arrow indicating the location of each site in terms of the number of kilobase pairs from the left hand end of the molecule. Numbers above the horizontal line indicate the size of each fragment in kilobase pairs, and correspond to the numbers in Fig. 2.2. The actual position of each fragment cannot be determined from the data in Fig. 2.2 alone; in practice, several separate digestions are done concurrently, each with a different restriction enzyme. The actual location of each fragment can then be determined from jointly considering the results of all digestions, as explained, for example, by Suzuki *et al.* (1981, Figs. 13.32, 13.33 and 13.34). (After Old and Primrose 1985.)

Prize, in this case to Temin and Baltimore, in 1975. Reverse transcriptase is important in recombinant DNA technology because there are many situations in which it is easier to isolate RNA than DNA from cells. For example, cells in the oviduct of chickens produce large quantities of ovalbumin and very little else. Consequently, it is relatively easy to isolate large quantities of ovalbumin mRNA from chicken oviduct cells. Then, by simply adding free nucleotides and the enzyme reverse transcriptase to the isolated mRNA, a complementary strand of DNA (called *complementary* DNA or *copy* DNA or cDNA) can be synthesized readily. Finally, by adding more nucleotides and the enzyme DNA polymerase (see Fig. 1.14), double-stranded cDNA can be formed from the single-stranded cDNA. The construction of cDNA is illustrated in Fig. 2.4.

2.4   Recombinant DNA, and DNA cloning

One of the main objectives of work with recombinant DNA is to produce an unlimited number of copies of a particular segment of DNA. This process is called *gene cloning* or *DNA cloning*. It is achieved by joining the DNA that is to be cloned (called *foreign* DNA) to a *vector* which is capable of replication within a particular *host*. This joining, which is called *splicing* or *ligation*, is accomplished with the aid of the enzyme DNA ligase, whose natural role was described in Fig. 1.14. The DNA molecule that results from joining the foreign DNA segment to the vector is called *recombinant DNA*. The most commonly used vectors are *plasmids*, which are circles of double-stranded DNA that exist in certain bacterial cells (hosts) independently of the main bacterial chromosome. A general outline of the process involved in DNA cloning using plasmids is given in Fig. 2.5.

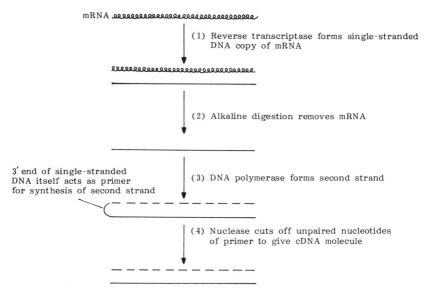

Fig. 2.4. The four steps involved in the construction of cDNA from mRNA. (From Australian Academy of Science 1980.)

In many cases, the efficiency of transferring foreign DNA into a host cell can be greatly increased by adding cohesive (*cos*) sites from λ phage DNA to a plasmid, forming a new type of vector called a *cosmid*. The presence of cos sites enables cosmids to be 'packaged' *in vitro* into phage particles which then obligingly follow their usual practice of attaching to bacterial cell walls and injecting their DNA (cosmids to which foreign DNA has been previously joined) into the bacterial (host) cell. In this way, much larger pieces of foreign DNA can be transferred to host cells.

2.5 Production of polypeptide from cloned DNA

Arranging for a DNA fragment to be cloned in a bacterial host is not the same thing as arranging for that DNA fragment to produce its corresponding polypeptide in that host.

In the former case, all that is required is for the foreign DNA to be inserted into the DNA of the vector. In the latter case, various molecular controls need to be present before a piece of DNA can be transcribed into mRNA and then translated into polypeptide.

The features that are common to most control mechanisms are illustrated in Fig. 2.6. The *promoter* region precedes, i.e. is located *upstream* from, the transcription initiation site. The promoter region contains specific sequences that are highly *conserved*, by which we

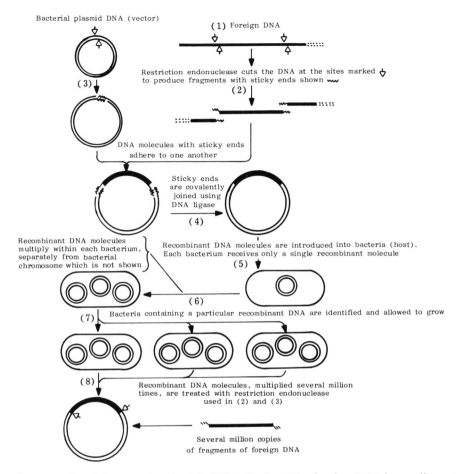

Fig. 2.5. The eight steps involved in DNA cloning. The foreign DNA is usually part or all of a cDNA molecule, or a restriction fragment of DNA obtained directly from foreign cells, or an artificially synthesized DNA segment. (From Australian Academy of Science 1980.)

mean that the same sequence occurs at the same site in most, if not all genes. Such highly conserved sequences are called *boxes*. The promoter region of prokaryote genes contains the *TATAAT box* and the *TTG box*, which are thought to be sites for recognition and binding of RNA polymerase. Sites that are likely to have a similar function in eukaryotic promoters are the *TATA box* and possibly the *CCAAT box*. Adjacent to the promoter is the *leader* sequence, which is transcribed but not translated. It contains a transcription initiation site (the *CAT box*), a ribosome binding site, and possibly other regulatory sites. The *structural gene* lies between the start and stop codons. In eukaryotes it is usually split into exons and introns, as described in Sections 1.3.3 and

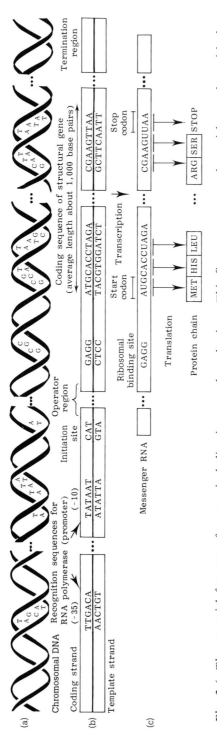

Fig. 2.6. The essential features of a gene, including its control mechanisms. This figure represents a bacterial gene, but with the exceptions described below, the features shown are common to most genes in most organisms. (a) The DNA double helix, showing certain base sequences. (b) The two strands of the DNA double helix aligned side-by-side. Only one strand acts as a template for mRNA. The TATAAT or Pribnow box (10 bases upstream from the transcription initiation site) and the TTG box (35 bases upstream) are thought to be sites for recognition and binding of RNA polymerase. The eukaryotic equivalent of the TATAAT box is the TATA or Goldberg–Hogness box, located about 30 bases upstream from the transcription initiation site. The only other highly-conserved sequence that may act as a recognition sequence in eukaryotes is the CCAAT box, located about 80 bases upstream. The operator region, which is a site for binding of a 'repressor' protein produced by a separate regulator gene, is not usually found in eukaryotes. The termination region stops transcription. If this gene were a eukaryotic gene, it would most likely be split into exons and introns. (c) mRNA, consisting of the leader sequence (upstream of the start codon), the structural gene (between the start and stop codons), and the trailer sequence (between the stop codon and the transcription termination region). Only the structural gene is translated into protein. (From 'The genetic programming of industrial micro-organisms' by Hopwood, D. A. Copyright © (1981) by Scientific American, Inc. All rights reserved.)

2.7, but in prokaryotes it is not split. *Downstream* from the stop codon is the *transcription terminator*.

In contrast to the above similarities in control mechanisms across species, there is one important difference between prokaryotic and eukaryotic control arising from the absence of split genes in the former. Having no split genes, prokaryotes have no need for the elaborate splicing mechanisms (see Fig. 2.14, below) required by eukaryotes for converting a primary mRNA transcript into mature mRNA. Because of this:

> *The only eukaryotic genes that can be transcribed and translated in prokaryotes are those consisting solely of exons.*

In practice, this is not a serious limitation. If mature mRNA for the desired polypeptide is available, then cDNA can be produced from the mature mRNA. Alternatively, if the amino acid sequence of the desired polypeptide is known, then the nucleotide sequence can be deduced using the genetic code, and the appropriate DNA molecule can be synthesized from single nucleotides. 'Gene machines' are now available for automatically synthesizing specific sequences of DNA (Fig. 2.7). Currently, only short sequences of DNA can be synthesized from nucleotides. However, in the future it will be possible to synthesize much longer sequences.

It is important to note that even if there are no introns in the DNA to be transcribed and translated, then there are still sufficient differences in control mechanisms between species for it to be generally agreed that:

> *Irrespective of the origin of the foreign DNA, the control mechanisms used for its successful transcription and translation should be the same as those of the host cell.*

In practice, this is usually achieved by ensuring that the foreign DNA is the structural gene only, and by inserting this structural gene immediately downstream from a promotor and a leader sequence in the vector DNA. This effectively 'tricks' the host into thinking that the foreign structural gene is, in fact, one of its own structural genes. In some cases, it is easier to insert the foreign structural gene into the middle of a vector structural gene, which results in the production of a hybrid polypeptide from which the foreign polypeptide can be released by appropriate chemical treatment.

Finally, it is worth noting that:

> *The production of polypeptide from foreign DNA in host cells is possible only because the genetic code is universal.*

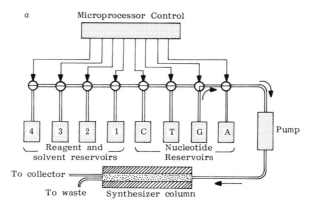

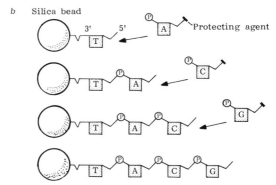

Fig. 2.7. 'Gene machines', several versions of which are now available, synthesize specified short sequences of single-strand DNA automatically and very quickly under the control of a microprocessor. A version made by Bio Logicals, a Canadian company, is illustrated. (a) The desired sequence of bases is entered on a keyboard. The microprocessor opens valves that allow successive nucleotides and the reagents and solvents needed at each step to be pumped through the synthesizer column. The column is packed with tiny silica beads (about the consistency of fine sand); each bead serves as a solid support on which the DNA molecules are assembled. (b) To make a given sequence a column is used in which many thousands of copies of the nucleoside (base plus sugar) that is to be at the 3′ end of the sequence (a T in this case) have already been fixed to each bead, leaving the nucleoside's 5′ side free. The microprocessor pumps millions of copies of the next nucleotide (A), with its 5′ side protected against unwanted reactions, into the column, and the A's bind to the T's. The protecting agent is removed, leaving the 5′ side free to accept the next nucleotide (C). In this way, chains of about 40 nucleotides have been synthesized at the rate of about one nucleotide every 30 minutes. The completed chains are cleaved from the beads and are eluted into the collector. (From 'The genetic programming of industrial micro-organisms' by Hopwood, D. A. Copyright © (1981) by Scientific American, Inc. All rights reserved.)

The triplet GAA, for example, means the same thing to a bacterium as it does to a lucerne plant or to a cow. There are, however, differences between species in 'preference' for certain codons in relation to those amino acids having more than one codon. And there is some suggestion that polypeptide production will be more efficient in the host if the foreign DNA consists of triplets most favoured by the host. Thus if the foreign DNA has to be synthesized on the basis of a known amino acid sequence, it may be worthwhile using host-preferred triplets in its construction.

This 'fine tuning' does not diminish the importance of the general conclusion reached above. Indeed, the success achieved in coaxing *E. coli* and other prokaryotic hosts to mass-produce polypeptide from a diversity of eukaryotes is remarkable testimony to the universality of the genetic code.

2.6   Gene banks or libraries

We have already seen that it is possible to clone a specific fragment of DNA once it has been isolated. We shall now take this idea one step further and consider how to clone all the DNA from, say, a chicken or a cow.

The basic processes involved are illustrated in Fig. 2.8. In the first step, the total genomic DNA is digested with one or more restriction enzymes into a large number of DNA fragments of various sizes.

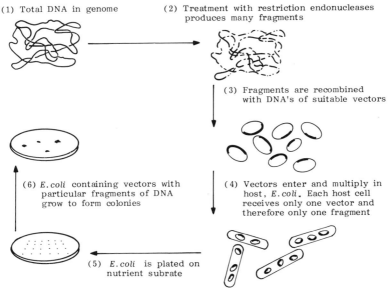

(1) Total DNA in genome

(2) Treatment with restriction endonucleases produces many fragments

(3) Fragments are recombined with DNA's of suitable vectors

(6) *E. coli* containing vectors with particular fragments of DNA grow to form colonies

(4) Vectors enter and multiply in host, *E. coli*. Each host cell receives only one vector and therefore only one fragment

(5) *E. coli* is plated on nutrient subrate

Fig. 2.8. Formation of a gene bank of all the DNA of an organism by use of the shotgun procedure. For a more detailed account, see Old and Primrose (1985, Chapter 6). (From Australian Academy of Science 1980.)

Because this process is somewhat analogous to firing a shotgun at a dinner plate, it is called the *shotgun* procedure. The fragments are recombined into suitable vectors which then enter host cells which can be maintained in colonies. A set of colonies from one shotgun procedure should contain all the DNA present in the organism from which the DNA was obtained initially. In this way, *gene banks* or *libraries* have been established for many different species ranging from *E. coli* to humans, and including several domestic animals. An example of the use of a gene bank is given in Section 2.7.

In order to obtain an idea of the size of gene banks, we need to know the total size of the haploid genome of various organisms, and the approximate size of fragments produced by the shotgun procedure. Fig. 2.9 indicates the range in genome size for most species. The size of fragments resulting from the shotgun procedure depends on the number of enzymes used, and whether there is a partial or complete digestion.

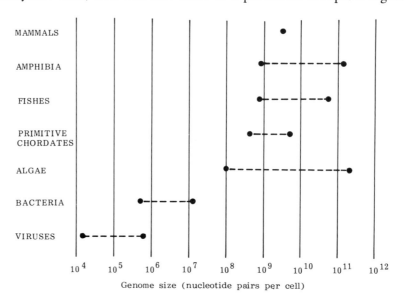

Fig. 2.9. The amount of DNA per haploid cell in various groups of organisms. All mammals have about the same number of nucleotide pairs, approximately 3×10^9, which equals 3×10^6 kbp. (From 'Repeated segments of DNA' by Britten, R. J. and Kohne, D. E. Copyright © (1970) by Scientific American, Inc. All rights reserved.)

For mammalian gene banks, a partial digestion with two enzymes (*Hae*III and *Alu*I) is often used, giving rise to fragments that are predominantly around 20 kbp long. With approximately 3×10^6 kbp in a mammalian haploid genome (Fig. 2.9), we would require a minimum of $3 \times 10^6/20 = 1.5 \times 10^5$ or 150 000 host colonies to have any chance of including all segments of a mammal's DNA in a gene bank. In order to have a high chance of including all segments, somewhere between 3×10^5 and 8×10^5 host colonies are needed. For a description of the construction of a gene bank in a domestic animal (chicken in this particular case), see Dodgson *et al.* (1979).

2.7 Split genes

Having now reviewed the most important techniques involving recombinant DNA, we shall see how they were used in the discovery of split genes. It should be obvious that split genes were not discovered during the course of a research programme designed specifically to find them, because at that time, no one imagined they existed! Instead, as so often happens, this unexpected discovery was made during the course of a research programme designed with a different aim in mind. In this case, the research workers were investigating the activation of the ovalbumin gene in the oviduct of laying hens. In order to do this, Chambon and his colleagues in Strasbourg decided to compare the ovalbumin gene isolated from oviduct cells (in which it is active) with the same gene isolated from other cells in which it is inactive.

They started by isolating ovalbumin mRNA from oviduct cells. This mRNA was then used as a template for the production of single-stranded cDNA, which was then copied to form double-stranded cDNA. This 'gene' for ovalbumin was then recombined with a plasmid and was cloned in *E. coli*, enabling the production of sufficient ovalbumin cDNA for restriction mapping. Among other things, the restriction map showed that ovalbumin cDNA happened not to contain any sites for the enzymes *Eco*RI and *Hind*III.

Having thus obtained considerable information about the structure of the ovalbumin gene in oviduct cells, they then set about investigating the structure of the same gene isolated from erythrocytes, in which the gene is not expressed. Remembering that ovalbumin cDNA is not cleaved by *Eco*RI or by *Hind*III, they digested all the DNA from erythrocytes with these two enzymes. In the case of *Eco*RI, for example, the digestion produced approximately 500 000 fragments, ranging in size from 1000 to 15 000 base pairs. As shown in the lower half of Fig. 2.10, these fragments were separated using electrophoresis.

The next step was to isolate those fragments carrying the ovalbumin gene. This was done by firstly using a technique called *Southern blotting*, which is a process (developed by E. M. Southern in 1975) whereby the double-stranded DNA is denatured into single strands, and then transferred to filter paper by placing the filter paper on top of the gel, somewhat analogous to soaking up ink with blotting paper. The next step involved pouring radioactively labelled single-stranded ovalbumin cDNA over the filter paper. Being single-stranded, this cDNA bound

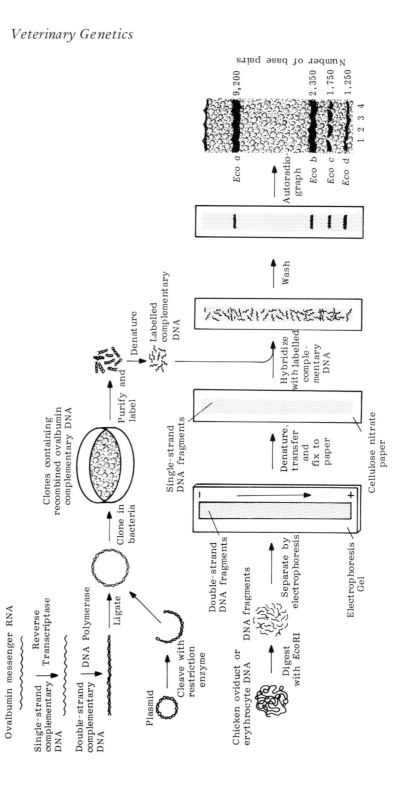

to any complementary fragment of DNA on the filter paper, and thus acted as a *probe* for the ovalbumin gene.

Because the DNA digestion had been done with restriction enzymes that did not cleave ovalbumin cDNA, it was expected that the probe would bind to only one fragment (the one containing the ovalbumin gene), and would thus be detected in only one band after autoradiography of the filter paper. But in the case of *Eco*RI, four distinct bands were detected (Fig. 2.10), indicating that DNA complementary to ovalbumin cDNA occurred in four different *Eco*RI fragments of DNA from erythrocytes. In the case of *Hind*III, three distinct fragments were found.

At first it was thought that errors had occurred in the various techniques. But checks and careful repeats of the experiments led to the same results. And, even more surprisingly, the same bands appeared whether the digestion was done on erythrocyte DNA (in which the ovalbumin gene is not expressed) or on oviduct DNA (in which the gene is expressed). It appeared as if ovalbumin DNA contained nucleotide sequences that were lacking in ovalbumin mRNA.

Final proof of this came when other workers searched through a chicken gene bank (see Section 2.6), and isolated a fragment of chicken DNA containing the whole of the ovalbumin gene. They then denatured this fragment of DNA and mixed it with purified ovalbumin mRNA, under conditions which encouraged the hybridization of complementary strands of RNA and DNA. This experiment provided a fascinating picture of a split gene, as shown in Fig. 2.11.

In order to ensure that no short exons had been overlooked by the above experiment, Chambon and his colleagues made use of another recent development in DNA technology, namely *DNA sequencing*. Chambon used the Maxam and Gilbert (1977) method of DNA sequencing, which is illustrated in Fig. 2.12. In this way, Chambon and his colleagues obtained the complete nucleotide sequence of ovalbumin cDNA from which they were able to deduce the nucleotide sequence of ovalbumin mRNA, which is shown in Fig. 2.13. By comparing this

---

Fig. 2.10. The experiment that led to the discovery of split genes. Lanes 1 and 2 of the autoradiograph are from chicken erythrocytes, while lanes 3 and 4 are from chicken oviduct cells. The bands are identical in all four lanes, indicating that ovalbumin DNA has the same structure whether it is being expressed (in oviduct cells) or is not being expressed (in erythrocytes). The presence of four bands indicates that the restriction enzyme *Eco*RI cuts ovalbumin DNA into four segments, whereas it does not cut ovalbumin cDNA copied from ovalbumin mRNA. The explanation for this is that the DNA is split into exons and introns, while the cDNA consists of exons only, and that the only *Eco*RI sites in the ovalbumin gene occur in the introns. (From 'Split genes' by Chambon, P. Copyright © (1981) by Scientific American, Inc. All rights reserved.)

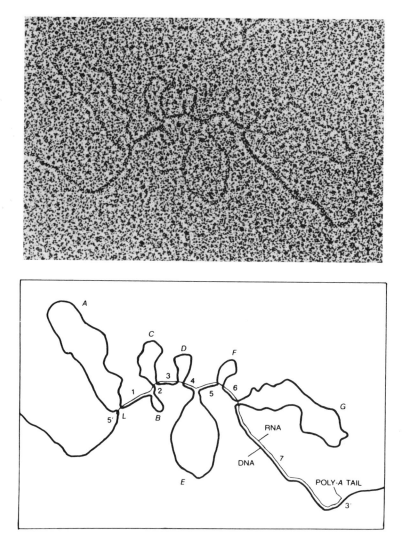

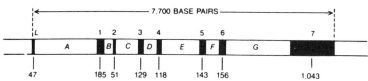

Fig. 2.11. Visual proof that the ovalbumin gene is split into introns and exons. Top: an electron micrograph showing hybridization between mRNA and cellular DNA. Centre: an interpretation of the micrograph, showing the eight exons (L for the leader sequence, and exons 1 to 7), and the seven introns (A to G) which form loops because they have no RNA with which to hybridize. Bottom: a map of the ovalbumin gene, showing the location and size (in base pairs) of each exon. The introns range in size from 251 to approximately 1600 base pairs. (From 'Split genes' by Chambon, P. Copyright © (1981) by Scientific American, Inc. All rights reserved.)

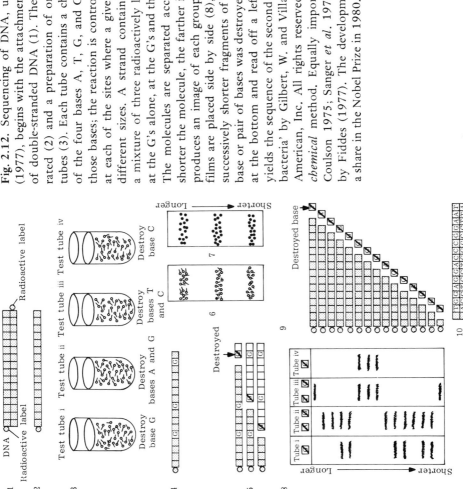

Fig. 2.12. Sequencing of DNA, using the method devised by Maxam and Gilbert (1977), begins with the attachment of a radioactive label to one end of each strand of double-stranded DNA (1). The strands of large numbers of molecules are separated (2) and a preparation of one of the two strands is divided among four test tubes (3). Each tube contains a chemical agent that selectively destroys one or two of the four bases A, T, G, and C, leading to cleavage of the strand at the site of those bases; the reaction is controlled so that only some of the strands are cleaved at each of the sites where a given base appears, generating a set of fragments of different sizes. A strand containing three G's (4), for example, would produce a mixture of three radioactively labelled molecules (5). The reactions break DNA at the G's alone, at the G's and the A's, at the T's and the C's, and at the C's alone. The molecules are separated according to size by electrophoresis on a gel; the shorter the molecule, the farther it migrates down the gel (6). The radioactive label produces an image of each group of molecules on an X-ray film (7). When four films are placed side by side (8), the ladderlike array of bands represents all the successively shorter fragments of the original strand of DNA (9). Knowing what base or pair of bases was destroyed to produce each of the fragments, one can start at the bottom and read off a left-to-right sequence of bases (10), which in turn yields the sequence of the second strand. (From 'Useful proteins from recombinant bacteria' by Gilbert, W. and Villa-Komaroff, L. Copyright © (1980) by Scientific American, Inc. All rights reserved.) The Maxam–Gilbert method is known as the *chemical* method. Equally important is the *plus-and-minus* method (Sanger and Coulson 1975; Sanger *et al.* 1977), the basic principles of which were reviewed by Fiddes (1977). The development of these methods earned Gilbert and Sanger a share in the Nobel Prize in 1980.

```
Cap ─── Exon L (Leader) ───▶        Exon 1 ───▶
5' m⁷G ACAUACAGCUAGAAAGCUGUAUUGCCUUUAGCACUCAAGAGUUCACC AUG GCUCCAUCGGCCAGCAAGCAU
                   │    │    │    │    │    │    │    │    │
                   10   30   50   70                        90

GGAAUUUUGUUUGAUGUAUUCAAGCUGUAUUGCCAUUGCUCAUCGUACUGCCCAUUGCUCAGCUCUAGCCAUGGUA
                                        Exon 2 ───▶              Exon 3 ───▶
        110        130        150        170        190

UACCUGGGUGCAAAAGACACAGGACACCACCAGGACUUGUUGCCUUUGAUAAACUUCCAGGAUUCGGAGACUCAGUGUGGCA
        210        230        250        270        290

CAUCUGUAAACGUUCACUCUUCACUUAGAGACAUCCUCAACCAAAUCACCAAAAUCAAAUGAUGUUUAUUCGUUCAGCCUUGCCUUGCUUUAUGCUGA
          Exon 4 ───▶
        310        330        350        370        390

AGAGAGAUACCCAAUCCUGCCAGAAUUACUUGCAGUGUGUGAAGGAACUGUAUAGAGGAGCUUGGAACCUAUCAUCAACAACAGCUGCAGAUCAAGCC
                      430        450        470        490

AGAGAGCUCAUCAAUUCCUGGGUAGAAAGUCAGAAAUGGCCUUCAACUGAAAAUGCCUGUCCGUGGAUCUCAAACUGCAAUGGUUCUGG
          Exon 5 ───▶                                              Exon 6 ───▶
        510        530        550        570        590

UUAAUGCCAUUGUCUCAAAGGACGUGUGGGAGAAACAUUUAAGGAUGAAGAGACACACAACAAGCAAUGCCUUUCAGAGUGAGCAAGAAAGCAAACCUGU
        610        630        650        670        690

GCAGAUGAUGUACCAGAUUGGGUUUAUUUAGAGUGGCAUCAAUGGCCUUCGACUCCUUGAGCAAUUGAAGAUCCUGGAGCUUCCAUUGCCAGUGGGACAAUGAGCAUG
          Exon 7 ───▶                                  770
        710        730        750                             790

UUGGUGCUGUUGCCUGAUGAAGUCUCAGGCCUUGAGCAGCUUGAGAGUAUAAAUCAACUUUGAAAAACUGACAUUGAAUCAACUGACUGAACUGACCAGUUCAACUGAUGGAAG
        810        830        850        870        890

AGAGGAAGAUCAAAGUGUACUUACCUCCGCAUGAAGAUGGGAGAAAAAUACACAACCUCACACUGUCUUAUGGGCAUUACUGACGUGUUAGCUC
        910        930        950        970        990

UUCAGCCAAUCUGUCUGGCAUCUCCUCAGCAGAGGCCUGAAGAGAUCUCAAGAGAGCCUGAGAAUCAAGCAGAUAUCUCAAGAAAUCAAUGAAGCACACAAUGAAGAGGCAGCAGAGGUG
        1010        1030        1050        1070        1090

GUAGGGUCAGCGAGGCUGGGAUGGCUGCAAGCGUCUCUCGAGAAUGCUGCUCAUUCUCUUCUGUAUCAAGCACAUCGCAACCAACG
        1110        1130        1150        1170        1190

CCGUUCUCUUCCUUUGGCAGAUGUGUUCCCUA UAA AAAGAAGAAAAGCUGAAAAACUCGUCCUUCCAACAAGACCCAGAGCACUGUAGUAUCAGGGGGUA
        1210        1230        1250        1270        1290
```

AAAUGAAAAGUAUGUCUCUGCUGCAUCCAGACUUCAUAAAAGCUGGAGCUUAAAUCUAGAAAAAAUCAGAAAUUACACUGUGACGAACAGGUGCA
    1310   1330   1350   1370   1390

AUUCACUUUCCUUUACACAGAGUAAUACUGGUAACUCAUGGAUGAAGGCUUAAGGGAUGAAUGAAAUUGGACUCACAGUACUGAGUCAUCACUGAAAAAU
    1410   1430   1450   1470   1490

GGAACCUGAUACAUCAGCAGAAGGUUUAUGGGGAAAAAUGCAGCUUCCAAUUAAGCCAGAUAUCGUAUGACCAAGCUGCCUCCAGAAUUAGUCACUCA
    1510   1530   1550   1570   1590

AAAUCUCAGAUUAAAUUAUCAACUGUCACCAACCAUUCCUAUGCUGACAAGGCAAUUGCUUGUGCUCUGUGUUCCUGAUACUACAAGGCUCUUCCUGA
    1610   1630   1650   1670   1680

CUUCCUAAAGAUGCAUUAUAAAAAUCUUAUAAAUUCACAUUUCUCCCUAAACUUUGACUCAAUCAUGGUAUGUGGCAAAAUGGUAUAUUAUCUAAUUCCAAA
    1710   1730   1750   1770   1790

UUGUUUCCUGUACCCAUAUGUAAUGGGUCUUGUGAAUGUAAUGGGUCUUGUGUCCUUUAAUCAUAAUAAAACAUGUUUAAGC——Poly-A
    1810   1830   1850   1870

Fig. 2.13. The nucleotide sequence of ovalbumin mRNA, consisting of a total of 1872 nucleotides. Molecules of ovalbumin, which consist of 386 amino acids, are produced from the $3 \times 386 = 1158$ nucleotides bounded by the start (AUG) and stop (UAA) codons which are boxed. The 64-nucleotide leader sequence at the 5′ end, and the 650-nucleotide trailer sequence at the 3′ end are not translated. The methylated guanine 'cap' at the 5′ end, and the poly-A 'tail' at the 3′ end are characteristic of functional mRNA, and are added after transcription, as shown in Fig. 2.14. (From 'Split genes' by Chambon, P. Copyright © (1981) by Scientific American, Inc. All rights reserved.)

sequence with that obtained from various fragments of cellular DNA, it became evident that the conclusion drawn from Fig. 2.11 was correct: the ovalbumin gene is split into eight exons and seven introns.

It remains now to explain how mRNA (consisting of exons only) is produced from cellular DNA which consists of exons and introns. What happens is that exons and all introns are transcribed initially into an RNA molecule, which then undergoes a series of processing steps in which the introns are excised and the exons are spliced together, as illustrated in Fig. 2.14. The mechanism by which introns are recognized and removed is not yet known, although the observation that most introns start with the nucleotide sequence GT and end with the sequence AG may be a key to an understanding of the mechanism.

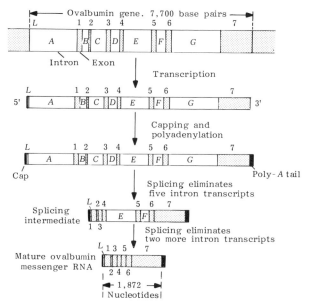

Fig. 2.14. The production of mature messenger RNA from cellular DNA. The capping and polyadenylation step is typical of eukaryotic mRNA (see Fig. 2.13). The excision of introns occurs in a number of stages, two of which are illustrated here. All intermediate stages occur in the nucleus; only mature mRNA is transferred to the cytoplasm. (From 'Split genes' by Chambon P, Copyright © (1981) by Scientific American, Inc. All rights reserved.)

Work in many laboratories has now confirmed that most eukaryotic genes are split into exons and introns, with the latter ranging in number from two to more than 50, and in size from 10 bases to 10 000 bases, and often accounting for at least as much DNA as the former. For example, a δ-crystallin gene in the chicken has 15 introns comprising approximately 80 per cent of the gene: the average length of introns in this gene is 2.3 kbp, while the average length of exons is only 0.6 kbp.

What is the purpose of so much DNA being 'wasted' in introns? Do introns have functions as yet undiscovered?

There has been much speculation on these and related questions. It appears that exons in some genes correspond to functional units of the final polypeptide, in which case introns may have some role as spacers separating functional units. But this is hardly an adequate explanation. Some people have suggested that introns and other un-translated segments of DNA have no function other than their own self-preservation within the genome. As such, they are said to be 'selfish DNA', which is an intriguing concept that has been the subject of much recent debate. We shall not elaborate on it here, but interested readers are referred to Crick (1979), Doolittle and Sapienza (1980), Orgel and Crick (1980), Orgel *et al.* (1980), Dover and Doolittle (1980), and Jain (1980).

2.8 Production of foot-and-mouth disease vaccine

Foot-and-mouth disease (FMD) is one of the most important of all animal diseases, and more FMD vaccine is produced throughout the world than any other vaccine. To give some idea of the scale of the problem, South America alone requires 500 million doses of vaccine per year. The conventional vaccine is based on inactivated viruses, and its large-scale production is not always successful; batches contain-ing live virus are sometimes inadvertently produced, leading to 'out-breaks' of the disease. This problem could be avoided if the vaccine were based solely on the antigenic portion (the portion that stimulates the production of antibodies: see Chapter 8) of the virus. Such a vaccine could never become infective.

As an illustration of the practical application of recombinant DNA, we shall now review briefly the use of this technology in the produc-tion of a non-infective FMD vaccine.

The genome of the FMD virus is a single-stranded molecule of RNA, approximately 8000 bases long. Among other things, this RNA molecule codes for four structural polypeptides, VP1, VP2, VP3 and VP4, which together form the protein coat (*capsid*) of the virus. The key to the use of recombinant DNA in producing a non-infective FMD vaccine is that VP1 is the antigenic portion of the virus. Thus it should be possible to mass-produce an effective but safe vaccine by cloning the gene for VP1. For convenience, we shall concentrate on the work of just one laboratory, that of Küpper *et al.* (1981), in Germany. Other laboratories have performed similar work, as described briefly at the end of this section.

Since the genome of FMD virus consists of RNA, the first step was to produce double-stranded cDNA from the RNA, using the techniques illustrated in Fig. 2.4. Not all of the resultant cDNA fragments were

8000 bases long; they ranged in size from around 200 bases up to 8000 bases as shown, for example, in lanes 1 and 3 of Fig. 2.15. Digestion of these fragments by various restriction enzymes (Fig. 2.15, lanes 4 to 9) led to the construction of a restriction map of the FMD virus, as illustrated in Fig. 2.16. Knowing the approximate molecular weights of the polypeptides produced by FMD RNA, and their order of translation, it was then possible to roughly align the restriction map with the position of viral genes, as also shown in Fig. 2.16.

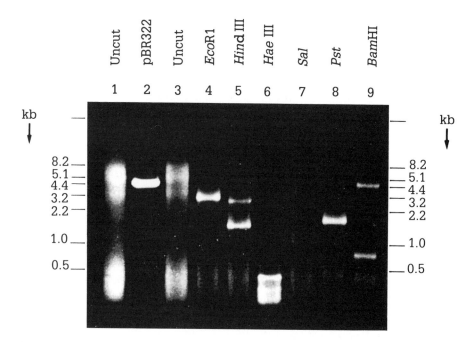

Fig. 2.15. Electrophoresis of fragments of cDNA copied from the RNA of foot-and mouth disease (FMD) virus. Lane 2: plasmid pBR322, used as a standard, because it is known to be 4362 bases long (written as 4.4 kb in the figure). Lanes 1 and 3: uncut cDNA fragments, as formed from FMD viral RNA. The two bands of approximately equal intensity indicate that around one half of the fragments ranged in size from 3 kb to 8 kb, while the other half were much smaller, being less than 0.5 kb. Lanes 4 to 9: results of digestion with the restriction enzymes shown. (From Küpper *et al.* 1981.)

The next step involved cloning various cDNA fragments in *E. coli*, using the general principles illustrated in Fig. 2.5. Then, by restriction-mapping a number of these clones, it was possible to align the cloned cDNA fragments with the whole viral genome (Fig. 2.16), and thus to identify fragments 144, 715, 1034, 1448, and 1824 as probable bearers of the VP1 gene. Since the amino acid sequence of VP1 was known

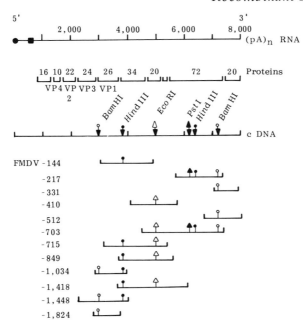

Fig. 2.16. Top line: the RNA genome of foot-and-mouth disease (FMD) virus. ● indicates a viral genomic protein linked covalently to the 5′ end. ■ is a poly (C) tract. $(pA)_n$ represents polyadenylation at the 3′ end. Numbers indicate the number of bases. Second line: the position of genes within the viral genome. Numbers indicate molecular weight $(\times 10^{-3})$ of proteins produced by the genes. Third line: a restriction enzyme map of a cDNA copy of the viral genome obtained from digestions like those shown in Fig. 2.15. Bottom: alignment of various cloned cDNA fragments. Symbols indicate cleavage sites corresponding to those in the restriction map above. (From Küpper *et al.* 1981.)

from previous work, it was then a matter of determining the DNA sequence of the relevant fragments, to see if their DNA sequence corresponded to the amino acid sequence of VP1. In the German laboratory, this was done for fragments 144, 715 and 1034, with the results as shown in Fig. 2.17. From this and other evidence it was concluded that cloned fragment 1034 would be suitable for the mass production of VP1.

Similar work to that described above has been conducted in the USA (Kleid *et al.* 1981), where VP1 is called VP3, and in England (Boothroyd *et al.* 1981). The English workers have concentrated on a cloned cDNA fragment that codes for all four coat proteins, whereas the Americans worked with a cloned fragment that coded only for VP1, which they were able to produce at the rate of approximately 10^6 molecules per *E. coli* cell. In initial trials, the Americans have shown that VP1 produced from *E. coli* could be an effective vaccine.

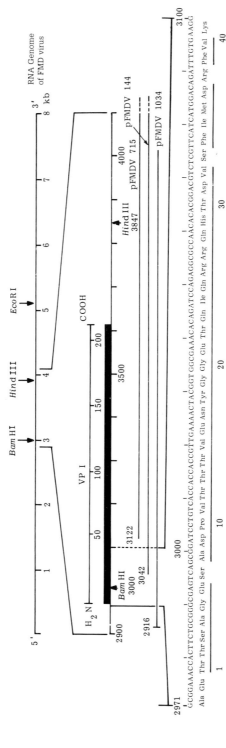

Fig. 2.17. A detailed map of the VP1 gene, shown in relation to the whole viral genome. Also shown is part of the nucleotide sequence obtained from cloned fragments 144 and 1034, and the deduced amino acid sequence of the VP1 protein. Only the few amino acids not underlined differ from the amino acid sequence that had been determined previously by other workers. (From Küpper *et al.* 1981.)

Work in this area is continuing at a great pace, especially in relation to field trials of the resultant vaccines, and also in relation to determining the genetic difference between immunologically different forms of the FMD virus (called *serotypes*). It is only a matter of time before the nucleotide sequences of cDNA for the VP1 protein of all important serotypes are determined. We will then have a much greater understanding of the nature of each serotype and of the differences between them. With this knowledge, it should be possible to produce large quantities of a set of non-infective vaccines, and to use these in a well-organized and soundly-based vaccination programme.

2.9   The future of recombinant DNA

There are three main areas in which recombinant DNA offers substantial prospects.

2.9.1 *Production of large quantities of particular molecules*

In principle, it should be possible to use recombinant DNA technology to mass produce any polypeptide, because polypeptides are the translation products of DNA. Principally, what is required is the isolation or construction of the relevant fragment of DNA, i.e. the relevant gene, and its insertion alongside or somewhere downstream from a strong promoter in a suitable host.

The list of polypeptides that could be usefully produced is very long. It includes naturally occurring microbial products such as enzymes (including the many enzymes used in genetic engineering) and many products of higher organisms such as growth hormone, insulin, epidermal growth factor (a defleecing agent), and interferon. It also includes the antigenic portion of many disease-causing organisms, which could be used as vaccines.

2.9.2 *Probes*

The use of DNA probes will have a major impact on detection of viruses and other disease organisms in plants, animals and humans. All that is needed is for a radioactively-labelled cloned fragment of DNA corresponding to a portion of the DNA or RNA of the pathogen to be mixed with DNA or RNA extracted from tissue of the individual being tested. It is then only a small step, at least in principle, to use DNA probes corresponding to plant or animal or human genes to identify individuals having particular genes. In the short term, this may enable detection of individuals carrying defective genes (see Section 11.3.8), or genes for disease resistance. In the longer term, it may provide a new tool for plant and animal breeders aiming to improve characteristics such as growth rate or productivity.

2.9.3 Modification of the genetic potential of cells and organisms

Recombinant DNA has opened the way for the controlled transfer of genes from one organism to another. There is still much work to be done, but the basic principles are clear. It is possible, for example, that micro-organisms used extensively in beverage, food, and antibiotic production, and those that operate in symbiosis with animals (such as rumen bacteria) could be made more efficient by the insertion of a gene or genes from other strains or species. In the longer term, it is possible that foreign genes could be introduced into plants and animals, either to correct genetic deficiencies, or to enhance performance.

2.10 The limitations of recombinant DNA

Before we get completely carried away with the implications of recombinant DNA, it is important to realize that:

> *Although the possibilities of recombinant DNA are considerable, there are many technical and basic scientific problems that need to be solved before many of the possibilities become realities.*

Some of these problems will be solved in the next few years, but others will require a much longer time.

2.11 The potential dangers of recombinant DNA

When recombinant DNA technology first became a reality in the early 1970s, fears were expressed by molecular biologists as to the potential dangers of this work. It was feared, for example, that if harmful strains of *E. coli* were inadvertently produced, unprecedented pandemics of novel animal and human diseases may result. Scientists working in the area agreed to moratoria on those types of experiments considered to be potentially the most dangerous, and guidelines were drawn up for the conduct of all work with recombinant DNA.

In the following years, it has become apparent that the initial worst fears of both scientists and the community in general are unlikely to be realized. Consequently, although considerable care is still taken in recombinant DNA work, the guidelines are now more relaxed.

2.12 Summary

Recombinant DNA technology is the result of more than thirty years of research in molecular genetics. Two of the most important stages in its development were the discovery of restriction enzymes and the discovery of the enzyme reverse transcriptase. Restriction enzymes cut DNA within specific nucleotide sequences, producing either blunt or sticky ends. The enzyme reverse transcriptase enables a strand of complementary DNA (called cDNA) to be produced from RNA.

Gene cloning is the production of a large number of copies of a particular gene. It is achieved by joining DNA that is to be cloned (called foreign DNA) to a vector which is able to replicate within a particular host. This joining, which is called splicing or ligation, is brought about by an enzyme called DNA ligase. The DNA molecule resulting from joining the foreign DNA segment to the vector is called recombinant DNA. The most commonly used vectors are plasmids, which are circles of double-stranded DNA that exist in certain bacterial cells (hosts) independently of the main bacterial chromosome.

The production of polypeptide from cloned DNA requires the presence of sequences of DNA that control transcription and translation. The success achieved in coaxing *E. coli* and other prokaryotic hosts to mass-produce polypeptide from a diversity of eukaryotes is remarkable testimony to the universality of the genetic code.

The gene bank or gene library of an organism is the whole of its DNA stored as fragments in a large number of colonies of the host (e.g. bacteria) into which the fragments have been cloned.

Recombinant DNA offers substantial prospects in the production of large quantities of particular proteins, in the use of probes for the detection of disease organisms and of genes in plants, animals and humans, and in the modification of the genetic potential of cells and organisms. Although the possibilities of recombinant DNA are considerable, there are many technical problems that need to be overcome before many of the possibilities become realities.

2.13 Further reading

Chambon, P. (1981). Split genes. *Scientific American* **244** (5), 48–59. (A more detailed review of the work on split genes described in this chapter.)

Freifelder, D. (Ed.) (1978). *Recombinant DNA: readings from Scientific American*. W. H. Freeman and Company, San Francisco. (A collection of review articles.)

Gilbert, W. and Villa-Komaroff, L. (1980). Useful proteins from recombinant bacteria. *Scientific American* **242** (4), 74–94. (A useful review of the techniques and applications of recombinant DNA.)

Malik, V. S. (1981). Recombinant DNA technology. *Advances in Applied Microbiology* **27**, 1–84. (A thorough review, with a lengthy discussion of the potential practical applications of recombinant DNA.)

Messel, H. (Ed.) (1981). *The biological manipulation of life*. Pergamon Press, Sydney. (A set of lectures primarily concerned with recombinant DNA, designed specifically as a basic introduction to the subject.)

Nobel lectures, published in the journal *Science*. Eight research workers have been awarded the Nobel Prize for their pioneering work in recombinant DNA. They are Baltimore and Temin (1975, reverse transcriptase), Arber, Nathans, and Smith (1978, restriction enzymes), Berg (1980, the first recombinant DNA), and Gilbert and Sanger (1980, DNA sequencing). The Nobel lectures delivered by each of these recipients are valuable reviews, particularly because of the insight they give into how some of the important discoveries in recombinant DNA were made. The abbreviated references for these lectures are Baltimore (1976) **192**, 632–6; Temin (1976) **192**, 1075–80; Arber (1979) **205**, 361–5; Nathans (1979) **206**, 903–9; H. O. Smith (1979) **205**, 455–62; Berg (1981) **213**, 296–303; Gilbert (1981) **214**, 1305–12, and Sanger (1981) **214**, 1205–10.

Nucleotide Sequence Data Library News. (A newsletter available from the European Molecular Biology Laboratory, Postfach 10 22 09, D-6900, Heidelberg, F.R.G., where an up-to-date library of all known nucleotide sequences in all species is maintained on magnetic tape. Access to the library is available to anyone, through purchase of a tape, or through on-line access from a computer terminal via a telephone line.)

Old, R. W. and Primrose, S. B. (1985). *Principles of gene manipulation* (3rd edn). Blackwell Scientific Publications, Oxford. (Probably the best introduction to recombinant DNA, but written at a more advanced level than the lectures in Messel's book.)

Rutter, J. M. (1981). Gene manipulation and biotechnology. *Veterinary Record* **109**, 192–4. (A brief summary of recombinant DNA.)

Science **196** (4286), **209** (4463). (These two issues of *Science* were devoted entirely to recombinant DNA. The first, published in 1977, is concerned mainly with methodology. The second, published in 1980, concentrates more on the various ways in which recombinant DNA can be applied to particular areas of research.)

Scientific American **245** (3). (This issue, published in 1981, contains a number of papers describing the practical applications of recombinant DNA.)

Sinsheimer, R. L. (1977). Recombinant DNA. *Annual Review of Biochemistry* **46**, 415–38. (Another useful review.)

Watson, J. D., Tooze, J., and Kurtz, D. T. (1983). *Recombinant DNA: a short course*. Scientific American Books, New York. (An excellent book, with an abundance of very helpful diagrams.)

Part II
Genetics and animal disease

Introduction

Part II of this book deals with genetic aspects of animal disease. It is primarily concerned with the principles of medical genetics, illustrated wherever possible with examples from domestic animals. In choosing animal examples, preference has been given to defects or diseases that are economically important or are well documented, or which illustrate a particular point clearly. No attempt has been made to present a complete catalogue of inherited defects and diseases. Those readers requiring information on a particular defect or disease not mentioned in this book should consult one or more of the catalogues listed below.

The last two chapters of this part cover topics that are particularly relevant to animal disease. The first of these, Chapter 10, is concerned with the genetical aspects of interaction between hosts on the one hand, and parasites and pathogens on the other. Particular attention is given to the development of resistance in parasites and pathogens. The final chapter examines the environmental and genetic options available for the control of inherited diseases in animals.

There are several books and reviews that deal with genetical aspects of animal disease. Readers requiring another point of view or further information on a particular topic may benefit from consulting some of the publications listed below.

Books

Hámori, D. (1983). *Constitutional disorders and hereditary diseases in domestic animals.* Elsevier, Amsterdam.

Popescu-Vifor, St., Sarbu, I., Ciupercescu, D. D., and Grosu, M. (1980). *Genetica si eredopatologie [Genetics and Inherited Pathological Conditions]* Editura Didactica si Pedagogica, Bucharest, Romania.

Wiesner, E. and Willer, S. (1974). *Veterinarmedizinische Pathogenetik [Veterinary Pathogenetics].* VEB Gustav Fisher Verlag, Jena, East Germany.

Reviews

Basrur, P. K. (1980). Genetics in veterinary medicine. In *Scientific Foundations of Veterinary Medicine* (Eds Phillipson, A. T., Hall, L. W. and Pritchard, W. R.) pp. 393–413, Heinemann, London.

Mulvihill, J. J. (1972). Congenital and genetic disease in domestic animals. *Science* 176, 132–7.

Wegner, W. (1976–9). Defekte und Dispositionen. (This is a series of articles published in *Tierärztliche Umschau*. As a whole, this series constitutes an extensive and thorough review of genetics in relation to animal disease. For bibliographic details, see the reference list at the end of this book.)

Catalogues

Some of the following catalogues are concerned with congenital defects. They are included here because a large number of congenital defects are genetic in origin. Full bibliographic details of the catalogues are given in the list of references at the end of this book.

General: Altman and Katz (1979); Andrews *et al.* (1979); Ballarini (1977); Blood *et al.* (1979, chapter 34); Cornelius (1969); Dezco (1974); Festing (1979); Foley *et al.* (1979); Gershwin and Merchant (1981); T. C. Jones (1978); Jones *et al.* (1972 *et seq.*); Patterson (1975); Rose and Behan (1980); WHO (1973–1982); Young (1967).

Laboratory animals: T. C. Jones (1978).

Cats: Patterson (1979); Robinson (1977, chapter 7); Saperstein *et al.* (1976).

Dogs: Erickson *et al.* (1977); Leipold (1977); Patterson (1977); Patterson (1979); Robinson (1982, chapter 8); Van der Velden (1979).

Sheep: Behrens (1979, pp. 255–60); Dennis and Leipold (1979); Saperstein *et al.* (1975).

Pigs: Huston *et al.* (1978); Queinnec (1975).

Cattle: Cho and Leipold (1977); Lauvergne (1968); Lauvergne (1978); Leipold *et al.* (1972); Predojevic (1973); Queinnec (1977); Leipold *et al.* (1983).

Horses: Antaldi (1980); Huston *et al.* (1977); Trommershausen-Smith (1980).

Wild and zoo mammals: Saperstein *et al.* (1977).

Birds: Arnall and Keymer (1975, pp. 367–78); Riddle (1975); Tudor (1979).

3
Biochemical genetics

3.1 Introduction

Ultimately, all simply inherited disorders will be understood in biochemical terms. At present, however, there is only a small proportion of inherited diseases for which a biochemical explanation is possible, and these are the subject matter of this chapter. Although simply inherited diseases currently represent only a small proportion of all inherited disorders, they do include some well-known and economically important diseases and hence are worthy of our consideration. A more important reason for studying them is that in learning about known biochemical disorders, we are learning about the basic principles that in future will be found to determine many other inherited diseases.

All the inherited disorders whose biochemical basis has been identified to date are the result of gene mutations. At one extreme, the simplest alteration in polypeptide structure that can result from such a mutation is the substitution of just one amino acid for another at a certain position along the polypeptide chain. At the other extreme, a mutation may render the cell incapable of synthesizing any polypeptide at all. If an altered (mutant) polypeptide is synthesized, then it may be just as effective in its specific biological role as the polypeptide it has replaced. Alternatively, the mutant polypeptide may be unable to carry out its specific function, or may carry out that function with reduced efficiency. Whether a polypeptide is present but unable to act properly, or is completely absent, the end result is the same: there is a deficiency of functional polypeptide and a consequent impairment of the physiological process in which that polypeptide is required. In discussions of biochemical disorders, therefore, whether a molecule is completely absent, or is present in a form that is unable to function properly, it is said to be *deficient*.

The aim of this chapter is to explain the main types of biochemical disorders that result from deficiencies of polypeptides brought about by mutation.

3.2   Inborn errors of metabolism

If a particular polypeptide acts as an enzyme or is part of an enzyme, then a mutation in the relevant gene sometimes results in a deficiency of that enzyme, with a consequent blockage in the relevant biochemical pathway at the point where that enzyme is required. Diseases that result from such blockages are called *inborn errors of metabolism*. A good example of an inborn error of metabolism in animals is congenital haemolytic anaemia associated with pyruvate kinase deficiency, which has been reported in Basenji and Beagle dogs.

3.2.1   *Congenital haemolytic anaemia due to pyruvate kinase deficiency*

In this disease, the affected biochemical process is anaerobic glycolysis in red blood cells. The relevant reactions in the pathway are illustrated in Fig. 3.1. It is clear from the pathway diagram that a deficiency of the enzyme pyruvate kinase (PK) will result in decreased glucose utilization, a build-up of intermediates between glucose and phosphopyruvic acid, a deficiency of pyruvic and lactic acids, and, most importantly, a decreased level of adenosine triphosphate (ATP) in red blood cells.

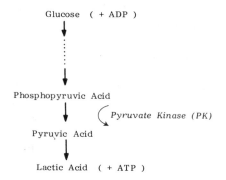

Fig. 3.1. A portion of the pathway describing anaerobic glycolysis in red blood cells, showing some of the intermediate steps by which glucose (plus adenosine diphosphate, ADP) is converted to lactic acid (plus adenosine triphosphate, ATP, which is an important source of energy for red blood cells). The enzyme pyruvate kinase (PK) is required for the conversion of phosphopyruvic acid to pyruvic acid. (After Patterson 1975.)

Since ATP is a major source of energy in these cells, the decreased ATP level is an important contributing factor to a decreased life span of red blood cells, and hence to what is seen as congenital haemolytic anaemia. Other clinical signs, many associated with a lack of energy, include syncope during exercise, weakness, excessive sleeping, pallor, tachycardia, splenomegaly, and occasionally orange faeces (Prasse 1977). It is a sobering thought that all these signs have probably resulted from a single base substitution in a DNA molecule. These diverse results of a single mutation provide a good example of *pleiotropy*, which is the existence of more than one end result from the action of one gene. A gene that influences a variety of characteristics is said to be *pleiotropic*.

If tests for erythrocytic PK activity are conducted among normal dogs within a population known to have produced animals with congenital haemolytic anaemia, then it is found that in some normal animals, including all those that are parents of affected animals, the level of PK activity is approximately 50 per cent of that observed in the remainder of normal animals. It is also found that those animals with the disease have essentially zero PK activity.

The explanation for these observations is that in affected individuals, both genes responsible for the production of PK are mutants. Affected animals are thus homozygous for the defective gene (*dd*). Those normal animals with normal PK activity have two normal PK genes (*DD*), and are thus homozygous for the normal gene. Those normal animals with approximately 50 per cent of normal PK activity are heterozygous (*Dd*), having one normal gene (producing normal PK) and one defective gene (producing very little or no active PK). Thus the level of PK activity is directly proportional to the number of 'doses' of the normal gene.

> *This is important from a practical point of view, because it provides a means of identifying heterozygotes or carriers.*

Obviously, the same conclusion applies to any simply-inherited, autosomal disease provided that the gene product is known and that its quantity or activity can be measured in live animals: carriers can be detected by a simple biochemical test, because they show approximately 50 per cent of normal enzyme activity. As discussed in Chapter 11, the detection of carriers is of major importance in many attempts at controlling genetic diseases.

As well as illustrating carrier detection by biochemical means, inborn errors of metabolism provide good illustrations of other important genetic concepts.

3.2.2 Type of gene action

In the case of congenital haemolytic anaemia discussed above, and many other inborn errors of metabolism, the *D* allele is said to be completely *dominant* to the *d* allele with respect to clinical signs of the disease. Another way of expressing this is to say that *d* is *recessive* to *D*, or that congenital haemolytic anaemia is a recessive disease. More generally, an allele is recessive for any characteristic if its effect with respect to that characteristic is not evident in the heterozygote. Likewise, an allele is dominant with respect to a particular characteristic if its effect is the same in heterozygotes as in homozygotes. Now, suppose that the characteristic with which we are concerned is the level of PK activity, rather than clinical signs. In this case, the *D* allele and the *d* allele are said to be each *co-dominant* or *incompletely dominant*, because the heterozygote exhibits the effect of both alleles. The terms recessive, dominant, co-dominant and incompletely dominant describe specific relationships between alleles, or specific *types of gene action*. From the above example, it is evident that:

> *Two alleles at one locus can exhibit more than one type of gene action, depending on the characteristic being considered.*

3.2.3   Genotype and phenotype

It is convenient to introduce two more terms at this stage. The first is *genotype* which refers to the genetical constitution of an individual at one or more loci. Carriers of congenital haemolytic anaemia, for example, have the genotype *Dd*, while affected animals have the genotype *dd*. The other term is *phenotype* which is an observable characteristic of an individual. In relation to PK activity in congenital haemolytic anaemia, there are three different phenotypes (normal activity, 50 per cent activity and zero activity) corresponding exactly to three respective genotypes (*DD*, *Dd*, and *dd*). With respect to clinical signs, however, the same three genotypes give rise to only two phenotypes: both *DD* and *Dd* animals are normal while *dd* animals are affected. There is not always, therefore, a one-to-one relationship between genotype and phenotype. In fact, situations in which there is a one-to-one relationship are the exception rather than the rule.

3.2.4   Inborn errors of lysosomal catabolism

Other good examples of inborn errors of metabolism are the inherited lysosomal storage diseases that result from defects in lysosomal catabolism. Lysosomes are small membrane-bound organelles found in the cytoplasm. They act as the digestive system of the cell and thus contain many enzymes which act in a step-wise manner to break down many different, complex molecules into monomeric units of simple lipids, amino acids, monosaccharides and nucleotides. If a particular enzyme is absent or inactive, then the step-wise degradation is halted, with a consequent build-up ('storage') in the lysosomes or elsewhere of the material that was to have been broken down by that enzyme. In other words, an inborn error of lysosomal catabolism produces a *lysosomal storage disease* (Jolly and Blakemore 1973). Not all lysosomal storage diseases are inherited, and not all inherited lysosomal storage diseases have been identified as owing their origin to an inborn error of lysosomal catabolism. In the following discussion we shall be concerned mainly with those storage diseases that are due to inborn errors of lysosomal catabolism, as they are the best understood.

The formal criteria for a disease to be classified in this category are listed in Table 3.1, while Table 3.2 lists diseases that satisfy these criteria. Although there are potentially a large number of different storage materials and hence potentially many different signs of disease, most lysosomal storage diseases have certain general clinical signs in common. According to Blakemore (1975), affected animals are usually normal at birth but fail to grow as rapidly as their litter mates or con-temporaries; many of the diseases manifest as neurological disorders in young animals; the disease is always progressive and has a fatal outcome, and the age of onset and speed of progression are variable but the two are usually directly related. Histologically, lysosomal storage diseases are characterized by cellular bodies containing the storage substance or, if the storage substance dissolves during histological preparation, by empty vacuoles.

Table 3.1. The formal criteria for a disease to be classified as an inborn error of lysosomal catabolism

---

1. The disease should be a storage disease.
2. It should be simply inherited.
3. The storage material, which need not be homogeneous, should be stored at least initially within lysosomes.
4. There should be a partial or absolute deficiency of one of the lysosomal enzymes.
5. This enzyme would normally hydrolyse the storage material.
6. Other enzymes should have normal or increased activities.

(Adapted from Hers 1965.)

The most economically important and extensively studied inborn error of lysosomal catabolism is mannosidosis in cattle. Prior to the introduction of a control programme in 1974 in New Zealand, where it was most prevalent, approximately 3000 Angus calves were affected each year. As it usually produces death within the first year of life, this disease represented a substantial problem to breeders of Angus cattle, and especially to owners of herds that happened to have a high incidence of the disease. We shall discuss the results of the New Zealand mannosidosis control programme in Chapter 11, when we consider the various means available for the control of genetic diseases.

The other inherited lysosomal storage disease that we shall consider is Chediak-Higashi disease, an autosomal recessive disease that is reported commonly in mink, and rarely in mice, cats, cattle, killer whales and humans. It is characterized by defective pigmentation leading to partial

Table 3.2. Inherited storage diseases due to inborn errors of lysosomal catabolism (adapted from Jolly 1977*a*)

Deficient enzyme	Storage substance	Disease synonyms	Species	Breed	References
α-glucosidase	Glycogen	Glycogen storage disease type II, Pompe's disease	cat dog sheep cattle	domestic Lapland Corriedale Shorthorn	Sandstrom *et al.* (1969) Mostafa (1970) Manktelow and Hartley (1975) Howell *et al.* (1981)
α-mannosidase	Oligosaccharides containing mannose and N-acetyl-glucosamine	Mannosidosis, Pseudolipidosis	cattle	Aberdeen Angus Murray Grey Tasmanian Grey	Jolly (1978) Jolly (1978) Jolly (1978)
β-mannosidase	As above	Mannosidosis	goat	Nubian Anglo-Nubian	Jones and Dawson (1981) Healy *et al.* (1981)
α-fucosidase	Glycoasparagines	Fucosidosis	dog	Springer Spaniel	Healy *et al.* (1984)
α-iduronidase	Glycosaminoglycans	Mucopolysaccharidosis I, Hurler syndrome, Scheiesyndrome	dog cat	Plott domestic	Shull *et al.* (1982) Haskins *et al.* (1979*b*)
Arylsulphatase B	Glycosaminoglycans	Mucopolysaccharidosis VI, Maroteaux-Lamy syndrome	cat	Siamese	Haskins *et al.* (1979*a*)

Enzyme	Substrate	Disease	Species	Breed	Reference
β-galactosidase	GM_1 ganglioside	GM_1 gangliosidosis	cat	domestic, Siamese	Barnes *et al.* (1981) Farrell *et al.* (1973)
			dog	Various breeds	Rittmann *et al.* (1980)
			cattle	Friesian	Donnelly and Sheahan (1981)
Hexosaminidase A	GM_2 ganglioside	GM_2 gangliosidosis, Tay-Sach's disease, Cerebrospinal lipodystrophy	cat	domestic	Cork *et al.* (1977)
			pig	Yorkshire	Baker, H. J. *et al.* (1976, p. 1199)
Sphingomyelinase	Sphingomyelin	Sphingomyelinosis, Niemann-Pick's disease	cat	Siamese	Wenger *et al.* (1980)
Glucocerebrosidase	Glucocerebroside	Glucocerebroside storage disease, Gaucher's disease	dog	Australian Silkie	Farrow *et al.* (1982)
β-galactocerebrosidase	Galactocerebroside	Globoid cell leucodystrophy, Krabbe's disease	cat	domestic	Johnson (1970)
			dog	various	Suzuki *et al.* (1974)

albinism, anomalous giant granules in leucocytes, an increased suscepti-bility to disease and premature death (Padgett 1968, 1979). It appears that the giant granules are lysosomes distended because of the storage of an incompletely broken-down lysosome metabolite, as yet unidenti-fied. Unlike many other lysosomal storage diseases, Chediak–Higashi disease is not associated with neurological disturbances. The most interesting aspect of this disease is that it occurs in all mink that are homozygous for a recessive coat colour allele known as Aleutian. The reason for this is that the Aleutian coat colour, which results in attrac-tive and valuable pelts, is the result of the same defective pigmentation that is part of the clinical signs of Chediak–Higashi disease. Thus in selecting mink for Aleutian coat colour, breeders have been unknow-ingly selecting for a lysosomal storage disease as well. Apparently, the Aleutian coats are sufficiently valuable for the breeders to tolerate the increased susceptibility to disease and early death of the animals that grow them. So serious is the susceptibility to disease that for many years a particular viral disease was seen only in Aleutian mink and in fact became known as Aleutian disease, a name that remains in use today (Porter *et al.* 1980). Although it is now known that all mink are susceptible to this disease, those mink that are homozygous for the Aleutian allele have a 5 to 8-fold higher mortality to the relevant virus than do other mink.

3.2.5   *Porphyria and protoporphyria*

Two interesting inborn errors of metabolism are associated with defects in the synthesis of haem. The first of these is porphyria, an autosomal recessive disease due to a deficiency of the enzyme uroporphyrinogen III cosynthetase, which leads to a build-up of the intermediates of haem biosynthesis and a lack of haem itself. This deficiency of haem gives rise to haemolytic anaemia which is one of the signs of the disease. The intermediates in the haem pathway, called porphyrinogens, are readily oxidized to porphyrins, which are aromatic compounds that absorb visible light and induce photosensitivity. The most common porphyrinogen involved in porphyria is uroporphyrinogen I, which when accumulated in excess, results in a characteristic red staining of teeth, bone and urine. Known as pink-tooth in cattle, this disease has also been observed in cats, pigs and humans (Levin 1974). In all of these species it is a rare disease in which heterozygotes have an enzyme activity intermediate between that of normal homozygotes and diseased animals.

In contrast, all fox squirrels (*Sciurus niger*) have extremely low activity of uroporphyrinogen III cosynthetase and hence all have red bones and teeth due to the deposition of excess uroporphyrinogen I. However, fox squirrels show neither photosensitivity nor haemolytic anaemia, which indicates that they have developed some physiological means of coping with low activity of uroporphyrinogen III cosynthetase.

Another enzyme involved in the biosynthesis of haem is ferrochelatase. If this enzyme is absent, or is present with markedly reduced activity, then there is a build-up of protoporphyrins which results in photosensitivity. Unlike porphyria, there is no associated colouration of teeth, bones and urine. This disease, which is called protoporphyria, has been reported in cattle (Ruth *et al.* 1977) and in humans.

3.2.6 *Dermatosparaxis*

The final inborn error of metabolism that we shall consider is a heritable disorder of connective tissue. Some animals are born with easily extendible and very fragile skin, a condition that in sheep (Fjolstad and Helle 1974) and cattle (Hanset and Ansay 1967) is known as dermatosparaxis. In animals suffering from this disease, severe lacerations result from the slightest scratch that in normal animals would cause only slight damage. In sheep and cattle, the cause of this disease is a build-up of an abnormal procollagen due to a deficiency of the enzyme procollagen peptidase. Normal cylindrical collagen fibrils cannot be formed in animals with this disease, and instead the abnormal procollagen forms itself into flattened twisted ribbons. In both sheep and cattle, this disease is inherited as an autosomal recessive condition.

3.2.7 *Genetical heterogeneity of disease*

A disease with the same clinical but different histo-pathological signs as dermatosparaxis has been reported in mink and dogs (Hegreberg *et al.* 1969), and in cats (Patterson and Minor 1977). In these three species, the disease is called cutaneous asthenia, and appears to be inherited as an autosomal dominant condition. However, its biochemical basis in these three species has not yet been identified.

> *The fact that a specific set of clinical signs can represent more than one disease in terms of inheritance is an indication of genetical heterogeneity of disease.*

In sheep, cattle, mink, dogs, and cats the clinical signs are similar; it is only when the biochemical and histo-pathological features are examined as well that at least two different diseases are recognized. The first disease (autosomal recessive in sheep and cattle) involves a deficiency of an enzyme involved in the processing of procollagen, giving rise to characteristic arrangements of abnormal procollagen, while the second disease (autosomal dominant in mink, dogs and cats) is characterized by abnormal collagen involved in packing defects in fibrils and fibres, and apparently no enzyme deficiency. The locus for the first disease is the structural gene for an enzyme involved in pro-collagen processing, while the locus for the second disease is probably the structural gene for collagen itself.

This is an example of genetic heterogeneity *between* species. Although such heterogeneity can cause confusion from time to time, a far greater source of confusion is genetic heterogeneity of disease *within* a species. For example, both forms of the above connective tissue disorder occur in humans.

If a set of clinical signs appears to have a genetic basis, but if genetic heterogeneity within a species remains undetected, it will be impossible to establish the exact form of inheritance. Indeed, as long as the heterogeneity remains, the available data will present a most confusing picture of inheritance. Although it is easy to be wise after the event, it is very difficult in practice to ensure that the available data on any syndrome all belong to the one disease entity and hence are homogeneous. Data are more likely to represent genetic homogeneity if they have been obtained from animals that not only present with the same clinical signs, but which also have the same histo-pathological and biochemical lesions as well.

3.3   Type of gene action and type of disease

In the connective tissue diseases described above, two types of gene action have been observed. One disease is inherited as an autosomal recessive condition and is associated with deficiency of an enzyme that processes procollagen. The other appears to be inherited as an autosomal dominant condition, and is associated with an abnormal molecule which may be due to mutation of the collagen structural gene.

> *There are reasonable grounds for expecting recessive diseases to be associated with enzyme deficiencies, and dominant or codominant diseases to be caused by defects in non-enzymatic polypeptides.*

Inborn errors of metabolism, for example, are inherited as recessive diseases, because enzymes are required in such small quantities that 50 per cent activity in heterozygotes is sufficient for normal functioning. On the other hand, if the mutant polypeptide has, for example, a structural role rather than an enzymatic role, then the structures incorporating that polypeptide will be defective in some way, and the heterozygote may well show some form of the disease. This is the most likely explanation currently available for the situation seen with cutaneous asthenia in mink, dogs and cats. It remains to be seen whether further research will substantiate this explanation.

If the polypeptide is a substrate in a particular process or is involved in transport, then a decrease by one-half in the production of normal polypeptide as occurs in heterozygotes might be expected to cause some clinical signs in the heterozygote because, like those with a structural role, polypeptides involved in transport and acting as substrate are also often required in relatively large quantities.

In general, therefore, non-enzymatic polypeptides should lead to dominant or at least incompletely dominant diseases, while enzyme defects should give rise to recessive diseases.

Whether or not this prediction holds for the inherited connective tissue diseases discussed above, there is increasing evidence, mainly from human inherited diseases, that it is quite useful as a general rule. It has certainly been well enough substantiated for there to be general agreement that it is a waste of time to search for an enzymatic deficiency in any disease that is inherited in a dominant manner. Likewise it is generally agreed that a search for enzymatic defects should have high priority in attempts to determine the biochemical basis of recessive diseases.

3.4   Inherited bleeding disorders

Haemostasis, or the arrest of bleeding, involves the blood vessel wall, blood platelets and coagulation factors. Among the many recognized bleeding disorders are those due to a deficiency in one of the coagulation factors. Table 3.3 lists the coagulation factors for which a simply inherited deficiency disease has been reported in animals. With the exception of haemophilia A and haemophilia B, heterozygotes for the bleeding disorders listed in Table 3.3 often exhibit a mild form of the disease.

In these cases, the type of gene action with respect to the disease is the same as the type of gene action with respect to the coagulation factor activity; the normal and the defective allele are co-dominant or incompletely dominant in terms of the disease and in terms of the biochemical lesion. It is obvious that detection of carriers for these conditions can be based initially on clinical signs, and subsequently confirmed with a coagulation factor assay. Since none of the factors deficient in these diseases is an enzyme, the co-dominant or incompletely dominant pattern of inheritance is consistent with the generalization made in the previous section. However, factors II, VII, IX, X, and XI are all precursors of enzymes. We must conclude that the presence of one-half of the normal concentration of enzyme precursor in heterozygotes is not sufficient to prevent some clinical signs appearing.

Unlike all the other disorders in Table 3.3, haemophilia A and haemophilia B are recessive diseases. Heterozygotes, therefore, exhibit no signs of the disease although they do show a reduced coagulation factor activity. Another interesting aspect of these two diseases is

Table 3.3. Inherited bleeding disorders associated with the deficiency of a specific coagulation factor (compiled mainly from Dodds (1975, 1981) and from Spurling (1980))

Deficient factor	Disease synonyms	Form of inheritance	Species	Breed	References
I	Hypofibrinogenaemia, Afibrinogenaemia	Autosomal	dog goat	St. Bernard Saanan	Kammermann et al. (1971) Breukink et al. (1972)
II	Hypoprothrombinaemia, Dysprothrombinaemia	Autosomal	dog	Boxer	Dodds (1977)
VII	Hypoproconvertinaemia	Autosomal	dog	Alaskan malamute, Beagle	Spurling (1980)
VIII	1* Haemophilia A, Classic Haemophilia	X-linked	cat dog	domestic most breeds	Cotter et al. (1978) Spurling (1980)
	2* von Willebrand's disease, Pseudohaemophilia, Vascular Haemophilia	Autosomal	horse rabbit pig dog	thoroughbred, standardbred crossbred several breeds numerous breeds	Archer and Allen (1972) Benson and Dodds (1977) Dodds (1981) Dodds (1981)
IX	Haemophilia B, Christmas disease	X-linked	dog cat	several breeds British shorthair	Spurling (1980) Dodds (1978)
X	Factor X deficiency, Stuart factor deficiency	Autosomal	dog	Cocker Spaniel	Dodds (1973)
XI	Factor XI deficiency, PTA deficiency	Autosomal	dog cattle	several breeds Holstein	Dodds (1981) Gentry et al. (1975)
XII[+]	Factor XII deficiency, Hageman trait	Autosomal	cat	domestic	Kier et al. (1980)

* While both these diseases involve factor VIII deficiency, they can be distinguished in terms of von Willebrand factor protein, VWF (normal in 1 and low in 2) and platelet retention or adhesiveness (normal in 1 and low in 2). For further distinguishing features, see Dodds (1977). † Deficiency of this factor does not result in a bleeding disorder.

that they are the only two that are X-linked. Because they are X-linked, heterozygotes are seen only in the homogametic sex and there are no carriers of these diseases among the heterogametic sex. Thus in species in which these diseases have been reported, all carriers are female.

The differences between the haemophilias on the one hand, and all the other bleeding disorders in Table 3.3 on the other, raise two important questions: why are the haemophilias recessive diseases when they appear not to be due to an enzyme deficiency, and how can a deficiency of the one coagulation factor (factor VIII) be inherited in two different ways, with haemophilia A being X-linked and von Willebrand's disease being autosomal? Although there is still much to be understood in this area, clues to the answers to these questions are provided by the fact that the product of the haemophilia A locus is a protein (called factor VIII procoagulant activity protein or VIIIC) that is involved in regulation of the coagulation cascade, and is required in relatively small quantities. On the other hand, the product of the von Willebrand locus is a protein (called von Willebrand factor protein or VWF) that acts as a carrier of VIIIC, and also plays an important structural role in clot formation; it is required in much larger quantities. The two different forms of inheritance of factor VIII deficiency can be explained by noting that factor VIII is a complex consisting of VWF and VIIIC, with VWF accounting for 99 per cent of the mass of the complex (Zimmerman *et al.*, 1983).

The reduced factor VIII activity that is often observed in heterozygotes for these two diseases is probably a consequence of random X-inactivation, which, as we saw in Chapter 1, occurs in all female mammals. In the case of haemophilia, the end result will be that, on average, one-half of the cells of a heterozygous female will express the normal allele and thus will produce normal quantities of factor VIII, while the other half of the cells will express the harmful allele, and consequently will fail to produce any factor VIII.

One final point should be noted in this regard. Since inactivation is a random process, not all females will have exactly one-half of each type of cell. In fact, a wide range of proportions of the two cell types can occur, from females having most or all cells with the normal allele active, to females having most or all cells with the haemophilia allele active. Females with a large proportion of normal cells may be indistinguishable from those who are homozygous for the normal allele, while females with a low proportion of normal cells may actually exhibit signs of the disease. It is evident that heterozygote detection in such cases is not straightforward. In general:

> *Heterozygote detection for X-linked recessive biochemical defects is not as effective as for autosomal recessive biochemical defects, because of random X-inactivation.*

The final aspect of bleeding disorders that we shall consider is the location of the two X-linked haemophilia loci. In humans, it is known that although both loci are on the X chromosome, the two diseases are inherited independently (unlinked), thus indicating that the two loci are widely separated on the same chromosome. In an attempt to see whether the same was true for dogs, Brinkhous *et al.* (1973) observed the results of matings involving females that were carriers of both haemophilia A and haemophilia B, having the recessive gene for haemophilia A on one X chromosome and the recessive gene for haemophilia B on the other X chromosome. Since only females have two X chromosomes, it is only in females that crossing-over can occur with respect to the two haemophilia loci. The four possible results of meiosis in such females (two crossover X chromosomes and two non-crossover X chromosomes) are shown in Table 3.4. Each of these four chromosomes is readily recognizable in male offspring of the above carrier females,

Table 3.4. Results of test matings conducted between carrier females and any type of male, in order to determine the linkage relationship between haemophilia A and haemophilia B in dogs

Maternal* X chromosome of offspring

Noncrossovers		Crossovers	
Chromosome	Number	Chromosome	Number
a + (●—●)	21	a b (●—●)	12
+ b (●—●)	16	+ + (●—●)	29
Totals	37		41

Frequency of crossovers = 41/(41 + 37) = 0.53, which is not significantly different from the crossover frequency of 0.5 expected with independent segregation.

At each locus, the symbol for the normal (dominant) allele is +, while the recessive alleles for haemophilia A and haemophilia B are indicated by a and b respectively. Since both diseases are X-linked, and since only females have two X chromosomes, crossing-over between the two haemophilia loci can occur only in females. Thus the only chromosomes shown are those from the females used as parents in the test matings. Each of these females was heterozygous at both loci, with the haemophilia alleles being in *repulsion*, i.e. on different chromosomes:

a + / + b
●—● / ●—●

(Adapted from Brinkhous *et al.* (1973) "Expression and linkage of genes for X-linked hemophilias A and B in the dog." *Blood* **41**, 577–85. Reprinted by permission of Grune and Stratton, Inc. and the authors.)

* The maternal X chromosome is the one inherited from the mother or dam.

irrespective of the genotype of the male parent, since the male off-spring have only one X chromosome. These four chromosomes can also be distinguished in female offspring, so long as the genotype of the male parent is known. The results of the matings described by Brinkhous *et al.* (1973), are presented in Table 3.4, and clearly indicate that the two haemophilia loci are unlinked in dogs, just as in humans. These results provide yet another confirmation of the generally accepted hypothesis proposed by Ohno (1967, 1973) that the X chromosome has been particularly conservative throughout mammalian evolution, so that the same loci occur in the same position on the X chromosome in most mammalian species.

3.5 Inherited haemoglobin disorders

Haemoglobin is one of the most intensively studied of all biological molecules. Amino acid sequences have been obtained for haemoglobins from a large number of different species, giving rise to a fascinating picture of evolution. Within many species, different forms of haemo-globin have been identified and, in many cases, explained in terms of several or even just one base substitution in the corresponding DNA. In humans, this work has identified haemoglobins whose function is defective and which give rise to a group of hereditary disorders known as the haemoglobinopathies. The most famous of these is sickle-cell anaemia, whose profound effects were shown in 1956 to be the result of a single substitution of valine for glutamic acid in position 6 of the β peptide chain. From the genetic code in Table 1.2, it can be deduced that this substitution must have been due to changing the relevant DNA triplet from CTT to CAT or from CTC to CAC, either of which involves just one base substitution, of adenine for thymine, in the DNA. The second of these alternatives was confirmed in 1977 when the nucleotide sequence of messenger RNA from the normal β chain was determined. It was found that glutamic acid in position 6 of the normal β chain is encoded by GAG in RNA, and hence by CTC in the corre-sponding DNA.

Another group of haemoglobin disorders in humans is known as the thalassemia syndromes. Their main clinical sign is anaemia which is due to reduced or complete absence of production of either α or β globin. Recent research in this area, using many of the recombinant DNA techniques described in Chapter 2, has been reviewed by Bank *et al.* (1980). It now appears that the deficiencies in production of α or β globin are due to deletions of certain nucleotide sequences in the DNA molecules that normally code for the globins, i.e. in the structural genes. Interestingly, some of the anaemias are also associated with deletions in DNA which is normally adjacent to the structural genes (the so-called flanking sequences), thus confirming that flanking sequences play a role in regulation of expression of structural genes.

Although many different haemoglobins have been identified in animals, no examples have been reported to date that involve a sufficient defect in function to result in clinical signs. Similarly, no cases of thalassemia syndrome have been reported in animals. This discrepancy between the situation with humans and with animals reflects the more intensive clinical supervision given to humans, and also the more intensive research effort in humans. There is every reason to believe that haemoglobinopathies will become known in animals as research proceeds in the future.

3.6   Inherited immunodeficiencies

As described in Chapter 8, the immune system of higher animals consists of two main sections: that which gives rise to cell-mediated immunity (concerned with transplantation immunity and delayed hypersensitivity) and that responsible for the production of antibodies. Figure 3.2 is a simplified representation of the immune system and indicates those points at which blockages have occurred, producing inherited immunodeficiencies in animals.

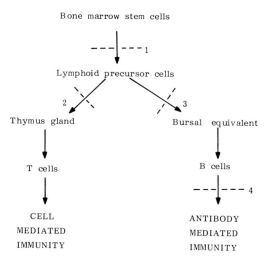

Fig. 3.2. A simplified representation of the immune system, showing points at which blockages are thought to occur in animals, leading to inherited immunodeficiencies.

The most widely studied of such diseases is combined immunodeficiency disease (CID) in Arab horses (Splitter *et al.* 1980). It is an autosomal recessive disease which appears to be due to a blockage between the stem cells and the lymphoid precursor cells (position 1 in Fig. 3.2). As a result, affected foals lack both T cells and B cells, and hence are unable to mount any form of immune response. Clinical signs include lymphopenia, immunoglobulin deficiency, absence of cell-mediated immunity, thymic hypoplasia and a severe reduction in splenic and lymph node lymphocytes. Consequently, foals show a markedly increased susceptibility to infection, and usually die within 5 months of birth. Infections are usually due to equine adenovirus, although the protozoan *Pneumocystis carinii* or various bacteria have been implicated in a number of deaths.

The biochemical defect resulting in CID is not yet known, but because it is a recessive disorder there is a good chance, as we saw earlier, that an enzyme deficiency will be found to be involved. Indeed, in some forms of CID in humans, a deficiency of adenosine deaminase has been detected, although foals with CID show no such deficiency. Recent evidence indicates that CID in foals may be associated with a defect of purine metabolism.

There are various other disorders of the immune system that appear to represent blockages at other points in Fig. 3.2. In Black Pied Danish cattle, for example, there is an autosomal recessive disease known as lethal trait A-46. It is characterized by thymic hypoplasia (position 2 in Fig. 3.2), which leads to a complete loss of ability to mount a cell-mediated response, while interfering only partially with the production of antibodies. Affected calves usually die by four months of age. A case of complete lack of immunoglobulins but with normal T cell function has been reported in a thoroughbred horse. Known as agammaglobulin-aemia, it probably reflects a blockage between the lymphoid precursor cells and the bursal equivalent (position 3 in Fig. 3.2). Selective deficiencies of particular immunoglobins such as IgG2 in Red Danish cattle, and IgM in horses may be due to a specific defect beyond the B cells (position 4 in Fig. 3.2). Defects at other positions have been reported in humans and will probably be found in animals in the future. For a general review of immunodeficiencies in domestic animals, see Perryman (1979).

While most inherited immunodeficiencies are quite rare and are thus not economically important, CID is a serious problem in Arab horses. For example, in 1977 it was reported that 2.3 per cent of 257 foals sampled from 9 states within the USA were affected. And one stud in Australia had the misfortune to import from England two stallions that were both subsequently shown to be carriers of CID. Extensive use of these stallions and their sons and daughters over the years before CID was shown to be an inherited disease, led to the birth of 17 affected foals out of a total of 204, or 8.3 per cent of all foals born in the stud. In Chapter 11 we shall consider alternative programmes that could be used to control CID in such a stud.

3.7 Summary

Only a small proportion of inherited disorders can be explained currently in biochemical terms. These few are worth studying because the principles that they illustrate must also apply to many other inherited disorders.

If a particular polypeptide acts as an enzyme or is part of an enzyme, then a mutation in the gene that codes for that polypeptide sometimes results in a deficiency of that enzyme, with a consequent blockage in the relevant biochemical pathway at the point where that enzyme is required. Diseases that result from such blockages are called inborn errors of metabolism.

Examples of inborn errors of metabolism occurring in animals include congenital haemolytic anaemia associated with pyruvate kinase deficiency, inborn errors of lysosomal catabolism, porphyria, protoporphyria and dermatosparaxis. In all inborn errors of metabolism, heterozygotes show a level of gene product intermediate between that of the two relevant homozygotes. This provides a relatively easy means of detection of carriers of inherited diseases whose biochemical basis is known.

Two alleles at one locus can exhibit more than one type of gene action.

Genetical heterogeneity of disease occurs when the same clinical signs are associated with more than one genetic disease. If it remains undetected, then such heterogeneity makes it impossible to determine the exact form of inheritance.

In general, recessive diseases are due to enzyme deficiencies while dominant diseases are caused by defects in non-enzymatic polypeptides.

Inherited bleeding disorders, inherited disorders of haemoglobin and inherited immunodeficiencies all provide further illustrations of the principles underlying inherited biochemical diseases.

3.8 Further reading

Andrews, E. J., Ward, B. C., and Altman, N. H. (Eds) (1976). *Spontaneous animal models of human disease* (Vols I and II). Academic Press, New York. (These books contain precise summaries of all the diseases discussed in this chapter.)

Baker, H. J., Mole, J. A., Lindsay, J. R., and Creel, R. M. (1976). Animal models of human ganglioside storage diseases. *Federation Proceedings* **35**, 1193–1201. (A detailed account of one type of lysosomal storage disease in both animals and humans.)

Dodds, W. J. (1977). Inherited hemorrhagic defects. In *Current veterinary therapy. VI. Small animal practice*. (Ed. Kirk, R. W.) pp. 438–45. Saunders, Philadelphia. (A concise review of inherited bleeding disorders, with references to other reviews in the same area.)

Harris, H. (1980). *The principles of human biochemical genetics* (3rd edn). North Holland, Amsterdam. (The standard text on human biochemical genetics, explaining in considerable detail many of the principles discussed in this chapter.)

Jolly, R. D. (1975). Mannosidosis of Angus cattle: a prototype control program for some genetic diseases. *Advances in veterinary science and comparative medicine* 19, 1–21. (A detailed account of the most economically important inborn error of lysosomal catabolism in animals, with a description of how this disease in particular, and biochemical diseases in general, can be controlled.)

Jolly, R. D. and Blakemore, W. F. (1973). Inherited lysosomal storage diseases: an essay in comparative medicine. *Veterinary record* 92, 391–400. (A very useful account of inherited lysosomal storage diseases in animals, with a comparison of their human counterparts.)

Jolly, R. D. and Hartley, W. J. (1977). Storage diseases of domestic animals. *Australian Veterinary Journal* 53, 1–7. (A broader review, including storage diseases of different aetiologies, and placing inherited storage diseases in their proper context.)

Spurling, N. W. (1980). Hereditary disorders of haemostasis in dogs: a critical review of the literature. *Veterinary Bulletin* 50, 151–73. (A thorough review of inherited bleeding disorders in dogs.)

Tizard, I. R. (1982). *An introduction to veterinary immunology* (2nd edn). Saunders, Philadelphia. (A textbook containing a review of inherited immunodeficiency diseases, and giving references to the original research in this area.)

4
Chromosomes
and chromosomal aberrations

4.1 Introduction

From time to time, mistakes occur in mitosis, in meiosis or in fertilization, and sometimes these mistakes give rise to aberrant karyotypes. The study of aberrant karyotypes has not only increased our knowledge of the genetic basis of certain diseases and abnormalities, but has also provided much valuable information on normal processes in which chromosomes are involved. One of the aims of this chapter is to discuss what is currently known about normal chromosomes. The other aim is to describe various types of chromosomal aberrations that occur, and to discuss the role that such aberrations play in determining abnormalities in animals.

The area of genetics concerned with chromosomes, both normal and aberrant, is called *cytogenetics*.

4.2 Normal karyotypes

As we saw in Chapter 1, a karyotype is the set of all chromosomes that exist in a particular cell. The karyotypes of hundreds of different species have now been investigated, and an illustrated catalogue of mammalian karyotypes has been assembled by Hsu and Benirschke (1967 *et seq.*). In order to describe karyotypes, chromosomes are classified into three groups (Fig. 4.1), depending on whether the centromere is at one end (*telocentric*), considerably closer to one end than to the other (*acrocentric* or *submetacentric*) or approximately in the middle (*metacentric*). In practice, it is often difficult to distinguish between a true telocentric chromosome and one which has a very small segment of chromosome on the other side of the centromere. For this reason, the terms acrocentric and telocentric are sometimes used interchangeably. A summary description of the karyotypes of common domestic animals is given in Table 4.1.

Avian karyotypes are very different from mammalian karyotypes, because birds have a ZW sex determining mechanism (which was described in Chapter 1), and, in addition to ordinary autosomes, they possess a number of very small chromosomes called *microchromosomes*.

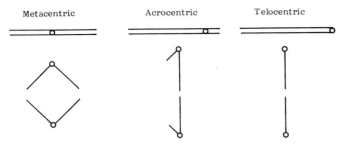

Fig. 4.1. The three main types of chromosomes, as determined by the position of the centromere (top), and the corresponding characteristic V, J and rod shapes (bottom) seen when the chromatids are separating during the stage of mitosis known as anaphase. (From Swanson C. P., Merz, T., and Young, W. J., *Cytogenetics: the chromosome in division, inheritance and evolution*, © 1981, p. 77. Adapted by permission of Prentice-Hall, Inc., Englewood Cliffs, NJ.)

Table 4.1. A summary description of the karyotypes of eight common domestic mammals

Species	Diploid number (2n=)	Autosomal pairs		Sex chromosomes	
		Metacentrics	Acrocentrics or telocentrics	X	Y
Cat, *Felis catus*	38	16	2	M	M
Dog, *Canis familiarus*	78	0	38	M	A
Pig, *Sus scrofa*	38	12	6	M	M
Goat, *Capra hircus*	60	0	29	A	M
Sheep, *Ovis aries*	54	3	23	A	M
Cattle, *Bos taurus*	60	0	29	M	M
Horse, *Equus caballus*	64	13	18	M	A
Donkey, *Equus asinus*	62	24	6	M	A

After Ohno (1968) and Eldridge and Blazak (1976).
    M = Metacentric; A = Acrocentric.

A typical karyotype of the domestic chicken (*Gallus gallus domesticus*) is illustrated in Fig. 4.2.

4.2.1   Banding

When karyotypes were first investigated, individual pairs of chromosomes could be identified only according to their shape and size. More recently, various methods of staining chromosomes have been developed to the stage where reproducible banding patterns on particular chromosomes can be obtained. The methods most commonly used are G, R,

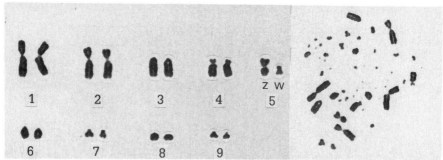

Fig. 4.2. (left) The 9 pairs of ordinary chromosomes (macrochromosomes) of a domestic chicken. (right) A complete genome of a domestic chicken consisting of 9 pairs of macrochromosomes and 30 pairs of microchromosomes. (Courtesy of M. Thorne.)

C, T, and Q-banding. The first four involve the use of Giemsa dye. G-banding produces a particular pattern of light and dark bands, and R-banding gives a pattern that is the reverse of G-banding. C-banding causes the constitutive heterochromatin (see Section 1.3.4) to stain darkly, while T-banding stains the telomeric (end) regions and the centromeres. In Q-banding, the use of quinacrine dye results in alternating brightly and dimly fluorescent bands which can be viewed through an ultraviolet fluorescent microscope, with the bright bands corresponding to the dark bands of G-banding. For a thorough review of banding techniques in domestic animals, see Gustavsson (1980*a*).

As an example of banding, the G-bands of cattle are illustrated in Fig. 4.3. Since the position, width and number of bands are generally different for each pair of chromosomes, it is now possible to be much more certain about the identification of particular chromosomes. By studying many cells treated in the same way, it is possible to draw up an *idiogram*, which is a representation of the bands that consistently appear on particular chromosomes. An idiogram of the G-bands of cattle is shown in Fig. 4.4. Karyotypes of other domestic species are illustrated in Figs 4.5 to 4.10.

Sometimes, as in Fig. 4.5, animals are discovered in which the two chromosomes of a pair have different banding patterns. Such animals are said to be heterozygous for banding pattern. When the offspring of such animals are investigated, it is found that each form of banding is passed on from the heterozygous parent in approximately equal frequency, just as for any other Mendelian character. The presence of two or more banding patterns in a population is called a *banding polymorphism*. The existence of such polymorphisms often enables particular chromosomes to be traced through many generations.

Two other cytogenetic techniques that should be mentioned are N-banding, which is done with Giemsa, and silver staining, both of which identify regions called *nucleolus organizer regions* (NORs).

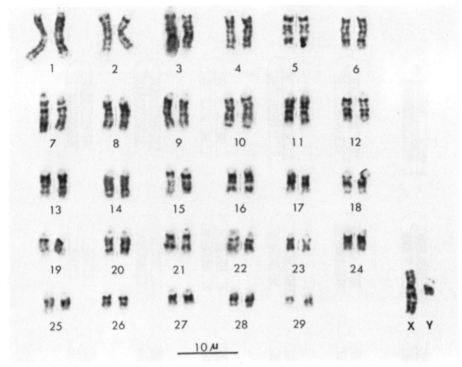

Fig. 4.3. A G-banded karyotype of a bull, consisting of 29 pairs of acrocentric autosomes plus an X and a Y chromosome. (From Ford *et al.* 1980.)

Each NOR is the site of production of a *nucleolus*, which is a discrete structure found within the nucleus (Fig. 4.11). Nucleoli are important because they are the assembly points for ribosomes, which, as described in Section 1.3.3, are the structures on which mRNA is translated into polypeptide. Among the major components of ribosomes are several different molecules of RNA, all of which are called *ribosomal RNA* (rRNA). Ribosomal RNA is not itself translated into protein, but being a component of ribosomes, it plays a major role in enabling mRNA to be translated into polypeptide.

Since ribosomes are composed partly of rRNA, and since ribosomes originate from nucleoli which in turn originate from NORs, it follows that NORs must contain DNA that codes for rRNA. A relatively large quantity of rRNA is required for the construction of sufficient ribosomes to satisfy each cell's requirements for translation. Indeed, rRNA constitutes between 80 and 90 per cent of total cellular RNA. It is not surprising, therefore, to find that there are several NORs in many species, and that within each NOR there are multiple copies of rRNA genes. In the horse, for example, an NOR is located on each homologue of chromosomes 1, 25, and 30, as illustrated in Fig. 4.12. The total

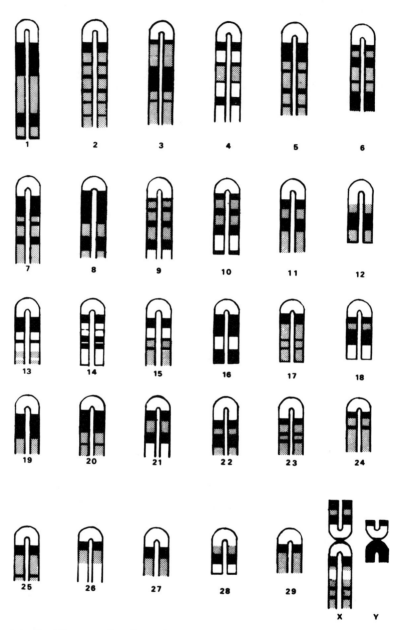

Fig. 4.4. An idiogram showing G-bands in cattle. (From Gustavsson *et al.* 1976.)

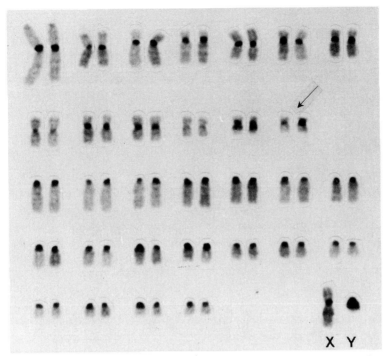

Fig. 4.5. A C-banded karyotype of a stallion, showing the typical dark staining centromeric regions. The first 13 pairs are metacentric and the remaining 18 pairs of autosomes are telocentric. Note that the two chromosomes of pair 13 (arrowed) have different C-band patterns, indicating a banding polymorphism. (From Buckland *et al.* 1976.)

number of copies of rRNA genes in higher eukaryotes is thought to be around 400 in a diploid nucleus (Lewin 1980, Chapter 27).

4.3   Abnormal karyotypes

Abnormal karyotypes arise from errors in chromosome replication, in fertilization or in early cleavage divisions of the fertilized egg. In many cases, abnormal karyotypes have been observed in healthy and/or highly productive animals, and likewise, many diseased and/or unproductive animals have completely normal karyotypes.

> *It follows that a normal karyotype is not in itself a guarantee of high production and freedom from disease, and neither is an abnormal karyotype certain to indicate a diseased or unproductive animal.*

The most important effect of abnormal karyotypes is their contribution to lowered reproductive performance, through (1) a decreased ability or complete failure to produce functional gametes, or (2) death of embryos.

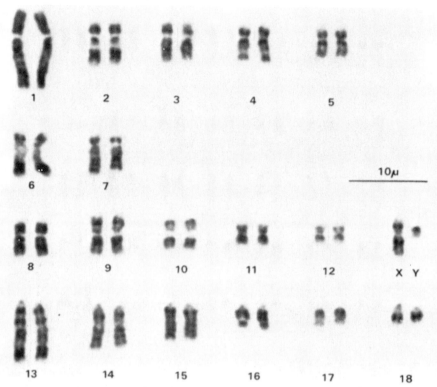

Fig. 4.6. A G-banded karyotype of a boar. Pairs 1 to 12 are metacentric and the remaining autosomes (pairs 13 to 18) are telocentric. (From Ford *et al.* 1980.)

4.3.1   *Abnormal chromosome number*

As an indication of the importance of chromosomal abnormalities in reproductive failure, we shall consider 29 cases of infertility in mares reported in the recent literature, and summarized by Blue *et al.* (1978). External appearance was generally normal in all of these mares. But oestrous cycles were either irregular or absent, the uterus and ovaries were often smaller than average, and ovaries often lacked follicles. When the karyotypes of these mares were examined, it was discovered that 20 (or 69 per cent) had abnormalities of the X chromosome. While this is likely to be a biased sample of all infertile mares, it is comparable with an incidence of X chromosome abnormalities of approximately 50 per cent among women who fail to menstruate.

The most common abnormality (occurring in 12 cases) was the lack of one X chromosome. Animals that lack a chromosome are said to be *monosomic* for that chromosome. These 12 mares are therefore said to be monosomic for the X chromosome. Their karyotype is written as 63,XO, with the O indicating absence of an X chromosome.

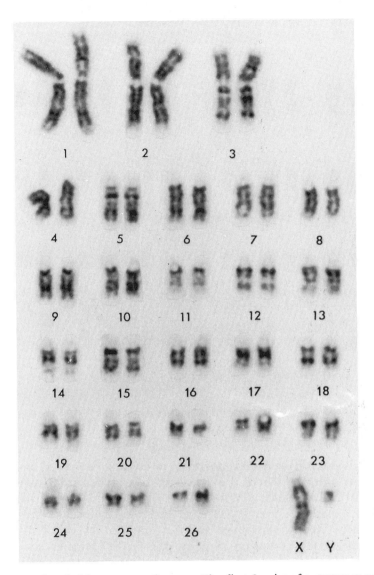

Fig. 4.7. A G-banded karyotype of a ram. The first 3 pairs of autosomes are meta-centric, and the remaining 23 pairs are acrocentric. (From Ford *et al.* 1980.)

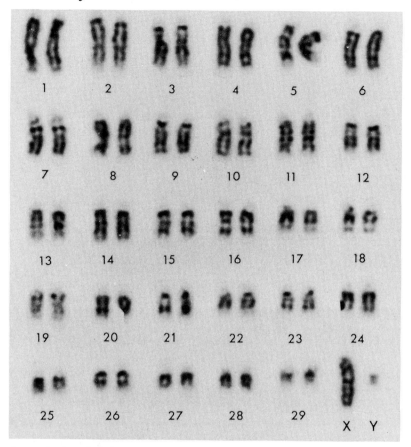

Fig. 4.8. A G-banded karyotype of a male goat. All 29 pairs of autosomes are acrocentric, as in cattle. (From Ford *et al.* 1980.)

XO individuals have been reported in mice, rats, cats, pigs, and horses, and also in humans, where the condition is called Turner's syndrome. All XO individuals have a more or less normal female external phenotype, but except for young XO mice and rats, are usually sterile. This difference between young mice and rats, and the rest, has led to speculation that the degeneration of germ cells in XO females may be progressive, such that the longer the delay before puberty, and the older the XO female in relation to puberty, the less chance there is of an XO female being fertile.

Another abnormality detected in one of the 29 infertile mares discussed above was the presence of three X chromosomes rather than the normal two. This is an example of *trisomy* for the X chromosome, and in the case of the mare with an extra X, the condition is written as 65,XXX. The karyotype of an XXX mare is shown in Fig. 4.13.

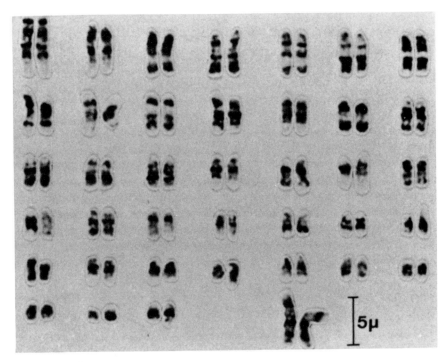

Fig. 4.9. A G-banded karyotype of a female dog. All 38 pairs of autosomes are acrocentric. (From Selden *et al.* (1975). The Giemsa banding pattern of the canine karyotype. *Cytogenetics and Cell Genetics* **15**, 380–7. S. Karger AG, Basel.)

When examined cytologically, non-dividing cells of X trisomics are seen to contain two Barr bodies, thus indicating that only one X chromosome in each cell remains active. Although the XXX mare in this study was infertile, XXX individuals are often fertile, and have been reported as giving rise to offspring with normal karyotypes in cattle and in humans.

XO and XXX individuals arise from *non-disjunction*, which is failure of the chromosomes or chromatids to disjoin during meiosis. Since there is normally one disjunction in each stage of meiosis (one during meiosis I and one during meiosis II), as described in Section 1.2.2, there are two opportunities for non-disjunction during the formation of a germ cell. Non-disjunction can occur with any of the autosomes or with the sex chromosomes. In the latter case, the results of non-disjunction differ depending on the sex of the individual in which meiosis is occurring and, in males, on whether non-disjunction occurs during meiosis I or meiosis II. In females, the results of non-disjunction of sex chromosomes are straightforward: irrespective of when non-disjunction occurs, some of the eggs will contain two X chromosomes while others

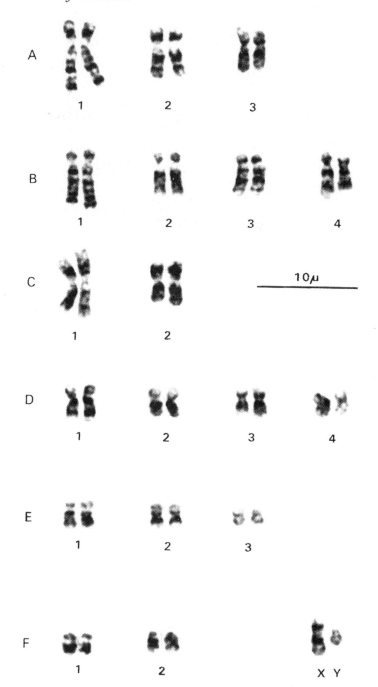

Fig. 4.10. A G-banded karyotype of a male cat, in which there are 16 metacentric and 2 acrocentric pairs of autosomes. (From Ford *et. al.* 1980.)

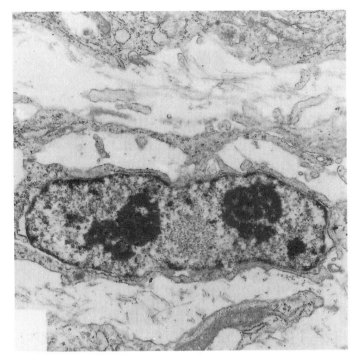

Fig. 4.11. A normal, diploid chicken cell with two nucleoli (the darkly stained structures), each being the product of a nucleolus organizer region (NOR) (not visible in figure). (From Macera and Bloom 1981.)

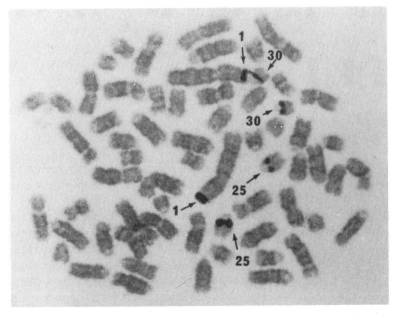

Fig. 4.12. A horse karyotype which has been N-banded and silver stained. Arrows indicate the sites of nucleolus organizer regions (NORs), and numbers identify the relevant chromosome. (From Kopp *et al.* 1981.)

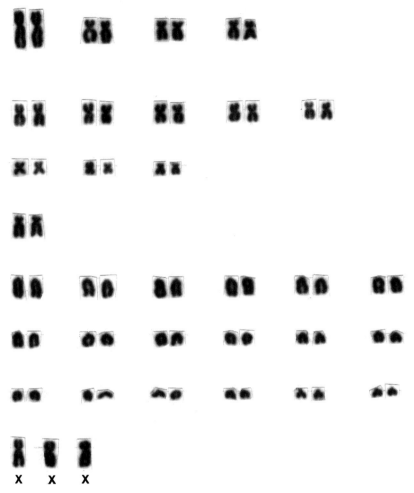

Fig. 4.13. X-trisomy in a mare. (Courtesy of I. A. Stewart Scott.)

will contain none. With males, however, four different types of unbalanced sperms can result, depending on when non-disjunction of sex chromosomes occurs (Fig. 4.14).

The unbalanced germ cells that result from non-disjunction are often still able to be involved in fertilization, and the possible outcomes are illustrated in Table 4.2. It is evident that XO and XXX individuals can arise from non-disjunction in either sex. It is also evident that many other types of unbalanced karyotypes can result from non-disjunction in one or both parents. In general, karyotypes with a small number of extra chromosomes, and karyotypes that lack a small number of chromosomes are said to be *aneuploid*, in comparison with those having

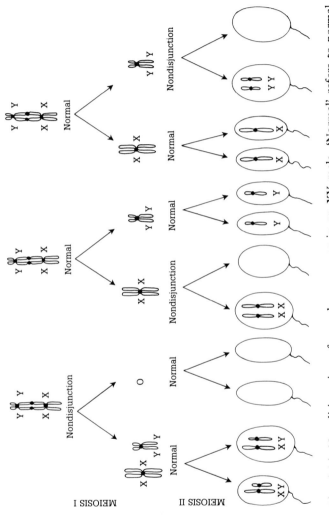

Fig. 4.14. Non-disjunction of sex chromosomes in an XY male. 'Normal' refers to normal disjunction, as illustrated in more detail in Fig. 1.5.

Table 4.2. The zygotes that can theoretically result from all possible combinations of normal and abnormal sperm and eggs, in relation to sex chromosomes. Since non-disjunction is the exception rather than the rule, abnormal sperms and eggs occur with a much lower frequency than normal sperms and eggs. The zygotes in the lower right-hand corner are expected to be very rare, because they are the result of the joint occurrence of two uncommon events

			SPERM					
			Normal		Abnormal			
			X	Y	XY	XX	YY	O
	Normal	X	XX	XY	XXY	XXX	XYY	XO
EGGS	Abnormal	XX	XXX	XXY	XXXY	XXXX	XXYY	XX
		O	XO	YO	XY	XX	YYO	O

the normal number which are said to be *euploid*. Thus monosomics and trisomics are both aneuploids. Some of the other aneuploids shown in Table 4.2 such as O, YO and YYO, have never been observed and are assumed to be lethal at the early embryonic stage. All the remaining aneuploids in Table 4.2 have been recorded in at least one species of mammal. Among those reported in domestic animals are XXY and XXXY, both of which have a predominantly male phenotype often associated with underdeveloped male sexual behaviour. Among domestic animals, XXY individuals have been reported in cattle, sheep, pigs, cats, and dogs. In humans, the condition is known as Klinefelter's syndrome. In all mammals, XXY individuals are usually sterile.

As with XO and XXX females, XXY males are generally not recognizable in domestic animals until they are karyotyped. One notable exception is male tortoiseshell cats. In Section 1.8, we saw that tortoiseshell cats are heterozygous at an X-linked coat colour locus, and since males have only one X chromosome, we would expect all tortoiseshells to be female. But male tortoiseshells do occur from time to time. They are usually sterile, and when examined cytogenetically, a majority of them are found to be XXY in at least a proportion of their cells. For example, of the 35 tortoiseshell males reported in the literature up to 1981 (Table 4.3), 10 were entirely 39,XXY and a further 13 had at least some XXY cells, being mixtures of various cell-lines such as 38,XY/39,XXY and 38,XX/38,XY/39,XXY/40,XXYY. In general, the most common cause of cell-line mixtures is non-disjunction during

Table 4.3. The karyotypes of 35 tortoiseshell male cats reported in the literature up to 1981*

Karyotype	Number	Sterile or fertile
39,XXY	10	Sterile
38,XY/39,XXY	5	Sterile
38,XY/39,XXY/40,XXYY	1	Sterile
38,XX/38,XY/39,XXY/40,XXYY	1	Sterile
38,XX/57,XXY†	4	Sterile
38,XY/57,XXY†	2	‡
Total with at least some XXY cells	*23*	
38,XX/38,XY	7	Some fertile, others sterile
38,XY §	4	Fertile
38,XY/39,XYY	1	Probably sterile

* Compiled from Loughman and Frye (1974), Centerwall and Benirschke (1975), Nicholas *et al.* (1980), Hageltorn and Gustavsson (1981), and Long *et al.* (1981). † The cells with 57 chromosomes contained three each of the 19 different chromosomes found in cats. This is an example of triploidy, which is discussed later in this section. ‡ Of the two cats having this karyotype, one was too young for fertility to be determined. The other was fertile, most likely because only 2% of its cells were abnormal, with the remaining 98% being 38,XY. § As explained in the text, these cats are most likely chimaeras, with a 38,XY/38,XY karyotype.

mitosis at an early stage of embryo development. For example, 38,XY/39,XXY individuals could arise from a normal XY embryo, as illustrated in Fig. 4.15. If non-disjunction during mitosis is the cause of XY/XXY individuals, then all cells have originated from a common

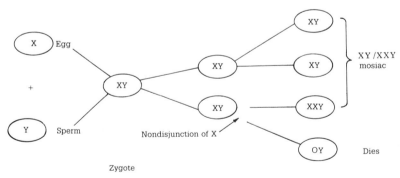

Fig. 4.15. Formation of an individual with an XY/XXY mixture of cells, as a result of non-disjunction of the X chromosome during mitosis. The OY cells are assumed to die, but the XXY cells will function more-or-less normally. Since all cells have originated from a common source, the individual is called an XY/XXY mosaic. (After *Genetics, evolution and man* by Bodmer, W. F. and Cavalli-Sforza, L. L. Copyright © (1976) by W. H. Freeman and Company. All rights reserved.)

source, in which case the individuals are called *mosaics*. In the case of tortoiseshell males, however, the above explanation is not adequate, since the X chromosomes in both cell lines in Fig. 4.15 have originated from a single X chromosome, in which case the resultant cat is expected to be either wholly orange or wholly non-orange. A possible explanation for some of the cell-line mixtures in tortoiseshell males is the fusion of two fertilized eggs. An XY/XXY individual could have arisen, for example, from the fusion of an XY egg and an XXY egg. In this case the two cell-lines have different origins and the resultant individual is called a *chimaera*. In many cases of naturally occurring cell-line mixtures, it is not possible to distinguish between these two alternatives. In some cases, however, one alternative is much more likely than the other. For example, the 38,XX/38,XY cats in Table 4.3 were almost certainly the result of the fusion of two fertilized eggs, one male (XY) and one female (XX).

The most interesting tortoiseshell males are those that appear to have a normal 38,XY karyotype. Being tortoiseshell, they must have two different types of X chromosome. The most likely explanation is that 38,XY tortoiseshell males are chimaeras resulting from the fusion of two fertilized male eggs, one carrying an orange allele on its X chromosome, and the other carrying a non-orange allele. Having a normal genome in every cell, 38,XY tortoiseshell males are fertile. Thus it is possible to find a fertile male tortoiseshell cat. But they constitute only a minority of tortoiseshell males, and are therefore very rare.

As an indication of the range of different types of sex chromosome aneuploidy that have been detected in domestic species, Table 4.4 lists various examples and describes the main phenotypic effect of each example. An illustration of the effect of sex chromosome aneuploidy is given in Fig. 4.16, which shows the external genitalia of an XX/XXY Arabian horse that was registered as a filly, but which had unusual vaginal folds, an enlarged clitoris resembling a glans penis, with a urethral process opening on its dorsal aspect, and an elongated anogenital distance. An internal examination revealed the absence of ovaries and uterus, and the presence of two small testicles. Not surprisingly, this 'filly' was sterile.

Bearing in mind all the information described above, we can conclude that:

Aneuploidy of sex chromosomes is usually associated with sexual abnormalities.

We have concentrated on aneuploidy of sex chromosomes because these are the most common type of aneuploidy seen in animals. However, aneuploidy of autosomes can also arise from non-disjunction, and some cases have been reported. In humans, where the majority of studies have been conducted, cases of trisomy for six different small autosomes are known, the most famous of these being trisomy for chromosome 21, which results in Down's syndrome. Human trisomics usually have severe abnormalities, including mental retardation and heart malformations, and often have a reduced life expectancy.

Table 4.4. Examples of sex chromosome aneuploidy in domestic animals. (From Gustavsson 1980b, who provides a reference for each example given)

Sex chromosomes	Species	Main phenotypic effects
XO	Pig	Intersexuality, ovarian hypoplasia
	Horse	Ovarian hypoplasia
	Cat	(died before puberty)
XO/XX	Horse	Ovarian hypoplasia
XO/XX/XY	Pig	Intersexuality
XXX	Cattle	No effects, ovarian hypoplasia
	Horse	Infertility
XXY	Cattle	Testicular hypoplasia
	Sheep	Testicular hypoplasia
	Pig	Testicular hypoplasia
	Dog	Testicular hypoplasia
	Cat	Testicular hypoplasia
XXY/XY	Cattle	Testicular hypoplasia
XXY/XX	Cattle	Intersexuality
	Pig	Intersexuality
	Horse	Intersexuality
	Cat	Testicular hypoplasia
XXY/XX/XY	Cattle	Testicular hypoplasia
XXY/XY/XO	Cattle	Testicular hypoplasia
XXY/XY/XX/XO	Horse	Cryptorchidism
XXXY	Horse	Intersexuality
XXXY/XXY	Pig	(No information)
XYY/XY	Cattle	No effects

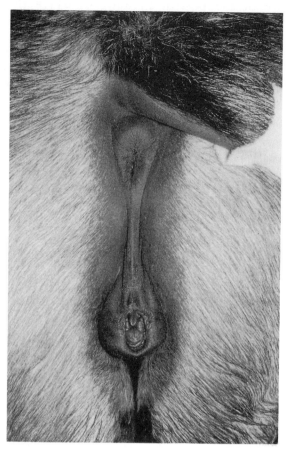

Fig. 4.16. Six-month-old Arabian 'filly' with unusual vaginal folds, an enlarged clitoris resembling a glans penis and an elongated anogenital distance. In addition to having 64,XX and 65,XXY cells, this 'filly' also had 3 per cent of 63,XO cells. These may represent a third population of cells, but it is more likely that they resulted from the loss of an X chromosome in a few cells during culturing and/or preparation. (From Fretz and Hare 1976.)

Isolated reports of autosomal trisomy in domestic animals have been made. For example, trisomy for chromosome 23 was observed in four calves by Gluhovschi *et al.* (1972). Since all four calves were dwarfs, it was concluded that the form of dwarfism shown by the four calves actually resulted from the trisomy. However, much more conclusive evidence is required before a cause-and-effect relationship can be established, especially in view of several other reports of dwarfs with normal karyotypes (Weaver 1975).

Trisomics for large autosomes and monosomics for any autosome have not been observed in living humans or animals. The absence of monosomics is somewhat surprising, because non-disjunction gives rise to equal numbers of gametes with an extra chromosome and gametes lacking a chromosome. Evidence in mice (Epstein and Travis 1979) indicates that equal numbers of trisomic and monosomic embryos are produced at fertilization and remain alive at least until the late morula or early blastocyst stage (day 3 in the mouse). Thereafter, however, the frequency of monosomics relative to trisomics decreases, and no monosomics remain alive beyond days 12 to 13 (Epstein *et al.* 1977).

The previous paragraph has introduced us to the second major effect of chromosomal abnormalities: the loss of embryos prior to birth. Several surveys of embryos in several species including poultry, rabbits, pigs and sheep have all indicated a surprisingly high frequency of chromosomal abnormalities, many of which are much less frequent or entirely absent in animals sampled after hatching or after birth.

One set of studies involved 9216 chicken embryos sampled 16 to 18 h after incubation (Fechheimer 1981). An examination of their macrochromosomes indicated that 5.2 per cent of embryos had abnormal karyotypes. Among these abnormal karyotypes, 81 per cent involved the presence of only one set of chromosomes (called *haploidy*) or the presence of more than two sets of chromosomes (called *polyploidy*). Haploid and polyploid karyotypes are compared with a normal diploid karyotype in Fig. 4.17. By studying the sex chromosomes present in haploids and polyploids, and by drawing on various other lines of evidence, conclusions were drawn as to the origin of the various abnormal karyotypes. Haploids arose entirely from sperm cells and not from unfertilized eggs. *Triploids* (three sets of chromosomes) arose mainly from errors of meiosis in females, with 75 per cent coming from suppression of meiosis II and 15 per cent coming from suppression of meiosis I. The remaining 10 per cent arose from *dispermy*, which is the fertilization of one egg by two sperms. There was a suggestion that the frequency of triploids was affected by the stage of lay, with the frequency decreasing as age of hen increased from 19 weeks to 27 weeks. All *tetraploids* (four sets of chromosomes) contained two maternal and two paternal sets of chromosomes, and were thought to have arisen from failure of the two products of mitosis to separate at the first cleavage division of a hitherto normal embryo. Only two *pentaploids* (five sets of chromosomes) were observed in the 9216 embryos studied, and both these appeared to arise from the fertilization of a tetraploid egg by a normal sperm.

These results should not be taken as indicating the way in which haploids and polyploids always arise; they are quoted here simply because they represent the only relevant data available from domestic animals, and because they give a good indication of at least some of the ways in which haploids and polyploids can arise.

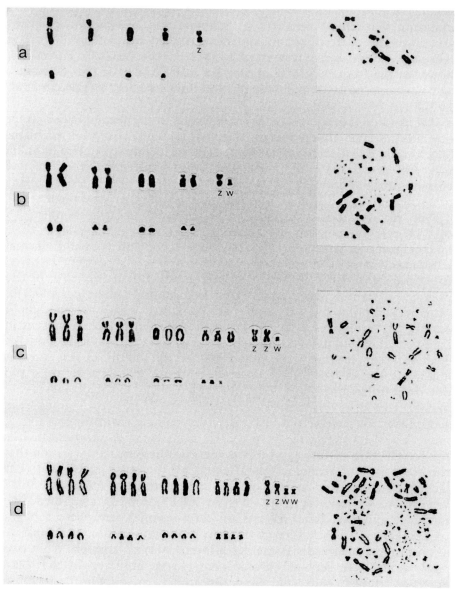

Fig. 4.17. A haploid chicken karyotype (a), compared with those from diploid (b), triploid (c), and tetraploid (d) chicken embryos. In each case, karyotypes of macrochromosomes are shown beside a cluster of all chromosomes from a cell. (Courtesy of M. Thorne.)

In many cases of haploidy and polyploidy reported in the above study, more than one cell-line was present in the one embryo. By noting the presence of sex chromosomes and by utilizing other evidence, mosaics could often be distinguished from chimaeras.

Apart from haploidy and polyploidy, the most frequent abnormality was aneuploidy, especially monosomy and trisomy, which together accounted for 11 per cent of all abnormalities. The frequency of chromosomal abnormalities in general, and of monosomics and trisomics in particular, was greater in some families than in others, and also differed in the three populations of chickens that were tested. This indicates that the production of abnormal karyotypes is itself under a degree of genetic control. A summary of the results of this study of chicken embryos, which indicates the differences in frequency of abnormalities between the three populations of chickens tested, is given in Table 4.5.

Table 4.5. Frequency of various chromosomal abnormalities in chicken embryos whose karyotypes were examined between 16 and 18 h after incubation (1n = haploid, 2n = normal diploid, 3n = triploid, 4n = tetraploid, 2n + 1 = trisomic, 2n − 1 = monosomic)

Abnormality	Incidence per 1000		
	Leghorn	Mixed*	Broiler
1n and 1n/2n	3	19	54
Pure polyploidy	6	9	21
2n/4n	1	5	16
2n/3n	0	1	2
2n + 1	0	6	2
2n − 1	2	3	5
Others	3	3	2
Totals	*15*	*46*	*102*
	=1.5%	=4.6%	=10.2%

* Consists of first cross (F1), second cross (F2) and first backcrosses.
(After Fechheimer 1981.)

Complementary to this information are the results of a karyotype survey of more than 4000 chickens which survived until three to six weeks after hatching (Fechheimer 1979). No chromosomal abnormalities were found among these chickens, despite the fact that they came from lines known to have between 5 per cent and 13 per cent chromosomal abnormalities in embryos. It appears, therefore, that chicken embryos with chromosomal abnormalities die either during incubation or in the first few weeks of life.

Although no chromosomal abnormalities were detected in the above survey, there are reports of abnormal karyotypes sometimes occurring in adult chickens. For example, adult chickens trisomic for a particular microchromosome have normal viability and are able to reproduce (Macera and Bloom 1981). Also, triploidy and tetraploidy have been reported on rare occasions in adult chickens. Indeed, one triploid chicken has been used experimentally to demonstrate that, in chickens at least, three homologous chromosomes can synapse at the same time (Comings and Okada 1971). In general, however, haploid and polyploid embryos are unable to survive to hatching.

Much smaller studies on embryos in other species, as reviewed by Hare *et al.* (1980), have yielded results similar to those in the chicken. In cattle blastocysts, for example, various types of polyploidy occur with frequencies usually ranging from less than 1 per cent to around 10 per cent. In pigs, the analogous range of frequencies is from zero up to 27 per cent. In contrast to the situation in chickens as described above, there is evidence in pigs that the incidence of polyploidy increases with the age of ova, so that delayed fertilization can increase the frequency of polyploidy. There is some evidence (Hancock 1959) that the increase in polyploidy is due to the increased tendency of older ova to accept more than one sperm (*polyspermy*).

The importance of abnormal karyotypes as a source of embryonic loss has been shown clearly in sheep, by Long and Williams (1980). Among 84 zygotes sampled two or three days after fertilization (at the one, two, four, or eight cell stage), one was a 1n/2n mosaic, four were trisomic, four were monosomic and one had a break in one chromatid.

Bearing in mind that the latter two types of abnormalities may have been artefacts of the cell culture technique, these results indicate between 6 and 12 per cent of zygotes with abnormal karyotypes. In contrast, between 10 and 18 days after fertilization in sheep (at the late blastocyst stage), no abnormal karyotypes were found (Long 1977). The most reasonable conclusion is that sheep zygotes with abnormal karyotypes do not survive beyond early pregnancy. In general, the abnormal karyotypes reported in studies of embryos are absent or are relatively rare in animals sampled after birth or hatching.

It can be concluded, therefore, that:

> *Chromosomal abnormalities are an important cause of embryonic loss.*

In pigs, for example, it has been estimated that around 30 per cent of all embryos die during the first half of gestation. And if approximately 8 per cent of all embryos have abnormal karyotypes that are rarely seen in pigs after birth (this being the overall frequency from studies in pigs, as summarized by Hare *et al.* (1980), neglecting one exceptionally high estimate), then around one-quarter of embryonic loss in pigs is associated with chromosomal abnormalities. In humans, where the data are much more extensive, the proportion is slightly higher: around one-third of spontaneous abortions are associated with chromosomal abnormalities. Since approximately 20 per cent of all conceptions in humans are spontaneously aborted, it appears that around 7 per cent of all human conceptions result in zygotes with chromosomal abnormalities sufficiently serious to produce spontaneous abortion. This corresponds well with the figures of between 5 per cent and 10 per cent obtained in the chicken and mammal studies described above.

So as to view such chromosomal abnormalities in their proper perspective, it is important to compare the above frequencies with those observed in individuals that survive to full term. Since good estimates are not yet available in animals, we must look to humans for relevant figures. In a survey of 24 468 consecutive hospital births in the United Kingdom, United States and Canada, only 126 children had chromosomal abnormalities, indicating an overall frequency of 5 per 1000 (Jacobs 1972). The most frequent abnormalities were XYY (0.9 per 1000), XXY (0.7 per 1000) and XXX (0.4 per 1000). There was only one case of an XO female in the whole survey, but other surveys have indicated an XO incidence of up to 0.2 per 1000. Since XYY and XXX individuals are usually fertile, the incidence of chromosomal abnormalities that result in sterility (XO and XXY) is probably no greater than 1 per 1000 newborn individuals, which is relatively low.

> *In general, while abnormal chromosomal numbers have a high incidence among newly conceived zygotes, they have a relatively low incidence among individuals sampled after birth.*

4.3.2   Abnormal chromosome structure

4.3.2.1 Reciprocal translocations   In Sweden in 1963, a Swedish Landrace boar served 21 sows and produced an average litter size of 5.6 piglets. In previous pregnancies when mated to other boars, the same sows had an average litter size of 12.7 piglets. The karyotype of this boar was examined by Henricson and Bäckström (1964), who found that part of one chromosome had been interchanged with part of another, by a process called *reciprocal translocation*.

> *A reciprocal translocation involves each of two non-homologous chromosomes breaking into two segments, followed by an exchange of segments between the two chromosomes.*

A more recent study of the same translocation by Hageltorn *et al.* (1973) utilized banding techniques to show that the translocation was between chromosomes 11 and 15, as shown in Fig. 4.18. In order to

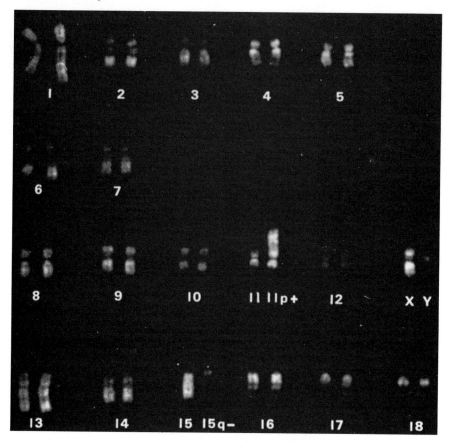

Fig. 4.18. The Q-banded karyotype of a Swedish Landrace boar showing a translocation between chromosomes 11 and 15. (From Hageltorn *et al.* 1973.)

standardize methods of describing translocations and other abnormalities, a convention has been adopted whereby the short arm of each chromosome is designated p (think of petite = small), and the long arm is designated q. In the translocation shown in Fig. 4.18, one chromosome 15 lacks most of its long arm and is therefore written as 15q−. And since the short arm of one chromosome 11 has been extended by the addition of the portion from 15, it is designated as 11p+. The translocation is written as t(11p+;15q−). Several other reciprocal translocations have been reported in pigs, including t(1p−;6q+), t(13q−;14q+), t(6p+;14q−), and t(4q+;14q−). In all cases that could be investigated, individuals carrying a reciprocal translocation have shown a significant reduction in fertility.

How do such translocations arise, and why do they lead to a substantial drop in fertility? We shall answer these questions using t(1p−; 6q+), for which good photographs are available (Fig. 4.19). The creation of

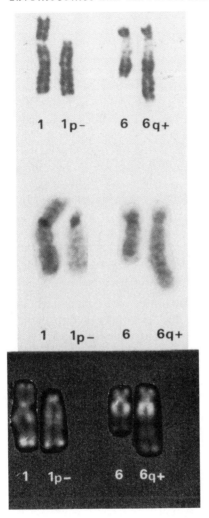

Fig. 4.19. Pairs of chromosomes 1 and 6 from a boar with t(1p—; 6q+). From top to bottom: G-banding, C-banding and R-banding. (From Ločniškar *et al.* 1976.)

t(1p—; 6q+), by the breakage and joining of non-homologous segments, is shown in Fig. 4.20. The breakage in chromosome 6 occurred very near to the end of the long (q) arm, and the small fragment so released became attached to the remaining portion of the short (p) arm of chromosome 1. Because only one member of each pair of chromosomes is affected, the individual in Fig. 4.19 is said to be a *translocation heterozygote*. Since no part of any chromosome was lost during the formation of the translocation, individuals carrying it still have a normal, balanced complement of genetic material and usually have normal phenotypes.

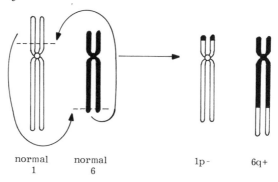

Fig. 4.20. The formation of a reciprocal translocation of the type illustrated in Fig. 4.19.

The only problem with translocations arises during meiosis.

Homologous chromosomes are attracted very strongly to each other during meiosis, and normally synapse or pair together to form what is called a *bivalent*. In translocation heterozygotes, however, if homologous chromosomes are to pair throughout their whole length, then four chromosomes must join together in a *quadrivalent*. With t(1p−; 6q+), for example, the quadrivalent consists of 1, 1p−, 6, and 6q+, as shown in Fig. 4.21. There are many possible outcomes from such a quadrivalent, depending on the position of cross-overs within the quadrivalent and on the type of segregation or disjunction. The effect of crossing-over within the quadrivalent depends on whether it occurs in an *interstitial segment* (between the centromere and the breakpoint), or in any of the other segments (called *distal segments*). If crossing-over occurs in distal segments, then the outcome is determined by which two of the four centromeres segregate together to the same end or pole of the cell. *Alternate segregation* occurs when diagonally-opposite centromeres segregate together. The other possibility is that adjacent centromeres segregate together, in which case we have either *adjacent-1 segregation* (if homologous centromeres disjoin) or *adjacent-2 segregation* (if non-homologous centromeres disjoin). Working from the quadrivalent shown in Fig. 4.21, and remembering that crossing-over involves the exchange of segments between two non-sister but homologous chromatids (as shown in Fig. 1.21), readers should be able to verify that after crossing-over in any distal segment, the two chromatids attached to a single centromere still have the same structure. For example, by drawing a distal cross-over between 1 and 1p−, readers will see that after crossing-over, the two 1p− chromatids are still attached to one of the white-coloured centromeres, and the two 1 chromatids

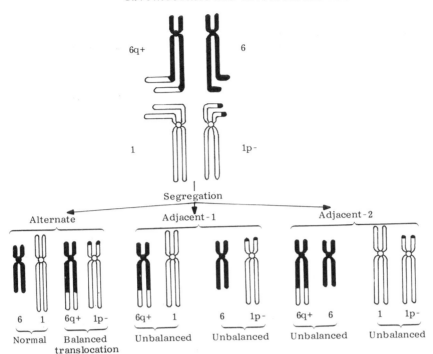

Fig. 4.21. Top: Quadrivalent formed by pairing of homologous segments of chromosomes 1 and 6 in an individual heterozygous for the reciprocal translocation t(1p−; 6q+). Below: The six possible gametes that can result from such a quadrivalent, assuming that there is no crossing-over in interstitial segments, i.e. between the centromere and the breakpoint.

are still attached to the other white centromere. It follows that the results shown in Fig. 4.21, in which cross-overs have been omitted so as to simplify the diagram, are exactly applicable to all cases of crossing-over in distal segments. It can be seen from Fig. 4.21 that all gametes arising from alternate segregation are balanced, whereas all those arising from adjacent-1 or adjacent-2 segregation are unbalanced. Most unbalanced gametes are able to function quite well as ova or sperm cells. However, when an unbalanced gamete combines with another gamete to form a zygote, that zygote is unbalanced, and dies before implantation. Thus unbalanced gametes result in embryonic death.

If a cross-over occurs within an interstitial segment, then the two chromatids attached to a single centromere have different structures. For example, if an interstitial cross-over is drawn between 6q+ and 6 in the quadrivalent shown in Fig. 4.21, then it can be deduced that each of the black centromeres will have one 6q+ chromatid and one 6 chromatid. In this situation, the concepts of adjacent and alternate segregation are no longer relevant, because it is impossible to say which centromeres are adjacent or alternate to each other. All that can be done is to consider whether homologous or non-homologous centromeres disjoin. By working through an appropriate meiosis, readers can verify that in the case of homologous centromeres disjoining, one-half of the gametes are balanced and the other half are unbalanced. If non-homologous centromeres disjoin, it is obvious that all gametes are unbalanced.

We can conclude that:

> *The level of fertility of an individual heterozygous for a reciprocal transloca-tion depends on the position of cross-overs within the quadrivalent formed during meiosis, and on the relative frequency of the types of segregation or disjunction.*

If the interstitial segment is short in both chromosomes involved in the translocation, then interstitial cross-overs will be rare. In theory, an individual heterozygous for such a translocation could range in fertility from normal (100 per cent), if all segregation is alternate, to zero, if all segregation is adjacent. In contrast, if interstitial segments are relatively large, then crossing-over within such segments will be common. Because such crossing-over gives rise to a maximum of 50 per cent balanced gametes, it follows that the maximum fertility expected from individuals heterozygous for reciprocal translocations with long interstitial segments is 50 per cent.

In practice, the observed fertility of pigs heterozygous for reciprocal translocations has ranged from 44 per cent (Henricson and Bäckström 1964) to 74 per cent (Ločniškar *et al.* 1976). The other domestic species studied in this regard are cattle and chickens. In the former, heterozygotes for reciprocal translocations have shown approximately 40 per cent of normal fertility (Mayr *et al.* 1983), while in the latter, a reduction in hatchability of fertile eggs of approximately 50 per cent has been observed (Blazak and Fechheimer 1981).

The importance of this reduction in fertility is that it is an inherited defect, being passed on from parents to offspring. This is because one-half of the balanced gametes resulting from meiosis contain transloca-tion chromosomes, which when combined with a normal germ cell, result in another translocation heterozygote. Thus, among the offspring produced by a translocation heterozygote, one-half will themselves be translocation heterozygotes.

Finally, if two germ cells carrying balanced translocated chromo-somes unite, then a *translocation homozygote* will result, with pairs of chromosomes that are exactly homologous and which will consequently form only bivalents during meiosis. Thus, individuals that are homo-zygous for reciprocal translocations will produce only balanced gametes and will show normal fertility.

4.3.2.2 Centric fusions The chromosome structures resulting from translocations between non-homologous chromosomes are many and varied. A type of translocation that has generated considerable interest is one in which the centromeres of two acrocentric chromosomes fuse to produce one metacentric chromosome which is known as a *centric fusion* (Fig. 4.22). It contains all of the genetic material previously present in the two separate chromosomes (Fig. 4.23). A slightly different translocation that also occurs between two acrocentric chromosomes is one in which a break occurs in one chromosome adjacent to the centromere which is then lost prior to the fusion of the two chromosomes. Known as *Robertsonian translocations*, after Robertson who first described them in 1916, they are often difficult to distinguish from centric fusion translocations, the only difference being the presence of one centromere instead of two fused centromeres. Because the difference between them is so slight and often so difficult to detect in practice, the terms centric fusion and Robertsonian translocation are often used interchangeably, and unless otherwise stated, we shall follow that practice here.

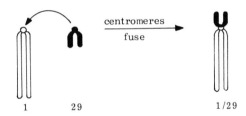

centromeres
fuse

1   29   1/29

Fig. 4.22. The origin of a centric fusion, in this case between chromosomes 1 and 29 in cattle.

Since centric fusions involve the replacement of two chromosomes by one, it follows that the total number of separate chromosomes in individuals heterozygous for a centric fusion will be one less than the normal number (Fig. 4.24). However, unlike monosomics, which also have one less than the normal number of chromosomes, individuals heterozygous for a centric fusion have a complete genome and hence have normal phenotypes. Individuals that are homozygous for centric fusion translocations will have two less than the normal number of chromosomes (Fig. 4.25), but they too have normal phenotypes because they have a complete genome.

Ever since centric fusions were discovered in animals, their effect on reproductive ability has been a controversial issue, in contrast to reciprocal translocations which clearly reduce fertility. Meiosis in individuals heterozygous for a centric fusion is certainly unusual, because three chromosomes have to synapse, forming a *trivalent*, as

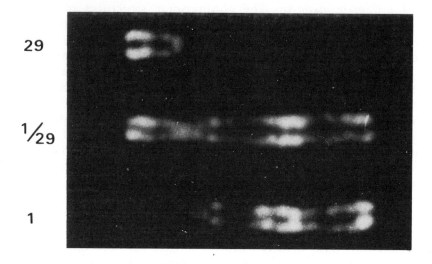

Fig. 4.23. An R-banded centric fusion (centre) between chromosomes 1 and 29 in cattle, compared with the two R-banded chromosomes (29: top, and 1: bottom) from which it is derived. (From Gustavsson and Hageltorn 1976.)

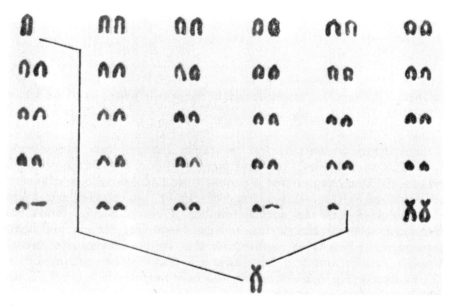

Fig. 4.24. An unbanded karyotype of a Swedish Red and White cow that is hetero-zygous for the 1/29 centric fusion translocation. The lines show the origin of the translocation. Note that there is a total of 2n = 59 chromosomes rather than the normal number for cattle, which is 2n = 60. (From Gustavsson and Rockborn 1964.)

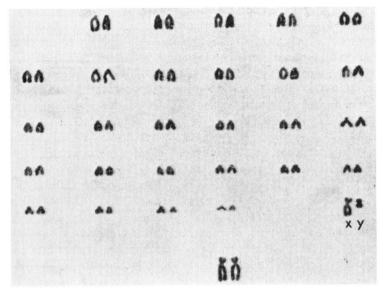

Fig. 4.25. An unbanded karyotype of a Swedish bull that is homozygous for the 1/29 centric fusion translocation. Note that there is a total of 2n = 58 chromosomes; two less than the normal number for cattle, which is 2n = 60. (From Gustavsson 1966.)

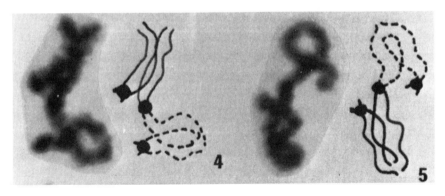

Fig. 4.26. Two examples of trivalents as seen during meiosis I in rams heterozygous for a centric fusion between chromosomes 9 and 12. For each trivalent, a suggested interpretation and the number of chiasmata are given. (From Chapman and Bruere 1977).

shown in Fig. 4.26. If the centric fusion chromosome disjoins from the other two, then balanced gametes will result. Any other type of disjunction will lead to unbalanced gametes which would be expected to lead to reproductive failure. The controversy concerning the effect of centric fusions on fertility arises from apparently different results observed in the two species of domestic animals most studied in this regard, namely sheep and cattle. Cytological evidence indicates that rams heterozygous for Robertsonian translocations produce about 5 per cent of unbalanced spermatocytes, but that these do not survive to become unbalanced gametes (Bruere *et al.* 1981). On the other hand, it appears that unbalanced gametes are capable of fertilization in cattle, giving rise to trisomic or monosomic embryos which subsequently do not survive (King *et al.* 1981). In support of the cytological evidence, extensive studies in sheep (as summarized by Bruere 1975) have shown that centric fusion heterozygotes have perfectly normal reproductive ability, while studies in cattle have shown heterozygotes to have reduced reproductive performance. Heterozygous cows, for example, show a reduction of around 5 per cent in non-return rate (Gustavsson 1969), and random samples of heterozygous bulls show a similar reduction in fertility, although their libido, serving ability and volume of ejaculate are all normal (Dyrendahl and Gustavsson 1979). Most importantly, bulls that are heterozygous for centric fusions are much more likely to be culled for low non-return rates than bulls with normal karyotypes.

There is no obvious explanation for the apparent difference in the effect of centric fusions in sheep and in cattle. It is to be hoped that further research will provide an explanation. In the meantime, the most important practical question is whether or not the 1/29 centric fusion should be removed from cattle populations because of its undesirable effect on reproductive ability. It has been argued that eradication of the 1/29 centric fusion in cattle would produce substantial financial benefits in some cattle populations, in terms of increased fertility (Gustavsson 1979).

> *The effect of eradicating a centric fusion such as the 1/29 in any breed of cattle will depend on its frequency in that breed; only if it is relatively common will its eradication have a noticeable effect on overall breed fertility.*

Substantial information has now accumulated on the frequency of the 1/29 centric fusion translocation in cattle (Table 4.6), and it is evident that the frequency varies considerably from region to region and from breed to breed. In the population in which eradication is having a significant effect on fertility (the Swedish Red and White), the frequency of the 1/29 centric fusion was relatively high, at around 13 per cent. Because of the effect of the 1/29 centric fusion on fertility, various artificial insemination centres throughout the world conduct regular karyotype screening programmes. Some centres prohibit the use of centric fusion heterozygotes. Given the expense of regular karyotype screening, and the apparent absence of the 1/29 centric fusion from certain popular breeds such as Herefords and Holstein-Friesians, and its low incidence in many other breeds, it is unlikely that regular screening

Table 4.6. Number of cattle investigated and incidences of the 1/29 centric fusion translocation in breeds extensively studied (over 100 cattle sampled at random) in different geographical areas

Geographical area	Breed	No. of cattle investigated	No. of translocation carriers	% 1/29
Australia	Holstein-Friesian	174	0	0
	Hereford	602	0	0
France	Limousin	231	13	5.6
	Aquitaine Blond	228	47	20.6
	Normandy	249	0	0
	Limousin	124	6	4.8
	Charolais	314	12	3.8
	F.F.P.N.	215	0	0
Great Britain	Holstein-Friesian	586	0	0
	Charolais	185	1	0.5
	Simmental	113	3	2.7
	Holstein-Friesian	330	0	0
Hungary	Hungarian Grey	106	4	3.8
Italy	Romagna	122	39	32.0
Norway	Norwegian Red	430	18	4.2
Sweden	Swedish Red and White	944	120	12.4
	Swedish Red and White	1173	164	14.3
	Swedish Holstein-Friesian	101	0	0
Switzerland	Simmental	654	21	3.2
	Swiss Brown	430	1	0.2
USA	Holstein-Friesian	743	0	0
	American Brown Swiss	224	3	1.3
West Germany	German Simmental	100	0	0

(After Gustavsson 1979.)

programmes of cows and natural service bulls would be economically justifiable in most breeds in most countries. If, however, a centric fusion is common in a particular breed in a certain country, then it may be worthwhile screening all bulls of that breed entering artificial insemination centres in that country.

One final point should be noted in relation to the 1/29 centric fusion in cattle. It was first reported in each of three Swedish Red and White cows that were suffering from lymphatic leukaemia. On the evidence presented, some investigators might have been tempted to suggest that the chromosomal abnormality was the cause of the leukaemia. However, the authors of the original report (Gustavsson and Rockborn 1964) were careful not to speculate on any such relationship, and their caution has subsequently been seen to be justified.

If a certain chromosomal abnormality is relatively common in a particular population, as the 1/29 centric fusion is in the Swedish Red and White breed, then it is quite likely that a small number of animals suffering a particular disease will by chance also carry the abnormality.

> *In fact, large-scale population surveys are required before cause and effect can be attributed in relation to chromosome abnormalities and disease.*

Apart from the 1/29 centric fusion in cattle, and the centric fusion in sheep discussed above, many other centric fusions involving various pairs of non-homologous chromosomes have been reported in domestic cattle, sheep, goats, pigs, and dogs, and in wild ungulates of various species. In all cases except the 1/29 in cattle, centric fusions have not been found to have any adverse effects on reproductive ability or on any other characteristic. However, until large-scale population surveys are conducted as described above, our understanding of the effect of most of these centric fusions will remain incomplete.

4.3.2.3 Inversions and deletions   The final types of structural abnormalities that we shall discuss are inversions and deletions. As their name suggests, *inversions* arise when a segment of chromosome becomes inverted following the breakage of a chromosome in two positions. If the segment includes the centromere, the inversion is said to be *pericentric*, and if the centromere is not included, the inversion is *paracentric*. *Deletions* arise if, following the breakage of a chromosome in two positions, the segment between the two break-points is lost.

> *Inversions give rise to a realignment of genes on a chromosome, and deletions result in a loss of genetic material.*

This is most easily illustrated by using letters to represent loci, as shown in Fig. 4.27. Inversions and deletions are the least frequently observed of chromosomal abnormalities in domestic animals, and only isolated cases have been reported. Deletions usually produce serious abnormalities, as the individual is effectively monosomic for the segment of chromosome deleted. In humans, for example, a deletion of

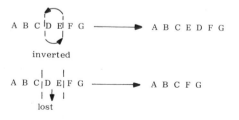

Fig. 4.27. The origin of inversions (top) and deletions (below).

about half the short arm of chromosome 5 results in a congenital syndrome called 'cri du chat', so-called because the cry of affected infants resembles the plaintive meow of a cat. In addition to this distinctive cry, affected children have various physical abnormalities, and are mentally retarded. Isolated cases of deletions have been reported in a pig blastocyst (McFeely 1967) and in sheep (Luft 1972). However, specific deletions have not been associated with particular abnormalities.

> *Because inversions involve the rearrangement of existing genes without any loss or addition, individuals carrying inversions have a normal phenotype.*

Indeed, since karyotypes are usually based on only mitotic cells, many inversions remain undetected. A chromosome with a paracentric inversion, for example, has exactly the same shape and size as a normal chromosome. It may, however, differ in banding pattern, and as banding techniques become more sophisticated, so too will more inversions be detected. Individuals heterozygous for an inversion are more easily detected during meiosis, because in order that all homologous segments should pair, the inverted segment must form a loop. Depending on the type of inversion (paracentric or pericentric) and on the number and distribution of cross-overs, unbalanced gametes may be formed (Fig. 4.28). Reports of inversions in domestic livestock are very rare, and so very little is known of their effect. In one case that has been reported, a pericentric inversion in a French bull converted an acrocentric chromosome into a metacentric.

4.4   Causes of abnormal karyotypes

Many chromosomal abnormalities occur without any apparent cause, and the frequencies of various abnormalities quoted in the previous section indicate how often some of these abnormalities arise in apparently normal circumstances. Although there is still much to be learnt about the causes of chromosomal abnormalities, one fact is beyond dispute.

> *Exposure to radiation, for example X-radiation, increases the frequency of chromosomal abnormalities.*

Indeed, a standard experimental method of obtaining chromosomal abnormalities for use in research is to expose sperm cells to X-rays. In domestic fowl, for example, Wooster *et al.* (1977) irradiated semen from 44 males with X-rays prior to its use in insemination. Among the 204 resultant chickens that hatched and were karyotyped, 18 (or 8.8 per cent) contained chromosomal rearrangements consisting of

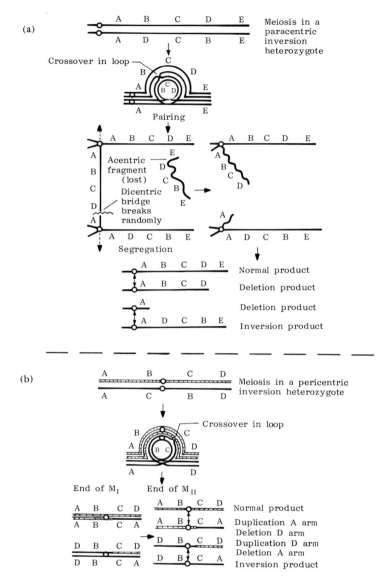

Fig. 4.28. The effect of crossing-over during meiosis in an individual heterozygous for a paracentric (a) and a pericentric (b) inversion. Neglecting multiple cross-overs, all recombinant gametes are unbalanced and fail to survive. Inversions are said to be *crossover-suppressors*, partly because the necessity of forming a loop inhibits proper chromosome pairing, and also because, with the exception of multiple cross-overs, the gametes resulting from crossing-over within an inversion are unfunctional. (From *An introduction to genetic analysis* (2nd edn) by Suzuki, D. T., Griffiths, A. J. F., and Lewontin, R. C. Copyright © 1981 by W. H. Freeman and Company. All rights reserved.)

translocations and one pericentric inversion. In a control group of chickens obtained from unirradiated semen from the same 44 males, no chromosomal rearrangements were recorded in 275 chickens examined. Furthermore, fertility of treated semen was approximately one-half that of control semen, and among the eggs that were fertile, 32 per cent of those from irradiated semen failed to hatch, compared with only 18 per cent from unirradiated semen. This difference of 14 per cent is a reflection of increased embryonic death, and is certain to be due at least in part to an increased frequency of aneuploidy and haploidy resulting from irradiation of semen.

More direct evidence on the effect of X-radiation on embryo death through chromosomal abnormalities comes from studies in mice in which whole-body irradiation of males (as distinct from irradiation of semen) produced aneuploidy in 10 out of 601 (or 1.7 per cent) foetuses, compared with none out of 211 foetuses from unirradiated controls (Hansmann *et al.* 1979).

Because of these striking effects of X-radiation of males and of their semen, and other equally striking results from irradiating ovaries in females, and also because of the effects of chromosomal abnormalities on embryo survival and fertility, it is obvious that exposure to X-rays should be avoided wherever possible, in both animals and humans. This warning is particularly relevant to veterinarians, who frequently use radiography as a diagnostic tool with animals from which the owner may intend to breed in the near future. One of the standard methods of diagnosing hip dysplasia in dogs, for example, is to take several radiographs of the hip region, which of course entails exposure of the gonads. All too frequently, radiographs are unsatisfactory for some technical reason and have to be retaken, thereby doubling the dose of X-radiation to the gonads.

Of course radiographs will continue to be a valuable diagnostic tool and there is no question at present of banning their use. The arguments raised here, however, should cause practising veterinarians to think twice before deciding on a radiograph, especially in relation to animals that are soon to be bred. They should also take all possible precautions to minimize the doses of radiation to which they, as radiographers, are exposed.

4.5 Evolution of karyotypes

Despite the wide range of chromosome numbers in different species of placental mammals, as illustrated in Table 4.1, the total amount of DNA in all placental mammals appears to be almost identical, as described in Section 2.6. The most logical explanation for this observation is that the evolution of modern placental mammals from their common ancestor was associated with much rearrangement and shuffling of the same total amount of DNA. One of the most common forms of shuffling appears to have been the fusion of two acrocentric chromosomes to form one metacentric, as was described in Section 4.3. Thus, two closely related species often differ by just one centric fusion. For example, two pairs of acrocentric chromosomes in Przewalski's horse (*Equus przewalskii*, 2n = 66) are present in the domestic horse (*Equus caballus*, 2n = 64) as one pair of metacentrics. Similarly, the metacentric chromosomes (1, 2, and 3) in the domestic sheep (*Ovis aries*, 2n = 54) appear to be the result of centric fusion of acrocentrics 1 and 3, 2 and 7, and 5 and 10, respectively, of the goat (*Capra hircus*, 2n = 60) (Hansen 1973*b*). That this is possible can be seen by comparing Fig 4.7 with Fig. 4.8.

Bearing this in mind, it is interesting to compare the total number of chromosome arms in different species. If chromosomal evolution has proceeded mainly by the accumulation of centric fusions, we would expect to find that the total number of major chromosome arms (called the *Nombre Fondamental* or *NF* by Matthey 1945) should be fairly constant across species. In many situations this is indeed the case. Within the superfamily *Bovoidea*, for example, 50 species have now been examined cytogenetically. Although the diploid number varies from 30 to 60 among these species, the NF in all but three cases varies only from 58 to 62. Table 4.7 illustrates this phenomenon for several species, including cattle, sheep and goats.

Table 4.7.   Relationship between diploid (2n) number, chromosome structure (acrocentric, A, or metacentric, M), and the Nombre Fondamental of Matthey (1945) in several members of the superfamily *Bovoidea*

Species	2n	NF†	Number of autosomes M	A	Type of X chromosome
Cattle, *Bos taurus*	60	62	0	58	M
Sheep, *Ovis aries*	54	60	6	46	A
Goat, *Capra hircus*	60	60	0	58	A
Musk ox,					
Ovibus moschatus	48	60	12	34	A
Rocky Mountain goat,					
Oreamnos americanus	42	60	18	22	A

* 2n = M + A + two sex chromosomes. † NF = 2M + A + the number of major arms in two X chromosomes. M is doubled because each metacentric has two major arms. The number of major arms in two X chromosomes is 4 if the X is metacentric, and 2 if it is acrocentric. (After Wurster and Benirschke 1968).

Bearing in mind the data in this table, it should be noted that the banding patterns on the acrocentric autosomes of goats, sheep and cattle appear to be almost identical, and that the two arms of each metacentric in sheep have banding patterns identical with particular acrocentrics in goats and cattle (Hansen 1973*a,b*). In other words, the chromosome structure within each arm is highly conserved. The concept of Nombre Fondamental and the hypothesis that evolution of karyotypes within *Bovoidea* has been almost solely by centric fusion do not help to explain how new species evolved within this superfamily. They do, however, provide an attractive means of bringing a sense of order into what would otherwise be apparently unrelated observations.

In contrast to the autosomes, the X chromosome appears to have altered very little throughout evolution. Indeed, in most placental mammals, including humans and many domestic mammals, the X chromosome is very similar in size and shape. In addition, genes that are X-linked in one species seem to be X-linked in all other species (Ohno 1967, 1973). In a similar manner, the Z chromosome found in birds and snakes appears to have been conserved intact throughout evolution. This evolutionary conservatism of the X and Z chromosomes can be particularly useful to those investigating inherited diseases. Thus if a certain disease is known to be X-linked in, say, mice, then it is a fairly safe bet that it is also X-linked in domestic animals and in humans. If the X chromosome has been so conservative throughout evolution, we might expect its banding pattern to be similar among most mammals. While this is true for certain species, such as humans and horses, it is not entirely true for others. In sheep, goats and cattle, for example, the number of bands on the X chromosome is the same but their sequence is different (Hansen 1973*b*). Presumably one or more inversions have occurred in the X chromosome since sheep, goats, and cattle diverged from a common ancestor.

4.6   Interspecific hybridization

An appreciation of the differences in karyotype among species is necessary for an understanding of the results of matings between members of different species. These matings are known as *interspecific hybridizations*, and, depending on which two species are involved, can produce a variety of results. At one extreme, the result is an unsuccessful fertilization which is reflected in a failure of the female parent to conceive. At the other extreme is the production of viable offspring which in certain cases are themselves fertile. This latter result is most likely when the two species being mated together not only have the same number, shape and size of chromosomes, but also have chromosomes with much genetic material in common and which can pair successfully during meiosis. The most common example at this end of the spectrum is the production of fertile cattle from the mating of *Bos taurus* (2n = 60) and *Bos indicus* (2n = 60). In this particular case, the only easily visible difference between the two karyotypes is the size of the Y chromosome. A slightly less successful cross is that between domestic cattle (*Bos taurus*, 2n = 60) and the American bison (*Bison bison*, 2n = 60), which produces sterile male offspring but occasionally fertile females, from which the Beefalo has been derived. A very common cross is that between the horse (*Equus caballus*, 2n = 64) and the donkey (*Equus asinus*, 2n = 62), producing mules (if the mating is mare × jackass) or hinnies (stallion × jennet), all of which have 63 chromosomes. A karyotype of a female mule is shown in Fig. 4.29. As with the *Bos taurus* × *Bison bison* hybrid above, it appears that male mules and hinnies are sterile, but that their female counterparts may be occasionally fertile.

However, reports of fertile female mules or hinnies must be treated with some caution. An interesting example of how appearances can deceive was given by Eldridge and Suzuki (1976) who investigated the apparent birth of a live horse-like foal to a female mule, which subsequently nursed the foal and showed all the normal maternal reactions to it. The foal was found to have a normal horse karyotype, and blood typing tests (like those described in Chapter 12) showed conclusively that the foal was in fact the offspring of a Shetland mare that had herself been seen to give birth in the same field to a foal three weeks after the alleged birth to the mule. The most plausible explanation is that the Shetland mare was in fact carrying twins which were born three weeks apart. The birth of the first foal presumably stimulated the mule to spontaneously commence lactation, a rare occurrence that has, nevertheless, been observed in horses.

Although the diploid chromosome number of horses and donkeys differs by only two, there are considerable differences in chromosome structure between the two species. For example, there are 24 pairs of metacentric autosomes in the donkey, but only 13 such pairs in the horse (Table 4.1). Banding studies by Ryder *et al.* (1978) have confirmed that horse and donkey karyotypes differ by a large number of complex chromosome rearrangements.

In contrast to these differences between horse and donkey chromosomes, the karyotypes of goats (2n = 60) and Barbary sheep (*Ammotragus lervia*, 2n = 58) differ by just one centric fusion: goats have 29 pairs of acrocentric autosomes and no metacentric autosomes, compared with 27 pairs of acrocentrics and one pair of metacentrics in Barbary sheep. Given the apparently smaller difference in karyotype between goats and Barbary sheep than between horses and donkeys, it is interesting to note that fertile male hybrids have been produced by mating Barbary sheep with goats (Moore *et al.* 1980), but never from mating horses with donkeys (Chandley 1981).

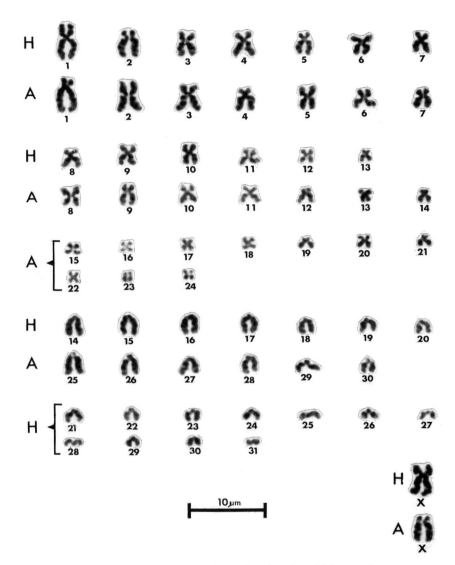

Fig. 4.29. The karyotype of a female mule, showing 32 horse chromosomes and 31 donkey chromosomes. The chromosomes cannot be paired, but are placed by length and conformation in comparison with horse and donkey karyotypes into rows marked H for horse and A for ass (or donkey). The first 5 and the 13th of the metacentric autosomes were assigned with confidence to the horse parent, and the first 5 and numbers 15 to 24 were assigned confidently to the donkey parent. Among the acrocentric chromosomes, 27 to 31 were assigned confidently to the horse parent. The two X chromosomes could also be identified with confidence. All other chromosomes were assigned arbitrarily to the horse or to the donkey parent. (From Eldridge and Blazak 1976.)

Although much work has been done in relation to interspecific hybridization (as summarized by Gray 1972), there is still much to be learnt about the factors that determine success or failure in such crosses. The results of future research in this area will shed considerable light on evolutionary relationships among species.

4.7 Biological basis of sex

We saw in Chapter 1 that sex is inherited in a simple Mendelian manner, determined by the presence or absence of a Y chromosome. It is only recently that a reasonable physiological explanation for this has begun to emerge (Ohno 1979; Wachtel and Koo 1981).

Embryological studies of the development of ovaries or testes from primordial germ cells and associated somatic cells, indicate that the Y chromosome exerts its effect through the ability of cells to recognize and interact with other cells. In most biological systems, this ability is determined by plasma membrane proteins. Because these proteins are involved in the determination of whether or not tissues are compatible, and because they can stimulate the production of specific antibodies, they are called *histocompatibility antigens*. It is now known that cells in normal individuals of the heterogametic sex in mammals (XY; males) exhibit a naturally occurring histocompatibility antigen that is inherited as if it were produced by a gene on the Y chromosome. It is called the *H-Y antigen*. Despite the extreme evolutionary divergence of the mammals so far tested for this antigen, in most cases their H-Y antigens are immunologically identical, indicating a remarkably similar biochemical structure, which points to an important and specific function. Furthermore, an antigen that is immunologically identical to mammalian H-Y antigen is also expressed in cells of the heterogametic sex (ZW; females) in chickens, and is inherited as if linked to the W chromosome.

The most significant evidence for the role of the H-Y antigen has come from studying abnormal individuals whose gonads do not correspond to their sex chromosomes; in most cases, the presence of testes or ovatestes is associated with the presence of H-Y antigen, and the lack of testes, and the consequent presence of ovaries, is associated with the lack of H-Y antigen.

> *This is a good example of how a study of abnormalities can lead to a better understanding of normal biological processes.*

An abnormality that has greatly extended our understanding of sex determination is *testicular feminization*, which has been reported in humans, horses, cattle, sheep, rats, and mice. Although they have normal female secondary sexual characteristics, including normal external female genitalia, individuals with this condition have undescended testes rather than ovaries. Also, instead of normally developed Müllerian duct derivatives (Fallopian tubes, uterus, cervix, and upper portion of vagina), they have underdeveloped Wolffian duct derivatives (epididymis, vas deferens, and seminal vesicle), as shown in Fig. 4.30. Cytogenetically, all of their cells are XY. In addition, all individuals with testicular feminization are positive for H-Y antigen. Testicular feminization is inherited as an X-linked recessive in mice, and available evidence indicates that it is probably inherited similarly in other species.

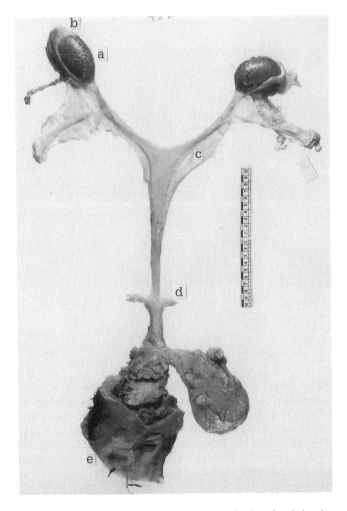

Fig. 4.30. The genital tract from a cow with testicular feminization. (a) Testis, (b) epididymis, (c) broad ligament, (d) seminal vesicle, (e) vulva. (From Long and David 1981.)

From the study of individuals with testicular feminization and with other abnormalities of sexual development, the mechanisms involved in normal sexual development are gradually becoming better understood. A simplified summary of current knowledge is given in Fig. 4.31. It must be emphasized that current knowledge is far from complete in this area. It is not yet certain, for example, whether the H-Y 'locus' on the Y chromosome is the structural gene for H-Y antigen or whether it is a regulatory gene for an autosomal or even X-linked structural H-Y gene. Also, there is some evidence that an H-Y receptor must be present in order for H-Y antigen to exert its effect. However, the

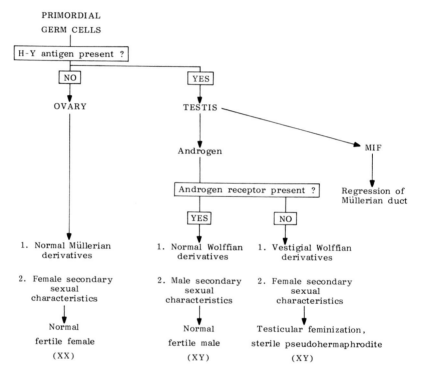

Fig. 4.31. A flow diagram illustrating the determination of sex in mammals. If H-Y antigen is present, the undifferentiated gonad becomes a testis, which secretes Müllerian Inhibition Factor (MIF) and androgen. MIF prevents differentiation of the Müllerian duct derivatives (Fallopian tubes, uterus, cervix, and upper portion of the vagina). Androgen induces development of Wolffian duct derivatives (epididymis, vas deferens, and seminal vesicle) and male secondary sexual characteristics. The X-linked *Tfm* (testicular feminization) allele causes absence of a receptor for androgen. Consequently, in XY embryos having the *Tfm* allele, the testis still produces MIF and androgen, but only the former can exert its effect. Thus, differentiation of Müllerian duct derivatives is inhibited as in normal males, but Wolffian duct derivatives remain underdeveloped, and the rest of the body follows its normal inherent tendency to develop as a female.

genetic origin of this receptor has yet to be determined. Further research in this area is certain to clarify these points, and to lead to a much fuller understanding of the biology of sex determination.

4.8 Freemartins

As long ago as the first century BC, it was known that most female calves born co-twin to a male are sterile. Such females are called *freemartins*. The same condition has since been recognized in other species, and the term freemartin is now used to describe sterile females born co-twin to a male in any species. In the following discussion, which is based on reviews by Marcum (1974) and Benirschke (1981), we shall concentrate mainly on cattle, because freemartins occur much more commonly in cattle than in any other species.

The processes that give rise to freemartins are:

1. Ovulation of at least two ova, and the subsequent fertilization of at least one of those ova by an X-bearing sperm and at least one by a Y-bearing sperm, to produce at least one male and one female zygote.
2. Fusion of chorions of a female zygote and a male zygote, and anastomosis of their blood vessels, giving rise to a circulation common to the two embryos.
3. Transfer of certain cells and possibly some other substance, or substances, between the two embryos.

Some of the results of these processes are common to both sexes, while other results are seen in only one sex.

4.8.1 Both sexes

If an individual receives cells from another individual, it then has two populations of cells, each derived from a different source. Such individuals are said to be *chimaeric* (cf. Section 4.3.1). In the case of freemartins, there is usually an exchange of haemopoietic cells which remain active for the remainder of each animal's life. Thus each member of the twin pair is chimaeric for erythrocytes and for leucocytes. This in turn means that each member exhibits its own blood groups plus those of its co-twin. Since chromosomes are readily visible in leucocytes (as described in Chapter 1), it follows that in unlike-sex twins, the two populations of leucocytes will be readily distinguishable according to their sex chromosomes: those derived originally from male haemopoietic cells will be XY, and those derived originally from female haemopoietic cells will be XX. The existence of these two readily observable different populations of leucocytes in unlike-sex twins is called *XX/XY chimaerism*. A wide range of proportions of XX and XY leucocytes in unlike-sex twins has been reported, ranging from 1 per cent to 100 per cent, with an average of approximately 50 per cent. Interestingly, there is a very high correlation (greater than 0.9) between the proportion of XY leucocytes in both members of a pair of twins, i.e. if one member has a low proportion of XY leucocytes then so does the other member, and similarly, if one member has a high proportion of XY leucocytes, then so does the other member.

We are accustomed to thinking of the karyotype of leucocytes as indicating the karyotype of all other cells in an individual. But in twins whose chorions were fused, this is not so. Indeed, with one possible exception mentioned below, erythrocytes and leucocytes are the only cells for which there is convincing evidence of chimaerism following vascular anastomosis: all other cells in female twins are XX, and all other cells in male twins are XY.

The possible exception mentioned above involves germ cells. It is well known that germ cells arise in a part of the embryo far removed from the site of the undifferentiated gonad, and that they subsequently migrate to the gonad. During this migration, it appears that some germ cells may enter the circulation (Ohno and Gropp 1965). It follows that if two embryos have a common circulation, then there is a good chance of their exchanging germ cells, and thus giving rise to *germ cell chimaerism*. There is limited evidence for the presence of XX cells in gonads of some male twins, and of XY cells in gonads of some female twins. And there is also some evidence to indicate that these 'foreign' germ cells undergo mitosis in the 'host' gonad. But there is no cytological evidence to indicate that these foreign germ cells produce gametes.

If germ cell chimaerism were a general phenomenon, and if the foreign germ cells did produce gametes, then we would expect a distortion in the sex ratio of the offspring of animals born as unlike-sex twins. Although there are several reports of males born co-twin to a female producing an excess of daughters, such results are the exception rather than the rule: most males born co-twin to a female do not have a distorted sex ratio among their offspring.

A more powerful test for the presence of germ cell chimaerism is to determine whether males born co-twin to females pass on any of their sisters' genes to their offspring. Extensive blood typing tests have so far failed to provide any such evidence. Further tests are likely in the future, using not only blood typing, but also the presence of appropriate 'marker' chromosomes, such as centric fusions (Stranzinger *et al.* 1981).

For the time being, we must conclude that:

While there is substantial evidence for the exchange of germ cells following vascular anastomosis, it appears that foreign germ cells do not produce gametes in the host's gonad.

The final result of vascular anastomosis that is seen in both males and females is *homograft tolerance*, which is the ability of a co-twin to accept a graft of skin or other tissue from its fellow co-twin without showing any signs of rejection. The practical application of this phenomenon is that in cattle, where vascular anastomosis is common, skin grafting cannot be used to distinguish between monozygous and dizygous twins.

4.8.2 Females

The major changes in females resulting from vascular anastomosis between unlike-sex twins are seen in the gonads and in the reproductive tract.

Until day 60 of foetal life in cattle, the female gonads appear to be developing normally. Thereafter, development of female gonads is 'masculinized' to an extent that varies considerably from female to female. At one extreme, the gonads sometimes develop into apparently normal ovaries that are capable of ovulation. At the other extreme, the gonads sometimes develop into miniature testes. In most cases, the end result is that one or both gonads are classified as *ovatestes*, containing both ovarian and testicular tissue.

The external genitalia are usually the same as those of normal females, except for the clitoris, which is often enlarged. Internally, the effect on females of vascular anastomosis between unlike-sex twins tends to be repression of Müllerian duct derivatives (Fallopian tubes, uterus, cervix, and upper portion of the vagina) and over-development of Wolffian duct derivatives (epididymis, vas deferens, and seminal vesicle). Depending on the extent of alteration from normal development, the internal reproductive tract of freemartins varies from more-or-less normal female to more-or-less normal male; some females have a normal vagina, cervix and uterus, while at the other extreme, some have a blind vagina together with vasa deferentia and seminal vesicles instead of a cervix and uterus. All intermediate combinations are possible.

4.8.3 Males

Vascular anastomosis between unlike-sex twins has little, if any, effect on the structure of male gonads and reproductive tracts; their structure is essentially normal. There is, however, an effect on reproductive ability. For example, in a study reported by Dunn *et al.* (1979), bulls born co-twin to a female had a 58 per cent chance of being culled for poor reproductive performance up to ten years of age, compared with only 5 per cent for single-born bulls. The poor reproductive performance of bulls born co-twin to a female was due to decreased sperm motility and sperm concentration.

4.8.4 Incidence of freemartins in cattle twins

The probability of vascular anastomosis occurring between cattle twins is very high, being approximately 90 per cent. Since vascular anastomosis almost invariably leads to sterility of the female in unlike-sex twins, it follows that:

> *Approximately 90 per cent of all female calves born co-twin to a male will be freemartins.*

4.8.5 Diagnosis of freemartins

Despite the considerable variation in the structure of reproductive tracts of females born co-twin to a male, as described in Section 4.8.2, it is possible to detect a large proportion of cattle freemartins as early as 3 to 6 weeks of age by a simple clinical examination (Kästli and Hall 1978). The two distinguishing features are: (1) short vagina (less than 12 cm); (2) absence of external opening of the uterus. This test is particularly valuable when young females are being offered as breeders with no indication as to whether or not they were twin-born.

   The other technique recommended for diagnosis of freemartins is to check for the presence of XX/XY leucocyte chimaerism. Because some freemartins have only a small proportion of XY leucocytes, it is obvious that the efficiency of this diagnosis will increase with the number of cells scored from each suspect female. Taking account of the observed distribution of XY leucocytes in a large number of freemartins, Dunn *et al.* (1981) showed that 168 cells must be scored from each suspect female, in order for the investigator to be 99 per cent confident that chimaerism will not be missed.

   Until 1977, it was thought that all females showing XX/XY leucocyte chimaerism were sterile. But in that year, there were two reports of chimaeric females born co-twin to a male giving birth to calves (Smith *et al.* 1977; Eldridge and Blazak 1977). However, since such cases are rare, testing for XX/XY leucocyte chimaerism is still regarded as a useful method of diagnosing freemartins. Although the presence of XX/XY leucocyte chimaerism indicates an exchange of cells between co-twins:

> *The extent of masculinization is not correlated with the proportion of XY leucocytes in the female co-twin. Neither is the degree of sterility in male co-twins correlated with the proportion of XX leucocytes in such males.*

   Indeed, both the fertile chimaeric females reported in 1977 had 26 per cent XY leucocytes, which is much higher than the percentage of XY leucocytes in some sterile chimaeric females. And in the study by Dunn *et al.* (1979), a bull with 85 per cent XX leucocytes had the highest fertility, while a bull with only 33 per cent XX leucocytes had the lowest fertility.

It should be noted that occasionally all the symptoms of freemartins will be found in a single-born female. The most likely explanation for such cases is that they actually commenced uterine life as a co-twin to a male which died and was resorbed prior to the birth of the single-born freemartin.

4.8.6 *Freemartins in other species*

Freemartins have been reported in sheep, goats, and pigs, but in each case, the incidence is much lower than in cattle. In sheep, for example, exchange of cells occurs in only about 1 per cent of twins. In other species such as horses and marmoset monkeys, exchange of cells between embryos is common, but the resultant chimaerism does not lead to infertility. The reasons for these differences are not known.

The most fascinating occurrence of freemartinism is in birds. Vascular anastomosis is common within doubled-yolked eggs, but the major end result is a 'feminization' of the male reproductive tract, in direct contrast to what happens in mammals. This fits in neatly with the fact that the sex determination mechanism in birds is the reverse of that in mammals (as described in Section 4.7).

4.8.7 *Causes of freemartinism*

Despite much research over a long period of time, there is still no adequate explanation for the occurrence of freemartins. Several theories have been advanced over the years, but none of them provides an adequate explanation.

Lillie's (1916) hormone theory proposed that the masculinization of freemartins was due to hormones transferred from the male co-twin, and the cellular theory of Herschler and Fechheimer (1967) suggested that freemartinism was due to the presence of cells bearing a Y chromosome, but neither theory is sufficient to explain the fact that not all females that share a common circulation with a male end up as freemartins.

It seems likely that the H-Y antigen (Section 4.7) is somehow involved in the creation of freemartins. For example, it appears that H-Y antigen is present on the undifferentiated gonads of embryos that become freemartins (Ohno *et al.* 1976). But the true meaning of this observation remains to be determined.

4.9 Classification of intersex

It is evident from Sections 4.3.1, 4.7, and 4.8, that although most individuals are either normal females or normal males, there is a continuing occurrence of *intersex* individuals, having a graded mixture of maleness and femaleness. In an attempt to clarify the classification of such individuals, Winter and Pfeffer (1977) suggested that intersexes should be classified according to the stage at which the abnormality in sexual development occurred. While this is certainly not the only method by which intersexes can be classified, it does provide a useful framework for the discussion of intersexes.

Before describing Winter and Pfeffer's classification, mention should be made of some terms that occur frequently in descriptions of intersex individuals. The first is *hermaphrodite*. In its broadest sense, this term is synonymous with intersex as defined above. However, it is sometimes used in a narrower sense to indicate the presence of both ovarian and testicular tissue. Such individuals are often called *true hermaphrodites*, to distinguish them from *pseudohermaphrodites*, which are intersexes having only ovarian or testicular tissue, but not both. If only ovaries are present, the intersex is sometimes called a *female pseudohermaphrodite*, whereas if only testicular tissue is present, the term *male pseudohermaphrodite* is used.

4.9.1 Chromosomal intersex

This category includes all animals whose abnormal sexual development is due to abnormalities in sex chromosomes. It includes all the cases of sex chromosome abnormalities discussed in Section 4.3.1.

4.9.2 Gonadal intersex

These are individuals having either a normal male or normal female karyotype, but with gonads that do not correspond to the chromosomal sex. Included in this category are XX or XY individuals with ovatestes, XX individuals with testes only, and XY individuals with ovaries only. Freemartins fit into this category because, as noted in Section 4.8.2, they are chimaeric in only blood cells and possibly germ cells: the remainder of their cells are XX. Also included in this category are cases of XX or XY individuals in which the gonads have failed to develop, a condition called *gonadal dysgenesis.*

4.9.3 Phenotypic intersex

Individuals in this category have normal chromosomal and gonadal sex, but some or all of their reproductive tract and other sexual characteristics are abnormal. Individuals with testicular feminization (Section 4.7) fit into this category.

4.10 Foetal sex diagnosis

The central role of chance in determining the sex of any particular offspring has been a source of frustration to humans for many years. In the absence of any useful methods for separating X-bearing sperm cells from Y-bearing sperm cells, the next best thing is to be able to determine the sex of embryos at a sufficiently early stage to enable abortion of those having the unwanted sex. A technique which assists in achieving this result is *amniocentesis*, in which a sample of amniotic fluid, which contains cells of foetal origin, is removed from the pregnant female.

If cells from the fluid are cultured *in vitro* for a period ranging from one to four weeks, then standard cytogenetic analysis of dividing cells (as described in Chapter 1) enables the karyotype and hence the sex of the foetus to be determined. Quicker methods that do not require the culturing of cells are also being investigated. These include testing for the presence of sex chromatin (Barr bodies), and radioimmunoassay of testosterone in allantoic fluid. Refinement of techniques in various species is certain to lead to greater accuracy and wider use of some or all of these methods in the future.

A somewhat different method of foetal sex diagnosis is being developed as an adjunct to multiple ovulation and embryo transfer (MOET), in which a number of fertilized eggs from a donor are placed singly or in pairs into recipient females, the eggs having been obtained following an appropriate hormonal treatment of the donor. Now a well-established technique in horses, cattle, sheep, goats and pigs, MOET usually involves the transfer of embryos at the morula or early blastocyst stage. The method of foetal sex diagnosis relevant to MOET involves the removal and immediate cytogenetic analysis of a small section of trophoblast prior to the transfer of the embryo to a recipient. Figure 4.32 illustrates a 14-day bovine embryo before and after the removal of a small section of trophoblast, together with the live calf

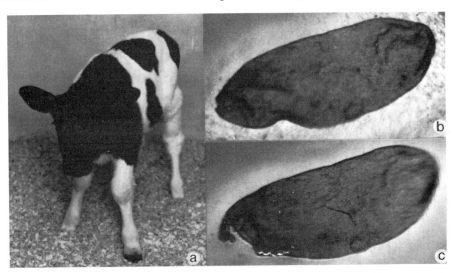

Fig. 4.32. A two-week bovine embryo before (b) and after (c) biopsy for the purpose of sex diagnosis using chromosome analysis. The calf that developed from the biopsied embryo is shown on the left (a). (From Mitchell 1977.)

that developed from it. Although accurate sex diagnosis is possible with this method, many of the techniques involved will require considerable refinement before the method becomes widely used.

4.11 Summary

Chromosomes can be classified into three groups, depending on whether the centromere is at one end (telocentric), considerably closer to one end than to the other (acrocentric or submetacentric), or approximately in the middle (metacentric). Further classification is based on banding patterns that often enable individual chromosomes to be identified. A normal karyotype is not in itself a guarantee of high production and freedom from disease, and neither is an abnormal karyotype certain to indicate a diseased or unproductive animal.

Abnormalities of chromosome number include monosomy (lack of one chromosome) and trisomy (presence of an extra chromosome), both of which arise from non-disjunction during meiosis. The most common forms of these abnormalities in living animals involve the sex chromosomes and can give rise to lowered reproductive performance. Occasionally trisomics for small autosomes are reported. Trisomics for large autosomes and monosomics for any autosome have not been observed in living humans or animals, presumably because they lead to embryonic death. The presence of only one set of chromosomes (haploidy) and the presence of more than two sets of chromosomes (polyploidy) are common chromosomal abnormalities among young embryos but are very rare among living animals, again because they lead to embryonic death. In general, while abnormalities of chromosome number have a high incidence among newly conceived zygotes (leading to a total loss of around 7 per cent of zygotes), they have a relatively low incidence (around 2 per 1000) among individuals sampled after birth.

Abnormalities of chromosome structure include translocations, inversions and deletions. Individuals heterozygous or homozygous for reciprocal translocations have a normal complement of genetic material and hence have normal phenotypes. Translocation heterozygotes usually, however, give rise to some unbalanced gametes and hence show reduced fertility. Centric fusions or Robertsonian translocations involve the joining together of two chromosomes into one, but involve no loss or gain of genetic material and hence give rise to normal phenotypes. Centric fusion heterozygotes in sheep have normal reproductive ability, but those in cattle show a reduction in fertility of around 5 per cent. The effect of eradicating a centric fusion such as the 1/29 in cattle will depend on its frequency; only if it is relatively common will its eradication have a noticeable effect on fertility.

While there is still much to be learnt about the causes of chromo-somal abnormalities, it is beyond dispute that exposure to radiation (for example, X-radiation) increases the frequency of chromosomal abnormalities.

One of the most common forms of karyotype evolution has involved centric fusion or centric fission. Thus, two closely related species often differ by just one centric fusion translocation.

Cells in normal individuals of the heterogametic sex exhibit a natur-ally occurring histocompatibility antigen that is inherited as if it were produced by a gene on the Y chromosome. It is called the H-Y antigen and its role is limited to the production of testes. All other sexual characteristics of males are the result of hormones produced by the testes: Müllerian Inhibition Factor causes regression of the Müllerian duct, and androgen induces development of Wolffian duct derivatives (epididymus, vas deferens, and seminal vesicle) and male secondary sexual characteristics. In the absence of H-Y antigen, the body follows its normal inherent tendency, which is to develop as a female.

A freemartin is a sterile female born co-twin to a male. It is charac-terized by masculinized gonads and internal genitalia, erythrocyte and leucocyte chimaerism, and homograft tolerance. Males born co-twin to a female sometimes have reduced fertility. Currently there is no adequate explanation for the freemartin syndrome.

Intersexes are individuals with a mixture of male and female charac-teristics, resulting from various abnormalities of sexual development. They can be classified according to whether the abnormality of sexual development occurred in the sex chromosomes (chromosomal intersex), in the gonads (gonadal intersex) or at a later stage of development (phenotypic intersex).

The sex of a foetus can be diagnosed by cytogenetic analysis of cultured cells or by testing for Barr bodies in cells obtained by amnio-centesis. If embryos are being transferred from a donor to a recipient, the sex of the embryo can be determined by an immediate cytogenetic analysis of a small section of trophoblast.

4.12 Further reading

Austin, C. R. and Edwards, R. G. (Eds) (1981). *Mechanisms of sex differentiation in animals and man*. Academic Press, London. (This book was written specific-ally for medical, veterinary and science students. It contains some very useful reviews that are directly relevant to this chapter.)

Eldridge, F. E. (1985). *Cytogenetics of livestock*. AVI Publishing Co., Westport, Connecticut. (A comprehensive review of the topics covered in this chapter.)

Fechheimer, N. S. (1979). Cytogenetics in animal production. *Journal of Dairy Science* **62**, 844–53. (A discussion of the application of cytogenetics to animal production.)

Gustavsson, I. (1980*a*). Banding techniques in chromosome analysis of domestic animals. *Advances in Veterinary Science and Comparative Medicine* **24**, 245–89. (A review of animal cytogenetics, with particular emphasis on banding techniques.)

Gustavsson, I. (1980*b*). Chromosome aberrations and their influence on the reproductive performance of domestic animals—a review. *Zeitschrift für Tierzüchtung und Züchtungsbiologie* **97**, 176–95. (Another review of animal cytogenetics, with particular emphasis on reproduction.)

Hare, W. C. D. and Singh, E. L. (1979). *Cytogenetics in animal reproduction.* Commonwealth Agricultural Bureaux, Farnham Royal, Slough, England. (A summary of the basic concepts of animal cytogenetics.)

Hsu, T. C. and Benirschke, K. (1967 *et seq.*). *An atlas of mammalian chromosomes* (Vol. 1 *et seq.*). Springer-Verlag, New York. (A catalogue of the karyotypes of all mammals that have been studied cytogenetically.)

Long, S. E. (1985). Centric fusion translocations in cattle: a review. *Veterinary Record* **116**, 516–18. (A review of centric fusions in cattle.)

Schulz-Schaeffer, J. (1980). *Cytogenetics.* Springer-Verlag, New York. (A comprehensive textbook, dealing with plants, animals and humans.)

Swanson, C. P., Merz., T., and Young, W. J. (1981). *Cytogenetics* (2nd edn). Prentice-Hall, Englewood Cliffs, New Jersey. (An updated edition of a well-known textbook.)

5
Single genes in populations

5.1 Introduction

Apart from those already discussed in Chapters 1 and 3, there are many other abnormalities and diseases that are due to the action of single genes. Most single gene conditions are rare and consequently are not cause for great concern. Occasionally, however, an abnormality or disease due to a single gene reaches a high frequency among the animals belonging to one or a few breeders, or sometimes within a breed as a whole.

The economic consequences of such an increase in frequency are sometimes quite severe, and breeders often ask for advice as to how the abnormality or disease in question can be decreased in frequency, if not eliminated. In order to give useful advice in cases such as this, we need to know more than just simple Mendelian genetics; we must also understand the way in which genes behave within a herd or flock or kennel or cattery, or within a breed as a whole. In other words, we need to understand the basic principles of population genetics. The aim of this chapter is to explain those basic principles.

5.2 Gene and genotype frequencies

Haemoglobin in sheep exists in two different forms (Hb^A and Hb^B), which are the products of two different alleles, A and B, at an autosomal locus.

With two different haemoglobin alleles, there are three different possible genotypes (AA, AB, and BB), each of which produces a distinctive electrophoretic pattern, as shown in Fig. 5.1. Thus for each genotype there is a distinguishable phenotype. Suppose that blood samples were taken from 175 sheep, and that after electrophoresis to determine haemoglobin type for each sheep, it was found that the numbers of genotypes were 91, 28 and 56 for AA, AB, and BB respectively. Given this information, it is possible to calculate the proportions of each genotype in the sample. These proportions are called *genotype frequencies*, and their method of calculation is shown in Table 5.1.

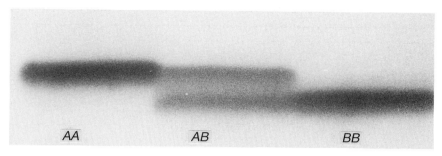

Fig. 5.1. The electrophoretic patterns, or phenotypes, corresponding to the three different haemoglobin genotypes *AA*, *AB*, and *BB* in Merino sheep. Note that the homozygotes have only one band, whereas the heterozygote, with two different gene products, has two bands. (Courtesy of K. Bell.)

It is also possible to calculate the proportions of each allele or gene in the sample. The calculation of these proportions, which are called *gene frequencies*, is also illustrated in Table 5.1.

The concept of gene frequency is basic to population genetics. In many situations, for example, differences between groups of animals are a reflection of differences in gene frequency at one or more loci.

Genotype frequencies and gene frequencies can be calculated for any group of animals. They are most useful, however, when calculated for a group of inter-mating individuals who share a common pool of genes which are transmitted from generation to generation according to the basic Mendelian laws as described in Chapter 1. We shall call such a breeding group a *population*. Populations can vary in size from all the members of a breed or even a species, to the members of just one flock or to an interbreeding group within, for example, a kennel or an aviary.

5.3   Random mating

Random mating exists within a population when each member of that population has an equal opportunity of mating with any individual of the opposite sex.

Bearing in mind that the breeding of domestic animals is largely under the control of humans, who often decide, for example, to mate a particular male to a particular female or group of females, it might be thought that the concept of random mating as defined above is largely irrelevant to domestic animals. However, in deciding which animals to mate to which, humans base their choice on a relatively

Table 5.1. Calculation of genotype and gene frequencies with respect to haemoglobin type in a hypothetical sample of sheep

Genotypes	*AA*	*AB*	*BB*	Total
Numbers	91	28	56	175
Genotype frequencies	91/175	28/175	56/175	1.00
	=0.52	=0.16	=0.32	1.00

Notice that the genotype frequencies sum to one.

Gene frequencies from numbers
One method of calculating gene frequencies is to count the total number of each gene, and divide by the total number of all genes, which is twice the number of animals in the sample (because each animal is diploid).

Each *AA* animal has two *A* genes.
Each *AB* animal has one *A* gene and one *B* gene.
Each *BB* animal has two *B* genes.

Therefore, the total number of *A* genes

$= 2 \times$ (number of *AA* animals) + number of *AB* animals
$= 2 \times 91 + 28$
$= 210$

And the total number of *B* genes

$= 2 \times$ (number of *BB* animals) + number of *AB* animals
$= 2 \times 56 + 28$
$= 140$

Now, since each animal is diploid, there is a total of $2 \times 175 = 350$ genes in the sample.
Therefore, the gene frequency of $A = 210/350$
$= 0.6$
Similarly, the gene frequency of $B = 140/350$
$= 0.4$

Gene frequencies from proportions
Another method is to calculate gene frequencies directly from genotype frequencies. The calculation involves multiplying each genotype frequency by the proportion of the relevant gene that it contains, and summing over genotypes. For gene *A*, we note that *AA* animals contain all *A* genes, *AB* animals contain $\frac{1}{2}A$ genes and *BB* animals contain no *A* genes.

Thus, the gene frequency of $A =$ frequency of $AA + \frac{1}{2}$ (frequency of *AB*)
$= 0.52 + \frac{1}{2}(0.16)$
$= 0.6$

Table 5.1 cont.

And the gene frequency of B = frequency of $BB + \frac{1}{2}$ (frequency of AB)
$$= 0.32 + \tfrac{1}{2}(0.16)$$
$$= 0.4$$

Some general points
Gene frequencies lie between zero and one, and, for any particular sample, must always sum to one. It follows that if we know the value of one gene frequency we can immediately calculate the other. For example, having calculated the gene frequency of A as 0.6, and knowing that the gene frequencies sum to one, it follows immediately that the gene frequency of $B = 1 - 0.6 = 0.4$.

small number of characteristics that are either measurable or at least easily visible, such as milk yield, fleece weight, coat colour, or various aspects of conformation.

> *For those characteristics that are taken into account by humans when matings are being planned, mating is often not random. For most other characteristics, however, mating is usually at random.*

In most situations, for example, mating is at random with respect to blood groups, because these are usually not known. For many simple genetic abnormalities and diseases, too, mating among those that survive to reproductive age is random, because for many such abnormalities and diseases, the different genotypes among survivors all give rise to the same phenotype. Thus the concept of random mating is very important in populations of domestic animals.

5.4   The Hardy–Weinberg law

A popular misconception among people being introduced to population genetics for the first time is that the type of gene action at a locus influences the gene frequencies at that locus. Some people believe, for example, that recessive genes will gradually decrease in frequency simply because they are recessive. Others believe that characteristics due to homozygosity for a recessive allele must occur in a population at a frequency of approximately 25 per cent. In order to illustrate why these beliefs are incorrect, we shall examine closely the implications of random mating within a population.

5.4.1 *The general case*
Consider, for example, the population of 175 sheep with known haemoglobin genotypes as initially described in Table 5.1. If genotype frequencies are the same in males as in females (as is often the case in practice), then the genotypic frequencies can be written as in Table 5.2.

Table 5.2. Genotype frequencies for the data in Table 5.1, assuming equal gene and genotype frequencies in males and females

Genotype	*AA*	*AB*	*BB*	Total†
Males	0.52	0.16	0.32	1.00
Females	0.52	0.16	0.32	1.00

† Frequencies have been expressed separately for each sex. Consequently, within each sex the genotypic frequencies sum to one.

> *If mating is random within a population with respect to a particular locus, then it follows that the probability or frequency of a mating between a male of a certain genotype and a female of a certain genotype equals the product of the frequencies of the respective genotypes.*

Thus, from Table 5.2, it can be seen that the frequency of matings between *AA* males and *AB* females equals $0.52 \times 0.16 = 0.0832$. Similarly, the frequency of matings between *AB* males and *AA* females equals $0.16 \times 0.52 = 0.0832$. Combining these two types of matings together, and describing them simply as *AA* × *AB* matings with sex not specified, it can be seen that the total frequency of *AA* × *AB* matings is $0.0832 + 0.0832 = 0.1664$.

For matings involving just one genotype, there is never any need to specify the sexes, since a mating between an *AA* male and an *AA* female is obviously the same as a mating between an *AA* female and an *AA* male. Therefore the frequency of *AA* × *AA* matings equals $0.52 \times 0.52 = (0.52)^2 = 0.2704$. Similarly, the frequency of *AB* × *AB* matings is $0.16 \times 0.16 = 0.0256$.

Table 5.3. Calculation of the total frequency of *AA* offspring

Mating	*AA* × *AA*	*AA* × *AB*	*AB* × *AB*
Frequency (1)	0.2704	0.1664	0.0256
Proportion of *AA* offspring (2)	1	$\frac{1}{2}$	$\frac{1}{4}$
Frequency of *AA* offspring {(1) × (2)}	0.2704	0.0832	0.0064

Total frequency of *AA* offspring $= 0.2704 + 0.0832 + 0.0064$
$= 0.3600$

Now, the three matings just described are the only source of AA offspring from this population, and according to Mendelian inheritance, the proportion of AA offspring expected from each type of mating is 1 from $AA \times AA$, $\frac{1}{2}$ from $AA \times AB$, and $\frac{1}{4}$ from $AB \times AB$. Knowing the frequency of each mating type and the proportion of offspring from each mating type that are AA, we can now calculate the total frequency of AA offspring, which as shown in Table 5.3, comes to 0.36. Similar calculations for AB offspring and for BB offspring give totals of 0.48 and 0.16 respectively, so that we now have among the offspring:

AA	AB	BB
0.36	0.48	0.16

which sum to one, as indeed they must if the calculations are correct. Thus one generation of random mating among parents has produced the above genotypic frequencies in the offspring.

Now comes the important step. Look back for a minute to Table 5.1 and notice that the gene frequencies in what we now call the parents were 0.6 for A and 0.4 for B. Can you see any simple relationships between the gene frequencies in the parents and the above genotype frequencies in the offspring? Yes, there are some simple relationships; the frequency of AA offspring, 0.36, is the square of the gene frequency of A, $(0.6)^2$, and the frequency of BB offspring, 0.16, is the square of the gene frequency of B, $(0.4)^2$, and the frequency of AB offspring (0.48) is twice the product of the parental gene frequencies $(2 \times 0.6 \times 0.4)$.

Of course these relationships may not appear to be of any real significance, and could in any case be a coincidence, or the result of some careful selection of data in the first place. However, let us imagine some other groups of parents in which the gene frequencies are the same, and let us look at the results of random mating within these groups.

Consider, for example, a population consisting of 105 AA and 70 BB, with no heterozygotes at all. Firstly, let us check that the gene frequencies are still 0.6 and 0.4. Using the same calculations as in Table 5.1, we have the gene frequency of $A = (2 \times 105 + 0)/(2 \times 175) = 0.6$, and hence the gene frequency of $B = 1 - 0.6 = 0.4$. Now, work through exactly the same set of calculations as shown in Tables 5.2 and 5.3 and you will obtain exactly the same proportions of offspring genotypes as before, namely 0.36 AA, 0.48 AB, and 0.16 BB. Take another group of parents, this time consisting of 35 AA and 140 AB with no BB's. The gene frequency of A is $(2 \times 35 + 140)/(2 \times 175) = 0.6$ and hence the gene frequency of $B = 1 - 0.6 = 0.4$ as before. Once again, if you do the calculations you will find that one generation of random mating produces offspring genotypes with frequencies of 0.36 *AA*, 0.48 *AB*, and 0.16 *BB*. In fact, random mating within any group of males and females in which the gene frequency of *A* is 0.6 and of *B* is 0.4 in both sexes, will produce offspring genotypes with frequencies of 0.36 *AA*, 0.48 *AB*, and 0.16 *BB*.

We can generalize our conclusions by repeating the above calculations for any group of males and females in which the gene frequencies of *A* and *B* are different from those used above, but are the same in each sex. In all cases, we will see that if a group of males and females is assembled in any way from any number of different sources, and if these males and females become parents by mating together at random:

> *The genotype frequencies of offspring are determined solely by the gene frequencies of parents, according to the following relationships:*
>
> (1) *the frequency of homozygotes equals the square of the relevant gene frequency;*
> (2) *the frequency of heterozygotes equals twice the product of the relevant gene frequencies.*

It must be emphasized that this conclusion is valid irrespective of how the parents were assembled in the first place. They could have all come from one herd or flock, or they could have come from any number of different breeds and/or different countries. The only requirements are that once the animals are assembled together, the gene frequencies in each sex are the same, and mating between them is at random with respect to the locus being considered.

Now, let us take any one of our offspring populations and determine the gene frequencies in it, using the appropriate method described in Table 5.1. For any offspring population, you will discover that:

> *The gene frequencies in the offspring equal the gene frequencies in the parents.*

Next, suppose that a particular population of offspring has reached sexual maturity and suppose that they become parents by mating at random amongst themselves. If you work through the appropriate calculations in the same manner as shown in Tables 5.2 and 5.3, you will discover that in the new generation of offspring, the genotype frequencies are the same as in the previous generation. This is exactly what we would expect from the two conclusions that we have just reached above: if genotype frequencies of offspring are determined solely by gene frequencies of parents, and if gene frequencies remain constant from one generation to the next, then it follows that genotype frequencies will remain constant from one generation to the next. And for as many generations as you care to repeat the process, the result will be the same: gene frequencies remain constant from generation to generation, and hence genotype frequencies remain constant from generation to generation.

In doing these calculations and reaching these conclusions, we have made certain assumptions, of which only one (random mating) has been clearly stated. A full list of the assumptions implicit in the above calculations is as follows:

1. mating is random;
2. each genotype has an equal opportunity to contribute offspring, and each offspring has an equal opportunity to survive until it, in turn, has an opportunity to mate. This is equivalent to assuming that there is no *selection*;
3. there is no mutation;
4. genes do not enter the population from outside the population. In other words, there is no *migration*;
5. the number of parents is sufficiently large, and the number of offspring that they produce is sufficiently large, that chance fluctuations in gene frequency are negligible.

The name given to changes in gene frequency that are due to chance is *genetic drift*. Using this term, we can say that the last of the above statements is equivalent to assuming that there is no genetic drift.

Taking account of these assumptions, we can now state our previous conclusions in the following general terms.

In a random mating population in which there is no selection, mutation, migration, or genetic drift,

  (1) *genotype frequencies in offspring are determined solely by gene frequencies in parents, such that*

   (a) *the frequency of homozygotes equals the square of the relevant gene frequency;*

   (b) *the frequency of heterozygotes equals twice the product of the relevant gene frequencies;*

  (2) *gene frequencies and genotype frequencies remain constant from one generation to the next.*

These statements are known as the *Hardy–Weinberg law*, being named after an English mathematician, Hardy, and a German physician, Weinberg, who were among the first to recognize the important implications of random mating within a population. The Hardy–Weinberg law can easily be proved, but only by recourse to algebra. Because of this, the proof has been placed in Appendix 5.1.

For reasons that will become evident later in this chapter, it is convenient to use a form of shorthand involving algebraic symbols when referring to gene and genotype frequencies. For the general case considered in this section, a popular convention is to let p and q represent the gene frequencies of, say, A and B respectively. Then, by applying the first part of the Hardy–Weinberg law as stated above, we can immediately write down the expected genotype frequencies as

AA	AB	BB
p^2	$2pq$	q^2

These frequencies are often called *Hardy–Weinberg frequencies* or *Hardy–Weinberg proportions*.

The part of the Hardy–Weinberg law that makes a specific prediction about the relationship between gene and genotype frequencies is often tested in the following manner. The data normally consist of numbers of individuals of each possible genotype at a particular locus, where all the individuals belong to a particular generation. In the example shown in Table 5.4, there are two alleles, and hence three different genotypes at the locus in question. The numbers of individuals are 8 *AA*, 78 *AB*, and 243 *BB*. We start the test by estimating the gene frequencies in these individuals, and then use these estimates of gene frequencies to predict what the genotype frequencies should be. We then convert the expected frequencies into expected numbers, and compare observed and expected numbers of each genotype, usually with the aid of a statistical test, as illustrated in Appendix 5.2. If there is sufficient agreement between observed and expected numbers, then we conclude that the prediction made from the Hardy–Weinberg law has been verified.

Usually, when we test a particular law, we are really testing the assumptions upon which that law is based; if the data have been chosen so as to provide a valid test of the law, and if the observations are not in conflict with the predictions arising from the law, then we conclude that the assumptions on which the law is based, are valid. If this were the case for the Hardy–Weinberg law, then for any population that has Hardy–Weinberg proportions of genotypes we would conclude that (1) there is random mating, and (2) there is no selection, mutation, migration, or genetic drift. In practice, however, the commonly-used test for the Hardy–Weinberg law, as illustrated in Table 5.4, is usually a test of only the first of the assumptions just listed; the test cannot detect mutation, and it fails to detect many cases of selection, migration, and genetic drift. The reasons for this are explained in detail in Appendix 5.3.

Because of these limitations of the test in detecting selection, mutation, migration, and genetic drift, and because mating is random with respect to most characteristics,

Most populations are observed to have Hardy–Weinberg proportions of genotypes.

Table 5.4. A test for Hardy–Weinberg proportions in a sample of Sonadi sheep in India, with respect to haemoglobin types A and B

	AA	*AB*	*BB*	Total
Observed numbers	8	78	243	329

1. Calculate gene frequencies:

    Frequency of $A = (2 \times 8 + 78)/(2 \times 329) = 0.14$
    Frequency of $B = (2 \times 243 + 78)/(2 \times 329) = 0.86$
    Check that the gene frequencies sum to one:

    $$0.14 + 0.86 = 1.00$$

2. Calculate the genotype frequencies expected from the Hardy–Weinberg law:

	AA	*AB*	*BB*
Expected proportions	$(0.14)^2$	$2 \times 0.14 \times 0.86$	$(0.86)^2$
	$= 0.0196$	0.2408	0.7396

3. Multiply the expected proportions by the total number of sheep in the sample, which in this case is 329, to obtain the expected numbers for each genotype:

	AA	*AB*	*BB*
Expected numbers	0.0196	0.2408	0.7396
	$\times$ 329	$\times$ 329	$\times$ 329
	$= 6.5$	$= 79.2$	$= 243.3$

    Check that the expected numbers sum to 329.

    $$6.5 + 79.2 + 243.3 = 329.0$$

4. Compare the observed and expected numbers:

	AA	*AB*	*BB*
Observed	8	78	243
Expected	6.5	79.2	243.3

Since the observed and expected numbers agree very closely, we can conclude that this sample of sheep comes from a population that has Hardy–Weinberg proportions of genotypes.

In the normal course of events, a comparison such as the one above should be conducted using a proper statistical test. An illustration of such a test is given in Appendix 5.2.

Data adapted from John and John (1977).

In fact, if mating among parents is random and if gene frequencies are the same in male and in female parents, then *all* populations have Hardy–Weinberg proportions of genotypes at the commencement of every generation, i.e. at the moment when zygotes are formed. In order to understand why this is so, consider a group of parents at the time of mating, and consider a single locus with two alleles, *A* and *B*. If the parents mate at random, then the probability of a resultant zygote having any particular genotype is the probability of the two relevant gametes uniting to form that zygote. And the probability of, say, a gamete carrying allele *A* uniting with another gamete carrying allele *A* is simply the frequency of *A* multiplied by the frequency of *A*, which is the frequency predicted by the Hardy–Weinberg law. Since the same argument applies to all genotypes, it follows that

> *At the moment of conception, zygotes are always expected to be in Hardy–Weinberg proportions, if their parents mated at random, and if gene frequencies were the same in male and in female parents.*

The only reason for us to see a departure from these proportions is if selection or migration or genetic drift act before the population is observed in any generation. It follows that a population could be undergoing a rapid change in gene frequency and yet still be observed to be in Hardy–Weinberg proportions every generation. This conclusion sometimes causes confusion, because people assume that Hardy–Weinberg proportions will be observed only when selection, mutation, migration, and genetic drift do not occur. From the above discussion, it should be evident that this is not necessarily so.

Most of the above discussion has been concerned with the first part of the Hardy–Weinberg law, the part that makes a prediction about genotype frequencies. The second part of the law is also important, because it tells us that:

> *the frequency of any gene will remain constant from generation to generation, unless selection, mutation, migration, or genetic drift act in such a way as to alter the frequency.*

We shall make practical use of this part of the Hardy–Weinberg law on several occasions in the remainder of this chapter, and in following chapters.

5.4.2 The special case for recessives

It is important to note that the Hardy–Weinberg law applies irrespective of the type of gene action at the locus in question.

> *In fact, so long as there is no selection, the type of gene action at a locus has no effect on the frequency of genes or of genotypes.*

The errors inherent in the two commonly held beliefs mentioned at the beginning of this chapter should now be clearly evident. For example, it is evident from the second part of the Hardy–Weinberg law that recessive characteristics will neither decrease nor increase in frequency from one generation to the next, unless selection, mutation, migration or genetic drift act in such a way as to alter the frequency of the recessive gene. And recessive characteristics can have a frequency anywhere in the range zero to one, depending solely on the frequency of the recessive gene.

Let us now examine the recessive case more closely, using coat colour in Angus cattle as an example. The typical black coat colour seen in Angus cattle is due to a dominant gene B, and the relatively rare red coat colour seen occasionally in Angus cattle is due to homozygosity for a recessive gene b. Because there are two genes at this locus in Angus cattle, there are three genotypes (BB, Bb, and bb). However, because B is completely dominant to b, the genotypes BB and Bb have exactly the same phenotype, namely black. Thus it is not possible to determine the genotype of all animals simply from their phenotype for characteristics determined by dominant and recessive genes. We cannot, therefore, calculate gene frequencies as for the case where all genotypes are identifiable.

We can, however, distinguish red (bb) from black ($B-$, where the dash indicates either B or b). In order to estimate gene frequencies, we make use of the general principle discussed in the previous section, namely that for the majority of loci, most populations have Hardy–Weinberg proportions of genotypes, which means that the genotypes at a single locus with two alleles are in the proportions p^2, $2pq$, and q^2. Applying this to the case of red versus black in Angus cattle, and letting q be the gene frequency of b, we have the following situation:

Genotype	BB	Bb	bb
Phenotype	Black	Black	Red
Frequency	p^2	$2pq$	q^2

It is obvious that the frequency of red Angus equals q^2, the square of the gene frequency of b. The gene frequency of b is then estimated as the square-root of the frequency of red calves.

For example, the frequency of red calves in pedigree Angus herds in the USA is approximately 5 per 1000. Assuming Hardy–Weinberg proportions, this leads to an estimate of the gene frequency of b as $\sqrt{(0.005)} = 0.07$. And since the only other gene at this locus is B, its frequency must be $1 - 0.07 = 0.93$.

We can now make one more interesting calculation. Since $p = 0.93$ and $q = 0.07$, it follows that the frequency of heterozygotes or carriers, which is $2pq$, equals $2 \times 0.93 \times 0.07 = 0.13$. Thus 13 per cent of Angus in the USA are carriers of red. This is a surprisingly high figure, but is indicative of the situation that holds for all *rare* recessive characteristics:

> *The frequency of carriers of rare recessive genes is much higher than the frequency of the recessive characteristic itself.*

Another example of a simple recessive character is yellow coat colour in Labrador dogs, which is recessive to black. Since many dog breeders prefer yellow to black, the frequency of the yellow genotype (*ee*) is quite high in many populations of Labradors. In one Australian population, for example, the frequency of yellow dogs is approximately 64 per cent. Assuming Hardy–Weinberg proportions, the gene frequency of *e* equals $\sqrt{0.64} = 0.8$, which means that the frequency of the dominant gene *E* must be only 0.2. This is a good illustration of the fact that recessive genes and recessive characteristics can be much higher in frequency than dominant genes and dominant characteristics.

5.5 Extensions of the Hardy–Weinberg Law

Two situations commonly encountered in practice were neglected in the discussion of the Hardy–Weinberg law given above. We shall now briefly consider these two situations.

5.5.1 *Multiple alleles*

Blood systems and enzymes in animals and humans provide the best illustrations of multiple alleles. For example, there are three different forms of glucose 6-phosphate dehydrogenase (G6PD) in horses, corresponding to three alleles, *D*, *F*, and *S* at the G6PD locus. Of course any one horse can have at the most only two different alleles. But in a large sample of horses, all three alleles will be found, and the frequency of each allele can be estimated.

If *p*, *q*, and *r* represent the frequencies of the three alleles, then the expected frequencies of the respective homozygotes are p^2, q^2, and r^2, and of the respective heterozygotes are $2pq$, $2qr$, and $2pr$. Using exactly the same principle, this prediction can be extended to any number of alleles at a locus.

5.5.2 *X-linked genes*

The most crucial assumption in the Hardy–Weinberg law does not hold for X-linked genes: mating is definitely not random with respect to sex!

To understand the implication of this, consider the X-linked coat colour locus in cats, for which each genotype in each sex has a distinguishable phenotype. Many population surveys have been conducted on cats in many countries, and the results for X-linked coat colour from two such surveys are shown in Table 5.5.

Table 5.5. Combined results of two surveys taken in Iceland of X-linked coat colour in cats, together with calculations of gene frequencies in males and in females

Sex	Female				Male		
Phenotype	Non-orange	Tortoiseshell	Orange	Total	Non-orange	Orange	Total
Genotype	*oo*	*Oo*	*OO*		*o*	*O*	
Numbers	117	53	3	173	149	28	177

Gene frequency of *o* in females = (2 × 117 + 53)/(2 × 173) = 0.83
Gene frequency of *O* in females = (2 × 3 + 53)/(2 × 173) = 0.17
Gene frequency of *o* in males = 149/177 = 0.84
Gene frequency of *O* in males = 28/177 = 0.16

Data adapted from Adalsteinsson *et al.* (1979).

Because males have only one X chromosome, the frequency of each phenotype in males equals the frequency of the respective gene, so that calculation of gene frequencies in males is very straightforward. In females, where there are three genotypes, gene frequencies can be calculated from first principles, as illustrated previously in Table 5.1. It is evident from Table 5.5 that the gene frequencies in males and females are essentially the same, with average values of 0.835 for *o* and 0.165 for *O*. Since the existence of three genotypes in females is analogous to the general case for autosomal genes discussed in Section 5.4.1, it is tempting to see whether the female genotypes are in Hardy–Weinberg proportions. Using the average gene frequencies, expected genotype frequencies are $(0.835)^2$, $2 \times 0.835 \times 0.165$, and $(0.165)^2$ for *oo*, *Oo*, and *OO* respectively. With a total of 173 females in the sample in Table 5.5, this gives *expected* numbers of 120.6, 47.7, and 4.7 which agree very closely with the *observed* numbers of 117, 53, and 3 respectively. Thus the female genotypes are in Hardy–Weinberg proportions.

These results are typical of those obtained with X-linked genes, which means that:

> *It is usually safe to assume that X-linked gene frequencies are the same in males and females and that X-linked female genotypes occur in Hardy–Weinberg frequencies.*

In other words, if the frequencies of two X-linked genes are p and q, then the frequencies of the two genotypes in males will be p and q, and the three genotypes in females will have frequencies of p^2, $2pq$ and q^2.

What are the practical implications of this conclusion? The most important implication is that X-linked characteristics are expected to occur with different frequencies in males and females.

> *This is most relevant to X-linked recessive characteristics, for which the frequency of the condition in males (q) is expected to be much higher than the frequency of the condition in females (q^2).*

Notice that the square of q is much smaller than q because q is always less than one. For example, if an X-linked recessive condition occurs with a frequency of 10 per cent in males ($q = 0.1$), then its expected frequency in females is 1 per cent ($q^2 = (0.1)^2 = 0.01$).

5.6 Selection and mutation

Selection acts on phenotypes, and occurs whenever some phenotypes contribute more offspring to the next generation than do other phenotypes. Selection may act at any stage during the life cycle of an individual from conception to mating. It sometimes also acts on germ cells between mating and conception, but this is not common, and need not concern us here.

Selection most commonly occurs through differential viability and/or differential reproductive ability, with reproductive ability including factors such as mating ability, fecundity and fertility. For convenience we shall refer to the combined effect of viability and reproductive ability as *fitness*. If selection occurs as a result of decisions by humans it is said to be *artificial selection*, while in all other situations it is referred to as *natural selection*. In either case, the principles by which selection operates are exactly the same. Although selection acts on phenotypes, we are mainly interested in its effect on genotypes and through them, on gene frequencies. Because of this interest, we often talk about selection acting on genotypes and on genes. Whenever we do this, however, we must bear in mind that selection really acts on individual animals according to their phenotype. The extent to which this affects genotypes and hence genes, depends on the extent to which particular phenotypes are associated with particular genotypes.

In order to understand how selection works in practice, we shall start by considering the population of Labrador dogs mentioned in the previous section. You will recall that approximately 64 per cent were homozygous for the recessive yellow gene, e, which corresponds to a gene frequency for e of 0.8, assuming Hardy–Weinberg proportions.

5.6.1 Selection against a lethal dominant

So as to illustrate some important principles, let us suppose that the Labrador breed society decides that from a certain day onwards, only yellow dogs and bitches can be used for breeding. This would amount to complete selection against black coat colour. Since black coat colour is due to the dominant gene, E, we can say that this amounts to complete selection against a dominant gene. Notice that this is equivalent to complete selection in favour of a recessive gene. Obviously black animals will continue to live and could reproduce, but since no black offspring can be registered, it is as if all black pups had zero viability. In cases such as this, where a gene is completely selected against, it is said to be *lethal*.

Since the only matings that are now permitted are yellow × yellow (*ee* × *ee*), and the only offspring eligible for registration are yellow (*ee*), it follows that all members of the Labrador breeding population are now homozygous for *e*, which means that the gene frequency for *E* has been altered by selection from 0.2 to 0.0. At the same time, the frequency of the recessive gene has increased from 0.8 to 1.0. This is an extreme, and yet still useful, illustration of how selection can alter gene frequencies. We conclude that:

> *The immediate effect of complete selection against a dominant gene is to remove it from the population altogether.*

5.6.2 Selection/mutation balance for a lethal dominant

With the black gene removed from the population, we might expect that black dogs would never appear again, unless the gene for black were introduced from another population by migration. However, mutation occurs from time to time at all loci, and the effect of mutation in the present context will be to alter occasionally a yellow gene, *e*, to a black gene, *E*. Because the *E* gene is dominant, any pup in which the new mutant *E* gene appears will have a black coat. Thus every time a mutation occurs from *e* to *E*, a pup with a black coat will appear in the population, even though all matings are between yellow dogs. However, the breed society rules described previously prevent any black dog from contributing genes to the next generation, so that each mutant gene is effectively selected out as soon as it appears.

We thus have two forces opposing one another: mutation is occasionally introducing dominant genes into the population, and selection is removing them. For a lethal dominant gene, the immediate result is that these two forces just balance each other; the number of mutant genes entering the population is the same as the number removed by selection, and the frequency of the dominant gene remains stable from generation to generation. This is called a *selection/mutation balance* or *equilibrium*.

The number of mutant genes entering the population each generation equals the frequency of mutation (which is called the *mutation rate*), multiplied by twice the number of offspring born each generation. (We multiply by two because each offspring has two genes.)

In order to calculate how many mutant genes are removed by selection, we first need to note that since mutation rates are very low, the chance of two uniting gametes both containing a new mutant is negligible. We can therefore assume that there will be no homozygotes for the mutant gene. Now, for every heterozygote removed by selection, one out of two (or one-half) of the genes eliminated is a mutant gene. Thus, the number of mutant genes removed by selection each generation equals one-half of the frequency of heterozygotes, multiplied by twice the number of offspring born each generation. (As above, we multiply by two because each offspring has two genes.) Now, because there are no homozygotes for the mutant gene, it follows from first principles (Table 5.1) that one-half of the frequency of heterozygotes equals the frequency of the mutant gene. Thus, the number of mutant genes removed by selection each generation equals the frequency of the mutant gene multiplied by twice the number of offspring born each generation.

Combining the information in the last two paragraphs, we conclude that in situations where selection against a dominant gene is complete, an equilibrium will exist when the frequency of the lethal dominant gene among newborn pups is equal to the frequency of new mutations occurring each generation. In general:

> *The equilibrium frequency for a lethal dominant gene equals the mutation rate.*

Since mutation rates are generally in the range of 1 in 10 000 to 1 in 1 000 000 (10^{-4} to 10^{-6}) per generation, it is obvious that the equilibrium frequency of lethal dominants is rather low.

If the equilibrium frequency of the dominant gene equals the mutation rate, what is the equilibrium frequency of the dominant phenotype? We have already noted that all new mutants appear in heterozygotes. It follows that the frequency of the mutant phenotype in the population equals the frequency of individuals that are heterozygous for the mutant gene. And we have just noted above that the frequency of the mutant gene equals one-half the frequency of heterozygotes. It follows that the frequency of heterozygotes is twice the frequency of the mutant gene. Recalling that the equilibrium frequency for a lethal dominant gene equals the mutation rate, we conclude that:

> *The equilibrium frequency of a lethal dominant phenotype equals twice the mutation rate.*

5.6.3 Less severe selection against a dominant

In many cases, a dominant gene is not completely selected against, but instead results in only a partial reduction in fitness. Not surprisingly, the frequency of such a gene decreases at a slower rate than with complete selection. Consequently, it will take longer than one generation for the gene to be removed completely from the population. The decrease in gene frequency resulting from selection against a dominant depends upon two factors. They are: (1) the strength of selection, expressed as the *selection coefficient*, which is the proportion by which the fitness of *EE* and *Ee* individuals is less than the fitness of *ee* individuals; (2) the frequency of the dominant gene.

A general expression for the change in gene frequency resulting from selection against a dominant gene is derived in Appendix 5.4.1, and has been used to illustrate the result of selection against a dominant gene in Fig. 5.2.

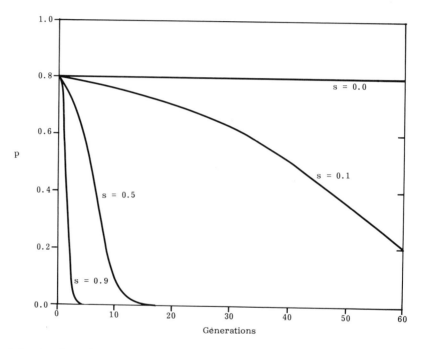

Fig. 5.2. Results of selection against a dominant gene, starting with an initial frequency of 0.8. Notice that as the strength of selection decreases (*s* decreases), it takes longer to remove the gene from the population. The curves in this figure were obtained from the expression $\Delta p = -sp(1 - p)^2 / \{1 - sp(2 - p)\}$, which is derived in Appendix 5.4.1.

Finally, we need to consider a selection/mutation balance when the dominant gene is selected against but is not lethal. In such cases, not all new mutants entering the population will be eliminated immediately, and those that remain will be passed on to the next generation. There will then be two types of dominant genes in the population: new mutants and those inherited from the previous generation. Since both types of mutant genes produce exactly the same phenotype, selection cannot distinguish one from the other.

In this situation, a stage will be reached at which the number of new mutants entering the population by mutation is exactly balanced by the total number of mutants (new and not new) eliminated by selection. We might expect that the smaller the effect of the dominant gene on fitness, the higher will be the equilibrium frequency. The exact relationship between equilibrium frequency and fitness for a dominant gene is derived in Appendix 5.4.2. It turns out that, for example, if the gene reduces fitness by one-half, then the equilibrium gene frequency is twice that for a lethal dominant, being equal to twice the mutation rate. If fitness is reduced by only one-tenth, the equilibrium frequency of the dominant gene is 10 times greater, being 10 times the mutation rate. Notice that this is still a relatively low rate of occurrence: if the mutation rate is 1 in 1 000 000, for example, then the frequency of the dominant gene in the last case cited above will be only 10 in 1 000 000. This leads us to an important point that should be noted.

> *Only very weak selection against a dominant gene is required to keep that gene at a very low frequency.*

5.6.4 Selection against a recessive

Suppose now that the Labrador breed society had made the opposite decision, and declared that only black dogs and bitches can be used for breeding. In this case the recessive gene for yellow is effectively lethal, and selection is completely in favour of the dominant gene for black.

Will the results of complete selection against a recessive gene be as effective as complete selection against a dominant? Will the recessive gene be immediately removed from the population? To both questions the answer is no. This is because selection cannot 'detect' all the recessive genes: it can detect only those that exhibit themselves in yellow dogs, leaving those in black heterozygous dogs undetected.

Let us examine the effects of this selection more closely. The immediate effect is the removal from the breeding population of all the yellow genes in the 64 per cent of the original population that had yellow coat colour. What effect will this selection have on gene frequency? Since the Hardy–Weinberg proportions among black dogs in the original population were 4 per cent EE and 32 per cent Ee, the genotype frequencies among the only animals that can contribute genes to the next generation are $4/(4 + 32) = 11$ per cent of EE and $32/(4 + 32) = 89$ per cent of Ee. Using one of the methods shown in Table 5.1, gene frequencies among these 'survivors' can be calculated as 0.11 +

$\frac{1}{2}$ (0.89) = 0.55 for E and 0.0 + $\frac{1}{2}$ (0.89) = 0.45 for e. Thus the effect of selection has been to decrease the frequency of the recessive gene from 0.80 to 0.45. Since the new gene frequency for e is the frequency of that gene in parents at the time of mating, it is also the frequency of the gene in the offspring resulting from these matings. The difference beween parents at the time of mating and the offspring that result from these matings is in genotype frequencies only; gene frequency is the same in each case. In fact, gene frequency will not alter again until selection occurs in the offspring generation.

Since both surviving genotypes among the parents have equally black coats, mating will be random for this coat colour locus, and consequently the expected genotype frequencies among the offspring are $(0.55)^2$ = 0.3 EE, 2 × 0.55 × 0.45 = 0.5 Ee, and $(0.45)^2$ = 0.2 ee, resulting in phenotypic frequencies of 80 per cent black and 20 per cent yellow. An important principle emerges from this last sentence: despite very strong selection against a particular gene, mating can still be, and usually is, random among those animals that remain in the population after selection. In other words:

> *Random mating is very common even in populations undergoing intense selection.*

What will be the effect of selection in the offspring just obtained? Obviously all yellow dogs will be culled, leaving 0.3/(0.3 + 0.5) = 0.38 EE and 0.5/(0.3 + 0.5) = 0.62 Ee among the black pups that will be the parents of the next generation. From first principles, we can calculate the gene frequencies after selection as 0.38 + $\frac{1}{2}$ (0.62) = 0.69 for E, and 0.0 + $\frac{1}{2}$ (0.62) = 0.31 for e. Thus two generations of complete selection against the recessive gene have changed its frequency from 0.80 to 0.45 to 0.31. We are making some progress, but nowhere near as rapidly as with selection against a dominant gene. Further generations of selection against the recessive gene will produce further decreases in its frequency, but, as shown in Fig. 5.3, at an ever decreasing rate. We can conclude that:

> *Selection against a recessive gene is a very inefficient means of removing that gene from a population.*

The reason for the decreasing effectiveness of selection against a recessive gene, as should now be clear, is that as the frequency of the recessive gene decreases, an increasing proportion of recessive genes are 'hidden' from the effects of selection by occurring in heterozygotes. It follows, therefore, that:

> *Selection against a recessive gene would be much more effective if heterozygotes could be detected.*

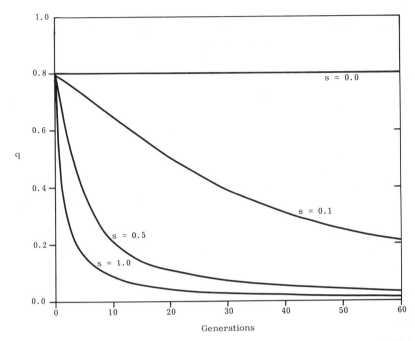

Fig. 5.3. Results of selection against a recessive gene, starting with an initial frequency of 0.8. Notice that once the recessive gene has reached a relatively low frequency, it remains in the population for a long time, despite continual selection against it. Compare this with Fig. 5.2 in which a dominant gene is removed from a population much more quickly once it has reached a relatively low frequency. The lowest curve shows complete selection against a recessive, and was obtained from the formula $q_t = q_0/(1 + tq_0)$, which is derived in Appendix 5.4.3, where the various terms are defined. If the recessive gene is not completely lethal, this simple formula no longer applies, and the response to selection must be calculated generation by generation, using the formula $\Delta q = -sq^2(1 - q)/(1 - sq^2)$, which is also derived in Appendix 5.4.3. When this is done, it is seen that response to selection against a recessive gene decreases as the strength of selection (s) decreases.

If all heterozygotes were detected and then not used for breeding, all recessive genes would be removed from the population at once, and the frequency of the recessive gene would fall to zero. This is the reason why so much research effort is being devoted to detection of heterozygotes in relation to recessive abnormalities and diseases. Notice that there is no requirement to destroy heterozygotes once detected. In fact, they could still be used for breeding, provided that no two heterozygotes were ever mated together. In this way, although the recessive gene remained in the population, the abnormality or disease that it causes would never occur, because homozygotes for the recessive gene would never occur.

5.6.5 *Selection/mutation balance for a recessive*

Even if we are successful at removing a recessive gene from the population or at least preventing its appearance in homozygotes, we still have to contend with the effect of mutation, which slowly but consistently, will be adding new recessive genes to the population.

If we are detecting heterozygotes and not using them for breeding, then we are in essentially the same situation as with a dominant gene. Each new recessive gene entering the population is immediately removed, and a selection/mutation balance is reached at which the frequency of the recessive gene equals the mutation rate. However, because homozygotes never occur, the frequency of the recessive characteristic at this selection/mutation balance is zero: affected individuals never appear.

If we are detecting heterozygotes and then using them for breeding, but avoiding matings between heterozygotes, then there is no selection operating against the recessive gene. In this case the frequency of the recessive gene will increase very gradually at a rate determined solely by the mutation rate, which is usually so low that we would be unlikely to detect an appreciable increase in frequency during a time period of say, 100 years, even in species with short intervals between generations.

The final case that we must consider is the conventional scheme of selection against a recessive gene, where heterozygotes are not detected. New mutants are entering the population at whatever the mutation rate is per generation, but are not eliminated until they occur in a homozygote. Since each offspring has two genes at a locus, the number of mutant genes entering the population each generation equals the mutation rate multiplied by twice the number of offspring born. On the other hand, the number of mutant genes removed each generation equals the frequency of homozygous offspring eliminated by selection multiplied by twice the number of offspring born. In this case, therefore, a selection/mutation balance occurs when the frequency of homozygotes eliminated equals the mutation rate.

If the recessive gene is effectively lethal so that all homozygotes are eliminated, it follows that:

> *The frequency of occurrence of a lethal recessive characteristic equals the mutation rate.*

This is exactly one-half the frequency of occurrence of a lethal dominant characteristic. If the recessive gene results in only a partial reduction in fitness of homozygotes, then the equilibrium frequency is proportionately higher, as in the case for a dominant gene.

The general case of selection against a recessive and of selection/mutation balance for a recessive is presented in Appendices 5.4.3 and 5.4.4.

5.6.6 *Selection favouring heterozygotes*

The Palomino is a very popular breed of horse in many countries, its main distinguishing feature being an attractive golden coat colour. With most breeds of animals, the aim of the relevant breed society is to register only animals that will breed true to type, which means that they will produce offspring that conform to the standards of the breed. Palominos, however, do not breed true to type, because the Palomino coat colour occurs only in horses that are heterozygous at a particular locus called the dilution locus; all Palominos are heterozygotes. This inconvenient fact of life has forced Palomino breed societies to maintain 'open' stud books, to which horses from other breeds are continually added, the main requirement for registration being a Palomino coat colour. Because non-Palominos cannot be registered, the rules of Palomino breed societies amount to complete selection in favour of heterozygotes: both homozygotes have zero fitness in relation to the Palomino breed societies. Exactly the same situation exists with a breed of domestic chicken called Blue Andalusian, in which the characteristic blue feather colour is seen only in heterozygotes. Both cases are examples of co-dominant gene action, in which the heterozygote has a phenotype that is intermediate between those of the two homozygotes. In the Palomino case, the homozygotes are chestnut (*DD*) and cremello (*dd*), which is creamy-white, and for Blue Andalusians they are black (*bl bl*) and white (*Bl Bl*). Both cases involve genes that dilute pigmentation; in the former case *d* and in the latter case *Bl*.

We saw in previous sections that the effect of selection against a dominant or against a recessive gene is to reduce the frequency of that gene to a relatively low level, at which stage a balance between mutation and selection maintains an equilibrium. Selection favouring heterozygotes results in a rather different type of equilibrium. In the extreme case exemplified by Palominos, two genes are maintained in the population at equal frequencies of 0.5 each, because the population consists solely of heterozygotes. In less extreme situations, in which both homozygotes have only a partially reduced fitness, the result is exactly the same so long as both homozygotes have equal fitness. Thus even if both homozygotes have only a 1 per cent reduction in fitness, selection is acting equally against each gene and will maintain an equilibrium gene frequency of 0.5. However, if one homozygote has a lower fitness than the other, then selection is less intense against the gene whose homozygote has the higher fitness.

Snorter dwarfism in cattle is a congenital abnormality due to a recessive gene. It reached a frequency as high as 10 per cent in some Hereford herds in the USA between 1945 and 1955. Why should a condition determined by a lethal recessive gene reach such high frequencies? Although not conclusively proven, the answer seems to be associated with the fashion at that time for breeding blocky, compact Herefords. If the dwarf gene has even a slight effect in heterozygotes such that on average heterozygotes are slightly more compact and blocky, then the fitness of the heterozygote will be relatively higher than that of the normal homozygote, because of the tendency for cattle breeders to favour the compact, blocky type of animal. If this argument is correct, then the end result is selection favouring the heterozygote, but in contrast to the Palomino and Blue Andalusian cases, the two types of homozygote in the dwarfism case have different fitnesses; the homozygote for the dwarfism gene has zero fitness, while that for the normal gene may have a fitness (as a result of artificial selection) only slightly less than that of heterozygotes.

In this case, selection is still favouring heterozygotes, which means that both genes will remain in the population, but it is acting more intensely against the dwarfism gene than against the normal gene. We might expect, therefore, an equilibrium at which the frequency of the dwarfism gene is less than 0.5. In fact, as shown in Appendix 5.4.5:

> *The equilibrium gene frequencies when selection favours heterozygotes are determined solely by the relative fitnesses of the two homozygotes.*

If, for example, one homozygote is only 10 per cent less fit than the heterozygote, while the other one is lethal (100 per cent less fit), then the equilibrium frequency of the lethal gene is $0.10/(0.10 + 1.00) = 0.09$. Thus only a small difference in fitness between heterozygotes and normal homozygotes can lead to an equilibrium frequency much higher than that expected under mutation/selection balance with a recessive lethal gene.

Although definite proof is usually lacking, selection favouring heterozygotes is often suggested as a reason for a recessive lethal condition reaching unusually high frequencies. In many situations, it is certainly the most plausible explanation available, and sometimes data are available to back it up. Syndactyly or fusion of digits in Holstein-Friesian cattle, for example, is an effectively lethal recessive abnormality for which there is some evidence available of selection favouring heterozygotes. In this case, there is a suggestion that heterozygotes produce higher milk and butterfat yields and thus have a higher fitness as a result of artificial selection (Leipold *et al.* 1973).

The term *polymorphism* is often applied to the maintenance in a population of at least two genes at a locus, at more or less intermediate frequencies, where intermediate is defined very broadly as anywhere in the range between, say, 0.05 and 0.95. Thus we can say that selection favouring heterozygotes often results in a polymorphism, whereas a selection/mutation balance usually does not.

One final point must be made in relation to selection favouring heterozygotes. If such selection occurs before genotypes are counted in a particular generation, we would expect a test for Hardy–Weinberg proportions to indicate an excess of heterozygotes. Because of this, situations in which an excess of heterozygotes has been observed in practice have often been taken as evidence of selection favouring heterozygotes. There are, however, several other possible explanations for an excess of heterozygotes.

As explained in Appendix 5.3, one very likely explanation is simply that gene frequency is unequal in males and females. For reasons that we shall not investigate here, any difference in gene frequency between the sexes inevitably results in an excess of heterozygotes at autosomal loci (Robertson 1965).

Other possible explanations are the recent admixture of two very different populations, and, at a more basic but no less likely level, the misclassification of genotypes (Archibald *et al.* 1979).

5.6.7 *Selection against heterozygotes*

Neonatal diarrhoea in piglets is of considerable economic importance. It is often caused by strains of *Escherichia coli* bacteria having a cell-surface antigen called K88, which combines with a receptor on the cell wall of a piglet's intestines, enabling the bacteria to attach themselves to the intestine (Fig. 5.4). Once attached, they proliferate, releasing

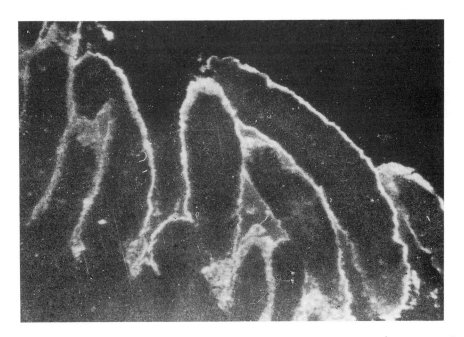

Fig. 5.4. Attachment of *E. coli* bacteria to piglet intestinal mucosa, demonstrated by a fluorescence technique. (From Jones and Rutter 1972.)

enterotoxins and thus producing diarrhoea which can lead to 90 per cent mortality. Some strains of *E. coli* lack the K88 antigen (K88-negative) and cannot therefore attach themselves to the intestinal mucosa (Fig. 5.5). Being thus unable to proliferate and release enterotoxins, such strains are non-virulent. Certain piglets, however, are not susceptible even to K88-positive bacteria, and it has been found that they lack the appropriate receptor for K88, thus preventing the attachment and subsequent proliferation of bacteria.

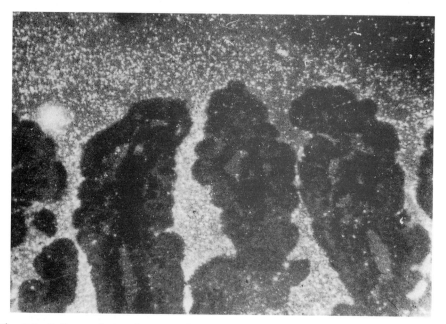

Fig. 5.5. Failure of attachment of *E. coli* bacteria to piglet intestinal mucosa, demonstrated by the same fluorescence technique as used in Fig. 5.4. (From Jones and Rutter 1972.)

This is the first example encountered in this book of a phenomenon that is of particular importance: variation in the host and variation in the pathogen. We shall discuss the matter in detail in Chapter 10, but for the time being we shall concentrate on the variation in the host, which for this particular disease has a simple genetic basis: resistance to the forms of neonatal *E. coli* diarrhoea caused by the lack of a K88 receptor is due to homozygosity for a recessive gene called *s*. Piglets having the receptor and therefore being susceptible to K88-positive bacteria are either homozygous for a dominant gene *S* or are heterozygous (Gibbons *et al.* 1977). The locus at which these alleles occur has not been mapped to a particular chromosome, but it appears to be linked to the locus that codes for transferrin.

Since K88-positive *E. coli* are fairly common, and since the susceptible gene is dominant, we might expect that the susceptible gene would be maintained at a very low frequency by a selection/mutation balance. However, in four English herds examined in one survey, the frequency of the dominant gene in three herds was much greater than 0.5, and in the other herd it was around 0.4. The most satisfactory explanation for these unexpectedly high frequencies involves a fascinating combination of basic principles from immunology and population genetics.

Consider a population into which the resistant gene *s* has only recently been introduced by mutation and/or migration, and suppose that a build-up of K88-positive bacteria occurs. Because nearly all animals are susceptible, there will be very strong selection against the dominant susceptible gene, i.e. in favour of the recessive, resistant gene. At the same time, however, the susceptible sows mount an immune response to the K88-positive bacteria, supplying antibodies to K88 via their colostrum to all their piglets. These antibodies are sufficient to prevent diarrhoea, and so selection against the dominant, susceptible gene becomes less intense not long after it commenced. Thus the previously expected rapid elimination of the dominant gene does not occur. It has, however, been reduced in frequency from 1.0 to, say, 0.7, which means that $(0.3)^2 = 0.09$ of all sows are now homozygous for the resistant gene. These sows certainly carry the K88-positive bacteria, but because attachment does not occur, they never develop antibodies to K88 and hence they provide no protection to their piglets. This is of no consequence to their *ss* piglets because they are naturally resistant anyway. However, not all offspring of *ss* sows are *ss*; some of the *ss* sows will mate with *Ss* or *SS* boars and will consequently produce some or all *Ss* piglets, all of which are susceptible. And these heterozygous piglets get the worst of both worlds; they have a receptor for K88, and they fail to receive antibody from their *ss* female parent.

Thus we have selection against heterozygotes, with both homozygotes having equal and normal fitness, in the case of *SS* because of antibodies received from the sow, and in the case of *ss* because of lack of a receptor for K88. Selection against heterozygotes is only partial, because only those born to *ss* sows are likely to be affected by diarrhoea. What are the likely effects of selection against heterozygotes?

A moment's reflection will indicate that for every heterozygote eliminated from a population, an equal *number* of both genes is removed, which will have a greater effect on the *less common* gene. Consider, for example, a herd of 100 pigs which has Hardy–Weinberg genotype frequencies corresponding to gene frequencies of 0.7 and 0.3 prior to selection. The numbers of the genotypes are 49 *SS*, 42 *Ss*, and 9 *ss*. Suppose that ten heterozygotes die from neonatal *E. coli* diarrhoea, leaving 49 *SS*, 32 *Ss*, and 9 *ss* in a total of 90 pigs after selection. The gene frequencies are now $(2 \times 49 + 32)/(2 \times 90) = 0.72$ for *S* and

$1 - 0.72 = 0.28$ for s, which represents a reduction in frequency of the *less common* gene.

> *In general, if homozygotes have equal fitness, then selection against hetero-zygotes results in a decrease in frequency of the less frequent gene.*

Thus, the gene that was at a frequency of less than 0.5 when selection against heterozygotes commenced, will gradually decrease in frequency so long as such selection continues.

In the piglet case, selection against heterozygotes will cease when the pathogenic *E. coli* disappear, leaving the frequency of the s gene at whatever level it had reached at that time. With the next outbreak of neonatal *E. coli* diarrhoea, the cycle of events described above will recommence.

If the s gene had increased in frequency to more than 0.5 before selection against heterozygotes became effective, then such selection would increase its frequency still further. If it were less than 0.5, then it would decrease in frequency towards its former low level. Thus if a large number of herds were surveyed, it might be expected that the s gene would be at quite a low frequency in some and at a high frequency in others. Results of the very limited surveys reported to date do not contradict this prediction.

For an algebraic treatment of selection against heterozygotes, see Appendix 5.4.6.

5.7   Genetic drift and the founder effect

There is yet another reason why deleterious and even lethal conditions can reach quite high frequencies in certain populations. It is called the *founder effect*, which is the change in gene frequency that usually occurs when a new population is founded with a small number of individuals.

The most extreme case of the founder effect in domestic animals can be illustrated by imagining that only one male and one female are chosen to found a new population. Consider a single locus with two alleles A and B. Irrespective of what the frequency of A was in the population from which the parents were chosen, the gene frequency in the actual parents must be either 0.0 (if both parents happen to be homozygous for B), 0.25 (if one parent is AB and the other is BB), 0.50 (if the parents are AB and AB, or AA and BB), 0.75 (AA and AB) or 1.00 (both parents homozygous for A). Suppose that the frequency of A was, say, 0.1, in the population from which the parents were chosen, and suppose that the parents chosen at random happened to be AA and AB. In this case, the gene frequency has altered from 0.1 to 0.75, a change of 0.65, which is very large and which is due entirely to chance or sampling. Since the gene frequency in the parents is now 0.75, it follows that the gene frequency in the offspring of these parents will also be 0.75, unless selection, mutation, or migration cause a further change. And even if the population size rapidly increases to many thousands, the gene frequency will remain at 0.75 in the absence of selection, mutation, or migration. Thus a population that is founded by just one male and one female can have a very different set of gene frequencies to that from which the founding parents were chosen.

The founder effect is an example of a *population bottleneck*, which is a situation in which the number of parents in a population becomes very small for one or more generations. As illustrated above, a population emerging from a bottleneck may be very different from the population that existed before the bottleneck.

The founder effect is a special case of the more general phenomenon of *genetic drift*, which, as we saw in Section 5.4.1, refers to changes in gene frequency due entirely to chance. These changes result from the sampling of finite numbers of genes that is inevitable in all finite populations. Because these changes in gene frequency are entirely due to chance, their direction is random and is completely outside the control of humans. Since all populations are finite, it follows that genetic drift occurs in all populations. However:

> *The smaller the population size, the larger is the magnitude of genetic drift.*

In order to proceed further with the general concept of genetic drift, we require a special definition of population size. Since it is most convenient to present that definition in Chapter 16, we shall postpone further discussion of the general concept until then. At present, it is sufficient to say that some populations of domestic animals are sufficiently large (hundreds of parents) for genetic drift to be negligible, while others are sufficiently small (less than ten parents) for genetic drift to be extremely important.

To return to the founder effect, consider the cases of combined immunodeficiency (CID) in Arabian foals discussed previously in Section 3.6. A particular Arabian stud in Australia was founded by the importation of two stallions from England. Estimates of the frequency of the recessive gene for CID are not available in England, but it is likely to be less than 0.1. Unfortunately, both of the foundation sires for the Australian stud were heterozygous for this gene, so that the frequency of the gene in the two males that founded the stud in Australia was 0.5. This is an example of the founder effect. The result, as described in Section 3.6, was an unacceptably high incidence of CID in this stud, at a level much higher than that observed in England from where the stallions were obtained.

There are other cases where increases in the frequencies of undesirable genes, apparently due to chance, have caused considerable trouble in domestic animals. In practice, it is often very difficult to distinguish between an increase due entirely to chance, and one due to heterozygote advantage. In many cases, the only data available simply indicate that a particularly fashionable sire (who consequently appears in the pedigree of many animals) was a carrier of an undesirable recessive gene. In some cases the gene may have arisen by mutation in the gonads of one of the sire's parents. The sire may then be used extensively throughout his lifetime without ever siring an affected offspring. Only when his sons, who are often also very popular, are mated to some of the original sire's female descendants does the defect begin to appear, because until then the opportunity for carriers to mate together may not have arisen. The defect soon appears with alarmingly high frequency in certain studs, especially those who have concentrated on the popular bloodline originating from the sire in question. Whether this sire became so popular because of some advantage conferred on him by the otherwise defective gene, or whether it was simply bad luck that a very popular sire happened to carry a defective gene, is often difficult to determine. In either case, the end result is an unacceptably high frequency of a defect in particular studs and often in the breed as a whole. Examples of this problem are encountered all too frequently, in all species of domestic animals. It is also encountered in humans, in which the classic example is that of Queen Victoria, who appears to have received a new mutant for X-linked haemophilia from one of her parents. Her relatively large family of nine children married into various other European royal families, into which they introduced the haemophilia gene. In domestic animals, a good example of the spread of a defective gene was described by Lamb *et al.* (1976), in relation to a defect called limber legs in Jersey cattle in the USA. Methods by which problems such as this may be overcome are discussed in Chapter 11.

In relation to genetic drift and the founder effect, it must be noted that there is an equal chance that an undesirable gene will decrease in frequency due to chance. For example, some populations have undoubtedly been founded by individuals completely lacking undesirable genes that are present in the original population. Such favourable results of the founder effect occur quite often, but because they result in the lack of a problem, they often remain unnoticed, or at least undocumented. Of course, it is unlikely that such a population will remain free of the particular gene forever, because mutation will probably introduce it at some time in the future.

5.8   Summary

A population is a group of inter-mating individuals who share a common pool of genes which are transmitted from generation to generation according to the basic Mendelian laws. For most loci in most populations, mating is random. The Hardy–Weinberg law provides predictions of genotype frequencies and hence of phenotype frequencies that often have considerable practical importance. The law shows for example, that the frequency of carriers of genes for rare recessive diseases is much higher than the frequency of the disease itself. In addition, the law explains why the frequency of characteristics due to X-linked recessive genes is much higher in males than in females.

Selection acts on phenotypes, and occurs whenever some phenotypes contribute more offspring to the next generation than do other phenotypes. To the extent that phenotypes result from the action of genes, selection may alter gene frequency.

The immediate effect of complete selection against a dominant gene is to remove it from the population altogether, while selection against a recessive gene is much less effective. The elimination of genes from a population as a result of selection is often opposed by mutation which continually introduces new versions of the same gene. A selection/mutation balance usually results, in which defective genes are permanently maintained at relatively low frequencies.

If selection favours heterozygotes, a stable equilibrium also results, but usually with gene frequencies less extreme than in the case of a selection/mutation balance. Selection favouring heterozygotes is a possible explanation for the rapid spread of an undesirable recessive gene. Another possible explanation is genetic drift, in which the frequency of a particular gene changes due to chance effects during sampling of genes. A particular case of genetic drift is the founder effect, the undesirable effects of which are seen when a new population is founded with a small number of individuals, some or most of which by chance carry undesirable genes that are relatively rare in the population from which they came.

5.9   Further reading

Cavalli-Sforza, L. L. and Bodmer, W. F. (1971). *The genetics of human populations*. Freeman, San Francisco. (A very clear account of the theory of population genetics, with most of the concepts well illustrated by examples.)

Falconer, D. S. (1981). *An introduction to quantitative genetics* (2nd edn). Longman, London. (The first two chapters of this classic textbook cover much of the same ground discussed in the chapter just concluded.)

Hartl, D. L. (1980). *Principles of population genetics*. Sinauer, Sunderland, Massachusetts. (Provides a thorough grounding in population genetics.)

Hartl, D. L. (1981). *A primer of population genetics.* Sinauer, Sunderland, Massachusetts. (For those who want a simple, straightforward introduction to population genetics.)

Jolly, R. D. (1977*b*). The founder effect and genetic disease of cattle. *New Zealand Veterinary Journal* 25, 109–10. (A discussion of the practical implications of the founder effect, followed by recommendations as to how its undesirable effects may be alleviated.)

Appendix 5.1

Proof of the Hardy–Weinberg law

Consider a locus with two alleles A and B with frequencies p and q respectively ($p + q = 1$). Let the genotype frequencies of AA be P, of AB be H, and of BB be Q in each sex, where P, H, and Q are positive fractions having any value between zero and one inclusive, providing that $P + H + Q = 1$. If mating is at random for this locus, the relative frequencies of the nine different mating types are the products of the relevant genotype frequencies:

		Males		
		AA	AB	BB
		P	H	Q
Females	AA P	P^2	PH	PQ
	AB H	PH	H^2	HQ
	BB Q	PQ	HQ	Q^2

Notice that matings involving different genotypes occur twice in the table, each time having the same frequency. Since these matings are simply reciprocals of the same mating type, we can combine them together and thus reduce the number of different mating types to six.

We now consider the proportions of each genotype resulting from each mating type. Matings of the type $AA \times AA$, for example, produce only AA offspring. Thus the frequency of AA offspring from this source is P^2, the frequency of $AA \times AA$ matings. Other sources of AA offspring are $\frac{1}{2}$ of the offspring from $AA \times AB$ matings, and $\frac{1}{4}$ of the offspring from $AB \times AB$ matings. Given the respective mating type frequencies of $2PH$ and H^2, the frequency of AA offspring from these matings is $\frac{1}{2} \times 2PH = PH$ and $\frac{1}{4}H^2$ respectively. The total frequency of AA offspring is then $P^2 + PH + \frac{1}{4}H^2 = (P + \frac{1}{2}H)^2$. Reference to Table 5.1 will indicate that one method of estimating gene frequency is to add half the frequency of heterozygotes to the frequency of homozygotes for that gene. In other words, $p = P + \frac{1}{2}H$. Thus the frequency of AA

offspring, which is $(P + \frac{1}{2}H)^2$, obviously equals p^2. By following the same argument for AB and BB offspring, we obtain:

Mating		Offspring		
Type	Frequency	AA	AB	BB
$AA \times AA$	P^2	P^2	—	—
$AA \times AB$	$2PH$	PH	PH	—
$AA \times BB$	$2PQ$	—	$2PQ$	—
$AB \times AB$	H^2	$\frac{1}{4}H^2$	$\frac{1}{2}H^2$	$\frac{1}{4}H^2$
$AB \times BB$	$2HQ$	—	HQ	HQ
$BB \times BB$	Q^2	—	—	Q^2
Totals		$(P+\frac{1}{2}H)^2$	$2(P+\frac{1}{2}H)(Q+\frac{1}{2}H)$	$(Q+\frac{1}{2}H)^2$
		$= p^2$	$= 2pq$	$= q^2$

It is evident that irrespective of the values of P, H, and Q, one generation of random mating results in genotype frequencies of p^2, $2pq$ and q^2.

What are the gene frequencies in the offspring? From first principles, the frequency of gene A in offspring $=$ frequency of AA plus $\frac{1}{2}$ frequency $AB = p^2 + \frac{1}{2} \times 2pq = p^2 + pq = p(p + q) = p$, since $p + q = 1$. Thus gene frequencies have remained unchanged from one generation to the next.

Appendix 5.2
A statistical test for Hardy–Weinberg proportions

From Table 5.4, we have

	AA	AB	BB	Total
Observed numbers	8	78	243	329
Expected numbers	6.5	79.2	243.3	329

The null hypothesis is that the sample of 329 comes from a population that has Hardy–Weinberg proportions of genotypes, in which case the observed numbers should equal the expected numbers, on average. A statistical test is required that will enable us to decide if the differences between observed and expected numbers shown above can be explained solely in terms of chance variations. If so, then the null hypothesis can be accepted as true.

An appropriate and simple test is the chi-squared test, in which the quantity

$$\chi^2 = \sum \frac{(O - E)^2}{E}$$

is calculated from the data as shown below, and then compared with a tabular value.

Genotype	Observed number (O)	Expected number (E)	O − E	(O − E)²	(O − E)²/E
AA	8	6.5	1.5	2.25	0.35
AB	78	79.2	−1.2	1.44	0.02
BB	243	243.3	−0.3	0.09	0.00

$$\chi^2 = 0.37$$

In order to compare this value with the appropriate tabular value, we must decide how many degrees of freedom we have. Since there are three classes (*AA*, *AB*, and *BB*), there would normally be two degrees of freedom (number of classes −1). However, in order to obtain the expected genotype frequencies, gene frequencies were estimated from the observed genotype frequencies. Because only one of these gene frequencies is independent (the second being equal to 1 − the first), we lose just one extra degree of freedom by estimating gene frequencies. Thus the number of degrees of freedom is the number of classes −1 −1, which equals one.

Reference to standard statistical tables shows that the tabular value of chi-squared with one degree of freedom at the 5 per cent level of significance is 3.84. Observed chi-squareds larger than this are expected by chance in only 5 per cent of tests, and so observed chi-squareds larger than 3.84 would normally lead to a rejection of the null hypothesis. In our case, the observed chi-squared is much smaller than 3.84, which indicates that the relatively small differences between observed and expected numbers can easily be attributed to chance.

Thus we conclude that the population from which this sample is drawn does have Hardy–Weinberg proportions of genotypes.

Appendix 5.3

Implications of the Hardy–Weinberg test

It was explained in Section 5.4.1 that at the moment of conception, zygotes are always expected to be in Hardy–Weinberg proportions, if their parents mated at random and if the gene frequencies were the same in male and in female parents. The conventional Hardy–Weinberg test is conducted on the observed genotype frequencies in those zygotes, once they have developed to the stage at which their genotypes can be determined. Let us consider the effect of selection, mutation, migration, and genetic drift on the observed genotype frequencies.

If selection occurs between the conception of zygotes and the stage at which the data are collected, then it may significantly alter the genotype frequencies from those expected under the Hardy–Weinberg law, if its effect is large enough. But selection that occurs *after* the data are collected, i.e. from the time of data collection until the time when gametes from the surviving individuals unite, will not be detected.

The effect of mutation is to alter gene frequency in the gametes that are uniting to form zygotes. Thus, even if mutation causes large changes in gene frequency, offspring genotypes will still be in Hardy–Weinberg proportions.

The situation with respect to migration is the same as with selection. If it occurs between the time when zygotes are formed and when the data are collected, then it may cause significant departures from Hardy–Weinberg proportions, if its effect is large enough. If it occurs at any other time, it will not be detected.

If genetic drift occurs when zygotes are being formed, then this may lead to a significant departure from Hardy–Weinberg proportions, if its effect is large enough. But if genetic drift occurs *within* a generation, between the conception of zygotes and the time when the next set of parents is chosen, it will not be detected.

In summary, the conventional test of the Hardy–Weinberg law is a test for random mating amongst parents. In addition, it may detect departures from Hardy–Weinberg proportions due to selection, migration, and genetic drift, but only if the effect of these factors is relatively large, and only if they occur at certain limited stages of the life-cycle. The test cannot detect mutation.

A further point should be noted. A departure from Hardy–Weinberg proportions could be due to selection or migration or genetic drift occurring at the appropriate stage of the life-cycle, or to any combination of two or more of these factors. Unfortunately, it is generally not possible to determine from the data used in the test which of these was responsible for any observed departure from Hardy–Weinberg proportions.

On several occasions in Section 5.4, it was assumed that gene frequencies were equal in both sexes. As this is generally true in practice, this assumption is usually valid. If gene frequencies are different in males and females, then it can be shown (Robertson 1965) that there will be an excess of heterozygotes in the offspring. If this excess of heterozygotes were sufficiently large, it could cause a significant departure from Hardy–Weinberg proportions. However, irrespective of how large the difference in frequency between male and female parents, the gene frequency in their offspring will be the same in each sex, being equal to the average frequency in the parents. It follows that when these offspring reach reproductive age and produce the next generation of offspring, then these latter offspring will have Hardy–Weinberg genotype frequencies. In other words, differences in gene frequency between the sexes delay the attainment of Hardy–Weinberg proportions by one generation.

Obviously, these conclusions apply only to autosomal loci; the special case of X-linked loci is discussed in Section 5.5.2.

Appendix 5.4
An algebraic treatment of selection and mutation

A5.4.1 Selection against a dominant

Consider a single locus with two alleles, E being dominant to e. Relative to the fitness of genotype ee, fitness of genotypes EE and Ee is reduced by a proportion s, where s is the *selection coefficient*. Relative fitnesses are then $1 - s$, $1 - s$, and 1 for EE, Ee, and ee respectively. To obtain the genotype proportions after selection, we multiply the genotype frequency prior to selection (which is the Hardy–Weinberg frequency) by the relative fitness.

	Genotype			
	EE	*Ee*	*ee*	Total
Frequency prior to selection	p^2	$2pq$	q^2	1
Relative fitness	$1 - s$	$1 - s$	1	
Proportion after selection	$p^2(1 - s)$	$2pq(1 - s)$	q^2	$1 - sp(2 - p)$

Notice that the sum of the proportions after selection is $1 - sp(2 - p)$, which is less than one. The relative frequencies of genotypes after selection can be obtained from these proportions as:

	EE	*Ee*	*ee*	Total
Frequency after selection	$\dfrac{p^2(1 - s)}{1 - sp(2 - p)}$	$\dfrac{2pq(1 - s)}{1 - sp(2 - p)}$	$\dfrac{q^2}{1 - sp(2 - p)}$	1

The gene frequency of E after selection, which we shall call p', is then estimated from first principles as

$$p' = \text{frequency of } EE + \tfrac{1}{2} \text{ frequency of } Ee$$

$$= \frac{p^2(1 - s)}{1 - sp(2 - p)} + \frac{1}{2} \times \frac{2pq(1 - s)}{1 - sp(2 - p)} \tag{1}$$

The change in gene frequency is $\Delta p = p' - p$, which after some algebra reduces to

$$\Delta p = -sp(1-p)^2/\{1 - sp(2-p)\}, \tag{2}$$

where the negative sign indicates that the change in gene frequency of E is a decrease rather than an increase.

If there is complete selection against the dominant gene E such that $s = 1$, then $\Delta p = -p$ and $p' = 0$, and the gene is eliminated after one generation of selection.

Expression (2) can be simplified by noting that if s is large, then there is strong selection against E, in which case p will be small. On the other hand, if p is large, then selection against E must be weak, in which case s will be small. Thus in most cases, the product sp will be small, and hence $1 - sp(2-p)$ will approximately equal 1. This gives

$$\Delta p \doteq -sp(1-p)^2, \tag{3}$$

which is sufficiently accurate for most purposes.

A5.4.2 *Selection/mutation balance for a dominant*

From the previous section, it can be seen that the loss of E genes as a result of selection is sp^2 from EE genotypes and $\frac{1}{2} \times 2spq$ from Ee genotypes, where the $\frac{1}{2}$ allows for only $\frac{1}{2}$ the genes in Ee genotypes being E. Thus the total loss of E genes is $sp^2 + spq = sp(p+q) = sp$.

In contrast, new E genes are entering the population as a result of mutation from e to E. Since the frequency of e is q, and the rate of mutation is μ, it follows that the gain of E genes resulting from mutation of e to E is μq, which is approximately μ, since q is very nearly 1. At equilibrium, therefore, the loss of E genes is sp and the gain is approximately μ, which gives $sp \doteq \mu$, or $\hat{p} \doteq \mu/s$, where $\hat{\ }$ indicates equilibrium gene frequency. Thus for lethal dominant genes ($s = 1$), the equilibrium gene frequency equals the mutation rate. Since μ is generally in the range 10^{-4} to 10^{-6}, even quite weak selection (small values of s) are sufficient to keep $\hat{p}$ relatively small. Thus at equilibrium, $\hat{p}$ is generally low.

The frequency of the dominant phenotype at any generation is $\hat{p}^2 + 2\hat{p}(1 - \hat{p})$ which approximately equals $2\hat{p}$, since $\hat{p}$ is generally low. Recalling that $\hat{p} = \mu/s$, this gives the frequency of the dominant phenotype at equilibrium as $2\mu/s$.

In summary, at equilibrium between selection and mutation in relation to a deleterious dominant gene, the equilibrium frequency of the dominant gene $= \mu/s$, and the frequency of the dominant phenotype $= 2\mu/s$.

A5.4.3 *Selection against a recessive*

Consider the same single locus as before, but this time with relative fitnesses of 1, 1, and $1 - s$ for EE, Ee, and ee respectively. We then have

	Genotype			
	EE	Ee	ee	Total
Frequency prior to selection	p^2	$2pq$	q^2	1
Relative fitness	1	1	$1 - s$	
Proportion after selection	p^2	$2pq$	$q^2(1-s)$	$1 - sq^2$

The frequency of the recessive gene after selection, which we shall call q', is estimated as

$$q' = \frac{q^2(1-s)}{1 - sq^2} + \frac{1}{2} \times \frac{2pq}{1 - sq^2} \qquad (4)$$

The change in gene frequency resulting from selection is

$$\Delta q = q' - q, \qquad (5)$$

which after some algebra reduces to

$$\Delta q = -sq^2(1-q)/(1-sq^2), \qquad (6)$$

where the negative sign indicates a decrease in frequency of the recessive gene.

Using the same argument as used when discussing selection against a dominant gene (Section A5.4.1), we note that in many cases either s or q is small, in which case $1 - sq^2$ is approximately 1, which gives

$$\Delta q \doteq -sq^2(1-q) \qquad (7)$$

as a useful approximation.

With complete selection against the recessive gene, we have $s = 1$. Noting that $(1-q^2) = (1-q)(1+q)$, it follows from (4) that

$$q' = q/(1+q), \qquad (8)$$

and from (6) that

$$\Delta q = -q^2/(1+q). \qquad (9)$$

For this particular case only (complete selection against a recessive), it is possible to express the gene frequency after t generations of selection (q_t) in terms of the original gene frequency (q_0), by noting from eqn (8) above that

$$q_1 = q_0/(1 + q_0), \qquad (10)$$

and thus

$$q_2 = q_1/(1 + q_1),$$
$$= \{q_0/(1 + q_0)\}/\{1 + q_0/(1 + q_0)\},$$

which simplifies to

$$q_2 = q_0/(1 + 2q_0).$$

Continuing these substitutions soon indicates that for t generations

$$q_t = q_0/(1 + tq_0). \tag{11}$$

Alternatively, we can rewrite (10) as

$$\frac{1}{q_1} = \frac{1}{q_0} + 1,$$

and thus

$$\frac{1}{q_2} = \frac{1}{q_1} + 1$$
$$= \frac{1}{q_0} + 2.$$

For the general case, we then have

$$\frac{1}{q_t} = \frac{1}{q_0} + t, \tag{12}$$

which is equivalent to equation (11) above.

A5.4.4 *Selection/mutation balance for a recessive*

From the previous section it can be seen that the loss of e genes resulting from selection amounts to sq^2. In contrast, new e genes are entering the population as a result of mutation from E to e. Since the frequency of E is p, and the rate of mutation is μ, it follows that the gain of e genes resulting from mutation from E to e is μp, which is approximately μ, since p is very nearly 1. At equilibrium, therefore, the loss of e genes is sq^2 and the gain is approximately μ, which gives $sq^2 \doteq \mu$, or $\hat{q} \doteq \sqrt{(\mu/s)}$, where, as before, ^ indicates equilibrium gene frequency. The frequency of the recessive phenotype at equilibrium is $\hat{q}^2$ which equals μ/s.

We can now compare the selection/mutation equilibrium frequencies for dominants and recessives:

Equilibrium frequency

	Gene	Phenotype
Dominant	μ/s	$2\mu/s$
Recessive	$\sqrt{(\mu/s)}$	μ/s

Notice that this table indicates a simple method for estimating mutation rates if relative fitnesses are known. If f is the frequency of the dominant or recessive phenotype, then mutation rate can be estimated as $fs/2$ for dominants, and as fs for recessives.

A5.4.5 *Selection favouring heterozygotes*

Since the heterozygote now has highest fitness, it is given a relative fitness of one. We then have

	A_1A_1	A_1A_2	A_2A_2	Total
Frequency prior to selection	p^2	$2pq$	q^2	1
Relative fitness	$1-s_1$	1	$1-s_2$	
Proportion after selection	$p^2(1-s_1)$	$2pq$	$q^2(1-s_2)$	$1-s_1p^2-s_2q^2$

It follows that the change in gene frequency of A_2 is given by

$$\Delta q = \frac{q^2(1-s_2)+pq}{1-s_1p^2-s_2q^2} - q \tag{13}$$

which can be simplified to

$$\Delta q = \frac{pq(s_1p-s_2q)}{1-s_1p^2-s_2q^2}. \tag{14}$$

At equilibrium, gene frequencies do not alter, and thus $\Delta q = 0$. Setting $\Delta q = 0$ in equation (14), the only non-trivial solution is

$$s_1p = s_2q.$$

Noting that $p = 1-q$, this can be rewritten as

$$\hat{q} = \frac{s_1}{s_1+s_2}. \tag{15}$$

Thus, when selection favours heterozygotes, equilibrium gene frequencies are determined solely by the relative fitness of the two homozygotes. If they are equal, then it follows from above that $\hat{q} = 0.5$. The equilibrium resulting from selection favouring heterozygotes is stable, in the sense that if q is less than $\hat{q}$, Δq is positive, and vice versa. Thus any departure from equilibrium results in selective forces that return the gene frequency to the equilibrium value.

A5.4.6 *Selection against heterozygotes*

We can draw conclusions about this type of selection by simply changing the signs of the selection coefficients in the previous section, so that the relative fitnesses now become $1 + s_1$, 1, and $1 + s_2$ for A_1A_1, A_1A_2, and A_2A_2 respectively.

By working through the expressions in the previous section, with the sign of s_1 and s_2 changed, we still arrive at exactly the same condition for equilibrium, namely

$$s_1 p = s_2 q, \tag{16}$$

or

$$\hat{q} = \frac{s_1}{s_1 + s_2}. \tag{17}$$

In contrast to the previous section, however, this equilibrium is unstable; by substituting appropriate values into the appropriate expression for Δq, the reader can verify that if q is less than $\hat{q}$, Δq is negative, and vice versa. Thus any departure from equilibrium results in selective forces that move the gene frequency towards zero or one; selection against heterozygotes by itself does not maintain a polymorphism.

In Section 10.5.2, there is discussion of an example of selection against heterozygotes in which the heterozygote has zero fitness. This is an extreme case which cannot be accommodated by the set of relative fitnesses given above. Instead, we need a set such as those shown below.

	A_1A_1	A_1A_2	A_2A_2	Total
Frequency prior to selection	p^2	$2pq$	q^2	1
Relative fitness	1	0	w	
Proportion after selection	p^2	0	$q^2 w$	$p^2 + q^2 w$

We then have

$$\Delta q = \frac{q^2 w}{(1-q)^2 + q^2 w} - q \tag{18}$$

which can be written as

$$\Delta q = \frac{q(1-q)\{q(1+w) - 1\}}{(1-q)^2 + q^2 w}. \tag{19}$$

At equilibrium, $\Delta q = 0$, which gives

$$\hat{q} = \frac{1}{1 + w} \tag{20}$$

as the only non-trivial solution. Once again, it is left to the reader to verify that this is an unstable equilibrium; if q is greater than $\hat{q}$, then Δq is positive which means that the population will become fixed for A_2; similarly, if q is less than $\hat{q}$, then Δq is negative, and A_2 is rapidly lost from the population.

If both homozygotes have equal fitness, then $w = 1$ and $\hat{q} = 0.5$, as we saw above. More generally, equation (20) tells us that as the fitness of one homozygote decreases relative to the other, then the higher becomes the equilibrium frequency of the allele corresponding to the less fit of the two homozygotes.

6
Familial disorders not due to single genes

6.1 Introduction

There are many important defects and diseases that are said to be *familial* because they tend to 'run in families', which means that the incidence among relatives of affected individuals is greater than the incidence in the general population. One possible reason for a defect or disease being familial is that members of the same family may share the same environment and may all, for example, be exposed to the same pathogen. Another possible reason is that members of the same family share the same genes, to an extent determined by their relationship to each other.

Thus if an abnormality or disease is familial, it may be due to a shared environment or to shared genes, or to a combination of shared environment and shared genes.

With single-gene defects and diseases of the type discussed in previous chapters, there is no difficulty in determining why they are familial: it is because of shared genes. There are, however, many familial defects and diseases that do not fit into any of the categories of simple single-gene diseases. And they are still familial even after all conceivable aspects of shared environment have been accounted for. This strongly implies that there is a class of diseases and abnormalities that are determined at least to some extent by genes, but in which the effects of the genes are not visible in any simple Mendelian manner.

The aim of this chapter is to explain what is known about such diseases and abnormalities.

6.2 Liability and threshold

Although the aim of this chapter is to discuss defects and diseases that are *not* due to single genes, the easiest way to understand the concepts relevant to such defects and diseases is to start by considering a defect that is due to a single gene.

A congenital syndrome known as internal hydrocephalus and retinal dysplasia (IHRD) is an autosomal recessive single-gene disease in Shorthorn cattle: all individuals homozygous for the recessive gene are affected, and all those having either one or two dominant genes are normal (Greene *et al.* 1978). Like most other recessive single-gene diseases, this disease is an example of an all-or-none trait: individuals are either completely affected (*all* of the clinical signs) or normal (*none* of the clinical signs).

For reasons that will soon become evident, it is convenient to think of each individual as having a certain *liability* to IHRD, where

> *Liability refers to the combined effect of all factors, both environmental and genetic, that render an individual more or less likely to develop a defect or disease.*

It must be clearly understood that the term 'environment' is used here and throughout this book in its widest sense, literally meaning non-genetic. Thus, an environmental factor is any factor that cannot be attributed to the action of the genes that determine the characteristic under consideration. Nutrition, housing, and climate are obvious environmental factors, but so too are factors such as sex and age of animal, size of litter into which the animal was born, and age of dam when the animal was born. A very important class of environmental factors are those which we cannot even identify let alone measure, such as the various non-genetic factors to which an embryo is exposed during gestation. We know that such factors operate, because identical twins (who share exactly the same set of genes) have different phenotypes at birth for a large number of characteristics. But apart from making a tentative list including factors such as position in the uterus and size of placenta, there is nothing we can do to control environmental factors such as these. All we can do is to recognize their existence and to remember that whenever the term environment is used in a genetics context, it means much more than just the wind and the rain.

Liability is a continuous variable that could, in principle, be measured on a continuous scale, in the same manner as we measure a characteristic such as body weight. But in this chapter we are discussing discontinuous characteristics in which there are often only two classes: affected and normal. We can accommodate discontinuous characteristics on a continuous scale by using the concept of a *threshold*, which is a certain level of liability above which all individuals develop the defect or disease, and below which all individuals are normal. In the case of IHRD referred to above, the position of the three genotypes for this single-gene disease can be represented on an arbitrary scale of liability as shown in Fig. 6.1. For single-gene defects or diseases, liability is usually determined solely by genotype, with environmental factors rarely exerting any effect. For example, in the case of IHRD, where affected individuals can produce offspring, matings in which both

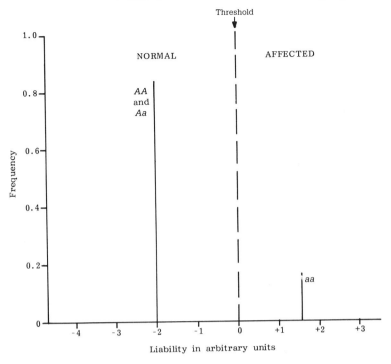

Fig. 6.1. A diagram illustrating the concepts of liability and threshold for an auto-somal recessive single-gene defect or disease. In this particular population, 16 per cent of individuals are homozygous for the recessive gene and are thus affected. The remaining 84 per cent are normal.

parents are affected (and hence are homozygous for the gene *a*) pro-duce nothing but *aa* offspring. Thus the mating of affected × affected produces 100 per cent affected offspring; liability to IHRD is deter-mined entirely by genotype.

6.3   Incomplete penetrance

Malignant hyperthermia syndrome (MHS) in pigs is a single-locus autosomal recessive defect characterized by a progressive increase in body temperature, muscle rigidity and metabolic acidosis leading rapidly to death. It can be induced in certain pigs by subjecting them to stress (such as loading and/or transport) or by exposing them to halothane vapour breathed through a tight-fitting mask. When given halo-thane, susceptible pigs (reactors) usually show signs of stiffening in hind-quarter muscles after two minutes, but if the mask is removed as soon as this happens, the majority of reactors show a complete recovery in five minutes. If there is no stiffening after three minutes of exposure to halothane, the animal is classed as a non-reactor, and is regarded as being resistant to MHS. The practical problems involved in administering halothane to pigs inevitably lead to variation in the efficiency of the test, even from pig to pig within a single test batch. It is to be expected, therefore, that the halothane test could produce the occasional false positive and false negative result. In other words, classification may not be completely accurate. A good illustration of this is given by data from the Animal Breeding Research Organization (ABRO) in Scotland, where halothane testing was commenced in 1974. There were no problems with detecting non-reactors but there were problems with detecting reactors. In the first full year of results (1975), when techniques were still being developed, the probability of misclassifying a reactor as a non-reactor was as high as 25 per cent. In 1976, however, by which time the operators had gained more experience, the probability of misclassification was down to as low as 2 per cent. Thus among one set of matings of affected × affected, reported by Smith and Bampton (1977), 98 per cent of offspring were classified as affected (reactors). Although the 2 per cent of non-reactors from these matings were almost certainly due to misclassification, the data as given do not correspond exactly to a simple recessive mode of inheritance. With results like this, the only way to explain the inheritance of MHS in terms of a recessive gene is to say that all the offspring of affected × affected matings are actually homozygous for the recessive gene, but only 98 per cent of them show the effect of the gene.

Results such as this have led to the use of the term *penetrance*, which is the proportion of individuals with a particular genotype that exhibit the phenotype normally associated with that genotype. As an example, the above results indicate a penetrance of the homozygous recessive genotype of 98 per cent. If penetrance is less than 100 per cent, it is said to be *incomplete*. In relation to liability, the existence of incomplete penetrance means that individuals with the same genotype (*aa*) can have different liabilities; those with MHS have a liability on the affected side of the threshold, and those that are not affected have a liability somewhere on the other side of the threshold. The most obvious cause for this difference in liability is misclassification as a result of various non-genetic (environmental) factors, such as the skill and experience of operators conducting the test. More recently, further research into the inheritance of MHS has clearly indicated that *aa* pigs are more likely to be misclassified as non-reactors if tested at three weeks of age than at eight weeks of age (Webb 1981). Thus, age is another non-genetic (environmental) factor that may give rise to incomplete penetrance of the *aa* genotype.

In cases such as this, where there is only a small departure from simple single-gene inheritance, and especially where that departure has obvious environmental causes, the most useful and sensible way in which to describe inheritance is to talk in terms of a single-gene defect with incomplete penetrance.

6.4 The multifactorial model

Now let us consider another familial defect, namely hip dysplasia in dogs, which in the past was thought by some authorities to be due to an autosomal recessive gene. In several different studies of this defect in German Shepherd dogs, the results of affected X affected matings have been an average of 86 per cent affected offspring (Willis 1976). Similar matings of affected X affected in Labradors produced an average of 63 per cent of affected offspring (Nicholas 1975). If we used the same arguments as were used previously for MHS in pigs, we could attribute hip dysplasia to a recessive gene (*a*) with 86 per cent penetrance in German Shepherds and 63 per cent penetrance in Labradors. In the latter case, for example, we would then have 63 per cent of *aa* Labradors with a liability sufficiently high to place them on the affected side of the threshold and 37 per cent with a liability sufficiently low to place them on the normal side. As with MHS in pigs, we could ask what factors are likely to be responsible for this difference in liability among *aa* animals. Since hip dysplasia is usually diagnosed by subjective evaluation of a radiograph (Fig. 6.2)

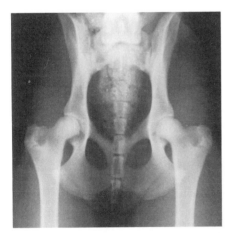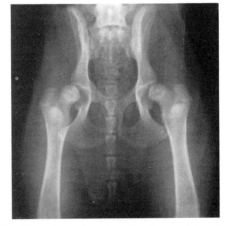

Fig. 6.2. Radiographs of (left) normal canine hip joints showing the femoral head sitting well into the acetabulum, and (right) hip joints showing bilateral coxo-femoral subluxation due to joint instability or laxity, which is the earliest sign of canine hip dysplasia, and which is usually followed by degenerative joint disease. In practice, radiographs of canine hip joints show all possible gradations between the two extreme cases shown here, but in many countries dogs are classified on the basis of a radiograph into one of only two categories: normal or dysplastic. The potential for misclassification is therefore quite high. (Courtesy of R. Zammit and G. Allan.)

or by palpation, misclassification certainly has to be listed as one of the likely causes. But there are other environmental factors that could be involved. For example, level of feeding and amount of exercise during early growth have been shown to affect an animal's liability to hip dysplasia. The difference in liability could also be due to the action of alleles at other loci that in various ways help to determine the way in which the hip joint develops. Or it could be due to the combined effect of various environmental factors and other genes, which is the most likely explanation. Thus, at this stage we are still saying that hip dysplasia is due to homozygosity for a recessive gene (*a*) at a particular locus, but that individuals who are homozygous (*aa*) for this gene do not all have the same liability to hip dysplasia; there are differences in liability among *aa* individuals, due to the combined effect of various environmental factors and alleles at other loci.

In fact, the situation is now like that shown in Fig. 6.3, where there is a bell-shaped or Normal distribution of liability values for *aa* individuals. Those animals that are overfed and overexercised, and happen to have alleles at other loci tending to produce ill-fitting hip joints, have a relatively high liability. Animals with only some of these predisposing factors will be more common than those with all the predisposing factors, which accounts for the increasing frequency of animals with liabilities closer to the mid-point of the distribution. A similar

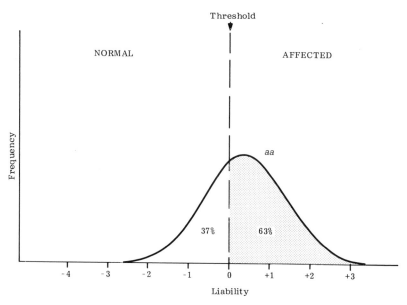

Fig. 6.3. A diagram illustrating the concepts of liability and threshold for an autosomal recessive single-gene defect or disease where the homozygote recessive genotype (*aa*) has incomplete penetrance, in this particular case at a level of 63 per cent.

trend is expected on the other side of the mid-point: those animals given the most favourable diet and most appropriate exercise regime, and who also happen to have alleles at other loci that tend to produce sound hip joints, will have much lower liabilities, but will be less frequent than those with only some of these factors operating to decrease liability.

The critical point to understand is that any animal that has a liability less than the threshold value, for whatever combination of environmental and genetic factors, will not have hip dysplasia.

When a characteristic is determined by the combined effect of several factors, both environmental and genetic, it can be said to be *multifactorial*. In general, if incomplete penetrance can be attributed to more than just one or two simple environmental effects such as misclassification, and especially if some other aspect of the simple, single-locus model appears to be not entirely appropriate (if dominance is incomplete, for example), then it is more sensible to describe the defect as being multifactorial than to talk in terms of a single locus with incomplete penetrance.

In order to understand the relationship between these two methods of describing the inheritance of an all-or-none characteristic, we shall expand upon the concepts illustrated in Fig. 6.3. If genes at other loci and/or environmental factors produce a Normal distribution of liability for *aa* homozygotes as in Fig. 6.3, then it is reasonable to expect a similar distribution in liability for genotypes *AA* and *Aa*, as shown in Fig. 6.4a. And if dominance is incomplete so that the *a* gene expresses itself in a small proportion of heterozygotes, we have a considerable overlap of distribution of liability among the three genotypes, leading to an overall distribution of liability (solid line in Fig. 6.4a) approaching that of a single Normal distribution, which is the distribution expected for a multifactorial defect or disease, as shown in Fig. 6.4b.

In the above discussion, we started by saying that hip dysplasia is a single-locus disease with incomplete penetrance, and finished by describing it is being multifactorial. Which of these descriptions or models is the most appropriate? It should be obvious from the above discussion that the latter model is much more appropriate than the former. Indeed, there are so many non-genetic and genetic factors contributing to liability for hip dysplasia that it is rather pointless to describe the inheritance of hip dysplasia in terms of a single locus with incomplete penetrance.

In general, if a defect or disease is familial but does not correspond closely to simple, Mendelian inheritance, then it is usually better to describe it as being multifactorial than to attribute it to a single locus with incomplete penetrance.

(a)

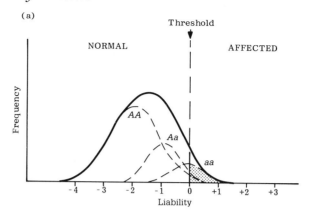

(b)

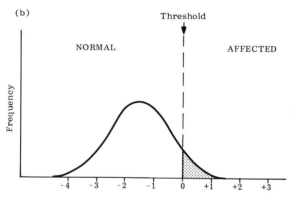

Fig. 6.4. Diagrams illustrating the similarity between (a) a single-gene model with incomplete penetrance and incomplete dominance, and (b) a multifactorial model. The solid line in (a) represents the total frequency of individuals with a particular liability, and is obtained by summing the frequency of relevant individuals from the three genotypes.

One virtue of the multifactorial model is that it enables a simple estimate to be made of the relative importance of genetic and environmental factors in contributing to the aetiology of a defect or disease. This is done by estimating a parameter called *heritability*, which for present purposes we shall define as the proportion of total variation in a population that can be attributed to variation in genetic factors. Variation in liability refers to the differences among individuals in liability to a particular defect or disease. Heritability of liability, then, is the proportion of those differences in liability that are due to genetic differences among individuals. If all members of a population are exposed to exactly the same set of environmental factors in relation to a particular disease or defect, then the differences in liability among members of that population must be due solely to genetic factors, and the heritability of liability is 100 per cent. At the other extreme, heritability is zero per cent if all individuals have exactly the same alleles at each of the loci that contribute to liability. Neither of these extreme situations is very common in practice, but they do serve to illustrate the meaning of heritability.

> *By far the majority of familial defects and diseases have an intermediate heritability, which indicates that both environmental and genetic factors contribute to their aetiology.*

6.5 An overall view of familial defects and diseases

A very useful method of investigating the inheritance of all-or-none characters is to compare the incidence in the general population with the incidence in relatives of affected individuals. Although not yet a common form of analysis for animal data, this method has been used on many defects and diseases in humans, and the results are shown in Fig. 6.5. There are two fairly distinct classes of disorders in this figure. Those corresponding to simple dominant or simple recessive inheritance tend to be in the upper left-hand corner, being relatively rare in the general population and much more frequent in full-sibs of affected individuals. (Full-sibs are individuals with both parents in common.) Those defects and diseases that are multifactorial are more common in the general population and relatively less frequent in full-sibs of affected individuals than single-gene defects and diseases.

6.6 More than one threshold

Congenital heart disease occurs in many different forms, the most important of which are familial in humans and in animals. In dogs, which are the animal species most studied in this regard, congenital heart disease occurs with an overall frequency of around five per thousand, which is of the same order of magnitude as reported in humans (shown as three per thousand in Fig. 6.5). In an extensive and very thorough set of experiments reviewed by Patterson (1976), the inheritance of congenital heart disease in dogs has been clearly demonstrated to be compatible with a multifactorial model.

The most common congenital heart disease is patent ductus arteriosus (PDA), resulting from defective closure of the ductus arteriosus. In the surveys reported by Patterson, PDA accounted for more than one-quarter of all congenital heart defects.

> *Like many other defects and diseases that are compatible with the multifactorial model, defective closure of the ductus arteriosus is a graded phenomenon, with increasing severity corresponding to increasing liability.*

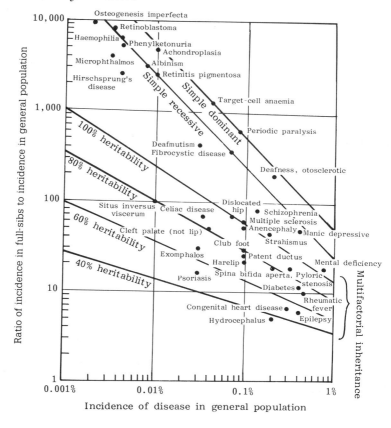

Fig. 6.5. Relationship between incidence in the general population and the relative incidence in full-sibs of affected individuals for a number of defects and diseases in humans. Full-sibs are individuals with both parents in common. The solid lines indicate the relationship expected for simple dominant, simple recessive and multifactorial inheritance. Although not all human defects and diseases are directly comparable with those in animals, at least some of those included in this diagram have analogues in animals that are inherited in a similar manner. One disease in the diagram that is not comparable with its animal analogue is dislocated hip, which in humans refers to clinical cases that are truly congenital (present at birth). The lines for simple dominant and simple recessive were obtained by assuming that the relevant genes are at a low frequency. This being so, the only significant mating type for the dominant case is $Aa \times aa$, from which one-half of the offspring and hence one-half of the full-sibs of an affected individual will be affected. The incidence in the general population is the frequency of AA plus the frequency of heterozygotes, which equals $p^2 + 2pq$, where p and q are the frequencies of the dominant and recessive genes respectively. Since p is low and q is close to one, $p^2 + 2pq$ is approximately $2p$. Thus the ratio of the incidence in sibs to the incidence in the general population is $\frac{1}{2} : 2p$, which corresponds to the line for simple dominants in this figure. By a similar argument, the frequency among full-sibs for a rare recessive condition is the frequency of affecteds from $Bb \times Bb$ matings, which is one-quarter.

This is clearly indicated in the results of various matings among Poodles. The degree of defect varied from normal closure, through partial closure (called ductus diverticulum or DD), to fully patent ductus arteriosus. The presence of these three graded categories of the one defect can be represented in the multifactorial model by two thresholds, as illustrated in Fig. 6.6, for a population of dogs resulting from PDA × PDA matings. An even better understanding of the multi-factorial model can be obtained from other types of matings, the results of which are given in Table 6.1 and in Fig. 6.7. The main conclusion to

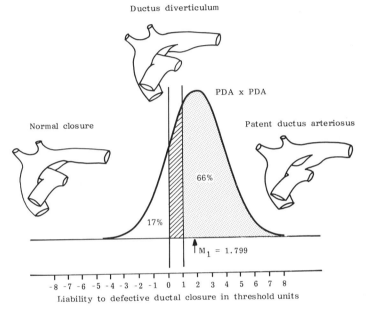

Fig. 6.6. A multifactorial model of defective ductal closure with two thresholds, corresponding to incidences of 17 per cent normal closure (white portion of distribution), 17 per cent DD (striped portion), and 66 per cent PDA (dotted portion), resulting from PDA × PDA matings in Poodles. Liability is measured in threshold units, where one threshold unit equals the distance between the two thresholds (see Falconer, 1981, p. 276). The position of the two thresholds relative to the distribution is determined solely by the above incidences and can be checked by reference to an appropriate table of the standard Normal curve. M_1 is the mean liability of all offspring from PDA × PDA matings. (From Patterson (1974). Pathological and genetic studies of congenital heart disease in the dog. *Advances in Cardiology* **13**, 210–49. S. Karger AG, Basel.)

Since the incidence in the general population is the frequency of the genotype *bb* which equals q^2, the appropriate ratio is $\frac{1}{4} : q^2$. Heritability with multifactorial inheritance, as shown in the diagram, is narrow-sense heritability which is defined in Section 6.7.1. (From Cavalli-Sforza and Bodmer 1971, whose diagram is a modification of one originally given by Newcombe 1964.)

Table 6.1. Results of test matings among Poodles having varying degrees of defective ductal closure

Mating type	Proportion of offspring		
	Normal	DD	PDA
PDA × PDA	0.17	0.17	0.66
PDA × N (1° Rel. PDA)	0.33	0.22	0.45
PDA × N	0.78	0.12	0.10

N = normal; DD = ductus diverticulum; PDA = patent ductus arteriosus; 1° Rel. PDA = first degree relative of dog having PDA (first degree relatives include parents, offspring, and full-sibs). Adapted from Patterson (1976).

be drawn from these results is that as the proportion of genes in common with PDA decreases in equal steps from $r = 1$ to $r = \frac{3}{4}$ to $r = \frac{1}{2}$, so too does the average liability decrease by roughly equal steps, from +1.8 to +0.8 to −1.5. Thus the mean liability is proportional to the percentage of genes for PDA.

From a practical point of view, the difference in incidence of affected offspring of the two types of matings involving normal dogs shown in Table 6.1 is particularly important; normal dogs that are closely related to affected dogs produce more than three times the incidence of affected offspring, and more than four times the incidence of severely affected offspring, when compared with normal dogs that are unrelated to affected dogs.

> *In general, the tendency for normal individuals to throw affected offspring, and the severity of the disorder among their affected offspring, depends on how closely the normal individual is genetically related to an affected individual.*

This important conclusion is a direct consequence of the multifactorial model. Another implication of the model is best illustrated by some data on hip dysplasia in Labradors, in which normal parents were classified according to how many of their parents were normal. The results are presented in Table 6.2, and the general trend can be summarized as follows.

> *The tendency for normal individuals to throw normal offspring depends on the number of their existing relatives that are normal.*

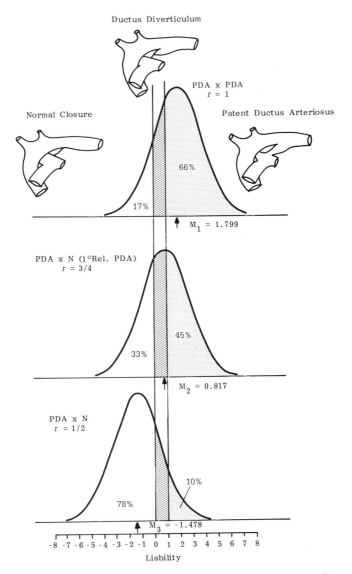

Fig. 6.7. Graphical representation of three different populations of Poodles, each having a different mean liability to defective ductal closure. The different mean liabilities are a consequence of the three different mating types detailed in Table 6.1, and the position of the distributions relative to the thresholds is determined solely by the incidences shown in Table 6.1. The top distribution is the same as in Fig. 6.6. Values of *r* indicate the proportion of genes derived from dogs having PDA. (After Patterson, D. F., Pyle, R. L., Buchanan, J. W. (1972). Hereditary cardiovascular malformations of the dog. In *The cardiovascular system* (ed. D. Bergsma) Part XV. Williams & Wilkins, Baltimore, for the National Foundation— March of Dimes, BD: OAS **8**(5), 160–74.)

Table 6.2. Effect of number of normal grandparents on the percentage of normal offspring from matings of normal × normal parents, in relation to hip dysplasia in Labradors

Number of normal grandparents*	Percentage of normal grandchildren†
3	100‡
2	71
1	68
0	50

* There were no cases of offspring with four normal grandparents in the available data. † The total number of grandchildren for which data were available on grandparents as well as parents was only 56, and thus there are no significant differences among the incidences shown in the table. The general trend, however, is sufficiently clear to justify presentation of the results. ‡ Since there were only two offspring in this class, this figure should not be taken as indicating that offspring with two normal parents and three normal grandparents will themselves always be normal. Adapted from Nicholas (1975).

In order to illustrate other aspects of the multifactorial model, we shall now return to heart defects. Although defective ductal closure was graded into three different classes, investigations into the incidence of left heart failure and severe pulmonary hypertension among dogs having the most severe defect, namely PDA, have indicated that further gradations in severity of heart malformation exist, and that these gradations have a very clear genetical basis. Among those animals having PDA, the incidence of left heart failure increases with the percentage of genes derived from dogs with PDA, as shown in Fig. 6.8. In contrast pulmonary hypertension shows no such trend. Other evidence summarized by Patterson indicated that the increased incidence of left heart failure was closely associated with increased size of ductus lumen, which when estimated by various physiological tests, including input conductance, was seen to be directly proportional to the percentage of genes derived from dogs with PDA.

Thus, increasing severity of defective ductal closure is directly proportional to the 'dose' of genes for liability to the defect.

Among the other forms of congenital heart disease studied by Patterson and colleagues, conotruncal septum defects (CSD) in Keeshonds clearly illustrate another important implication of the multifactorial model.

The frequency and severity of a defect or disease will be greatest among relatives of more severely affected individuals.

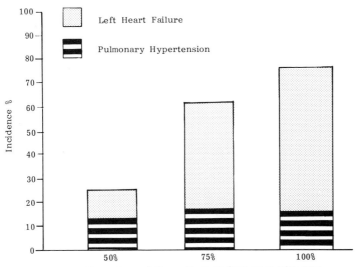

Fig. 6.8. Incidence of left heart failure and severe pulmonary hypertension among dogs having PDA, as a function of the percentage of genes derived from dogs with PDA. It must be emphasized that only dogs with PDA are included in these data, so that the increasing severity of heart defects shown in this diagram is occurring to the right-hand side of the second threshold in Fig. 6.7. (From Patterson and Pyle 1971.)

Conotruncal septum defects proved to be particularly useful in this context, as they can be divided into four different grades of increasing severity, as illustrated in Fig. 6.9. If CSD were due to a single gene, then the existence of more than one grade of defect would be said to indicate *variable expressivity* of that gene. But the possible causes of variable expressivity are the same as those for incomplete penetrance: environmental factors, alleles at other loci, or a combination of environmental and genetic factors. In general, therefore, single-gene defects with variable expressivity can be more usefully thought of as being multifactorial.

From an extensive set of matings among Keeshonds with various grades of CSD, the overall incidence of CSD and the incidence of each grade of CSD in offspring was determined for various mating types, according to the average CSD grade of parents. The results for overall CSD incidence and for incidence of grade 3 CSD are shown in Fig. 6.10. It is evident that the overall frequency of CSD and the frequency of the most severe form (grade 3) are almost directly proportional to the average severity in parents. These results are entirely consistent with the multifactorial model, in which frequency and severity in relatives of affected individuals are expected to increase as liability increases.

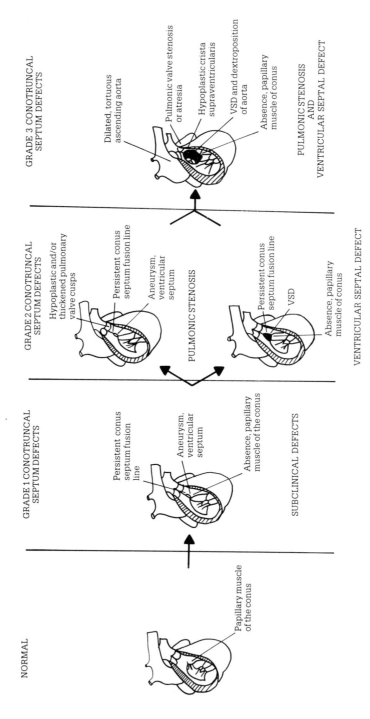

Fig. 6.9. A graded series of defects of the conotruncal septum in Keeshond dogs, indicating increasing severity of defect from grade 0 (normal) to grade 1 (sub-clinical defects) to grade 2 (grade 1 plus either pulmonic stenosis *or* ventricular septal defect) to grade 3 (grade 1 plus pulmonic stenosis *plus* ventricular septal defect). (From Patterson *et al.* 1974.)

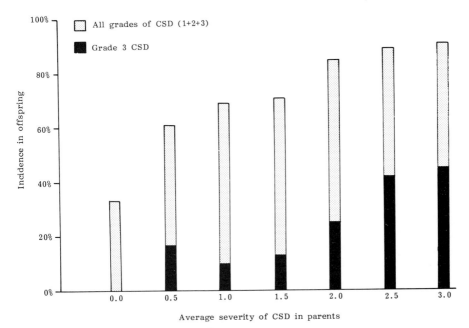

Fig. 6.10. Overall incidence of conotruncal septum defects or CSD (grades 1, 2 and 3) and incidence of grade 3 CSD (the most severe form) in offspring of matings in Keeshonds, as a function of average severity of CSD in parents. (Drawn from data presented by Patterson *et al.* 1974.)

6.7 Some final points

6.7.1 Heritability

Since the concept of heritability was not introduced until the multi-factorial model was introduced, it might appear that heritability is a valid concept only in relation to multifactorial characteristics. But this is not so.

> *Despite the fact that heritability is usually only mentioned in relation to multifactorial characteristics, it is an equally valid parameter for single-gene characteristics.*

As defined earlier in this chapter, heritability is 100 per cent for all simple single-gene characteristics, because all the differences in phenotype between individuals in relation to those characteristics are due to different genotypes at the locus in question. When heritability is defined in this sense, we refer to it as *heritability in the broad sense* or the *degree of genetic determination*.

There is, however, another sense in which the term heritability is used, and in practice it is used in this second sense much more frequently than in the sense of degree of genetic determination. Known as *heritability in the narrow sense*, it expresses the extent to which phenotypes are transmitted from parents to offspring. In other words, it expresses the extent to which offspring resemble their parents, or in more general terms, it determines the extent to which relatives resemble each other.

It follows that if we can measure resemblance between relatives, then we should be able to estimate narrow-sense heritability.

One easy way to measure resemblance between relatives for defects and diseases is to calculate the ratio of incidence of the defect or disease in relatives of affected individuals to incidence in the general population. If narrow-sense heritability is zero, then the defect or disease is no more frequent in relatives of affected individuals than in the general population, and the ratio equals one. If narrow-sense heritability is 100 per cent, then the degree to which relatives resemble each other, measured as a correlation coefficient, equals the proportion of genes that they have in common. For first-degree relatives (parents, full-sibs, offspring) this is $\frac{1}{2}$, and for second-degree relatives (grandparents, half-sibs, grandchildren) it is $\frac{1}{4}$. Unfortunately, the relationship between (1) the correlation among relatives, and (2) the ratio of incidence in relatives to incidence in the general population is mathematically complex and beyond the scope of this book. However, the end result of this relationship is illustrated in Fig. 6.5, by the line labelled 100% heritability. It should now be evident that we can use Fig. 6.5 to estimate heritability for defects whose inheritance is compatible with the multifactorial model. All we need to know is the incidence in the general population and the incidence in relatives (first-degree relatives if we are to use Fig. 6.5). For example, Fig. 6.5 indicates that club foot in humans has an incidence of around 0.1 per cent in the general population, and 30 times that in full-sibs, giving an estimate of heritability equal to 80 per cent.

Although this method of estimating heritability seems simple and straightforward, there are many potential problems and sources of bias that can be easily overlooked when using this method. Before proceeding to estimate heritability in this manner, therefore, the reader should carefully read the relevant references listed at the end of this chapter.

For reasons that are explained in Section 14.4.1, narrow-sense heritability may equal broad-sense heritability or it may be less, but it will never be greater. In practice, the difference between the two is often very difficult to detect. Because we are most interested in understanding and predicting how characteristics are transmitted from one generation to the next, narrow-sense heritability is much more useful than broad-sense heritability. So much so, in fact, that whenever we see the word 'heritability' on its own, we can take it to mean narrow-sense heritability, unless otherwise stated.

6.7.2 Penetrance and expressivity

We have seen in this chapter that the concepts of incomplete penetrance and to a lesser extent variable expressivity are often invoked to explain any departure from simple single-gene inheritance. Patterson (1976) has described them aptly as 'escape clauses'. With a little ingenuity, most data on most defects can be made to fit a single-gene model by choosing the appropriate values of penetrance for each genotype, by postulating a certain amount of incomplete dominance and if necessary by invoking variable expressivity as well. But having done this, are we any the wiser? The disadvantage of continuing to think in terms of a single-gene model is that we continue to think in terms of carriers of a single gene, and we are therefore inclined to think in terms of trying to remove that gene from particular populations. On the other hand, by thinking in terms of a multifactorial model, we can forget about detecting carriers or removing a gene from a population, and concentrate instead on more fruitful tasks. Our willingness to use an affected animal in a mating will depend solely on the severity of its own defect and on its relationship to other defective animals. We will not waste our time searching for Mendelian ratios but will simply proceed with various matings, using as our guide the incidence of affected animals achieved in previous matings of that type.

In addition, by thinking in terms of the multifactorial model, we will be spared the confusion caused by the very common habit people have of saying that if the inheritance of a defect or disease is not compatible with simple dominant or simple recessive inheritance, then its mode of inheritance is not completely understood, or worse still, not known. In fact, the mode of multifactorial inheritance is known and understood, and appropriate breeding plans for the reduction in frequency of a particular defect or disease can easily be drawn up and put into practice on the basis of that understanding.

6.7.3 Recurrence risks

On many occasions in this chapter, we have spoken about the percentage of affected offspring resulting from a certain type of mating. Such percentages can be called *recurrence risks*, because they give an indication of the probability or risk of a particular defect occurring again, should that type of mating be repeated. The recurrence risks presented in this chapter are called *empirical recurrence risks*, because they have been obtained by observing the results of certain types of matings. Empirical recurrence risks can be very useful in planning and conducting breeding programmes aimed at reducing the frequency of certain defects or diseases. For example, the figures for incidence of CSD shown in Fig. 6.10, which are really empirical recurrence risks, clearly show that the frequency of all grades of CSD will be decreased most rapidly by breeding only from parents that have no signs of CSD. If this is not possible, for example because the incidence of CSD is very high in a particular population of dogs, then the empirical recurrence risks in Fig. 6.10 indicate that breeders should aim for matings in which the average severity of CSD is as low as possible.

There are, however, limitations to the usefulness of empirical recurrence risks; they cannot indicate the risk for a type of mating from which no offspring records are currently available, and they may not take account of sex of offspring or severity of defect.

These disadvantages can be overcome by using *theoretical recurrence risks*, which are predictions arising directly from either a single-locus model or a multifactorial model, whichever is appropriate. Because they are predicted from the model, theoretical recurrence risks can be calculated for any imaginable type of mating, whether or not such matings have ever occurred before. In the simplest cases, theoretical recurrence risks equal segregation ratios for single-locus models, and, for the multifactorial model, they can be calculated from the population incidence and heritability. In most cases, however, the calculation of theoretical recurrence risks involves complications that are beyond the scope of this book. Readers requiring further information should consult Emery (1976) or any good book on genetic counselling, such as Stevenson and Davison (1976).

6.7.4 *Future resolution of multiple factors*

A major aim of research into defects and diseases whose inheritance is compatible with the multifactorial model is to identify the important factors, both environmental and genetic, that determine liability. Considerable progress has been made in this area, especially in regard to environmental factors. Much more progress can be expected in the future. On the genetic side, for example, extensive research in mice and in humans has identified a surprisingly large number of alleles at several closely-linked loci that give rise to so-called histocompatibility antigens (see Section 8.4). Some of these antigens show very strong associations with certain diseases that are presently classified as multifactorial. These antigens thus become identifiable genetic factors contributing to liability. Work in this area has already expanded into domestic animals, and is very likely to lead ultimately to the identification of several important genetic factors that play a role in determining liability to various diseases in domestic animals.

To describe a defect or disease as being multifactorial is therefore really only the beginning. The immediate virtue of such a description is that it clears the ground for a fruitful mating and selection programme. The longer-term virtue of the multifactorial approach is that it leads logically to research into identifying the various environmental and genetic factors that play a key role in determining liability.

6.8 Summary

Many familial defects and diseases do not follow simple Mendelian patterns of inheritance. One method of describing such disorders is to attribute them to a single gene with incomplete penetrance and/or incomplete dominance and/or variable expressivity. In many cases, however, it is more useful to describe such disorders in terms of a multifactorial model, in which liability to the disorder is determined by the combined effect of at least several factors, some of which are environmental and some of which are genetic.

Liability varies on a continuous scale from low to high, and although it can not be measured directly on individuals, it is a very useful concept. Individuals are affected by the disease or defect if their liability is above a certain threshold.

From a practical point of view, the implications of the multifactorial model can be stated as follows:

(1) the more severely an individual is affected, the more frequent and severe will be the disorder in the relatives, including offspring, of that affected individual;
(2) among normal individuals, the lower their genetic relationship with affected individuals and the larger the proportion of their relatives that are normal, the less frequent and severe will be the disorder in future relatives, including offspring.

Among the many familial defects and diseases that can be thought of in terms of the multifactorial model are hip dysplasia and various forms of congenital heart disease.

6.9 Further reading

Curnow, R. N. and Smith, C. (1975). Multifactorial models for familial diseases in man. *Journal of the Royal Statistical Society* A **138**, 131–69. (A detailed review of various models proposed for non-Mendelian familial diseases, with an appendix including the comments and criticisms of several other workers in this field, and a reply by Curnow and Smith.)

Falconer, D. S. (1965). The inheritance of liability to certain diseases, estimated from the incidence among relatives. *Annals of Human Genetics* **29**, 51–76. (This is the paper that rekindled interest in the multifactorial model, by presenting a very clear account of its implications and its use in the analysis of familial data. For an updated review, see chapter 18 of Falconer (1981).)

Inouye, E. and Nishimura, H. (Eds) (1977). *Gene-environment interactions in common diseases*. University Park Press, Baltimore. (This book contains useful reviews of common familial diseases and of the multifactorial model, together with the results of many studies of non-Mendelian familial diseases in humans and in animals.)

Mikami, H. and Fredeen, H. T. (1979). A genetic study of cryptorchidism and scrotal hernia in pigs. *Canadian Journal of Genetics and Cytology* 21, 9–19. (An illustration of how to estimate heritability of liability for two non-Mendelian familial disorders in pigs.)

Patterson, D. F. (1976). Congenital defects of the cardiovascular system of dogs: studies in comparative cardiology. *Advances in Veterinary Science and Comparative Medicine* 20, 1–37. (A very informative review of a detailed genetic analysis of congenital heart disease in dogs, including descriptions of the fitting of appropriate multifactorial models.)

7
Is it inherited?

7.1 Introduction

'Is it inherited?' is a common question asked in relation to a wide range of defects and diseases. Unfortunately, the answer for the majority of defects and diseases is 'we don't know', because insufficient data have been collected and analysed to enable a decision to be made.

This situation will gradually improve as more studies are conducted. In order to conduct such studies, an understanding is required of Mendelian inheritance (Chapter 1), population genetics (Chapter 5), the multifactorial model (Chapter 6) and elementary statistics. The aim of this chapter is to show how knowledge in these areas can be put to practical use in investigating the mode of inheritance of a defect or disease.

It is important that such studies be conducted, not only from the immediate veterinary point of view, but also because medical research workers are increasingly on the lookout for inherited diseases of animals which could serve as models of human disease.

7.2 General evidence for a genetic aetiology

If genes make any contribution to the aetiology of a defect or disease, then, as discussed in Chapter 6, it follows that there will be a positive relationship between the chance of an individual being affected and the extent to which that individual has genes in common with affected individuals.

The most important practical implications of this relationship are seen with respect to families and breeds. Since members of the same family have more genes in common than members of different families, and members of the same breed have more genes in common than members of different breeds, it follows that:

> *A genetic contribution to the aetiology of a defect or disease is indicated if incidence is higher in some families than in others within a breed, and also if incidence is higher in some breeds than in others.*

Of course, evidence of variation in incidence between families and/ or between breeds does not constitute proof of a genetic contribution to aetiology, because environmental factors common to members of certain families and/or common to members of certain breeds may be sufficient to account for the observed variation in incidence. The first step, then, in attempting to disentangle possible environmental causes from possible genetic causes, is to remove or allow for the effects of possible environmental factors. This can be done only after a detailed investigation of all environmental factors that are thought likely to affect the occurrence of the defect or disease. The trouble with this approach is that the investigator often has no prior indication as to which factors should be examined: each one that comes to mind as being possibly important must be investigated, in order to determine which ones (if any) are important.

If the effects of certain environmental factors cannot be removed, then in some cases they can be allowed for by the use of *matched controls*, which are unaffected animals chosen from the same population as affected individuals, so as to match the latter in relation to any non-genetic factors that might be thought to be important. As an example of the use of matched controls, imagine that you were investigating a defect such as epistaxis (bleeding) in thoroughbred horses, and found that certain sires produced many more offspring that bleed than other sires. While this might be taken as evidence favouring a genetic aetiology for epistaxis, it could simply be a reflection of the fact that some sires produce many more starters than do other sires. In order to check this, a matched control could be chosen for each recorded case of bleeding, by picking at random a non-bleeder from amongst the starters in races similar to those races in which bleedings were recorded. If the distribution of sires is the same in the matched controls as in the horses that bleed, then there is no evidence for a genetic contribution to the aetiology of bleeding, even though there is variation in incidence between families.

While differences between breeds are less likely to be due to environmental factors unless breeds are usually kept in very different environments, it is still vital that possible environmental causes of variation between breeds be removed or allowed for, before concluding that observed variation in incidence between breeds indicates a genetic contribution to aetiology. The average age of a random sample of Greyhounds, for example, is likely to be much less than the average age of a random sample of Labradors. This is not, however, an indication of any genetic difference between the two breeds in relation to life expectancy; it is simply a reflection of the fact that Labradors are mostly kept as pets and thus usually survive for at least several years, whereas the majority of Greyhounds are destroyed at a relatively young age because of poor performance on the track.

There are some further points that must be noted in relation to variation between and within breeds.

In many cases the two types of variation are directly related. Malignant hyperthermia syndrome (MHS: see Section 6.3), for example, has a completely genetic aetiology *within* pig breeds, in each of which it is due to homozygosity for a recessive allele. There is also considerable variation in incidence *between* breeds, from well over 50 per cent in the Pietrain breed down to less than 5 per cent in Large Whites, and this between-breed variation can be explained in terms of different frequencies of the recessive allele in different breeds.

But such a relationship need not necessarily exist. In fact:

> *The existence of a genetic contribution to aetiology within each of several breeds does not necessarily indicate that there will be a genetic difference between breeds.*

To continue with MHS as an example, it is evident that if the recessive allele occurs with the same frequency in two breeds, then there will be no variation between those two breeds but there will still be considerable variation within each breed.

The final piece of general evidence suggesting a genetic contribution to aetiology for a particular defect or disease is when the same or a very similar defect or disease is definitely inherited in another species of animal, or in humans. Biochemical disorders of the type discussed in Chapter 3 illustrate this point particularly well, and so too do inherited coat colours and patterns. Of course there will be exceptions to the rule, and we must be continually on the lookout for them; but the general principle holds up very well. In fact, as Ohno (1967, 1973) has pointed out (see Section 4.5), we can go one step further with X-linked loci, and predict fairly confidently that if a particular locus is X-linked in one species of mammal then it will be X-linked in them all. Thus a disease that is X-linked in one species is likely to be X-linked in other species.

7.3 The four models of simple, Mendelian inheritance

If we still find that a defect or disease is familial, even after removing or allowing for all conceivable environmental factors, then the next step is to determine if the available data correspond in general terms to any of the four simple, Mendelian models of inheritance: autosomal dominant, autosomal recessive, X-linked dominant, and X-linked recessive.

In many cases it is convenient to start by drawing pedigrees or family trees. Pedigrees are most informative if drawn in a standard format, using the symbols illustrated in Fig. 7.1. An example of a

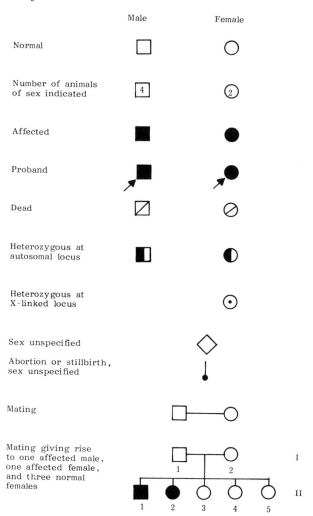

Fig. 7.1. The symbols used in pedigrees. A *proband* is an affected individual through whom the family came to the notice of an investigator. Note that generations are identified by roman numerals, and individuals within a generation are identified by arabic numerals from left to right.

pedigree is given in Fig. 7.2, in this case for hereditary multiple exostosis in a family of horses. Constructing pedigrees in this manner can be helpful in providing an initial impression of how a particular defect or disease is transmitted from one generation to the next. From Fig. 7.2, for example, it is evident that males and females are affected in approximately equal proportions (6 males and 5 females in generation IV), that matings of affected × affected can produce normal females and that matings of affected × unrelated normals give rise to

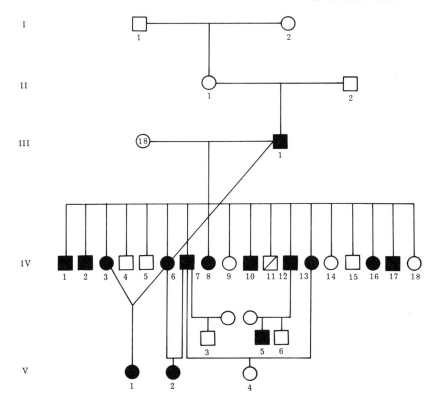

Fig. 7.2. Pedigree showing the pattern of inheritance of multiple exostosis in horses. This is a disease in which benign tumours occur in many different bones, sometimes being present at birth. Note that stallion III-1 was mated with his daughter IV-3, a surprisingly common practice in most species of domestic animals. Notice also the matings among half-sibs (IV-6 × IV-7, and IV-7 × IV-13). (Redrawn from Gardner *et al.* 1975. Copyright 1975 by the American Genetic Association.)

quite high proportions of affected offspring. The second observation above is not compatible with either form of X-linked inheritance or with autosomal recessive inheritance, but all three observations are compatible with autosomal dominance. Having examined a pedigree such as the one shown in Fig. 7.2, a tentative conclusion could be drawn that the disease in question, multiple exostosis in horses, is an autosomal dominant condition, assuming that III–1 is a new mutant. This conclusion could then be tested on further data.

The problems associated with drawing a pedigree as in Fig. 7.2 are firstly that it is a time-consuming task, and secondly that in the case of animals, the occurrence of matings between parent and offspring, and between full-sibs or half-sibs, often result in a maze of intercrossing lines that are difficult to interpret. Despite these disadvantages, pedigrees like the one in Fig. 7.2 can be quite useful, and are often drawn during the initial stages of investigation into the inheritance of a defect or disease.

With or without the aid of a pedigree, it is possible to draw up a list of criteria that taken together suggest a particular form of inheritance. Drawing on a knowledge of Mendelian inheritance (Chapter 1) and of population genetics (Chapter 5), the relevant criteria for each model are summarized below. One criterion is concerned with the degree of genetic relationship, the measurement of which is described in Chapter 13.

7.3.1 *Autosomal dominant*

(1) The defect or disease is transmitted from generation to generation without skipping any generations.
(2) Every affected offspring has at least one affected parent, except in the case of a new mutant.
(3) Normal offspring from affected parents produce only normal offspring when mated to normals, and the same is true for all their descendants.
(4) Approximately equal numbers of males and females are affected.
(5) If the defect or disease is rare but not lethal, then most matings producing affected offspring will be normal × affected ($aa \times Aa$), in which case the expectation is that one-half of each sex among the offspring will be affected. Thus the segregation frequency is $\frac{1}{2}$.*
(6) If the defect or disease is lethal, then it will be very rare, occurring sporadically with an incidence equal to twice the mutation rate.

7.3.2 *Autosomal recessive*

(1) The defect or disease may skip generations.
(2) All offspring of two affected parents are affected.
(3) Approximately equal numbers of males and females are affected.
(4) If the defect or disease is rare, then:

    (i) most affected individuals will have both parents normal;
    (ii) most matings producing affected offspring will be $Bb \times Bb$, for which the segregation frequency is $\frac{1}{4}$;*
    (iii) carriers (Bb) will usually mate with homozygous normals (BB), producing one-half carriers among their offspring. If, then, a carrier sire is mated to his own daughters, or to the daughters of another carrier sire, it follows that one-half of such matings will be $Bb \times Bb$, in which case the segregation frequency is $\frac{1}{2} \times \frac{1}{4} = \frac{1}{8}$;

---

* See Section 7.4 for a discussion of a bias that arises in data collected from such matings.

(iv) matings between an affected animal and an unrelated normal animal usually produce only normal offspring;

(v) affected × normal matings that do produce affected offspring must be $bb \times Bb$, in which case the segregation frequency is $\frac{1}{2}$;*

(vi) the average genetic relationship between normal parents of affected individuals is greater than between normal parents that have not produced affected individuals. (The reason for this is that the greater the genetic relationship between two individuals, the more likely they are to be carrying the same mutant gene; see Section 13.3.)

7.3.3 X-linked dominant

(1) Affected males when mated to normal females transmit the defect or disease to all their daughters but not to their sons.

(2) Unless the defect or disease is very common, affected females when mated to normal males transmit the defect or disease to an average of one-half of their sons and one-half of their daughters.

(3) If the defect or disease is rare, then its incidence in females is approximately twice that in males, in the general population.

(4) Every affected offspring has at least one affected parent, except in the case of a new mutant.

7.3.4 X-linked recessive

(1) The defect or disease may skip generations.

(2) All offspring of two affected parents are affected.

(3) Incidence is lower in females than in males, with the incidence of the defect or disease in females being approximately the square of the incidence in males, in the general population.

(4) If the defect or disease is rare, then:

(i) most affected individuals are males, and result from matings among normal parents;

(ii) most matings producing affected offspring will be $X^D X^d \times X^D Y$, for which the segregation frequency is zero in females and $\frac{1}{2}$ in males;*

(iii) affected males when mated with normal unrelated females transmit the defect or disease to none of their offspring, but all of their daughters are carriers;

(iv) affected females when mated with normal males transmit the defect or disease to all their male offspring but to none of their female offspring. All female offspring, however, are carriers.

* See Section 7.4 for a discussion of a bias that arises in data collected from such matings.

7.4 Simple segregation analysis

If the data available on a particular defect or disease appear to correspond in general terms to one of the four sets of criteria outlined above, then the next step is to determine more specifically whether the data are compatible with the respective simple Mendelian model. This is done by means of a segregation analysis. If specially planned matings can be arranged, then it is a simple matter to test agreement with a particular model by comparing the expected segregation frequencies, which can be obtained from Tables 1.3 and 1.4, with those observed. For example, if the defect or disease is thought to be autosomal recessive, then all normal individuals that have produced affected offspring must be carriers. If matings among such known carriers are specifically arranged, then we expect one-quarter of all offspring to be affected with the defect or disease, i.e. the expected segregation frequency is 0.25.

In many situations, however, it is not possible to arrange matings among known carriers and to observe all resultant offspring. Instead, it is much more usual for matings to come to the notice of the investigator after the event, and then only if an affected offspring is produced. This immediately introduces a bias into the data, because those matings of carrier × carrier that by chance do not produce any affected offspring will be automatically excluded from the data. The result is that even if the defect is really due to an autosomal recessive allele, a segregation ratio greater than 0.25 is expected. This is because all the affected offspring have been included in the data, but not all of the normal offspring.

The existence of this bias was recognized as long ago as 1912, and since then many different methods of segregation analysis have been developed in order to cater for it. As a result of a development by Davie (1979), there is now just one very straightforward method of simple segregation analysis that can be used in place of all other methods that have been used in the past. Known as the Singles Method, it provides an efficient and unbiased estimate of segregation frequency, and enables a very simple statistical test of whether the data are compatible with a particular Mendelian model. An illustration of the Singles Method is given in Appendix 7.1.

Although the above procedure has been described as simple and straightforward, in practice there are often considerable difficulties encountered by investigators attempting to determine whether a defect or disease is Mendelian. For example, it is quite commonly found that while the majority of available observations agree very closely with the general criteria for a particular model, there are one or two observations that are incompatible with that model. Misclassification of phenotypes, as discussed in Section 6.3 for the case of malignant hyperthermia syndrome in pigs, is one common cause of this problem. So too are the occurrence of mutation and the occurrence of *phenocopies*, which are phenotypes caused by environmental factors that mimic the effect of a gene. Misclassification, mutation, and the occurrence of phenocopies give rise to what are known as *sporadic cases*, which by definition will not necessarily fit any model. Methods of segregation analysis have been developed to allow for sporadic cases, and for other complicating factors as well. The problem is that as more and more complicating factors are accommodated into segregation analysis, the methods of analysis become more and more complex.

7.5 Complex segregation analysis

If an investigator really wishes to utilize all available data as efficiently as possible, then in most cases complex segregation analysis on a computer is essential. One form of analysis that has been used on animal data utilizes information on the relative incidence of different mating types, and can easily incorporate incomplete penetrance. Examples of its use are given by Smith and Bampton (1977) and by Wijeratne and Curnow (1978). Various packages of computer programmes have been developed for even more sophisticated analyses, as reviewed for example by Elston *et al.* (1978) and by Lalouel (1978). The problem with these computer packages is that they require considerable mathematical and statistical skill on the part of the investigator, in order not to be misused. Even then, the results of a properly-conducted complex segregation analysis are often open to debate, because, as discussed in Chapter 6, the available data may be compatible with more than one model. This is not to say that complex segregation analysis has no place in veterinary genetics. On the contrary; in the hands of a skilled investigator, complex segregation analysis can be very useful. If such an analysis is properly conducted, it will provide an estimate of the heritability of liability for the defect or disease, and will also indicate whether or not the segregation of a single gene with relatively large effect makes any contribution to the variation in liability.

7.6 Summary

A genetic contribution to the aetiology of a defect or disease is indicated if incidence is higher in some families than in others within a breed, and also if incidence is higher in some breeds than in others, but only if all possible environmental effects have been taken into consideration. The existence of a genetic contribution to aetiology within each of several breeds does not necessarily indicate that there will be a genetic difference between breeds.

Drawing on a knowledge of Mendelian inheritance and of population genetics, it is possible to compile a list of general criteria for each of the four simple, Mendelian models of inheritance, namely autosomal dominant, autosomal recessive, X-linked dominant, and X-linked recessive. If the available data agree in general terms with one of these models, then the Singles Method of simple segregation analysis can be used to test for specific agreement between the data and the model. If the data do not fit any simple, Mendelian model, then complex segregation analysis may be required.

In conclusion, the steps involved in an investigation designed to answer the question 'Is it inherited?' will usually be:

(1)   check on general evidence for a genetic aetiology;
(2)   check on general criteria for each simple, Mendelian model;
(3)   perform a simple segregation analysis, using the Singles Method;
(4)   if the simple segregation analysis rejects a simple, Mendelian form of inheritance, but the defect or disease is obviously familial, perform a complex segregation analysis.

7.7   Further reading

Leipold, H. W., Dennis, S. M., and Huston, K. (1972). Congenital defects of cattle: nature, cause and effect. *Advances in Veterinary Science and Comparative Medicine* 16, 103-50. (This review contains a useful section on genetic diagnosis which describes the use of segregation analysis in cattle.)

Morton , N. E. and Chung, C. S. (Eds) (1978). *Genetic epidemiology*. Academic Press, New York. (This book contains several useful accounts of the theory and practice of segregation analysis. Although concerned solely with humans, many of the concepts and examples are equally relevant to animals.)

Nicholas, F. W. (1984). Simple segregation analysis: a review of the methodology. *Animal Breeding Abstracts* 52, 555-62. (A review of simple segregation analysis as it applies to animals, illustrating the application of Davie's (1979) extension of the Singles Method.)

Pidduck, H. (1985). Is this disease inherited? A discussion paper with some guidelines for canine conditions. *Journal of Small Animal Practice* 26, 279-91. (A review of the material covered in this chapter.)

Appendix 7.1

An example of simple segregation analysis (adapted from Nicholas 1984)

As a result of an extension by Davie (1979), the Singles Method is now the easiest, quickest, and most efficient form of simple segregation analysis available. In the present context, the only requirement for this method to be valid is that *all* members of each reported family have been included in the data.

The most straightforward use of the Singles Method can be made when the investigator is certain that either:

(1) all families with affected offspring are included in the data, or
(2) a random sample of families with affected offspring has been included in the data.

If either of these assumptions can be satisfied, then the segregation frequency can be estimated as

$$\hat{p} = \frac{A - A_1}{T - A_1},\tag{1}$$

and its estimated variance is given by

$$\text{Est. Var. } (\hat{p}) \doteq \frac{(T - A)}{(T - A_1)^3}\left| A - A_1 + 2A_2 \frac{(T - A)}{(T - A_1)}\right|,\tag{2}$$

where A is the total number of affected offspring in the available data, T is the total number of all offspring in the available data, A_1 is the total number of families with just one affected offspring, and A_2 is the total number of families with two affected offspring.

In many situations, the investigator may not be certain that one or other of the assumptions listed above has been satisfied. For example, families with several affected offspring may be more likely to be included in the data than families with just one affected offspring. In such a case, the best that can be done is to obtain an upper and a lower estimate of p. The upper estimate is obtained exactly as shown above, and the lower estimate is obtained from the same two equations by replacing A_1 with n, the number of families, and by setting A_2 equal to zero.

A7.1.1 *Hypothesis testing*

The aim of simple segregation analysis is to test whether or not the data are compatible with a simple Mendelian form of inheritance. This test most commonly involves a statistical comparison of the estimated p with the hypothesized value (p_0) arising from the particular model of inheritance being tested. For example, if a recessive mode of inheritance is suspected, and if both parents in each of a set of full-sib families are normal, then the null hypothesis is that the true value of p is $p_0 = 0.25$.

If the assumptions in relation to family sampling have been satisfied, then just one estimate of p is obtained, and the null hypothesis can be tested by calculating $Z^2 = (\hat{p} - p_0)^2/\{\text{Est. Var. } (\hat{p})\}$, which is approximately distributed as χ^2 with one degree of freedom.

If upper and lower estimates have to be calculated, then testing the null hypothesis is less straightforward. In practice, the best that can be done is to conclude that the data are consistent with the above null hypothesis provided $\hat{p}_U + \sigma_{\hat{p}_U} > 0.25 > \hat{p}_L - \sigma_{\hat{p}_L}$, where subscripts U and L refer to upper and lower limits respectively, and where σ is the square root of the estimated variance (Steinberg 1959).

A7.1.2   Number of observations required

It is useful to have a simple guideline to indicate the number of families of a particular size that would be required in order to obtain a particular level of accuracy for the estimate of segregation frequency. For the Singles Method, a simple guideline can be obtained by writing the expected variance of $\hat{p}$ when estimated from n families of size s, as $p_0 q_0 / \{n(s-1)\}$, where p_0 is the hypothesized segregation frequency, and $q_0 = 1 - p_0$.

As an example, imagine that an investigator wishes to test a simple recessive hypothesis ($p_0 = 0.25$) on data from n families of size 6. The expected variance of $\hat{p}$ is approximately $(0.25)(0.75)/\{n(6-1)\} = 0.0375/n$, which gives expected standard errors of approximately 0.19, 0.09 and 0.06 for one, five and ten families respectively. Suppose that an investigator requires the coefficient of variation of $\hat{p}$ to be no greater than, say, 20%. If the null hypothesis is true, the number of families required to satisfy this criterion is the value of n for which $\sqrt{[p_0 q_0 / \{n(s-1)\}]} / p_0 \leqslant 0.2$. This gives $n \geqslant 25 q_0 / \{p_0(s-1)\}$ as the required number of families, for which $p_0 = 0.25$ and $s = 6$, gives $n \geqslant 15$.

Implicit in the above approach is the assumption that all families are of the same size, which of course in practice is rarely true. If desired, an expected variance could be calculated using the same approximate formula but taking account of the expected or actual distribution of family size. In most cases, however, the additional accuracy may not be worth the additional effort. Although they are approximate, calculations such as those above indicate that an investigator can obtain a rough idea as to whether there are sufficient data available for a satisfactory segregation analysis.

A7.1.3   An example

We now present a practical example of segregation analysis, using the data of Hartley *et al.* (1978), given in Table A7.1. These data were collected over a period of time, as individual breeders notified the investigators of the occurrence of litters containing at least one affected puppy. Since none of the parents of affected puppies were themselves affected, and since puppies of both sexes were affected, the most likely null hypothesis is that cerebellar degeneration is inherited as an autosomal recessive condition. Since in all 28 families at least one affected offspring was produced from normal parents, all parents must be heterozygous if the null hypothesis is true. Thus the expected segregation frequency under the null hypothesis is 0.25, and the null

Table A7.1. Data on the occurrence of cerebellar degeneration in the 28 families of rough-coated collies described by Hartley *et al.* (1978)

Family number	Total affected offspring	Family size
1	2	3
2	2	4
3	5	5
4	3	5
5	1	5
6	2	6
7	3	6
8	3	6
9	2	6
10	4	7
11	1	7
12	1	7
13	1	7
14	2	7
15	2	7
16	3	8
17	1	8
18	4	8
19	3	8
20	3	8
21	3	8
22	2	9
23	1	9
24	4	9
25	2	9
26	3	10
27	5	15
28	2	19
$n = 28$	$A = 70$	$T = 216$
	$A_1 = 6$	
	$A_2 = 9$	

hypothesis can be stated as $H_0 : p = 0.25$. However, since heterozygote-by-heterozygote matings that by chance failed to produce affected puppies are obviously not included in the data, the 28 families in Table A7.1 are a biased sample of all heterozygote-by-heterozygote matings. Thus the proportion of affected offspring in the available data ($A/T = 70/126 = 0.324$) is a biased estimate of segregation frequency. We shall now use the Singles Method of segregation analysis to remove the effect of this bias.

A7.1.4   *Assumptions 1 and 2 not satisfied*

After considering how the data were obtained, Hartley *et al.* (1978) concluded that the probability of a litter coming to their notice and hence being included in the data 'is certain to be partly determined by the number of affected offspring in a litter'. Thus neither assumption 1 nor assumption 2 is satisfied, and so the appropriate form of analysis is to obtain upper and lower estimates of p.

The upper estimate is obtained directly from equations (1) and (2):

$$\hat{p}_U = \frac{A - A_1}{T - A_1} = \frac{70 - 6}{216 - 6} = 0.305$$

$$\text{Est. Var.}(\hat{p}_U) = \frac{(T - A)}{(T - A_1)^3}\left| A - A_1 + 2A_2\frac{(T - A)}{(T - A_1)}\right|$$

$$= \frac{(216 - 70)}{(216 - 6)^3}\left| 70 - 6 + 2(9)\frac{(216 - 70)}{(216 - 6)}\right|$$

$$= 0.001206$$

$$\sigma_{\hat{p}_U} = 0.035$$

The lower estimate is obtained from the same two equations, by replacing A_1 with n, and by setting A_2 equal to zero:

$$\hat{p}_L = \frac{A - n}{T - n} = \frac{70 - 28}{216 - 28} = 0.223$$

$$\text{Est. Var.}(\hat{p}_L) = \frac{(T - A)}{(T - n)^3}(A - n)$$

$$= \frac{(216 - 70)}{(216 - 28)^3}(70 - 28)$$

$$= 0.000923$$

$$\sigma_{\hat{p}_L} = 0.030.$$

Using Steinberg's (1959) criterion, we have

$$\hat{p}_U + \sigma_{\hat{p}_U} = 0.305 + 0.035 = 0.340,$$

and

$$\hat{p}_L - \sigma_{\hat{p}_L} = 0.223 - 0.030 = 0.193$$

Thus, the condition $\hat{p}_U + \sigma_{\hat{p}_U} > 0.25 > \hat{p}_L - \sigma_{\hat{p}_L}$ is satisfied, and it is therefore concluded that the data are consistent with a simple recessive mode of inheritance.

A7.1.5   Assumptions 1 or 2 satisfied

For the purpose of illustration, we shall now assume that assumption 1 or assumption 2 is satisfied in the data of Hartley *et al.* (1978).

A single estimate of p is obtained directly from equations (1) and (2) (using exactly the same arithmetic as used for the upper estimate obtained above). Thus:

$$\hat{p} = \frac{A - A_1}{T - A_1} = \frac{70 - 6}{216 - 6} = 0.305,$$

$$\text{Est. Var. } (\hat{p}) = \frac{(T - A)}{(T - A_1)^3}\left| A - A_1 + 2A_2\frac{(T - A)}{(T - A_1)}\right|$$

$$= \frac{(216 - 70)}{(216 - 6)^3}\left| 70 - 6 + 2(9)\frac{(216 - 70)}{(216 - 6)}\right|$$

$$= 0.001206$$

$$\sigma_{\hat{p}} = 0.035.$$

Then the null hypothesis is tested using:

$$Z^2 = \frac{(\hat{p} - p_0)^2}{\text{Est. Var. } (\hat{p})}$$

$$= 2.508,$$

which is not significant at $\alpha = 0.05$. Thus the data are consistent with a simple recessive mode of inheritance, and $\hat{p} = 0.305 \pm 0.035$.

8
Immunogenetics

8.1 Introduction

If a foreign substance enters an animal or sometimes even if it just touches the animal's skin, then that animal automatically and unconsciously responds to the 'attack' by attempting to inactivate or destroy the foreign substance. Such a response is called an *immune response*, and the protection it provides is called *immunity*. There are two types of immune response, each of which is brought about primarily by the action of a different type of *lymphocyte* (a type of white blood cell). As shown in Fig. 8.1, lymphocytes originate in the bone marrow stem cells. Those that migrate to the thymus become *T-lymphocytes* (often called T-cells), while those that migrate to the Bursa of Fabricius in chickens, or to the 'bursal equivalent' (which is not clearly defined) in mammals, become *B-lymphocytes* (often called B-cells).

In one type of immune response (the *humoral* immune response), B-cells mature into plasma cells, which produce large quantities of *antibody* (*anti*-foreign *body*) against the foreign substance, which is called an *antigen* (*anti*body-*gen*erating). The maturation of B-cells into plasma cells, and their subsequent production of antibody, is triggered by the presence of antigen. There is almost an infinite range of potential antigens, including viruses, bacteria, foreign molecules of any kind, and blood cells from other animals. If the foreign substance is a cell or a particle of reasonable size, then the antibody produced against it is usually directed against specific structures on the surface of the cell or particle. In such cases, the term antigen is applied to the surface structure itself rather than to the whole cell or particle. Despite the huge potential number of different possible antigens, the individual under attack soon produces a quantity of antibody that is specifically directed against the antigen concerned, and the antibody binds with the antigen to form an antigen–antibody complex. The end result is agglutination (in which the antigen–antibody complex gives rise to clumps of inactivated cells or particles), or precipitation (clumping of soluble antigens) or cell death (in which the antigen–antibody complex gives rise to a series of reactions that lead to cell lysis). The protection that arises from the production of antibodies is called *humoral immunity*.

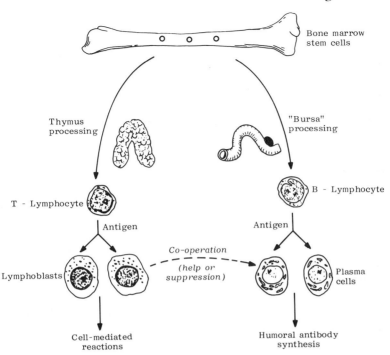

Fig. 8.1. The origin of the two types of immune response. Humoral immunity occurs when the presence of an antigen stimulates B-lymphocytes to mature into plasma cells, which produce antibody. Cell-mediated immunity occurs when the presence of antigen stimulates T-lymphocytes to develop into various types of mature T-cells (lymphoblasts) including cytotoxic T-cells, helper T-cells, suppressor T-cells, and lymphokine-producing T-cells. Figure 3.2 is a simplified version of this diagram. (After Roitt 1984.)

The other type of immune response (*cell-mediated* immune response) occurs when T-cells, after stimulation by antigen, develop into various types of mature T-cells, including cytotoxic T-cells (which are directly responsible for the death of viral-infected cells or foreign cells), helper T-cells and suppressor T-cells (which help or hinder the action of B-cells and other T-cells), and T-cells that give rise to *lymphokines*, which are soluble factors that can greatly enhance the destructive action of other white blood cells such as *macrophages*.

The general area of knowledge associated with the genetic basis of immunity is called *immunogenetics*. The aim of this chapter is to review those aspects of immunogenetics that are most relevant to animals.

8.2 Antibodies

Antibodies are protein molecules that belong to a class of proteins called *immunoglobulins*. The basic immunoglobulin molecule consists of four chains of amino acids, two identical light (L) chains and two identical heavy (H) chains, joined by di-sulphide bonds. The L chains are approximately 220 amino acids long, while the H chains range from approximately 450 to 700 amino acids in length. Each chain consists of a variable (V) region and a constant (C) region, with the variable regions, as their name implies, differing from one antibody to the next. The constant regions, on the other hand, are usually the same in a large number of different antibodies. The main features of an antibody molecule are illustrated in Fig. 8.2, and some remarkable photographs of an antibody are shown in Fig. 8.3.

Being polypeptides, antibodies are obviously the product of structural genes.

In fact, light and heavy chains are produced from different clusters or families of genes located on different chromosomes. Much of what is now known about these gene clusters is the direct result of the application of recombinant DNA technology, in particular the cloning of various parts of the clusters, and the subsequent sequencing of the DNA. In the following account, only the general features of current knowledge are presented. For a more detailed description, readers should consult the reviews by Leder (1982) and Tonegawa (1983).

8.2.1 Light chain assembly

There are actually two different types of light chains called κ and λ, each produced by a different cluster of genes. These clusters are located on different chromosomes, but are so similar that they must have arisen by duplication of an ancestral gene, followed by translocation. Because they are similar, and because only one of them is ever switched on in any given cell, we shall consider only one of them, κ, in the following account.

In undifferentiated cells, the DNA of a light-chain gene cluster consists of several hundred V genes, four J genes, and one C gene. As can be seen in the top left-hand corner of Fig. 8.4, each of these genes is separated from its neighbour by intervening sequences of DNA. These sequences range in length up to 24 kilobases.

During differentiation of the embryonic B-cell into an antibody-producing B-cell, all of the DNA between a particular V gene and a particular J gene is deleted from the *template* strand of DNA (the strand that will be transcribed), in a process called *V-J joining*. As shown in Fig. 8.4, the resultant template strand of DNA consists of

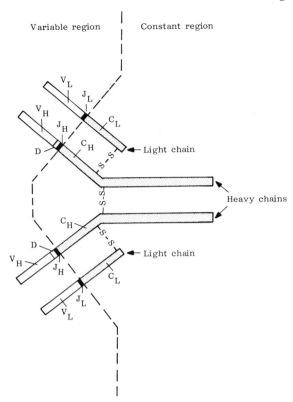

Variable region | Constant region

Fig. 8.2. The main features of an antibody molecule, showing two identical heavy (H) chains, and two identical light (L) chains, with each chain consisting of a constant (C) and a variable (V) region. In light chains, the V region consists of two segments: the variable segment (V_L) and the joining segment (J_L). In heavy chains, the V region contains a diversity (D) segment (see Section 8.2.2) in addition to the V_H and J_H segments. In both types of chains, the V segment contains approximately 95 amino acids, and the J segment approximately 15 amino acids. In H chains, the D segment consists of approximately 5 amino acids. Although not shown in the diagram, the V region consists of three hypervariable or complementarity-determining regions (CDR1 to CDR3) flanked by four framework regions (FR1 to FR4). The CDRs are the sequences of amino acids that actually touch the relevant antigen, and which are therefore primarily responsible for antigen recognition.

a V gene and a J gene adjacent to each other, but with an intervening sequence still separating J and C. This DNA is transcribed into mRNA in the nucleus of the lymphocyte, but by the time the mRNA has reached the cytoplasm, the intervening sequence between J and C has been removed by *RNA splicing*. The end result, as shown in Fig. 8.4, is an mRNA molecule that can be translated into a light chain consisting of three adjacent segments; V, J, and C.

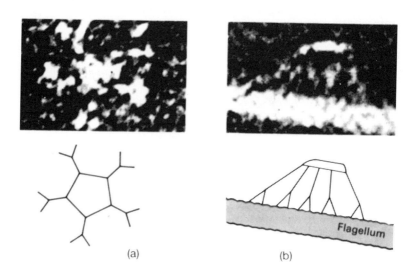

Fig. 8.3. Electron micrographs (above) and line drawings (below) of an antibody consisting of five of the units shown in Fig. 8.2. (a) A dorsal view, showing the regular arrangement of the five units. (b) A lateral view of a sheep antibody attached to a flagellum of *Salmonella paratyphi*. (From Roitt 1984.)

8.2.2 Heavy chain assembly

The heavy-chain gene cluster in undifferentiated cells is similar to the two light-chain clusters. It seems, therefore, that all three clusters arose by duplication and subsequent translocation of the one ancestral antibody-producing cluster. The difference between the heavy chain cluster and the light chain clusters is due to mutation and further duplication within the cluster following the initial duplication. As shown in Fig. 8.4, the heavy chain cluster consists of several hundred V genes, eight D genes, four J genes, and five C genes.

(The existence of five C genes is not directly relevant to the present discussion. For the remainder of this section, therefore, we shall consider only one of them, namely C_μ.)

During differentiation into an antibody-producing lymphocyte, only one each of the V, D, and J genes in the heavy chain gene cluster is joined together following deletion of appropriate pieces of the template strand of DNA, in a process called *V–D–J joining*. The remaining processes are the same as described above for light chains. Thus, the resultant template strand of DNA is transcribed into nuclear mRNA, which is spliced before it reaches the cytoplasm, giving rise to an

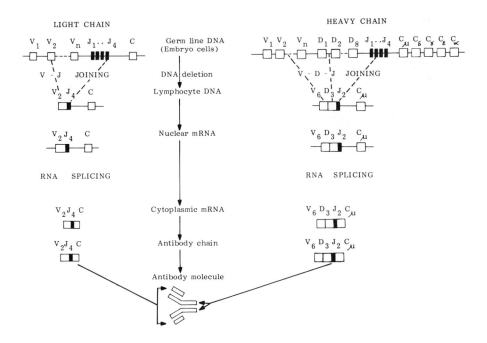

Fig. 8.4. The assembly of an antibody molecule. In general, the boxes represent coding sequences (exons) and the lines joining them represent intervening sequences. (The C_μ heavy chain gene is an exception: see Fig. 8.5.) The drawings are not to scale. For a detailed account of the length of coding sequences and of intervening sequences, see Leder (1982).

mRNA molecule that is translated into a heavy chain with adjacent V, D, J, and C_μ segments.

8.2.3 *The mechanism of V–J joining and V–D–J joining*

The mechanism by which V, D, and J genes are joined together is not understood at present. It is known, however, that:

> *The intervening sequences separating each V, J, C, and D gene have highly conserved inverted repeats located at specific positions.*

For example, the palindrome

CACAGTG

GTGTCAC

occurs adjacent to the right-hand end of every V and every D gene, and its inverse occurs adjacent to the left-hand end of every J and every D gene. Also, a highly conserved sequence of nine bases is located a further 11 or 22 bases to the right of every V and every D gene, and its inverse is located a further 11 or 22 bases to the left of every J and every D gene. The existence of such highly-conserved inverted repeats suggests that the deletion of DNA between particular V, D, and J genes involves the same mechanism by which insertion sequences (Section 1.3.4) and transposons (Section 10.4.4) excise themselves from DNA molecules.

8.2.4 Antibody diversity

As soon as an animal becomes immunocompetent, i.e. becomes capable of producing antibodies, and before being challenged by any antigen, it produces about one million (10^6) different antibodies. Each different antibody is produced by a different clone of B-lymphocytes, i.e. a particular clone of B-lymphocytes produces only one type of light chain and one type of heavy chain. For many years it was not evident how such a diversity of antibodies could be produced. The above account of antibody production provides an obvious answer to this question.

Suppose, for example, that there are 300 V genes and that they are all slightly different from each other. And suppose that the D genes and the J genes are also slightly different from each other. Recalling that there are 4 J genes and 8 D genes, it follows that there are $300 \times 4 = 1200$ different light chains that could be produced by V–J joining, and $300 \times 8 \times 4 = 9600$ different heavy chains that could be produced by V–D–J joining. Noting that an antibody molecule consists of two identical light chains and two identical heavy chains, it follows that a total of $1200 \times 9600 = 11.52 \times 10^6$ different antibody molecules could be produced. Thus, the so-called *combinatorial diversity* generated by V–J joining, V–D–J joining, and the combination of light and heavy chains is sufficient to explain the figure of one million different antibodies quoted above. However, these three processes are not the only source of antibody diversity. There is evidence, for example, of imprecision during V–J joining and V–D–J joining, such that the joining ends of particular V or D or J segments vary by up to ten nucleotides, giving rise to *junctional site diversity*. Another cause of diversity is the occasional insertion of one or several nucleotides at V–D and D–J junctions (called *junctional insertion diversity*). The final cause of diversity is *somatic mutation*, which, in the present context, refers to changes in nucleotide sequences that occur in the relevant V gene either during or after V–J and V–D–J joining. These somatic mutations are about three times more frequent in the three so-called *hypervariable* or *complementarity-determining regions* (CDRs) of the V segment than in the four *framework regions* (FRs) that flank the CDRs. Not surprisingly, it appears that the three CDRs correspond to the regions of an antibody molecule that actually come into contact with the relevant antigen, and which must, therefore, be primarily responsible for antigen recognition. Thus, variation in one or more of the CDRs would be more likely to give rise to a change in antibody specificity than variation in some other region.

We can conclude that:

> *Antibody diversity arises from:*
>
> *(1) combinatorial diversity;*
> *(2) junctional site diversity;*
> *(3) junctional insertion diversity;*
> *(4) somatic mutation.*

8.2.5 The heavy chain switch

We have already seen that prior to being challenged by any antigen, an animal produces about one million different antibodies, with each different antibody being produced by a different B-lymphocyte. On the surface of each B-lymphocyte, there are about 100 000 molecules of the particular antibody produced by that lymphocyte. These molecules all belong to the so-called IgM class of antibodies, and they are all attached to the cell in a particular manner: the end of the heavy chain furthest from the V region is embedded into the lipid layer of the cell membrane. This is achieved by having lipid-soluble amino acids at the C end of the heavy chain.

These membrane-attached antibody molecules act as antigen receptors. When an appropriate antigen enters the body, it eventually encounters a B-lymphocyte with surface antibodies (antigen receptors) that are able to bind to that antigen. This binding occurs at the end of the antibody molecule furthest from the cell, i.e. at the V region end.

> *Obviously, the variability of the V region reflects the vast number of different antigens that have to be detected, while the constancy of the C region reflects its main function, which is the same for all antibodies, namely attachment of the antibody molecule to the lymphocyte surface.*

B-lymphocytes undergo further changes after initially producing membrane-bound IgM molecules. If stimulated by the presence of antigen, they produce IgM molecules that differ slightly in amino acid sequence at their C end, such that this end is not lipid-soluble and hence does not enable attachment of the antibody molecule to the lymphocyte surface. This second type of IgM antibody molecule is released into the serum and is thus called the *secreted* form, as distinct from the *membrane-bound* form produced initially. In addition, as B-lymphocytes mature, some of them change from producing IgM antibodies alone, to producing IgM and another class of antibodies, called IgD, at the same time. Finally, each B-lymphocyte becomes committed to producing any one of the five different classes of antibody molecules that are known, namely IgM, IgD, IgG, IgE, or IgA.

> *The different forms and classes of antibodies produced by a particular B-lymphocyte at different stages of its development differ only in the C region of their heavy chain; all of the light chain, and all of the V region of the heavy chain, are the same in all classes of antibody produced by a particular B-lymphocyte.*

The switch from production of the membrane form of IgM to the secreted form, and the subsequent switch to production of a different class of antibody, is called C_H *switching*. In the case of the two different forms of IgM, the mechanism of C_H switching can be better understood by taking a closer look at the C_μ gene, as shown in Fig. 8.5.

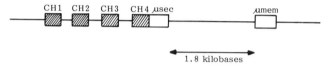

Fig. 8.5. A more detailed structure of the C_μ gene, showing the four main exons (CH1 to CH4) which code for the four main sections of the C_H chain of IgM, together with the two terminal exons, *μsec* for the secreted form, and *μmem* for the membrane form. Not to scale.

It can be seen that the C_μ gene consists of four main exons (CH1 to CH4) plus two other exons, one of which (*μsec*) is immediately adjacent to CH4, with the other (*μmem*) being about 1.8 kilobases away. The *μsec* exon produces the amino acid sequence found at the C end of the lipid-insoluble (secreted) form, while the *μmem* exon produces the amino acid sequence found at the C end of the lipid-soluble (membrane) form. It is thought that the C_μ gene gives rise to two different types of nuclear RNA, one of which terminates at the 3′ end of the *μsec* exon, while the other terminates at the 3′ end of the *μmem* exon. Two different forms of RNA splicing then bring either *μsec* or *μmem* adjacent to CH4 in the cytoplasmic RNA that is translated into a heavy chain.

The key to the production of more than one class of antibody by the one B-lymphocyte is the existence of five different C_H genes (see Section 8.2.2). These genes are called C_μ, C_δ, C_γ, C_ϵ, and C_α, and correspond respectively to the C_H regions of IgM, IgD, IgG, IgE, and IgA.

When a B-lymphocyte first starts to produce heavy chains, it does so by deleting from the template strand of DNA all C genes and associated intervening sequences except the C_μ gene, which is closest to the J genes. After splicing, the resultant mRNA gives rise to IgM antibodies. If only the last three C genes (C_γ, C_ϵ, and C_α) are deleted, then it seems that two different forms of RNA splicing can give rise to the production of IgM and IgD molecules at the same time, as illustrated in Fig. 8.6. If the B-lymphocyte finally becomes committed to producing, say, IgA molecules, then it is thought that this results from a deletion of all but the C_α gene, as also illustrated in Fig. 8.6. The mechanism by which this C_H *switching* occurs is not yet fully understood, but it is known to involve so-called *switch-sites*, which are repetitive nucleotide sequences located between most of the C genes.

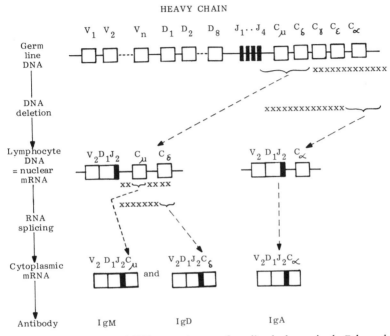

Fig. 8.6. The production of different classes of antibody by a single B-lymphocyte, due to C_H switching. Crosses indicate those portions of DNA or RNA that are deleted. Not to scale.

8.2.6 *Genetic control of antibody production*

> *Not only are antibodies encoded by genes, but the extent of their production is also under genetic control.*

This has been clearly shown by selection experiments in which, for example, chickens were selected for high and low production of antibody to injected sheep red blood cells (Siegel and Gross 1980). In just three generations, selection produced a significant difference between the high and low lines in terms of antibody titre, which is a standard measure of the concentration of antibody present. (It equals the highest serial dilution at which antibody activity can still be detected.)

> *The fact that antibody titre can be altered by artificial selection indicates that it is under genetical control.*

Furthermore, although selection was solely in terms of immune response to sheep red blood cells, the high line was found to be more resistant to a number of infectious agents including *Mycoplasma gallisepticum* and feather mites. It was less resistant than the low line, however, to *E. coli* (Gross *et al.* 1980). This latter result indicates, as might be expected, that having the ability to produce large volumes of antibody against a particular antigen (sheep red blood cells) does not necessarily increase ability to mount immune responses against all other potential antigens, including pathogenic ones.

8.3   Red-cell antigens

Red-cell antigens are antigens occurring on the surface of red blood cells. The first evidence of genetic control of any cell-surface antigens was presented by Landsteiner in 1900, in relation to the ABO blood group system in humans. The structure of the ABO antigens and their genetic basis are now well known, and are worth describing here because they probably bear a resemblance to red-cell antigens in animals, the exact details of which are much less well understood.

Neglecting the finer details of the ABO story, we can say that there are four distinct blood groups recognized, corresponding to the presence of one or two different antigens (A or B, or A and B) on the surface of red blood cells, or to their absence (O). Each antigen consists of a combination of sugars attached to a protein backbone, with antigens A and B differing only by the terminal sugar that is added to the common chain of four sugars. The addition of this final sugar is determined by different forms of the enzyme transferase, which are produced by different alleles at an autosomal locus on chromosome 9. One allele codes for a transferase that adds galactose as the final sugar, giving rise to the B antigen. Another allele at the same locus produces a transferase that adds N-acetyl galactosamine, a derivative of galactose, giving rise to the A antigen. The third allele results in no sugar being added to the four-sugar chain, and the consequent lack of a detectable antigen, which is called O. The allele giving rise to each antigen is given the same symbol as the antigen. The genetic basis of each antigen is illustrated in Table 8.1.

Thus we have three alleles, *A*, *B*, and *O* at a locus, and six different genotypes *AA, AB, BB, AO, BO,* and *OO,* corresponding to all possible pairwise combinations of the three alleles. Knowing the biochemical basis of the three antigens, it follows that these six genotypes will give rise to only four different phenotypes, because *AA* and *AO* will both exhibit just one antigen, namely A, and the same applies to *BB* and

Table 8.1. The genetic basis of the ABO blood group in humans. N = α-D-N-acetyl galactosamine, G = α-D-galactose

Allele	Allele product (enzyme)	Antigen	Detectable?
		4-sugar chain	
O	—*	O–O–O–O–[Protein backbone] $\mid$ O	No
A	N-acetyl galactosaminyl transferase	N–[O–O–O–O–[Protein backbone]] $\mid$ O	Yes
B	galactosyl transferase	G–[O–O–O–O–[Protein backbone]] $\mid$ O	Yes

* The O allele does in fact produce an enzyme, but a discussion of that enzyme would involve a discussion of two related blood group systems, Lewis and Secretor, which are beyond the scope of this book.

BO. The heterozygote *AB*, on the other hand, exhibits both antigens. Thus the four phenotypes are A, B, AB, and O. Alleles *A* and *B* are said to be *codominant* with each other because both are expressed in *AB* heterozygotes. Allele *O* is referred to as the *null* allele, as it gives rise to no detectable antigen. Like all null alleles, it is recessive; when present with any other allele in a heterozygote, the effect of the null allele cannot be detected.

There are many different red-cell blood groups in animals. How are they detected? The usual technique involves taking red blood cells from one individual (the donor) and injecting them into another individual (the recipient).

As an example, suppose the donor has an antigen called A_1, on the surface of its red blood cells. Suppose also that the recipient lacks the A_1 antigen, and hence recognizes the donor's cells that carry the A_1 antigen as foreign or *non-self*. The recipient mounts an immune response, and some of its B-lymphocytes produce antibody to the A_1 antigen. These antibodies accumulate in the recipient's serum, which is then called anti-A_1 antiserum. Some of this antiserum is collected from the recipient, and is then added to red blood cells taken from each of a number of other individuals who collectively constitute a *test panel*. If any of these individuals have the A_1 antigen on their red blood cells, then the anti-A_1 antibody in the recipient's serum will bind with the antigen, resulting in agglutination or haemolysis, which is said to be a *positive* test result. In this way, all individuals carrying the A_1 antigen can be detected, and similar procedures can be used to detect other antigens. Of course, at first the name of the antigen being detected is not known; all that is known is that antiserum from a certain recipient gives a positive test when combined with red blood cells from certain members of a test panel. As more donors are injected with red blood cells from other individuals, the number of antisera increases. Sometimes the antisera from two recipients give a positive test for exactly the same members of a test panel, thus indicating that those two antisera have the same antibody. Different antibodies will cause positive tests in a different group of members of the test panel.

The situation in practice can be complicated by many factors which we will not discuss here. In general, however, a number of laboratories proceed in the above manner until they each have a bank of antisera. Then, every few years, laboratories exchange samples of their antisera, and all sera are tested on one or more local panels of animals. The patterns of positive tests are analysed for all antisera, and names or symbols are agreed upon for antisera, and hence for antigens, that all give exactly the same reaction pattern.

The inheritance of each antigen is then studied. The general picture that emerges is that:

(1) some antigens correspond to particular alleles at a particular locus;
(2) the remaining antigens occur only in conjunction with certain other antigens, in which case the group of antigens (called a *phenogroup*) corresponds to a particular allele at a particular locus;
(3) these alleles are usually, but not always, codominant;
(4) each locus corresponds to a different blood group system.

In order to provide an indication of the present state of knowledge about red-cell blood groups, resulting largely from procedures like those outlined above, Table 8.2 lists the currently known blood group systems in domestic animals, together with the minimum number of alleles detected to date at each locus. It can be seen from Table 8.2 that there are usually multiple alleles at each locus; indeed, the B system in cattle has more than 600 different alleles! However, it is possible that this large number of 'alleles' is really the result of segregation within a cluster of closely-linked genes which collectively constitute the B 'locus'. For an account of the debate on this unresolved question, see Bell (1983).

By comparing antisera across species, it has been shown that the B system in cattle has the same antigens as the B system in sheep, that antigens of the C system are also the same in both species, and

Table 8.2. Blood group systems in domestic animals

Horses Locus	Min. No. alleles	Cattle Locus	Min. No. alleles	Sheep Locus	Min. No. alleles	Pigs Locus	Min. No. alleles	Dogs Locus	Min. No. alleles	Cats Locus	Min. No. alleles	Chickens Locus	Min. No. alleles
A	11	A	10	A	3	A	2	A	3	AB	2	A	5
C	2	B	>600	B	52	B	2	B	2	C	2	B	35
D	11	C	77	C	4	C	2	C	2			C	5
K	2	F	4	D	2	D	2	D	2			D	5
P	3	J	4	M	4	E	15	F	2			E	9
Q	5	L	2	R	2	F	3	Tr	3			H	3
U	2	M	3	X	2	G	3	J	2			I	5
		S	15			H	7	K	2			J	3
		Z	2			I	2	L	2			K	4
		T	2			J	3	M	2			L	2
						K	6	N	2			P	10
						L	6					R	2
						M	18						
						N	3						
						O	2						

Total number of blood group systems

7	10	7	15	11	2	12

Adapted from Bell (1983) and Rasmusen (1975).

that the M system in sheep is the same as the S system in cattle. In addition, it has been shown that systems C and J are linked in pigs.

For most red-cell blood groups, antibodies are produced only following a challenge with the appropriate antigen. The exceptions to this generalization are the ABO system in humans, the J system in cattle and the AB system in cats. In these systems, antibodies to antigens not carried by a certain individual occur 'naturally', without any obvious challenge. For example, anti-A antibody occurs in almost all cats that have the B antigen (Auer and Bell 1981). Apart from these exceptions:

> *Individuals do not normally carry antibodies to red-blood-cell antigens, unless they have been specifically challenged with the appropriate foreign red blood cells.*

Since animals do not normally carry antibodies to red-blood-cell antigens, it has been thought commonly that blood transfusions in animals can be conducted quite safely with any available blood, and that there is normally no need for blood typing before transfusion. However, transfusion with randomly-chosen, untyped blood may lead to an immediate transfusion reaction if, unbeknown to the practitioner, the recipient has been transfused previously with blood containing the same antigens. Even if this does not occur, then it is quite possible that a random, untyped transfusion will sensitize the recipient to subsequent transfusions, or to the blood cells of the recipient's future offspring, if the recipient is a female (see Section 8.3.1 below). Thus, whenever possible:

> *It is advisable to obtain blood for transfusion from donors who have been typed as being compatible or negative for red blood cell antigens that are known to evoke strong antibody responses.*

The most clinically important of such antigens are A in dogs, B in cats, A_a and Q_a in horses, and A, F, and some B antigens in cattle (Best 1983). If untyped donors must be used, then a simple cross-match should be conducted, in which a drop of plasma from the recipient is mixed on a slide with a drop of red-cell suspension from the donor. If agglutination occurs, it would be wise to find another donor. However, since the cross-match test is not always effective, a lack of agglutination does not guarantee that a transfusion reaction will not occur. Thus, care should be taken during transfusions, even if the cross-match test is negative.

8.3.1 *Neonatal isoerythrolysis*

Occasionally newborn foals that appear perfectly normal at birth, become weak and dull within 24 hours of birth, and develop acute anaemia, jaundice and haemoglobinuria. Their heart and respiratory rates become elevated, and they usually die within a few days. This disease is known as neonatal isoerythrolysis, NI, or haemolytic disease of the newborn. Judging from the above clinical signs, NI is associated with destruction of red blood cells. Why should this happen?

In the case of horses, the answer lies in foeto-maternal haemorrhage that occurs sometimes during pregnancy or birth, releasing red blood cells from the foetus into its dam's blood circulation. Consider the A blood group system, which happens to be the most important in relation to NI, and consider the A_a antigen within that system. Suppose that the foetus has inherited the A_a antigen from its sire, i.e. both sire and foetus are positive for A_a (written as A_a+). Suppose also that the dam lacks the A_a antigen (A_a-). Now, when the foetus' cells enter the dam, she will recognize antigen A_a as non-self, because she does not have that antigen. She will therefore produce anti-A_a antibodies in her serum. These anti-A_a antibodies will be transferred along with all other antibodies into the dam's colostrum, which the foal drinks (Fig. 8.7). The reason for the above symptoms of NI should now be evident. The anti-A_a antibodies are absorbed through the foal's gut and pass into its blood stream, where they rapidly destroy all cells with A_a antigen on their surface.

Fortunately, not all blood group systems give rise to this problem. Indeed, only the A and Q systems are regularly implicated in horses, with A being the more important of the two. Also, NI is rare in first foals, because the initial immune response is usually too slow to cause any trouble; the mare has not yet been sensitized. If challenged a second time, however, the mare will quickly mount an immune response and give rise to NI.

There are certain actions that can be taken to alleviate NI. Treatment of affected foals involves either transfusion with whole blood from a suitable donor, or transfusion of washed red blood cells from the dam. The main requirement is that the cells given to the affected foal must not carry any antigens that are carried by the sire but not the dam of the foal, as it is these antigens against which the dam has produced antibodies. Thus the sire is *not* a suitable donor.

In contrast, the dam should be a suitable donor, because none of the antibodies that the foal obtained from her colostrum will be directed against her own cells. But her serum contains the offending antibodies, produced by her against the foal's antigens inherited from its sire. Consequently, if the dam's cells are to be used for transfusion, they must first be washed with sterile saline, with the aim of removing all plasma and hence all offending antibody.

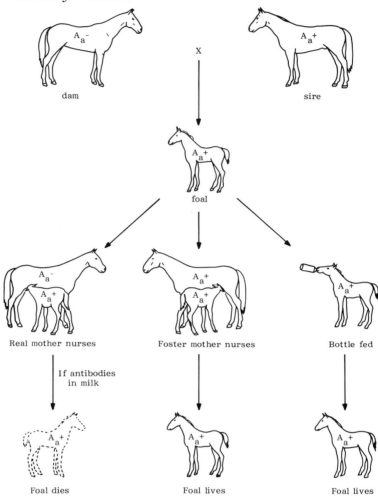

Fig. 8.7. The cause of neonatal isoerythrolysis in foals, and two methods of preventing it. A_a- indicates absence of A_a antigen, and A_a+ indicates presence of A_a antigen. (From *The horse* by Evans, J. W., Borton, A., Hintz, H. F., and Van Vleck, L. D. Copyright 1977 by W. H. Freeman and Company. All rights reserved.)

NI in foals can be prevented very simply by not allowing the foal access to its dam's colostrum for the first 24 to 36 hours, until protein molecules are no longer absorbed by the small intestine of the foal. Such action should be taken only if the foal is known to be at risk. This can be determined by screening for the presence of antibody to the foal's cells, in the pregnant mare's serum at four weeks, two weeks, and one week prior to the expected parturition date. The cells used in this screening test can be from the sire of the foal or from a panel of horses known to be positive for antigens involved in NI. However, because NI occurs with a frequency of less than one per cent, this would be a very wasteful procedure if done on all mares. A less wasteful procedure would be to blood-type all mares, and then to screen during the last four weeks of pregnancy only those that are A_a negative, i.e. that lack antigen A_a. The reason for concentrating solely on A_a is that antibodies to this antigen are thought to be the cause of more than 80 per cent of all cases of NI in horses. At the very minimum, all mares that are thought to have previously produced an NI foal should be screened. If this is not possible, then all subsequent foals of such mares should be given colostrum from a source other than their dam, as a precaution against NI (Fig. 8.7).

NI is best known in horses, but has been reported in dogs, cattle, and pigs as well. In the case of cattle and pigs, however, its cause appeared to be rather different to that described above. For example, cows with affected calves had antibodies directed against red-cell antigens in the calves, but so did bulls and steers from the same population. In addition, affected calves occurred with equal frequency in all parities, first parity included. Taken together, this evidence suggested that foeto-maternal haemorrhage was not implicated, but rather that there was an external stimulus to antibody production, received equally by both male and female cattle.

Investigations in the USA and in Australia, where the problem of NI in cattle was most acute, revealed that the external stimulus was the use of blood-based vaccines against babesiosis (tick fever) in Australia and against anaplasmosis in the USA. In the Australian case, for example, the vaccine was a mixture of whole blood from several animals infected with the protozoan *Babesia argentina*. Being a mixture of whole blood from several animals, the vaccine was certain to contain many red-cell antigens that were likely to stimulate production of antibodies by any animal receiving a dose of the vaccine. The anaplasmosis vaccine used in the USA had a similar origin and a similar effect. They both gave rise to calf losses as high as 20 per cent, and were thus potentially an important source of calf loss. A similar problem occurred for a time in pigs, associated with the use of whole-blood vaccines for swine fever (hog cholera). However, these problems have largely been overcome. In cattle vaccines, for example, most of the whole blood is now replaced with a cell-free diluent, and in pigs whole-blood vaccines are no longer used.

Finally, it must be noted that NI may occur in any parity (including the first) in any species of domestic animals, if the dam has been sensitized previously as a result of receiving a blood transfusion from a donor that was positive for clinically important antigens. Although this type of human-made NI is unlikely to be a common problem, it should be taken into account when transfusing blood into any female.

NI also occurs in humans, but in a different context. This is because in humans there is a two-way exchange of cells through the placenta, which means that the mother's antibodies can be transferred to the foetus prior to birth, resulting in severe anaemia and sometimes even death of the foetus. The blood group system most involved with NI in humans is the Rh system.

It is worth noting that the genetical effect of naturally-occurring NI in all species is selection against heterozygotes.

This is because the only situation in which the foetus has an antigen not carried by the dam is if the foetus inherited a different antigen from its sire, and must therefore be heterozygous.

8.4   The major histocompatibility complex (MHC)

It is common knowledge that organ and tissue transplants and skin grafts are usually rejected by the recipient. It is also well known that the chance of rejection is considerably reduced if the donor is a close relative of the recipient. For example, the chance of a successful transplant or graft is much greater with a full-sib donor than with an unrelated donor, and the chance of success is 100 per cent if the donor and recipient are identical twins. Obviously, there is a genetic basis to transplant rejection.

It is now known that rejection is determined by naturally occurring cell-surface antigens called *histocompatibility antigens*. The inheritance of these antigens is exactly the same as that for other blood group antigens: autosomal and codominant.

In the species where most work has been done in this field, namely mice, there are more than 30 loci whose antigens play a role in rejection. But there is one group of loci, the H-2 group, that plays a much more important role than the others. Because this group plays the most important role, it is known as the *major histocompatibility complex*, MHC. All higher vertebrates have an MHC. With the exception of mice, rats and chickens, nomenclature for the MHC is standard across species, as indicated in Table 8.3.

In all species studied to date, the MHC consists of a number of closely linked loci (Fig. 8.8), with the recombination frequency between loci usually being less than 1 per cent.

Some of the loci code for antigens that are detected serologically, i.e. by the use of antisera containing antibodies to those antigens. Such antigens are called SD (serologically defined) or *class I antigens*. Although they occur on virtually all nucleated cells in the body, typing for class I antigens is generally conducted on lymphocytes, which are

Table 8.3. Nomenclature for the MHC in various species

Species	Name	Symbols
Mouse	H-2*	H-2
Rat	RT1*	RT1
Dog	dog lymphocyte antigens	DLA
Pig	swine lymphocyte antigens	SLA
Goat	goat lymphocyte antigens	GLA
Sheep	ovine lymphocyte antigens	OLA
Cattle	bovine lymphocyte antigens	BoLA
Horse	equine lymphocyte antigens	ELA
Rhesus monkey	rhesus lymphocyte antigens	RhLA
Human	human lymphocyte antigens	HLA
Domestic chicken	B†	B

* The mouse and rat names were not changed to a standard form because H-2 and RT1 had become widely accepted prior to agreement being reached on a standard nomenclature. † The chicken MHC is the B blood group system, known for many years as a red-cell system.

readily available and easy to handle in the laboratory. Other loci within the MHC produce antigens that were detected initially following the observation that when lymphocytes from certain pairs of individuals are mixed to form a *mixed lymphocyte culture* (MLC), there is a mutual or one-way stimulation of cell division. It is now known that the cells from any particular individual are stimulated in an MLC only if the cells from the other individual have an LD (lymphocyte-defined) or *class II antigen* that is absent from the cells of the first individual. Class II antigens have a restricted distribution, occurring mainly on B-lymphocytes and macrophages. They can now also be detected serologically.

Because the MHC loci are so closely linked, the set of alleles (one per locus) that happen to be on a particular chromosome are usually inherited as a single unit known as a *haplotype*. Since chromosomes occur in pairs, each animal has two MHC haplotypes, one inherited from its dam, and the other inherited from its sire. If a cross-over occurs within the MHC region during meiosis, then two new haplotypes are formed. But because the MHC loci are so closely linked, crossing-over within the MHC is infrequent.

> *Thus, haplotypes are usually passed from parents to offspring in exactly the same form for many generations.*

In fact, for as long as crossing-over does not occur within the MHC region, each haplotype is inherited as if it were an allele at a single

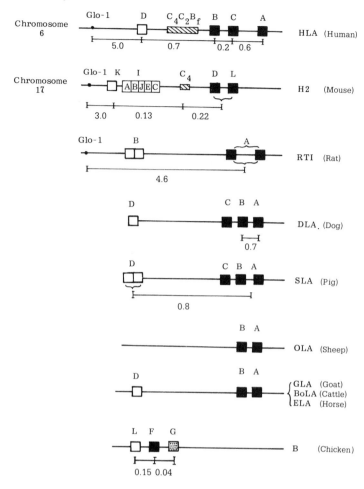

Fig. 8.8. A summary of loci within the MHC of various mammalian species. Solid squares indicate loci coding for class I antigens. Open squares represent loci coding for class II antigens. Shaded areas represent loci coding for various components of complement (see Section 8.4.2). Numbers under a particular system are percent recombination frequency; the drawings are not to scale. The I region in mice has five subsections: A, B, J, E, and C. The corresponding regions in rats and pigs each have two subsections. In humans, mice, and rats, the locus for the enzyme glyoxylase-1 (Glo-1) is closely linked to the MHC. The location of loci within HLA and H-2 is known with considerable certainty. These two systems have been positioned in the figure so as to emphasize the considerable similarity between them (following Bodmer 1981). Given this similarity between the MHC of humans and mice, it is most likely that the MHC of other mammals will be found to fit a similar pattern. Thus, although very little is known about the location of loci within the other systems, the loci that have been identified to date are drawn according to the same general pattern. Not included in this diagram are the MHC's of non-human primates, which are very similar to HLA (see Balner 1981, for a review). The

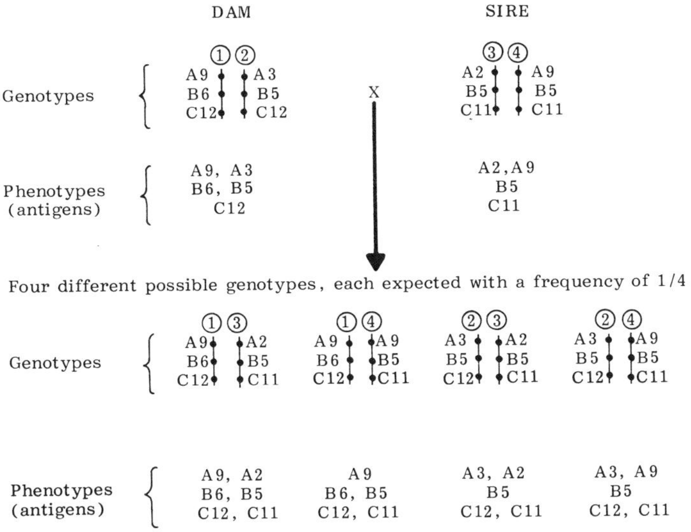

DAM   SIRE

Genotypes $\left\{\begin{array}{c} \text{①②} \\ \text{A9 } \updownarrow \text{ A3} \\ \text{B6} \updownarrow \text{ B5} \\ \text{C12} \updownarrow \text{ C12} \end{array}\right.$ X $\left.\begin{array}{c} \text{③④} \\ \text{A2 } \updownarrow \text{ A9} \\ \text{B5} \updownarrow \text{ B5} \\ \text{C11} \updownarrow \text{ C11} \end{array}\right\}$

Phenotypes $\left\{\begin{array}{c} \text{A9, A3} \\ \text{B6, B5} \\ \text{C12} \end{array}\right.$ $\left.\begin{array}{c} \text{A2, A9} \\ \text{B5} \\ \text{C11} \end{array}\right\}$
(antigens)

Four different possible genotypes, each expected with a frequency of 1/4

Genotypes $\left\{\begin{array}{cccc} \text{①③} & \text{①④} & \text{②③} & \text{②④} \\ \text{A9} \updownarrow \text{A2} & \text{A9} \updownarrow \text{A9} & \text{A3} \updownarrow \text{A2} & \text{A3} \updownarrow \text{A9} \\ \text{B6} \updownarrow \text{B5} & \text{B6} \updownarrow \text{B5} & \text{B5} \updownarrow \text{B5} & \text{B5} \updownarrow \text{B5} \\ \text{C12} \updownarrow \text{C11} & \text{C12} \updownarrow \text{C11} & \text{C12} \updownarrow \text{C11} & \text{C12} \updownarrow \text{C11} \end{array}\right.$

Phenotypes $\left\{\begin{array}{cccc} \text{A9, A2} & \text{A9} & \text{A3, A2} & \text{A3, A9} \\ \text{B6, B5} & \text{B6, B5} & \text{B5} & \text{B5} \\ \text{C12, C11} & \text{C12, C11} & \text{C12, C11} & \text{C12, C11} \end{array}\right.$
(antigens)

Fig. 8.9. Inheritance of MHC haplotypes, where there are three identifiable loci, A, B, and C. Haplotype numbers are shown in circles, and alleles and their corresponding antigens are identified by means of a standard locus/number code. Since each genotype in the offspring occurs with a frequency of $\frac{1}{4}$, there is a chance of $\frac{1}{4} \times \frac{1}{4} = \frac{1}{16}$ that any two full-sibs will each have a particular set of antigens. Also, for any particular individual, there is a chance of $\frac{1}{4}$ that the individual and any one of its full-sibs will have the same set of antigens.

locus. An illustration of the inheritance of haplotypes is given in Fig. 8.9.

To illustrate the importance of the MHC in graft rejection, transplantations have been conducted amongst different classes of individuals with respect to the MHC, and the survival of the grafts has been monitored. In one such experiment, dogs were typed for antigens at four DLA loci, A, B, C, and D, and then kidney transplants were performed in which the donor and recipient belonged to one of the following groups: (1) DLA-identical littermates; (2) DLA-identical nonlittermates; (3) littermates differing by one haplotype; (4) unrelated dogs differing at both haplotypes.

All recipients were given a standard daily immunosuppressive therapy of azathioprine (2 mg/kg body wt) and prednisolone (1 mg/kg

---

chicken MHC is much smaller than mammalian MHCs, but still has one class I locus (F) and one class II locus (L) producing antigens with a molecular structure very similar to that of mammalian class I and class II antigens respectively. The G locus codes for antigens detected only on red blood cells: it does not have a mammalian counterpart.

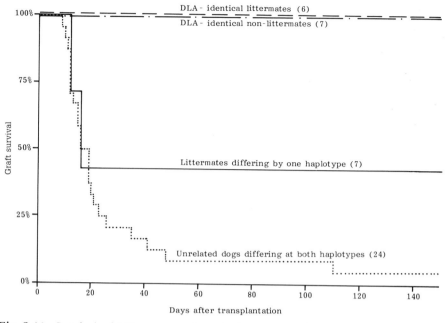

Fig. 8.10. Survival of kidney transplants in dogs, when the donor and the recipient are either DLA-identical littermates, DLA-identical non-littermates, littermates differing by one haplotype, or unrelated dogs differing at both haplotypes. Numbers in brackets are the numbers of recipients in each group, excluding four recipients that died from causes other than kidney rejection. (Drawn from data presented by Bijnen *et al.* 1980.)

body wt) for the first 100 days following transplantation. This therapy was then gradually withdrawn until day 150, after which no further therapy was given. The results up to day 150 are shown in Fig. 8.10. The importance of DLA antigens in relation to histocompatibility is clearly evident. When the donor and recipient were DLA-identical (i.e. they each had two DLA haplotypes in common), transplantation was 100 per cent successful. If they had only one DLA haplotype in common, transplantation was 43 per cent successful, and if they had no DLA haplotypes in common, the success rate was only 4 per cent. These results are particularly clear-cut, and indicate that in the presence of immunosuppressive therapy, the four known DLA loci are the sole determinants of histocompatibility in the dog. However, other results are not so clear-cut. For example, in a similar experiment in which no immunosuppressive therapy was given, the success rate for DLA-identical non-littermates was lower than that of DLA-identical littermates, and was about the same as that of littermates differing by one haplotype. This indicates that there are other loci in dogs, as yet undiscovered, that play a role in determining histocompatibility but whose effect can be overcome by immunosuppression. We can conclude that:

> The MHC certainly does have a major histocompatibility role, but there are other loci that also play some role.

In chickens, the MHC is the B blood group system, which was originally identified in terms of antigens on the surface of red blood cells, but which is now known to be detectable on lymphocytes as well. It accounts for approximately 75 per cent of variability in graft survival time (Marangu and Nordskog 1974), and so is certainly the major determinant of histocompatibility in that species. Although originally designated as a single locus with multiple alleles, it now appears to be a complex of several tightly-linked loci, just as in all other species studied to date.

8.4.1 Structure of MHC antigens

> The structure of major histocompatibility antigens appears to be very similar across a broad spectrum of species including chickens and all mammals.

As shown in Fig. 8.11, each class I antigen consists of two polypeptide chains. The chain that passes through the cell membrane is the product of an MHC locus; it carries the antigen specificities at its exposed end, on either side of the sugar (CHO) side-chain. It has a

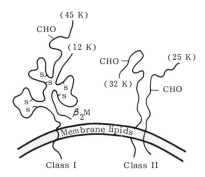

Fig. 8.11. The basic structure of class I and class II antigens, showing the relationship of the antigens to the cell membrane. CHO is a sugar side-chain, and S–S represents an intrachain di-sulphide bond. β_2M is β_2-microglobulin, which does not pass through the cell membrane. Numbers in brackets are molecular weights, e.g. 45K = 45 000 daltons. In each type of antigen, the two chains are held together by non-covalent bonds. (After Benacerraf 1981, *Science* **212**, 1229–38. Copyright 1981 by the AAAS.)

molecular weight of around 45 000 daltons. The smaller chain is β_2-microglobulin, with a molecular weight of approximately 12 000 daltons, and which is the product of a locus that is separate from the MHC region (in humans it is on a different chromosome). Interestingly, the amino acid sequence of the β_2-microglobulin shows considerable homology with a segment of the MHC-coded chain, and with the constant region of the heavy chain of certain immunoglobulins. This rather fascinating observation takes us almost the full circle; both antigen and antibody now appear to be at least partly composed of a very similar polypeptide chain, which is the product of separate pieces of DNA that probably arose by duplication from a single piece of ancestral DNA.

Apart from the above homologies, there are also considerable amino acid homologies between antigens from different class I loci within a species, and between species. For example, there is a 60 per cent to 80 per cent similarity in amino acid sequence between antigens from the D and K loci of the H-2 system. Analogous figures for the A and B loci in the HLA system range from 80 per cent to 90 per cent. Finally, there is a 45 per cent to 70 per cent homology between products of the class I loci in humans and mice. This evidence strongly indicates that all class I loci in all mammalian species have evolved from a single, ancestral piece of DNA (Jonker and Balner 1980).

Also shown in Fig. 8.11 is the general structure of a class II antigen. Once again there are two chains, but in this case they are more similar in size, and both pass through the cell membrane. Although it is only the smaller of the two chains that carries the antigenic determinant, available evidence indicates that both chains are the product of segments of DNA within the D 'locus'.

8.4.2 Complement

In certain circumstances, the formation of an antigen–antibody complex on the surface of a cell initiates an amplifying cascade of enzymic activations, the final result of which is fracture of the cell wall and cell death. The components of this cascade that undergo activation are serum proteins, which collectively are called *complement*. In mediating cell death, complement plays an important role in the body's defences.

While there is still much to be learned about the various roles of complement, the genetic basis of at least some of the components of complement is now well understood. In particular, loci coding for several of the components of complement are located within the MHC region, as shown in Fig. 8.8. The significance of this is not yet known, but it does indicate that the MHC is involved in the genetic control of the body's defences.

8.4.3 Disease associations

Marek's disease in chickens is a neoplastic disease in which the growth of tumour cells is caused by a DNA virus. Because of its economic importance, it has been the subject of considerable research effort. One of the main areas of research has involved attempts to alter the susceptibility of populations by means of artificial selection. For example, Cole (1968) reported a selection programme conducted in the Cornell randombred control strain of chickens. Before the selection programme commenced, mortality due to Marek's disease was 51 per cent. After only four generations of selection in one line for resistance and in another line for susceptibility, the lines showed respectively 7 per cent and 94 per cent mortality. In terms of the concepts discussed in Chapter 6, selection during only four generations produced large changes in liability to Marek's disease.

From the point of view of the present chapter, the interesting result of the above selection experiment, as shown in Table 8.4, was a marked difference between the resistant and susceptible lines in the frequency of two alleles at the B locus, which, as we saw above, is the MHC of chickens. Did this indicate an association between Marek's disease and the MHC? More specifically, did the very high frequency of the B^{21} allele in the resistant line indicate that this allele conferred, or was in some way actively associated with, resistance to Marek's disease? And was B^{19} conferring susceptibility?

Table 8.4. Results of selection for resistance and susceptibility to Marek's disease in the Cornell randombred control flock of chickens

Incidence of Marek's disease†		Frequency of B blood group alleles*	
Before selection	After four generations of selection	B^{21}	B^{19}
51% ➤ Resistant line	7%	1.00	0.00
➤ Susceptible line	94%	0.00	0.97

*After two more generations of selection and one further generation without selection.
† Deaths to eight weeks of age, plus lesions present at eight weeks of age. (Compiled from data presented by Briles *et al*. 1977 *Science* **195**, 193–5. Copyright 1977 by the AAAS.)

In order to investigate these questions, chickens were sampled from the *unselected* Cornell randombred control flock, and were typed for B antigens. Matings were then arranged amongst these unselected chickens to enable comparisons to be made between different B genotypes *within* families, so as to randomize all other possible genetic effects. The results are presented in Table 8.5. They show beyond doubt that there is a very strong association between the B^{21} allele and resistance to Marek's disease, and that B^{21} is almost completely dominant in its effect. They also show that B^{19} is not the only antigen associated with susceptibility. Thus, the increase in frequency of B^{19} in the susceptible selection line (Table 8.4) may have no special significance, but the increase of B^{21} in the resistant line (Table 8.4) certainly does have significance.

Table 8.5. Mortality resulting from a standard dose of Marek's disease virus, in chickens of different genotype obtained from the unselected Cornell randombred control flock ($-$ indicates an allele other than B^{21} or B^{19})

Genotype		Number of chickens			Genotype with respect to B^{21}	
		Exposed	Dead	% Dead		
B^{21}	B^{21}	65	0	0%	B^{21}	B^{21}
B^{21}	$-$	76	2	3% ⎱ 4%	B^{21}	$-$
B^{21}	B^{19}	60	4	7% ⎰		
B^{19}	$-$	50	19	38% ⎱	$-$	$-$
B^{19}	B^{19}	13	6	46% ⎱ 56%		
$-$	$-$	93	63	68% ⎰		

(Adapted from Table 9 of Stone *et al.* 1977, using the nomenclature of Briles *et al.* 1982, and Briles and Briles 1982.)

Among other disease-association studies that have been conducted are several in chickens (Guyre *et al.* 1982) and one in dogs (Bennett *et al.* 1975) which all show an association between MHC antigens and regression of experimentally-induced tumours. Many more studies are now under way in domestic animals, largely because of the many strong associations that have been detected between HLA antigens and various diseases in humans.

It should be noted that disease association studies in humans consist mainly of comparing the frequency of HLA antigens in healthy controls with the frequency in patients having a certain disease. Thus, antigens that are found to be associated with the disease are really associated with susceptibility to the disease. With animals, on the other hand, as illustrated in the Marek's disease example described above, it is possible to search for antigens associated with resistance to certain diseases.

Among the many human studies that have been conducted, the diseases that show strongest associations are those with a presumptive or suspected auto-immune origin.

Auto-immune diseases are those that are due to a breakdown in the body's ability to distinguish between self and non-self; the body produces antibodies against its own cells (called *autoantibodies*). Diseases known, presumed, or suspected of being in this category include juvenile onset diabetes, multiple sclerosis, and various forms of arthritis; all show quite strong associations with certain HLA antigens.

An auto-immune disease that has been studied in animals is auto-immune thyroiditis which has been reported in rats, dogs, chickens, and monkeys. It has been the subject of considerable research in chickens, including investigations of associations with B antigens. The results to date are not as clear-cut as with Marek's disease, but they do indicate that the B locus does influence the occurrence of auto-immune thyroiditis in the chicken (Wick *et al.* 1982).

Auto-immune thyroiditis in chickens is of interest from another point of view. Like Marek's disease, its incidence has been significantly altered by artificial selection, in one population from less than 1 per cent to over 90 per cent in females and from zero per cent to over 80 per cent in males, in a ten-year period (Cole *et al.* 1968). The way in which the disease incidence responded to selection is compatible with a multifactorial model for liability to auto-immune thyroiditis. And exactly the same is true for the way in which Marek's disease has responded to selection in a number of selection experiments, including the one described by Cole (1968). The existence of associations between B antigens and these two diseases indicates that the B locus is one of the many loci that contribute to the variation in liability to these diseases.

Why should MHC antigens be associated with resistance or susceptibility to disease? Certainly MHC antigens are concerned with recognition of self and non-self, and are thus in some way tied up with the body's ability to mount an immune response. But what is the actual mechanism? There is still much to be learned in answer to this question, but substantial evidence has been accumulated in relation to a phenomenon known as *MHC restriction*, whereby a T-lymphocyte can recognize a foreign antigen *only in the presence of an MHC antigen that it recognizes as 'self'*. In fact, the foreign antigen is recognized as being foreign only when it is 'presented' to the T-lymphocyte by an *antigen presenting cell*, which is often a macrophage. Because the macrophage has 'self' MHC antigens on its surface, the T-lymphocyte can directly compare the 'self' MHC antigens with the foreign antigen. The importance of MHC restriction arises from the fact that recognition of foreign antigens by T-lymphocytes is usually a prerequisite for a successful immune response. Since MHC antigens are likely to differ in the extent to which they enable a T-lymphocyte to recognize a particular antigen as being foreign, we would expect certain MHC antigens to be associated with a more effective immune response to that foreign antigen. If this foreign antigen were, for example, a disease-causing virus, then the end result would be that an individual's resistance or susceptibility to that viral disease would depend on the particular set of MHC antigens possessed by that individual.

The exact mechanism of MHC restriction differs for class I and class II antigens. For example, the only 'self' MHC antigens recognized by cytotoxic T-cells are class I MHC antigens. Since these antigens are characteristic of nearly all cells, it follows that cytotoxic T-cells, after being activated by recognition of foreign antigen in conjunction with 'self' class I antigen presented on macrophages, will kill any cell showing the relevant foreign antigen in conjunction with 'self' class I antigen, e.g. a virus-infected cell from any part of the body. In contrast, helper T-cells recognize only class II MHC antigens, which, as we saw above, have a very narrow distribution, being found mainly on macrophages and B-cells. This narrow distribution makes sense, because the main role of helper T-cells is to stimulate B-cells to produce antibody more effectively. Thus, having been activated by recognition of foreign antigen in conjunction with 'self' class II antigens on macrophages, helper T-cells have only one other type of cell with which they can interact, namely B-cells, which are the very cells that they are expected to stimulate. The narrow distribution of class II MHC antigens thus greatly increases the efficiency with which helper T-cells can find and stimulate B-cells to produce antibody.

We can conclude that:

> *Class I MHC antigens play a key role in cell-mediated immunity, while class II MHC antigens are important in humoral immunity.*

8.4.4 Mortality and production associations

Apart from being associated with specific diseases as described above, MHC antigens appear to be also associated with mortality in general. Once again, evidence comes from the chicken, which is the only species of domestic animal investigated to date in this regard. During the period 1965 to 1977, for example, B^1B^1 homozygotes in a non-inbred flock had the highest average mortality out of all B locus genotypes (approximately 33 per cent), compared with an average of around 12 per cent for all B^1- heterozygotes (Nordskog *et al.* 1977). And during the first eight years of this period, B^1B^1 homozygotes had the lowest egg production out of all B locus genotypes (Nordskog *et al.* 1973).

The mortality results verified earlier results obtained by Briles and

Allen (1961), who found a similar effect with B^1B^1 homozygotes in various inbred lines. The practical importance of these associations is obvious.

8.4.5 *Immune response genes*

We will now consider some further research done with chickens. This time, it involves the injection into chickens of a synthetic polypeptide called GAT, which is a linear random copolymer of glutamic acid, alanine, and tyrosine (Pevzner *et al.* 1978). Following immunization with GAT, chickens were classified as high or low responders, according to their ability to produce antibody to GAT.

With one exception, B^1B^1 homozygotes were low responders; indeed, some of them failed to produce any antibody at all. In contrast, B^1-heterozygotes and those not carrying B^1 were high responders. Ignoring the single exception for the time being, these results tie in with the mortality results outlined above. However, we now have to explain why the B^1 allele when homozygous confers inability to mount an immune response to GAT.

A clue to the answer to this question lies with the one exception mentioned above, which was a high responder within the B^1B^1 group. In order to investigate this individual, matings were arranged between him and three B^1B^1 low responders, producing a total of 27 non-responders and 19 high responders all of whom were B^1B^1. The ratio of $27:19$ is not significantly different from a $1:1$ ratio. The most likely explanation for this result is that the exceptional B^1B^1 male was heterozygous at a closely-linked locus that actually controls the immune response to GAT, and arose from a cross-over between the B locus and the locus controlling the immune response. This was subsequently confirmed by other studies.

The locus controlling the immune response to GAT is now called an immune response (Ir) gene, and its two alleles are given the symbols Ir^L for low responder, and Ir^H for high responder. The most likely interpretation of the above results, in terms of this Ir gene, is given in Fig. 8.12.

One final experiment must be mentioned. Pevzner *et al.* (1978) took some B^1B^1 chickens, divided them into high responders, Ir^H-, and low responders, Ir^LIr^L, and observed mortality in each group. During the first six months after hatching, 39 per cent of low responders died, compared with 19 per cent of high responders. Recalling that B^1B^1 chickens as a whole have around 33 per cent mortality, it is evident that the Ir alleles have a considerable effect on mortality. Presumably, inability to mount an immune response to GAT is to a certain extent indicative of inability to mount an immune response to other antigens, including pathogens. Further evidence in relation to

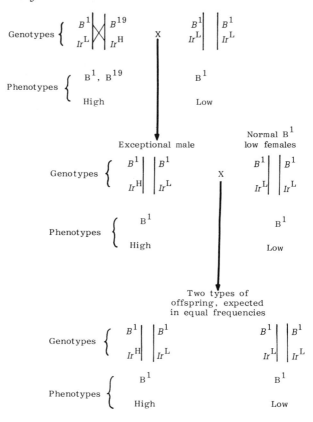

Fig. 8.12. The most likely explanation for the results obtained by Pevzner *et al.* (1978). The B locus and the Ir locus are closely linked, and until a cross-over occurred in the parent of the exceptional male, there were just two haplotypes, namely B^1Ir^L and $B^{19}Ir^H$. The cross-over (in the individual in the top left-hand corner of the diagram) produced a B^1Ir^H haplotype, which gave rise to the exceptional male. Being heterozygous at the Ir locus, and being mated to females homozygous for Ir^L which is recessive, the exceptional male produced approximately equal numbers of high and low responders, all of whom were homozygous for B^1.

the Ir gene for GAT was provided by Pevzner *et al.* (1981), who showed that average mortality of B^1B^1 low responders over two further years was more than double that of B^1B^1 high responders. They also showed that the incidence of Marek's disease in B^1B^1 low responders was more than double that in B^1B^1 high responders.

There is now evidence of immune response genes in mice, guinea pigs, rats, monkeys, chickens, and pigs. They have been detected in relation to a variety of synthetic and naturally occurring antigens. Most are located in the MHC region, but some are located elsewhere. The mode of action of immune response genes has been clarified greatly by evidence that many of them actually code for class II MHC antigens. Indeed, the main class II locus in the H-2 system of mice (see Fig. 8.8) is called the I region because a cluster of immune response genes is located there. Thus, in many cases, high-responder Ir genes are genes that code for class II antigens that are very effective at enabling helper T-cells to recognize foreign antigen, whereas low-responder Ir genes are genes that code for class II antigens that do not aid helper T-cells to recognize foreign antigen.

In Section 8.4.3, we saw that associations between MHC antigens and various diseases may involve class I antigens and/or class II antigens. And we have now seen that in many cases, immune response genes are in fact the genes that code for class II antigens. Of course, some MHC-disease associations may be due to genes within the MHC that have not yet been identified. Whatever the situation, it should be noted that if one gene within an MHC haplotype is associated with a particular disease, then all other identifiable genes in that haplotype will appear to be associated with the disease, because all genes in an MHC haplotype remain together for many generations, before a crossover separates them. This means that some MHC antigens may simply be convenient markers for another gene within the MHC that actually determines resistance or susceptibility.

The tendency for alleles at different loci to be associated with each other more often than would be expected if they were independent of each other is called *linkage disequilibrium*. Although it is an important concept in relation to disease associations, it is a difficult concept and we shall not pursue it further in this book. Readers seeking further information should consult the relevant references given at the end of this chapter.

8.5   Summary

Antibodies are proteins that occur in blood serum following stimulation of the immune response by an antigen. They have a specific structure, consisting of two light (L) chains and two heavy (H) chains, joined together by di-sulphide bonds. Light and heavy chains are produced from different clusters or families of genes located on different chromosomes. The light-chain gene cluster consists of several hundred V (variable) genes, four J (joining) genes, and one C (constant) gene, while the heavy-chain gene cluster contains eight D (diversity) genes in addition to approximately the same number of V and J genes as in the light-chain gene cluster. More than one million different antibodies can result from the joining of one each of the V and J genes in light chains, and one each of the V, D, and J genes in heavy chains, followed by combination of two copies of each chain to form an antibody molecule. These processes involve DNA deletion and RNA splicing. The different forms of antibody (membrane or secreted) and the different classes of antibody (IgM, IgD, IgG, IgE, or IgA) produced by a particular B-lymphocyte at different stages of its development differ only in the C region of their heavy chain, and result from a process called C_H switching, which involves both DNA deletion and RNA splicing.

Not only are antibodies encoded by genes, but the extent of their production is controlled by other genes, as indicated by successful artificial selection for increased titre of antibody in mice and chickens.

Red-cell antigens are structures, often glycoproteins, occurring naturally on the surface of red blood cells; they give rise to the red-cell blood group systems. Each blood group system corresponds to a separate locus, and each blood type within a blood group system corresponds to an antigen or to a combination of antigens, which in turn correspond to particular alleles at that locus.

Neonatal isoerythrolysis (NI, or haemolytic anaemia of the newborn) occurs naturally in horses and dogs. The destruction of red blood cells in this disease is due to concentration in the colostrum of antibodies to the offspring's red-cell antigen(s). These antibodies are produced in the dam after a flow of red blood cells from foetus to dam following haemorrhage between foetal and maternal tissue. Treatment of NI involves removal of the offspring from access to the dam's colostrum, and transfusion of whole blood from a suitable donor. If the foal's dam is used as a donor, then only saline-washed red blood cells should be transfused, because plasma from the dam contains the antibodies that are causing the disease in the foal. Prevention involves monitoring antibody levels late in pregnancy and if necessary using a substitute colostrum. Human-made NI has occurred in cattle and pigs after the use of blood-based vaccines.

The major histocompatibility complex (MHC) is a set of closely linked loci having several important roles in relation to the immune response. Some of the loci produce the larger of two polypeptide chains that together constitute class I antigens. Other loci produce both chains of class II antigens, and yet other loci produce at least some of the serum proteins of complement. There is a remarkable similarity between the MHC of all mammals and birds. The MHC is the most important factor determining the success of a tissue transplant, and more generally, it plays a vital role in enabling immune responses to be mounted against foreign antigens. Strong associations between MHC antigens and diseases have been reported in several species. These associations raise the possibility of identifying single genes for disease resistance in domestic animals.

8.6 Further reading

Bell, K. (1983). The blood groups of domestic mammals. In *Red blood cells of domestic mammals*. (Eds Agar, N. S. and Board, P. G.) pp. 133–64. Elsevier, Amsterdam. (A thorough review of red-cell blood groups in animals.)

Biozzi, G., Mouton, D., Heumann, A. M., and Bouthillier, Y. (1982). Genetic regulation of immunoresponsiveness in relation to resistance against infectious diseases. *Proceedings of the Second World Congress on Genetics Applied to Livestock Production* 5, 150–63. (A review of five classic experiments in mice.)

Festenstein, H. and Démant, P. (1978). *HLA and H-2: basic immunogenetics, biology and clinical relevance*. Edward Arnold, London. (An excellent introduction to the major histocompatibility complex, largely in terms of the human, HLA, and mouse, H-2, complexes.)

Gahne, B. (1980). Immunogenetics: a review and future prospects. *Livestock Production Science* 7, 1–12. (A review of immunogenetics, with particular reference to domestic animals.)

Götze, D. (Ed.) (1977). *The major histocompatibility system in man and animals*. Springer-Verlag, Berlin. (A review of the MHC in humans, subhuman primates, dogs, cattle, pigs, rabbits, Syrian hamsters, guinea pigs, rats, mice, chickens, and ectothermic vertebrates.)

Hala, K., Boyd, R., and Wick, G. (1981). Chicken major histocompatibility complex and disease. *Scandinavian Journal of Immunology* 14, 607–16. (A precise review of the chicken MHC and its role in disease resistance.)

Leder, P. (1982). The genetics of antibody diversity. *Scientific American* 246(5), 72–83. (An excellent review.)

Longenecker, B. M. and Mosmann, T. R. (1981). Structure and properties of the Major Histocompatibility Complex of the chicken. *Immunogenetics* 13, 1–24. (A good review of the chicken MHC.)

Newman, M. J. and Antczak, D. R. (1983). Histocompatibility polymorphisms of domestic animals. *Advances in Veterinary Science and Comparative Medicine* 27, 1–76. (A thorough review of histocompatibility systems.)

Nobel lectures. The 1980 Nobel Prize in Physiology or Medicine was awarded to Baruj Benacerraf, George Snell, and Jean Dausset for their pioneering work on the MHC of various species. Their Nobel lectures (*Science* 212, 1229–38, *Science* 213, 172–8, and *Science* 213, 1469–74, respectively) are valuable reviews.

Nordskog, A. W. (1983). Immunogenetics as an aid to selection for disease resistance in the fowl. *World's Poultry Science Journal* 39, 199–209. (An assessment of the possible role of the MHC in selection for disease resistance.)

Rapaport, F. T. and Bachvaroff, R. J. (1978). Experimental transplantation and histocompatibility systems in the canine species. *Advances in Veterinary Science and Comparative Medicine* 22, 195–219. (A detailed review of the role of various histocompatibility systems in transplantation in dogs.)

Roitt, I. M. (1984). *Essential immunology* (5th edn). Blackwell Scientific Publications, Oxford. (Rapidly becoming a classic textbook on basic immunology, and probably the best beginning for anyone who has trouble remembering the difference between an antigen and an antibody.)

Scott, A. M. and Jeffcott, L. B. (1978). Haemolytic disease of the newborn foal. *Veterinary Record* **103**, 71–4. (A review of naturally occurring neonatal isoerythrolysis in horses, providing details on prevention and treatment.)

Stormont, C. (1975). Neonatal isoerythrolysis in domestic animals: a comparative review. *Advances in Veterinary Science and Comparative Medicine* **19**, 23–45. (A review of neonatal isoerythrolysis, emphasizing the difference between naturally occurring and human-made cases, and paying particular attention to how the latter can be avoided.)

van Dam, R. H. (1981). Definition and biological significance of the Major Histocompatibility System (MHS) in man and animals. *Veterinary Immunology and Immunopathology* **2**, 517–39. (A general review of the MHC.)

9
Pharmacogenetics

9.1 Introduction

The action of drugs often depends upon the length of time before they are activated or inactivated in the patient. Since both these processes are determined at least in part by enzymes, and since enzymes are the products of genes, it follows that reaction to drugs is at least partly under genetical control. Pharmacogenetics is concerned with the nature and implications of this control.

The aim of this chapter is to provide an overview of pharmacogenetics. Because it is a relatively new science, our total knowledge in this area is rather limited. The best that can be done is to illustrate what is known by means of examples.

9.2 Strain differences in drug response

As mentioned in Section 7.2, differences between populations for any characteristic often indicate a genetic contribution to variation in that characteristic, providing of course that all possible environmental causes of differences between the two populations have been removed or taken into account. Thus, strain differences in drug response often indicate a genetic contribution to the variation in drug response.

Several strain differences in response to drugs have been reported in animals. In an experiment with two strains of chickens, for example, the frequency of malformations produced by administering sulphanilamide and carbachol at the end of the fourth day of incubation differed between strains (Landauer *et al.* 1976). Other examples include differences between strains of rabbits in ability to inactivate isoniazid, and differences between strains of mice in relation to chloroform toxicity.

More is known about inability to inactivate isoniazid in humans, where it has been shown that slow inactivation is due to homozygosity for an autosomal recessive allele that produces a relatively inactive form of hepatic acetyl-transferase. Moreover, slow inactivators for isoniazid are also slow inactivators for several other drugs that are chemically related to isoniazid. This is important because slow inactivators are more likely to develop side effects to these drugs than fast inactivators. The frequency of the slow-inactivation allele varies from population to population, but in both black and white Americans its frequency is around 70 per cent, which means that approximately one-half ($0.7 \times 0.7 = 0.49$) of Americans are slow inactivators. Given the similarity of drug metabolism in all mammals, it is likely that a slow-inactivation allele exists in various populations of domestic animals, and it could be such an allele that determines the strain differences reported in rabbits.

9.3   Genetics and anaesthesia

One of the best understood cases of the genetic basis of drug response in animals concerns reaction to halothane in pigs, which was discussed in Section 6.3. Reactors usually show signs of stiffening in the hind-quarters after two minutes exposure to halothane, associated with rapid hyperthermia which soon leads to death if halothane is not removed. Reaction to halothane is due to homozygosity for a recessive allele. The frequency of reactors varies considerably between breeds, from zero in many Large White populations to 100 per cent in one particular strain of the Pietrain breed. The same potentially fatal reaction to halothane occurs in humans, but the frequency is very low, being around 1 in 10 000.

An associated problem in humans concerns sensitivity to succinyl choline which is used during anaesthesia as a muscle relaxant. Normally, succinyl choline is short acting, requiring the use of an artificial respirator for only a short time following administration of the drug. But some individuals are particularly sensitive to succinyl choline, and require the artificial respirator for many hours. Sensitivity is due to an autosomal recessive allele that codes for an inactive form of pseudo-cholinesterase, the normal form of which is required to inactivate succinyl choline. The allele for low activity has a frequency of around 2 per cent in Caucasians. Again it is likely that a similar defect occurs in at least some domestic animals.

9.4   Warfarin resistance

Warfarin is an anticoagulant that is used widely in medicine and which is also used as a rodent poison. In humans, the standard dose of warfarin is between 5 and 10 mg per day. But some humans show no response to the drug when administered at this rate; they need up to 150 mg per day in order to achieve the same results as are normally obtained with the low dose. These individuals are showing warfarin resistance, which appears to be due to an autosomal dominant allele in humans.

Much more is known about warfarin resistance in rodents, because of the extensive use of the drug as a rodenticide. The mechanism of resistance can be best understood by firstly considering the normal clotting process which, as we saw in Section 3.4, involves a number of clotting factors in a complex series of biochemical reactions. Some of these factors, namely II, VII, IX, and X, require vitamin K for their activation. In fact, the carboxylation phase of activation of these four clotting factors occurs as a result of the oxidation of vitamin K. In the normal course of events, vitamin K oxide produced during activation of the above four clotting factors is transported to the liver, where it is reduced to its original form by a reductase enzyme. The reduced form of vitamin K is then used in subsequent activation of clotting factors.

The effect of warfarin is to inhibit reductase activity in the liver. The exact mechanism is not yet known, but since warfarin has a molecular structure similar to that of vitamin K, it is possible that the presence of warfarin 'tricks' a feed-back mechanism into thinking that there is sufficient vitamin K available, and thus leads to an inhibition of reductase activity. Whatever the exact mechanism, the presence of warfarin leads to a lack of reduced form of vitamin K, which in turn leads to insufficient activation of clotting factors II, VII, IX, and X, which in turn leads to a failure of the clotting process.

Warfarin resistance in rodents is due to an autosomal allele, R, which codes for a slightly different reductase which is less sensitive to the normal inhibitory effect of warfarin. However, this slightly different reductase is also less effective in the absence of warfarin, with the result that animals having the R allele require up to 20 times more vitamin K in their diet, to compensate for the decreased activity of the enzyme produced by this allele. If this increased demand can not be met, then homozygotes for the resistant allele suffer from the same bleeding disorders as those induced by warfarin in susceptible animals, namely disorders due to deficiencies in vitamin K-dependent blood clotting factors. Several studies of the relative fitness of each genotype resulting from the action of these factors have been conducted. Using the results from two such studies on rats in Great Britain, the following somewhat simplified account will be given in order to illustrate the main aspects of the warfarin story.

The two forms of reductase enzyme described above are the products of two alleles at an autosomal locus. Allele S codes for the susceptible form of reductase, and allele R codes for the resistant form. Individuals with the R allele have an increased requirement for vitamin K. Because this increased requirement sometimes cannot be met, and because RR homozygotes have a higher requirement than RS heterozygotes, the genotype RR has the lowest relative fitness in the absence of warfarin, with the RS genotype having a fitness somewhere in between the two homozygotes. For example, in a study by Partridge (1979), the relative fitnesses in the absence of warfarin were found to be:

SS	RS	RR
1.00	0.77	0.46

in which case the R allele would be either removed from the population, or maintained at a low frequency by a selection/mutation balance.

When warfarin is introduced, both RS heterozygotes and RR homozygotes are equally resistant. But RR homozygotes still have a lower fitness than RS heterozygotes because of their greater requirement for vitamin K. And SS homozygotes have a lower fitness than RS heterozygotes because they are susceptible to warfarin. In the presence of warfarin, therefore, there is heterozygote superiority for fitness. In one study, for example, the relative fitnesses in the presence of warfarin were:

SS	RS	RR
0.68	1.00	0.37

(Greaves *et al.* 1977). It is left to the reader to verify, using the concepts discussed in Chapter 5, that the expected consequence of this change in relative fitness is an increase in the frequency of the R allele to a new equilibrium of 0.34, which should correspond to a frequency of resistant rats of approximately 0.56 at birth in any generation.

This is a classic example of a stable polymorphism which should remain while ever warfarin continues to be used. It is an adequate explanation for the observation that in areas where warfarin was used extensively over many years, the frequency of resistant rats rose to an intermediate value and remained there (Fig. 9.1). The above account also provides an adequate explanation for the gradual decrease in the frequency of resistant rats observed in areas where warfarin use has been discouraged after previous extensive use.

The warfarin story is perhaps the most dramatic example of the practical implications of genetic variation in response to drugs. Although the use of warfarin as a rodenticide is only indirectly relevant to the use of it or any other drug in domestic animals or in humans, the warfarin story does emphasize that differences between individuals do exist in relation to drug response, and that these differences are sometimes under genetic control.

9.5   Multifactorial pharmacogenetics

In the examples described above, alleles at a single locus have a large effect on drug metabolism. With many drugs, however, there is no single locus with a noticeably large effect. Instead, metabolism is determined by the action of an unknown number of genes and by an unknown number of non-genetic effects. The end result of this

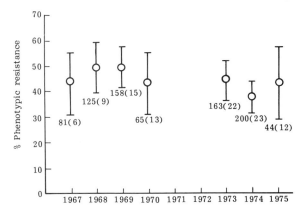

Fig. 9.1. The frequency of resistant rats over a period of nine years in an area where warfarin was regularly used as a rodenticide. Notice that the frequency has remained at an intermediate level. The fact that the observed frequency is a little lower than the figure of 56% predicted from the heterozygote superiority model, indicates that other factors such as immigration may also be acting. But the general conclusion is very clear: *selection favouring heterozygotes maintains a stable polymorphism*. The lines indicate two standard errors on either side of the mean, while the numbers indicate sample size and (in brackets) the number of infestations from which samples were drawn. (From Greaves, Redfern, Ayres, and Gill (1977) 'Warfarin resistance: a balanced polymorphism in the Norway rat.' *Genetical Research* **30**, 257–63. Cambridge University Press, Cambridge.)

combination of genetic and non-genetic factors is that when many individuals are each given a standard dose of certain drugs, a continuous distribution of response (or metabolism), as measured by drug concentration at some particular time, is seen (Fig. 9.2). At one extreme, the concentration may be so high as to produce a toxic effect. At the other extreme, the concentration may be so low that the drug is ineffective. This variability in response to a standard dose of certain drugs has obvious practical implications. Since there is usually no way

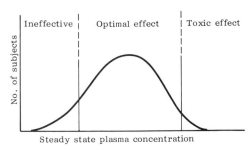

Fig. 9.2. The continuous distribution of drug concentration resulting from the action of many genes and many non-genetic factors. (From Vogel and Motulsky 1979.)

of knowing where a particular individual lies in the overall distribution, before administration of the drug, there is little that can be done in terms of predetermining the appropriate dose for a particular individual. It is important to remember, however, that there is variability between individuals in response to most drugs, and that this may lead occasionally to toxic or ineffective results from the administration of a standard dose of a drug.

9.6 Summary

The action of drugs often depends upon the length of time before they are activated or inactivated in the patient. Since both these processes are determined at least in part by enzymes, and since enzymes are the products of genes, it follows that reaction to drugs is at least partly under genetic control. Thus we expect to observe genetic differences in the ability of individuals to cope with or respond to a particular drug. The best example of this in domestic animals is malignant hyperthermia syndrome in pigs, which is a single-locus defect arising from reaction to halothane, with reaction being recessive to non-reaction. A more complex example is warfarin resistance in rodents, which is a single-locus condition with resistance being dominant to susceptibility, but also with homozygotes for the resistant allele having greatly reduced fitness because of a requirement for up to 20 times more vitamin K, which normal diets cannot fulfil. This gives rise to selection favouring heterozygotes, which accounts for the continued polymorphism observed at this locus.

With many drugs, reaction is due to the combined effect of an unknown number of genes and non-genetic effects. This gives rise to a continuous distribution of drug concentration, which may range from toxic to ineffective, following administration of a standard dose.

9.7 Further reading

Bishop, J. A. (1981). A neo-Darwinian approach to resistance: examples from mammals. In *Genetic consequences of man made change*. (Eds Bishop, J. A. and Cook, L. M.) pp. 37–51. Academic Press, London. (A review of the warfarin story.)

Evans, D. A. P. (1980). Genetic factors in adverse reactions to drugs and chemicals. In *Pseudo-allergic reactions. Involvement of drugs and chemicals. Vol. 1.* Dukor, P., Kallos, P., Schlumberger, H. D. and West, G. B. (Eds). pp. 1–27. S. Karger, Basel, Switzerland. (A review of pharmacogenetics.)

Motulsky, A. G. (1957). Drug reactions, enzymes and biochemical genetics. *Journal of the American Medical Association* 165, 835–7. (An early paper on genetic control of drug response, written before the term pharmacogenetics was coined.)

Propping, P. (1978). Pharmacogenetics. *Reviews of Physiology, Biochemistry and Pharmacology* 83, 124–73. (Another review.)

10

Hosts, parasites, and pathogens

'. . . can we doubt that individuals having any advantage, however slight, over others, would have the best chance of surviving and of procreating their kind?'   (Darwin 1859)

10.1 Introduction

Veterinarians and others involved with animal production would have a much easier life were it not for parasites and pathogens, which exert a profound influence on animal production throughout the world. It is therefore very important that we understand as much as possible about them. Even more importantly, we must understand the implications of attempts by humans to bring them under control.

The aim of this chapter is to provide such an understanding. We shall start by considering host-pathogen interactions, and shall then discuss the genetic basis of resistance in hosts and in parasites and pathogens. We shall then finish up with some illustrations of a variety of ways in which attempts are now being made to control parasites and pathogens, together with some comments on the likely chance of success.

10.2 Host–pathogen interactions

We shall consider three very different examples of interaction between hosts and parasites or pathogens in this section, with the aim of providing an indication of the variety of ways in which such interactions can occur.

10.2.1 Myxomatosis in rabbits

In 1859, a homesick Englishman who had grown a little tired of seeing nothing but kangaroos on his property near Geelong in Australia, imported some rabbits from England. He thought it would improve the hunting. Little did he know that this innocent importation would give rise to a plague of rabbits so serious that it would threaten the very existence of the pastoral industries in Australia.

Nothing done by humans had much effect on the rabbit population until the CSIRO released a virulent strain of the myxoma virus into the rabbit population in 1950. Having never before been exposed to the virus, the rabbits were very susceptible; their liability to myxomatosis was very high, with almost 100 per cent of infected rabbits dying. The combination of susceptible host and virulent pathogen assured the scheme of initial success; myxomatosis spread very rapidly and rabbits died by the thousands. But this very success of the scheme in its early days ensured that it could not be a success in the longer term. The CSIRO appreciated this because they realized that:

> *The death of large numbers of rabbits imposed very strong natural selection for resistance in the host, and also very strong natural selection for avirulence in the pathogen.*

If there had been absolutely no genetic variation for liability to myxomatosis in the rabbit, then this very strong natural selection would have had no effect. But there was genetic variation (the heritability of liability to myxomatosis is around 35 per cent), and those animals with genes for low liability had a greater chance of surviving to reproduce and hence to pass on their 'resistance' genes to their offspring. Similarly, had there been no genetic variation in the virus and had there been no mechanisms for new variation to be generated in the virus, then virulence would not have changed. But either there was genetic variation in the virus, or (as is more likely) new variation was created by mutation. And as soon as a mutant occurred that lessened the virulence, then the rabbits infected with the less virulent virus stayed alive longer and thus enabled more of the less virulent strain to be produced.

It is not surprising, therefore, that strains of rabbits emerged showing varying degrees of resistance to the virus, and strains of virus emerged with varying degrees of virulence (Fig. 10.1).

New, virulent strains of virus were developed in the laboratory and released, with considerable effectiveness in the period immediately following their release. But once again, the very success of the new virus imposed very strong natural selection for decreased virulence in the virus, and fewer and fewer rabbits died.

In a situation like this, there is obviously a very dynamic interaction between host and pathogen, with each side continually attempting to get into an optimum position. Humans can and do tamper with this interaction from time to time, but their ability to exert a lasting effect in the direction they desire is limited by biological realities of which we all should be very much aware.

That is not to say that the myxomatosis programme should never have been launched by the CSIRO; on the contrary, the programme has been very successful and continues to play a major role in rabbit control. It is simply to say that in order to avoid disappointment, the

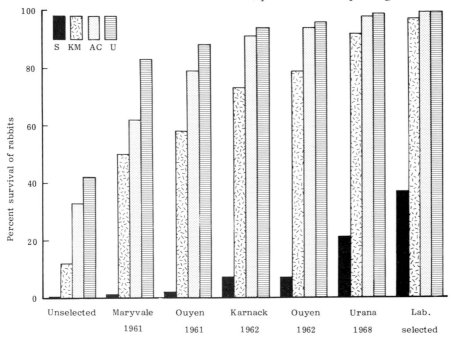

Fig. 10.1. Per cent survival in seven strains of rabbits when exposed separately to each of four strains of myxoma virus. The rabbit strains are listed in order of increasing resistance, from unselected (least resistant) to laboratory selected (most resistant). Field strains of rabbits are indicated by the site and date of capture. KM, AC, and U are field strains of virus, while S is the standard laboratory strain, similar to that first released in 1950. It can be seen that all field strains of rabbits are more resistant than the unselected strain, with resistance increasing over time. Artificial selection for resistance under controlled laboratory conditions (lab. selected strain) has been more effective than natural selection in the field (Sobey 1969; Sobey *et al.* 1970), and indicates likely future trends in the field strains of rabbits. Field strains of the virus are much less effective than the S strain on unselected rabbits, and are virtually useless against the more resistant rabbits. (Drawn from data presented by Rendel 1971.)

limitations imposed by biological realities must be appreciated before such a programme is commenced.

Myxomatosis is not confined to Australia; it was also introduced to France, from where it spread to Great Britain, just a few years after its release in Australia. Although there have been certain differences between these three countries in terms of the progression of the disease and its effects on rabbits, the overall results in all three countries have been much the same (Ross 1982).

There is still much to be learnt about the population genetic aspects of host–parasite interactions such as the one discussed above. Traditionally, population geneticists have been preoccupied with the effects of selection, migration, mutation, and drift in single populations, and have avoided the more complex problems presented by two populations interacting with each other. Recently, this previously neglected area has begun to receive some attention. The theoretical studies by Clarke (1976), Bremermann (1980), and Lewis (1981) are beyond the scope of this book, but should be read by people with a particular interest in this area. For a review of the way in which ecologists, population biologists, epidemiologists, and veterinarians are joining forces to study more effectively the interaction between hosts and parasites, see Anderson and May (1982).

The host–pathogen interaction exemplified by the rabbit and the myxoma virus is a relatively typical one. We shall now discuss an example of host–pathogen interaction that is just as important but which is somewhat less typical. Indeed, in this particular example, there is debate as to whether a pathogen is involved at all.

10.2.2 Scrapie in sheep

Scrapie is a disease of adult sheep that involves a progressive degeneration of the central nervous system. It is characterized by incoordination, a bewildered expression, grinding of teeth, compulsive rubbing (scraping) against fixed objects, and death a few weeks after the appearance of clinical signs.

Although the cause of the disease is still subject to debate, there is overwhelming evidence for the existence of an infectious agent (Kimberlin 1979*a*). For example, when normal sheep from flocks with no history of the disease are penned with affected sheep, some of the normal sheep become affected. And the same result is obtained if a sample of brain from an affected sheep is injected into normal sheep from flocks with no history of the disease.

Experiments have shown that the infective agent is about the size of a medium-sized virus, and that it increases in quantity when passaged through animals. But it does not behave like a conventional virus, in that it is remarkably resistant to various physical and chemical treatments, and it does not stimulate an immune response. Furthermore, it has defied all attempts to determine its chemical structure (Kimberlin 1982).

Whatever it is, one of its most notable features is its extremely long incubation period. For example, in one experiment, injections of brain tissue from affected sheep were given to 294 six-month-old lambs (Hoare *et al.* 1977). No clinical signs were observed in any of the lambs until 93 days after inoculation. Between then and 365 days, 123 lambs became affected, and by 495 days this had risen to 156 lambs affected. After day 495, no more cases were observed until the 870th day. Between then and the end of the experiment at 1310 days, a further 30 lambs became infected, with the last infection being recorded 1049 days (two years and ten months) after inoculation. Thus there were two distinct groups of animals: those with a 'short' incubation period (93 to 495 days), and those with a 'long' incubation period (870 to 1049 days).

Because of the length of incubation, scrapie is called a *slow virus disease*. The extraordinarily long incubation periods lead to the fascinating speculation that in certain circumstances the incubation period of the pathogen may be longer than the life span of the host (Dickinson *et al.* 1975). As we shall see below, this raises interesting problems for quarantine authorities.

Despite the evidence for a transmissible agent of one form or another, Parry (1979) argues strongly and presents evidence in support of his thesis that the disease is determined by an autosomal recessive gene. He admits that an agent does exist, but believes that it is a product of the recessive gene and has nothing to do with transmission of the disease.

Whatever the correct answer is, the evidence on scrapie presents an unusual picture of interaction between a host and an agent that may or may not be causative.

From a practical point of view, scrapie has become increasingly important in recent years because of the desire of scrapie-free countries like Australia and New Zealand to import sheep from countries in which scrapie does occur. The quarantine problems posed by the long incubation time for slow virus diseases are almost insurmountable; how long, for example, should the quarantine period be for a disease whose incubation time may be longer than the life span of the sheep?

This and other questions will continue to puzzle veterinarians for some time into the future.

10.2.3 *African trypanosomiasis in various animal species*

African trypanosomiasis is the most important of all animal diseases in Africa. It kills many thousands of cattle each year, and decreases production in hundreds of thousands of other cattle suffering chronic infection. In addition to cattle, the disease occurs also in sheep, goats, camels, horses, pigs, various wildlife species, and in humans, where it is known as sleeping sickness. It is a protozoan disease caused by various species of trypanosomes that are transmitted primarily via the tsetse fly.

The most interesting feature of trypanosome infection is that it is characterized by regular fluctuations in the numbers of trypanosomes in the infected host, ranging from virtually zero up to approximately 1500 per ml of blood. The reason for these regular fluctuations is a phenomenon called *antigenic variation*, which is the occurrence of a sequence of different antigenic variants all arising from the single population of pathogens that originally entered the host.

Trypanosomes are encapsulated by a glycoprotein coat, and the antigenic determinant of this coat is the protein portion, which consists of a single chain of approximately 600 amino acids. Being a single chain of amino acids, the antigenic determinant of a trypanosome is obviously the product of a gene in that trypanosome.

When a population of trypanosomes enters a host, all the members of that population exhibit a basic antigen, which is one of the frequently occurring antigen types. The host mounts a strong humoral immune response, producing antibodies directed against this basic antigen. Consequently, most of the trypanosomes that originally entered the host are destroyed. But by this time, some trypanosomes have 'switched off' the gene for the basic antigen, and have 'turned on' a gene for another antigen, which differs by many amino acids from the basic antigen. Trypanosomes carrying this second antigen multiply rapidly until the host's immune response system produces antibodies to this second antigen, by which stage some trypanosomes are now producing yet another different antigen that the host has not previously encountered. And so the regular fluctuations in numbers of trypanosomes continue for many cycles, with the pathogen regularly 'changing its spots' in order to keep one step ahead of the host's immune response system.

The trypanosome genome contains more than 100 genes that each code for a different type of antigen. Although there is some debate as to the exact mechanisms for antigenic variation, there is general agreement that rearrangement of antigen genes is involved. One popular suggestion, for which there is considerable evidence arising from research using recombinant DNA techniques, is that the 'switching on' of a particular gene involves the duplication of that gene, and the subsequent *transposition* (movement and insertion) of that duplicate gene into another region of the genome called the *expression site*. Once incorporated into the expression site, the duplicate gene is transcribed and translated into antigen. When the time comes for the next antigen to be produced, the corresponding gene is duplicated and the duplicate copy is inserted into the expression site, in place of the previous duplicated gene, which by this time has been removed and lost.

The above account is a very brief summary of the duplication/transposition model for antigenic variation. For more information on this and other models, and for details of the way in which recombinant DNA techniques are being used to increase our understanding of antigenic variation, readers should consult the review by Donelson and Turner (1985).

There are three other aspects of antigenic variation that should be mentioned. The first is that it occurs even in the absence of antibody, which indicates that trypanosomes are somehow 'pre-programmed' to produce a sequence of antigens even in the absence of external stimuli. The second point is that if trypanosomes are removed from a host

Table 10.1. A list of hosts for which there is evidence of genetic variation for resistance to the parasite shown alongside the host (after Wakelin 1978, who provides a reference for each example given)

Parasite	Host	Parasite	Host
PROTOZOA		*Ascaridia galli*	Fowl
Trypanosoma congolense	Cattle	*Cooperia oncophora*	Sheep
T. vivax	Cattle	*Haemonchus contortus*	Sheep
T. brucei	Mouse	*Nematodirus battus*	Sheep
T. cruzi	Mouse	*Oesophagostomum radiatum*	Sheep
Plasmodium falciparum	Man	*Ostertagia circumcincta*	Sheep
P. vivax	Man	*Trichostrongylus axei*	Sheep
P. cynomolgi	Man	*Trichostrongylus colubriformis*	Sheep
P. bastianelli	Man	*Haemonchus contortus*	Goat
P. berghei	Mouse	*Trichostrongylus* spp	Goat
P. vinckei	Rat	*Ascaris suum*	Pig
Leishmania donovani	Mouse	*Strongyloides ransomi*	Pig
Eimeria tenella	Fowl	*Aspiculuris tetraptera*	Mouse
E. brunetti	Fowl	*Brugia pahangi*	Rat
E. maxima	Fowl	*Litomosoides carinii*	Mouse
E. mivati	Fowl	*Nematospiroides dubius*	Mouse
E. necatrix	Fowl	*Nippostrongylus brasiliensis*	Rat
		Nippostrongylus brasiliensis	Mouse
DIGENEA		*Strongyloides ratti*	Rat
Schistosoma mansoni	Mouse	*Trichostrongylus colubriformis*	Guinea pig
Schistosoma mansoni	Rat	*Trichuris muris*	Mouse
Schistosoma mansoni	Hamster	*Trichinella spiralis*	Mouse
Schistosoma mansoni	Rhesus monkey		
		ARTHROPODA	
Fasciola hepatica	Rat	Arachnida	
		Ixodid ticks	Cattle
CESTODA		*Boophilus microplus*	Cattle
Echinococcus multilocularis	Mouse	*Ornithonyssus sylviarum*	Fowl
Hymenolepis citelli	Deer mouse	Insecta	
Hymenolepis nana	Mouse	*Melophagus ovinus*	Sheep
Taenia taeniaeformis	Mouse	*Polyplax serrata*	Mouse
Taenia taeniaeformis	Rat		
NEMATODA			
Enterobius vermicularis	Man		
Necator americanus	Man		

and are recycled through a tsetse fly, they revert to producing one of the basic antigens when they enter a new host, irrespective of which antigen was last being produced in the previous host. This also indicates a degree of 'pre-programming' of the trypanosome. Finally, it is obvious that:

> *The practical implication of antigenic variation is that it is very difficult to produce a vaccine that will be effective against the large number of different antigens that each host is likely to encounter.*

The phenomenon of antigenic variation is an important example of host–pathogen interaction. And it is all the more important because it occurs not only in trypanosomes but in other protozoa including members of the genus *Plasmodium*, which causes malaria, and the genus *Babesia*, which causes babesiosus or tick fever, an important disease of cattle that was mentioned in Section 8.3.1.

10.3 Resistance in hosts

There is no shortage of well-documented examples of genetic variation in hosts for resistance to pathogens or parasites. Indeed, the list of examples shown in Table 10.1, which covers parasites only, is so extensive that we must conclude that this type of genetic variation is present in most, if not all, populations of hosts.

In the following discussion, we shall briefly review several of the more important examples of genetic variation for resistance in hosts.

10.3.1 Resistance to Marek's disease in chickens

We have already seen in Chapter 8 that there is sufficient genetic variation in liability to Marek's disease to enable resistance and susceptibility to be substantially altered by selection. And in Section 8.4.3 it was shown that much of the response to selection for resistance in one particular population was due to a strong association between the B^{21} allele and resistance (Stone *et al.* 1977); this allele alone accounted for a considerable proportion of the total genetic variation in liability. It is now known that this same allele accounts for a significant proportion of genetic variation in resistance in other populations.

10.3.2 Resistance to neonatal scours in pigs

As we saw in Section 5.6.7, a major cause of neonatal scours in pigs are strains of *E. coli* having a cell-surface antigen called K88. But not all piglets are susceptible to K88 *E. coli*. In particular, only those piglets with a K88 receptor on the walls of their intestines are susceptible; those that lack the receptor are resistant. As reported in Section 5.6.7, the presence or absence of the K88 receptor is determined by two alleles at an autosomal locus, with the allele for presence of the receptor being completely dominant to that for lack of the receptor. Thus, resistance to neonatal *E. coli* scours in pigs is under the control of two alleles at a single locus, with resistance being recessive to susceptibility (Gibbons *et al.* 1977).

10.3.3 Resistance to worms in sheep

As reviewed by Le Jambre (1978), there is ample evidence of variation between breeds, and of genetic variation within breeds in relation to worm resistance in sheep. Genetic differences within breeds are particularly important, as they open the way for selection programmes aimed at increasing resistance. After infection with a standard dose of *Haemonchus contortus*, for example, the heritability in Australian Merino sheep was found to be 0.35 for maximum worm eggs per g of faeces, and 0.16 for haematocrit deflection, which is a measure of red blood cell loss due to worm burden. With heritability values of this order, selection for resistance should be reasonably effective. And there is good evidence to suggest that the situation is similar for other species of worms as well. For example, Windon and Dineen (1981) have had considerable success in selecting Merinos for response to *Trichostrongylus colubriformis*, following vaccination with irradiated larvae.

The evidence in relation to the role of any single locus contributing to resistance is equivocal. In view of the importance of the host's blood in the diet of so many worms, several studies have searched for associations between resistance and various types of haemoglobin. Although certain studies have indicated some useful associations, these associations have not always been confirmed by other studies.

10.3.4 Resistance to blowflies in sheep

It is very difficult to measure resistance to blowflies directly, because seasonal factors such as rainfall play such an important role in determining the incidence of strike. For this reason, research workers have tended instead to measure the incidence of fleece rot, which has a very high correlation with body strike by blowflies, and which is relatively easy to measure under controlled conditions that include artificial wetting of the fleece. Under standard conditions such as these, there is ample evidence of genetic variation in resistance to blowflies in the Australian Merino, as indicated by the following information taken from Atkins and McGuirk (1979).

There is, for example, variation between three of the major strains of Australian Merino (fine, medium, and strong) and also variation between flocks within strains, in the latter case ranging from 13 per cent up to 39 per cent incidence of fleece rot under standard conditions. And there is also variation within flocks, with the heritability of liability to fleece rot being approximately 15 per cent. That this is ample variation for selection is indicated by the result of an experiment in which lines of sheep were selected for resistance or susceptibility to fleece rot; the incidence has fallen to 15 per cent in the resistant line and risen to 43 per cent in the susceptible line.

10.3.5 Resistance to ticks in cattle

It is well known that *Bos indicus* cattle are much more resistant to ticks than *Bos taurus* cattle raised under the same circumstances. Even crosses between *Bos indicus* and *Bos taurus* cattle, e.g. Brahman ×

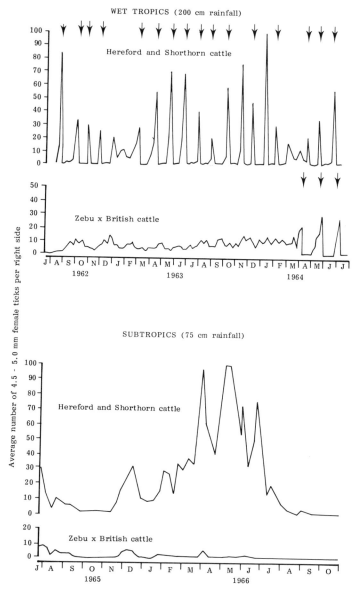

Fig. 10.2. The levels of resistance to ticks in British cattle (Herefords and Short-horns) and in *Bos indicus* (Zebu) × British crosses, assessed at monthly intervals in the wet tropics (above) and in the subtropics (below). Resistance has been assessed using the standard technique of taking several counts of the number of mature female ticks on one side of each animal. Arrows indicate the timing of acaracide treatment. (From Wharton 1976. Reproduced by permission of the Food and Agricultural Organisation of the United Nations.)

British and Africander × British, carry only 40 per cent of the ticks carried by British breeds (Seifert 1971). The extent of the difference in resistance between British cattle and *Bos indicus* × British crosses is clearly indicated in Fig. 10.2.

In addition to variation between breeds, there is genetic variation for tick resistance within both *Bos indicus* and *Bos taurus* breeds, and, within second and subsequent crosses between the two types of cattle, heritability is very high (around 80 per cent). This indicates that selection within a population formed from the crosses would be very effective. This is exactly what has been done with the development of new breeds of cattle suitable for tropical and semitropical conditions, as described, for example, by Hayman (1974), Turner (1975) and Utech and Wharton (1982).

10.4 Resistance in parasites and pathogens

We have seen above that genetic variation in resistance of hosts is almost universal, and that in some cases this variation has been utilized in the form of selection of hosts for increased resistance. We shall now turn to parasites and pathogens, in order to discover the extent of genetic variation in resistance in them, and to see the extent to which selection has exploited that variation.

10.4.1 Resistance to insecticides in sheep blowflies

The Australian sheep blowfly, *Lucilia cuprina*, is a major cause of loss of income in the Australian wool industry, through damage done by larvae to live sheep, and the consequent loss of wool production and death of sheep. Attempts by humans to control the blowfly have until recently concentrated mainly on the use of insecticides for spraying (jetting) the live sheep.

The use of insecticides soon gave rise to a predictable pattern of response in the blowfly. In 1955, for example, an organo-chlorine called dieldrin was released for use, and was extremely effective at first, giving protection for around eight weeks or longer after jetting. Within three years, however, its effectiveness had dramatically fallen, and sheep were being struck by blowflies only a week or two after jetting. It was said that the blowfly had become resistant to the chemical. In 1958 a new insecticide was released, this time an organo-phosphate called diazinon. It, too, was very effective at first, but, as with dieldrin, it soon lost its effectiveness. Other chemicals were released in an attempt to keep one step ahead of the blowfly. But in each case it was only a matter of time before the blowfly became resistant (Fig. 10.3).

It is now known that the blowflies became resistant to each chemical in turn because of very strong natural selection favouring alleles for resistance at one or more loci. These resistance alleles usually operate

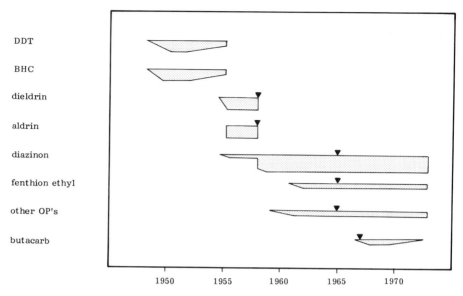

Fig. 10.3. The use of insecticides against the sheep blowfly (*Lucilia cuprina*) in Australia from 1947 to 1973. Band width indicates the extent of use, and ▼ indicates the diagnosis of resistance. The first four insecticides are all organo-chlorines. Dieldrin and aldrin are both cyclo-dienes, having a similar structure, a similar (but unknown) mode of action, and showing resistance at the same time. The next three are organo-phosphates, with a mode of action different to those above but similar within the group; they inhibit acetyl cholinesterase and consequently interfere with the transfer of signals across nerve synapses. All organo-phosphates showed resistance at the same time. Butacarb is a carbamate, with a mode of action very similar to that of organo-phosphates, even though it is a different type of chemical. Not surprisingly, resistance to butacarb developed very quickly. (From Shanahan and Roxburgh 1974. Reprinted from *PANS* **20**(2): 190–202 with permission of the controller of Her Britannic Majesty's Stationery Office. Crown copyright © 1974.)

by coding for an enzyme that is able to detoxify the insecticide, or by coding for a variant of the enzyme against which the insecticide acts, such that the variant is still able to function in the presence of the insecticide. Prior to the introduction of a particular chemical, the relevant resistance alleles are usually maintained at a low frequency in the overall population of flies by a balance between mutation and selection, which arises because resistance alleles, in the absence of the chemical, are usually disadvantageous. The introduction of a new insecticide produces an immediate change in the relative fitnesses of the three genotypes RR, RS, and SS (where R is the resistance allele and S is the susceptible allele), with the result that RR now has the highest relative fitness and SS the lowest. The inevitable consequence of this change in relative fitness is an increase in the frequency of the R allele.

And since the genotype *RR* now has the highest fitness, the theoretical end result of this selection is that the population will become homozygous for the *R* allele, at which stage all flies will be resistant.

Switching from one insecticide to another within the same class of chemical compound will produce very little change if the mode of action of the chemical, and hence nature of resistance, is the same in each case. On the other hand, replacing one type of compound with another may cause equally strong selection in favour of a resistance allele at a different locus, in which case the above story is repeated. If the first resistance allele also still exerts some effect on the new insecticide (if there is *cross-resistance*), then it will remain at a high frequency even after the original insecticide is withdrawn from use. This is what sometimes happens in practice. If there is no cross-resistance, then the withdrawal of the original insecticide will cause the relative fitnesses of the three genotypes to revert to their original values, and the resistance allele will decrease in frequency to its original very low value. This last alternative is what happened with the allele for resistance to dieldrin. When dieldrin was withdrawn in 1958, the frequency of the resistance allele was at least 0.42 and was possibly much higher (exact estimates were not made at that time). Ten years later, its frequency had fallen to around 0.02, where it has remained ever since. The fact that resistance to organo-phosphates was developing during the period in which dieldrin resistance was declining indicates that there is no cross-resistance between dieldrin and organo-phosphates.

By using many of the classical gene-mapping techniques developed in *Drosophila melanogaster*, researchers have determined the location of resistance loci in the blowfly genome. There are, for example, two different loci with alleles that confer resistance to organo-phosphates, one on chromosome 4 and another on chromosome 6. The resistance alleles are incompletely dominant to the susceptible alleles, and have been recorded at a very high frequency (0.98 to 1.00) in several natural populations of blowflies, which fits in well with the theoretical expectations outlined above. Dieldrin resistance is determined by an allele at a third locus, located on chromosome 5.

The above account indicates that the reason for blowflies becoming resistant to insecticides is basically very simple; it represents an excellent illustration of the principles of population genetics.

Not surprisingly, the real situation is a little more complex than outlined above. There are, for example, many modifying genes of relatively small effect that play a role in insecticide resistance, and the longer a particular insecticide is used, the more will these modifying genes be subject to selection, with the result that the major resistance gene becomes less and less likely to decrease in frequency when the insecticide is eventually withdrawn. Despite complications such as this, the above picture highlights the important general principles involved

Table 10.2. The occurrence of resistance to acaracides in various populations of ticks. The names within countries in the last column refer to different strains, each of which has a distinguishable level of resistance which is a reflection of differences in the frequency of various resistance alleles at a small number of loci (from Wharton 1976)

Species	Arsenic 1900–	DDT 1946–	Toxaphene-BHC-dieldrin group 1947–	Organophosphorus-carbamate group 1955–
Amblyomma hebraeum	Rhodesia 1975 South Africa 1975		South Africa 1975	South Africa 1975
Amblyomma variegatum	Zambia 1975		Tanzania 1975 Zambia 1975	
Boophilus decoloratus	Kenya 1953 Malawi 1969 Rhodesia 1963 South Africa 1938 Zambia 1975	South Africa 1954	Kenya 1964 Malawi 1975 Rhodesia 1969 South Africa 1948 Uganda 1970 Zambia 1975	Rhodesia 1975 South Africa 1966 'Berlin' = 'Ridgelands'
Boophilus microplus	Argentina 1936 Australia 1936 Brazil 1948 Colombia 1948 Jamaica 1948 South Africa 1976	Argentina 1953 Australia 1953 Brazil 1953 Venezuela 1966	Argentina 1953 Australia 1953 Brazil 1953 Colombia 1966 Ecuador 1966 Guadeloupe 1961 India 1964 Madagascar 1963 Malaysia 1967 Martinique 1961 Trinidad 1969 Venezuela 1966	Argentina 'Goya' 1970 'Las Guerisas' 1964 Australia 'Ridgelands' 1963 'Biarra' 1966 'Mackay' 1967 'Mt. Alford' 1970 'Gracemere' 1970 'Bajool' 1972 'Tully' 1972 'Ingham' 1973

Species			
Haemaphysalis leachii		South Africa 1976	Brazil 'Ridgelands' 1963, 'Minas Gerais' 1969; Colombia 'Ridgelands' 1967, 'Guaimarito' 1970; Venezuela 'Ridgelands' 1967, 'Guaimarito' 1970
Hyalomma marginatum		Spain 1967	
Hyalomma sp. (*rufipes, truncatum*)	South Africa 1975	South Africa 1975	
Ixodes rubicundus	South Africa 1976		
Rhipicephalus appendiculatus	Malawi 1975, Rhodesia 1975, South Africa 1975, Zambia 1975	Kenya 1968, Rhodesia 1966, South Africa 1964, Tanzania 1971, Uganda 1968, Zambia 1975	South Africa 1975
Rhipicephalus evertsi	Kenya 1975, Rhodesia 1975, South Africa 1975, Zambia 1975	Kenya 1964, Rhodesia 1966, South Africa 1959, Tanzania 1970, Uganda 1970, Zambia 1975	South Africa 1975

Reproduced by permission of the Food and Agriculture Organisation of the United Nations.

in the development of insecticide resistance in many species of insect, and of resistance to chemicals in general. The extent of insecticide resistance is indicated in Fig. 10.4, which illustrates the number of insect species throughout the world that have become resistant to various insecticides by the basic processes outlined above, and the time scale of the development of resistance. The extent of the problem is very evident.

We can conclude that:

> *The inevitable consequence of the widespread use of an insecticide is that insects will become resistant to it; and the more effective the insecticide is initially, the faster will resistance develop.*

10.4.2 Resistance to acaracides in ticks

The same basic principles as outlined above apply to ticks in cattle. Thus, as a result of the widespread use of a range of acaracides to combat ticks, there are now many populations of ticks resistant to various acaracides, as shown in Table 10.2.

10.4.3 Resistance to anthelmintics in worms

Once again the same principles apply. For example, just three years after the release in the USA of drenches based on the compound thiabendazole (TBZ), resistant strains of *Haemonchus contortus* were detected. With the continual widespread use of this and other benzimidazoles (BZ) in several major sheep-producing regions of the world, resistance to various forms of BZ has now become a major problem. Worse still, worms that are resistant to BZ sometimes show resistance to other compounds used as drenches, including organo-phosphates, salicylanilides, and substituted nitrophenols.

In an attempt to investigate the genetic basis of this resistance, various selection experiments have been conducted in the laboratory. For example, Le Jambre *et al.* (1976) conducted a selection programme in a strain of *Haemonchus contortus* that showed a relatively low level of resistance, with 20 per cent surviving the standard LD_{95} of TBZ (the LD_{95} is the dose that kills 95 per cent of worms in a standard, unselected strain). Selection was conducted by placing a standard number of worm larvae into previously worm-free sheep, drenching the sheep with the standard LD_{95} of TBZ, and collecting eggs from surviving worms. Larvae from these eggs were raised in the laboratory until ready to be placed into another group of worm-free sheep. After six generations, the selection line was split into two groups. The first group was selected as before until the tenth generation, while the second group was selected for a further four generations using TBZ together with another related compound called morantel, MOR.

After ten generations, the TBZ line was 100 per cent resistant to TBZ, with all worms surviving. The TBZ + MOR line showed 91 per cent resistance to TBZ and 25 per cent resistance to MOR. However, both lines showed complete lack of resistance to levamisole, LEV, which is a compound closely related to MOR: more than 95 per cent of worms from both lines were killed by a low dose of LEV.

Similar selection experiments on other species of worms and using different combinations of anthelmintics have produced the same type of result, indicating the presence of genetic variation for resistance in worms to all compounds. In some cases, resistance is due to a particular allele with a relatively large effect at a single locus, while in other cases, resistance appears to be *polygenic*, being determined by alleles of smaller effect at a number of loci. The main difference in results from one selection experiment to the next concerns resistance to other compounds; some lines become resistant to other compounds while others do not. Although it is difficult to see a logical pattern clearly emerging, it does seem as if there is a greater tendency for the development of resistance to more than one drench if the drenches belong to the same group of compounds than if they belong to different groups. In order to distinguish between these two situations, the term *cross-resistance* is often confined to the latter (resistance to two or more

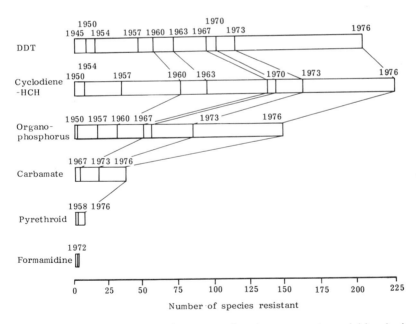

Fig. 10.4. The time-scale of development of resistance to insecticides in insect species. HCH (which is also known as BHC) shows strong cross-resistance with the cyclo-dienes. (From Wood 1981, after Brown 1977.)

chemicals from different groups). The term *side-resistance* is then used for the existence of resistance to two or more chemicals from the same group.

10.4.4 Resistance to antibiotics in bacteria

In the years following the introduction and widespread use of penicillin, strain after strain of resistant bacteria was isolated. Other antibiotics were introduced, but resistant strains soon emerged. Worse still, many such strains were resistant to more than one antibiotic, as shown, for example, in Fig. 10.5. The emergence of resistance was too rapid and widespread to be explained solely by the conventional process described for other species of pathogens and parasites in earlier sections, namely the increase in frequency of resistance genes resulting from selection during the *vertical* transmission of those genes from parent to off-spring. In fact, the rapid and widespread emergence of antibiotic resistance in bacteria has been mainly due to the ability of bacteria to

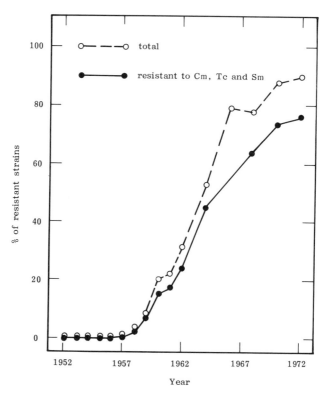

Fig. 10.5. Increase in the incidence of strains of *Shigella* showing multiple resistance to antibiotics in Japan. Cm = chloramphenical, Tc = tetracycline, Sm = streptomycin. (From Glass 1982.)

transfer genes *horizontally* (between contemporary individuals, within generations) as well as vertically.

There are three methods by which bacteria can transfer genes horizontally. They are *transformation* (the release of naked DNA from one cell, and its uptake by another cell), *transduction* (the transfer of DNA from cell to cell by means of a bacteriophage), and *conjugation* (the transfer of DNA from one cell to another, following the joining together, i.e. mating, of the two cells). Antibiotic resistance genes can be transferred horizontally by all three methods, with conjugation probably being the most important. As shown in Fig. 10.6, the transfer of antibiotic resistance by conjugation involves the transfer from one cell to another of a duplicate copy of a plasmid, which, as we saw in Section 2.4, is a circular segment of DNA that functions independently of the bacterial chromosome. The importance of conjugation lies in the fact that many important genes for antibiotic resistance are located on plasmids. Those plasmids that carry one or more resistance genes are called *R factors*.

Horizontal transfer of resistance can be demonstrated, for example, in chickens (Smith 1970); if donor *E. coli* with an R factor are added to drinking water, followed by another strain of *E. coli* or of *Salmonella typhimurium*, both of which lack the R factor, then before long, the R factor can be found in cells of the second *E. coli* strain or of *Salmonella typhimurium* isolated in the chicken. Moreover, if this experiment is done using chickens receiving antibiotics regularly in their feed, then the frequency of recipient cells carrying the R factor is increased, if the donor strain contains the R factor for that particular antibiotic. The reason for this is that, by destroying cells that lack the R factor, the antibiotic exerts a strong selective force in favour of recipient cells that have received an R factor.

The importance of horizontal transfer of resistance arises because of the large number of bacteria present in the external and internal environment of animals and of humans. The use of antibiotics creates an environment favouring the survival of strains that possess R factors. These strains then act as a reservoir for the transfer of R factors to any other strains (including pathogenic strains) that happen to be present from time to time. The remarkable extent to which plasmids can be transferred horizontally between bacterial species is shown in Fig. 10.7.

Many different types of R factors have now been identified, including those that have a single gene for resistance to almost any of the antibiotics currently in use. But there are also plasmids known to have genes for resistance to more than one antibiotic. At first it was hard to see how such plasmids could arise, because the chance of more than one relevant mutation occurring in the same plasmid is very small. The answer to this puzzle came with the discovery in the mid-1970s of *transposons*, which are a type of transposable genetic element (TGE).

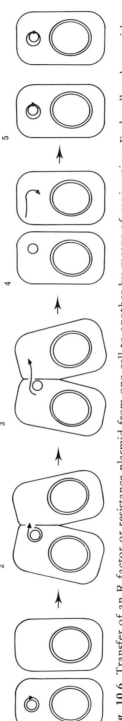

Fig. 10.6. Transfer of an R factor or resistance plasmid from one cell to another by means of conjugation. Each cell ends up with a copy of the original plasmid and in turn may become a potential donor cell. (From 'The molecule of infectious drug resistance' by Clowes, R. C. Copyright © 1973 by Scientific American, Inc. All rights reserved.)

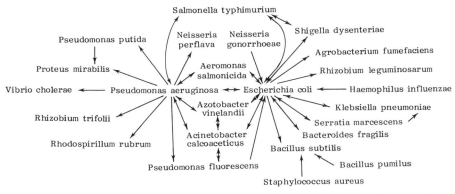

Fig. 10.7. The many different routes by which plasmids may be transferred horizontally from one bacterial species to another. (From Young and Mayer, 1979, *Reviews of Infectious Diseases* **1**, 55–62. The University of Chicago Press, Chicago. © 1979 by the University of Chicago. All rights reserved.)

As described in Section 1.3.4, TGEs are segments of DNA that can move from one site to another within and between chromosomes. In bacteria, transposons can move within and between plasmids and the main bacterial chromosome. They can also insert themselves into and remove themselves from bacteriophage DNA.

Structurally, transposons resemble insertion sequences, which were described in Section 1.3.4. Thus, a transposon is a segment of DNA flanked by inverted repeats. The main difference between the two types of TGE is that transposons: (a) are much larger, and (b) carry a gene or genes producing a phenotypically identifiable effect.

In relation to size, insertion sequences generally range between 500 and 1400 base pairs, whereas transposons range in size from around 2000 to 50 000 base pairs. The most common type of genes carried in transposons are those conferring resistance to antibiotics.

A list of transposons is given in Table 10.3, and the typical structure of one of them, Tn3, is shown in Fig. 10.8. Although most transposons carry only one gene for antibiotic resistance, it is now easy to see how plasmids with multiple resistance genes have arisen so quickly; they simply accumulate the appropriate transposons. In fact:

> *A plasmid carrying genes for resistance to several different antibiotics is really just a plasmid containing the appropriate transposons.*

An example of the way in which such transposons occur within plasmids with multiple resistance is given in Fig. 10.9. Notice that most of the transposons and hence most of the resistance genes have been incorporated into one large transposon called a *resistance-determinant segment*. Obviously, the transposition of such segments enables the simultaneous transfer of multiple resistance.

Table 10.3. A list of transposons, showing their size, the genes they carry, and the nature and size of their terminal repeated sequences. Note that the terminal repeated sequence of Tn9 and Tn1681 is actually an insertion sequence called IS1. In the case of Tn1681 it is inverted at opposite ends, but in Tn9 it is repeated in the same direction (a *direct repeat*) at each end (after Foster and Kleckner 1980, who give references for each transposon)

Transposon	Size	Markers	Repeated sequences
Tn1	4.8 kb	Ap	Short, inverted (40 bp)
Tn2	4.8 kb	Ap	Short, inverted (40 bp)
Tn3	4.8 kb	Ap	Short, inverted (40 bp)
Tn4	20.5 kb	ApSmSu	Short, inverted (40 bp)
Tn402	7.5 kb	Tp	None detected
Tn5	5.2 kb	Km	Long, inverted (1.4 kb)
Tn501	7.8 kb	Hg	Short, inverted
Tn551	5.2 kb	Em	Short, inverted (70–100 bp)
Tn6	4.1 kb	Km	Short
Tn7	12.8 kb	TpSm	Short, inverted
Tn9	2.5 kb	Cm	Long (IS1, 700 bp)
Tn903	3.1 kb	Km	Long (1050 bp)
Tn917	5.1 kb	Em	Short, inverted (280 bp)
Tn951	16.6 kb	Lac	Short, inverted (100 bp)
Tn10	9.3 kb	Tc	Long, inverted (1.4 kb)
Tn1681	2.9 kb	Ent	Long, inverted (IS1, inverted)

The abbreviations for the antibiotic resistance and other markers carried by transposons are: Ap (ampicillin resistance), Sm (streptomycin), Su (sulphonamides), Tp (trimethoprim), Km (kanamycin), Hg (mercuric salts), Em (erythromycin), Sp (spectinomycin), Cm (chloramphenicol), Tc (tetracycline), Lac (lactose fermentation), Ent (enterotoxin).

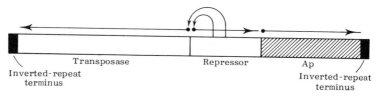

Fig. 10.8. The typical structure of a transposon, in this case transposon Tn3. Within the inverted repeats, there is a gene for the enzyme transposase which is required for transposition (insertion or removal of the transposon), a gene Ap, which confers resistance to ampicillin and related β-lactam antibiotics, by coding for the enzyme β-lactamase, and a gene for a repressor protein that controls transcription of the transposase gene and of the repressor gene itself. Arrows indicate the direction of transcription. (From 'Transposable genetic elements' by Cohen, S. N. and Shapiro, J. A. Copyright © 1980 by Scientific American Inc. All rights reserved.)

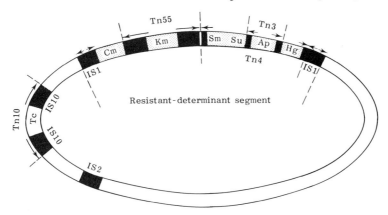

Fig. 10.9. A hypothetical plasmid carrying genes for resistance to chloramphenicol (Cm), kanamycin (Km), streptomycin (Sm), sulphonamide (Su), ampicillin (Ap), mercury (Hg), and tetracycline (Tc). The resistance-determinant segment, in which all but one of the resistance genes occur, is a large, composite transposon, flanked by the insertion sequence IS1. Transposon Tn3 is located within Tn4. Arrows indicate the direction of nucleotide sequences within inversion sequences. (From 'Transposable genetic elements' by Cohen, S. N. and Shapiro, J. A. Copyright © 1980 by Scientific American Inc. All rights reserved.)

There is abundant evidence of the role played by transposition in the creation of plasmids with multiple resistance, and of the role played by conjugation in enabling the rapid spread of such plasmids, and hence of multiple resistance, both within and between species of bacteria. There is also abundant evidence to indicate that the extensive use of antibiotics throughout the world, confers a strong selective advantage for bacteria carrying multiple resistance, thus favouring the rapid spread of plasmids carrying multiple resistance. For a review of this evidence, see Saunders (1981).

10.4.5 *Other mechanisms of resistance*

The examples of resistance discussed in the previous sections are the most important ones in relation to animals. There are, however, other mechanisms of resistance that should be mentioned briefly, as they may become relevant in the future.

The first is *gene duplication* or *gene amplification*, to which we have already referred in relation to similarities between antigen and antibody in Section 8.4.1. In the present context, there is some evidence associating gene duplication with increasing insecticide resistance in the peach–potato aphid, *Myzus persicae* (Devonshire and Sawacki 1979; Bunting and Van Emden 1980). Resistance to insecticides in this insect apparently arises from the increased production of a particular esterase, E4. Evidence is presented in the above papers to indicate that this increased production of enzyme is achieved by duplicating the relevant gene. For a general review of the role of gene duplication in drug resistance, see Schimke (1980).

Chromosomal rearrangements are another potential source of resistance. Translocations, for example, may bring together two genes for resistance that were previously on different chromosomes. And their proximity could generate an interaction that enhances their previously independent roles. Once again, the evidence for this mechanism of resistance comes from *Myzus persicae* (Blackman *et al.* 1978).

10.5 Control of parasites and pathogens

Having illustrated the fact that parasites and pathogens show widespread genetic variation, we shall now consider some of the ways in which humans are attempting to eradicate them, or at least to bring them under control.

10.5.1 Screw-worm flies

There are two types of screw-worm fly, the Old World fly (*Chrysomya bezziana*) and the New World fly (*Cochliomyia hominivorax*). They are both parasites of warm-blooded animals, with the damage being caused by the predilection of the larval stage for open wounds. The Old World screw-worm larvae, for example, cause up to 30 per cent loss of calves in New Guinea due to navel strike, and, at the height of their activity, the New World larvae resulted in millions of dollars worth of lost cattle production in the southern states of the USA. Because of its economic importance, considerable effort has been directed at eradicating the screw-worm fly from the USA, with a large degree of success.

The biggest contribution to this success has been a method of biological control known as the *Sterile Insect Release Method*, SIRM. This method involves rearing large numbers of flies of both sexes in the laboratory, and exposing these flies to doses of radiation sufficiently high to render them sterile. The sterile flies are then dropped from aeroplanes at rates of between 1600 and 4000 per square mile per week. The principle of the method is simply that if sufficient sterile flies can be released, then most (and preferably all) wild-type flies will mate with a sterile fly rather than with a wild-type (fertile) fly. Obviously this will occur only if very large numbers of sterile flies are released.

A judicious combination of SIRM and dipping of cattle resulted in the complete eradication of the screw-worm fly from south-eastern USA by 1959, only two years after the programme had begun. The south-western eradication programme was not started until 1962 and was a much more ambitious task. By 1971 a fly-free barrier 200 to 300 miles wide had been established along the USA–Mexico border

(Fig. 10.10), and all areas north of this barrier were free of the screw-worm fly. The maintenance of the barrier was a monumental task, involving the release of 200 million sterile flies per week! Despite this effort, the screw-worm fly reappeared north of the barrier in 1972, and it appeared as if the barrier was breaking down.

Fig. 10.10. The status of the USA–Mexico screw-worm eradication programme in 1977. By the mid 1980s, the site marked 'Future barrier', which is the Isthmus of Tehuantepec, had been reached. (From Williams *et al.* 1977. Reproduced by permission of the Food and Agricultural Organization of the United Nations.)

It now appears that the cause of this outbreak lies in the existence of several *non-interbreeding* populations of wild-type screw-worms, differing substantially in karyotype, phenotype, and behaviour (Richardson *et al.* 1982). Obviously, if the laboratory-reared strain is derived from a different population to the one that presents the major problem in the field at a particular time, then the release of sterile flies will have very little effect on the wild-type population, because there will be relatively few sterile X wild-type matings, irrespective of how many sterile flies are released. It took several years for the nature of this problem to be worked out to the stage where it could be acted upon, with the result that there were several major outbreaks of screw-worm fly in the USA between 1972 and 1976, and again in 1978. Since then, however, the USA has been effectively free of the screw-worm fly, because the barrier has been maintained with laboratory populations corresponding more closely to the various wild-type populations known to exist in Mexico.

Despite the success of the programme, which has shown a benefit : cost ratio of greater than 113 : 1, there is a continual search for ways to increase its effectiveness and/or to decrease its cost. A substantial improvement was achieved in the mid 1980s, by moving the fly-free barrier gradually southwards to the Isthmus of Tehuantepec, which, as shown in Fig. 10.10, is much narrower, thereby enabling the barrier to be maintained with less effort, and at a lower cost.

The use of SIRM in helping to eradicate the screw-worm fly is commonly regarded as one of the most successful of all human attempts at biological control. The method has also been used, with considerable success, against the Mexican fruit fly and the Codling moth. Until 1981, similar success had been achieved against the Mediterranean fruit fly (Medfly) in California. What happened in California in 1981 is a good example of how successful biological control programmes can fail if insufficient care is taken. There is still argument over exactly what happened, but it appears that 50 000 supposedly sterile flies released as part of the control programme were actually fertile. It is not clear to what extent this accidental release was the cause of the massive outbreak of Medfly that occurred in 1981; but the outbreak was so severe that aerial spraying of insecticide was eventually carried out amidst considerable political controversy.

Although it has come to be regarded as a form of genetic control, the SIRM itself involves no genetics. It does, however, have genetical implications in relation to problems caused by adaptation of laboratory flies, and also in relation to issues such as the possibility of selection for mating preference. In addition, the sterilization procedure can be viewed as giving rise to dominant lethal alleles.

10.5.2 Blowflies

The use of the conventional SIRM for the control of the sheep blowfly was considered by the relevant Australian authorities, but the benefit : cost ratio expected from rearing and sterilizing sufficient flies for a national control programme was found to be unacceptably low. Consequently, various alternatives have been considered. Some of these are modifications of the conventional SIRM, while others involve different techniques.

One of the disadvantages of the conventional SIRM is that in many insect species, females require a much higher radiation dose than males, in order to be made sterile. In screw-worms, for example, females require 6 krads whereas males require only 2.5 krads (La Chance 1979). Thus in order to ensure that all flies are sterile, males must be exposed to much higher doses than necessary, resulting in them being less competitive in the field. If males could be separated automatically from females prior to irradiation, then they could be given the lower radiation dose and consequently would be more competitive when released. In fact, the aim of the SIRM could be achieved by releasing sterile males only, because if sufficient sterile males are released, then most matings will occur between sterile males and fertile (wild-type) females. Bearing this in mind, Australian research workers set out to investigate various ways of automatically separating the sexes in the sheep blowfly.

One technique for the automatic sexing of larvae involves generating by irradiation a reciprocal translocation between a Y chromosome and a section of chromosome 5 carrying a dieldrin resistance gene, *R*. As shown in Fig. 10.11, strains carrying this translocation in conjunction with a normal chromosome 5 with the susceptible gene, *S*, consist of only resistant males and susceptible females. Exposure of all larvae to dieldrin will therefore kill all the females, leaving the males to be released as part of an SIRM programme. The fact that the released males are all carrying a dieldrin resistance gene does not matter, because they have all been sterilized by radiation prior to release, and thus cannot pass on the resistance gene, or any other gene, to subsequent generations in the field.

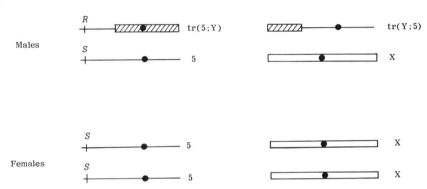

Fig. 10.11. The difference between males and females in a strain of blowflies carrying a reciprocal translocation between the Y chromosome and an autosome 5 carrying an allele *R* for resistance to dieldrin. Since *R* is dominant to susceptibility, *S*, all males are resistant while all females are susceptible.

The conventional SIRM involves the maintenance of laboratory-reared flies right up to the adult stage (including irradiation at the late pupal stage), and the release of adults from aeroplanes or other vehicles. If a genetic source of sterility could be used as an alternative to radiation, then the provision of labour and facilities for irradiation would be unnecessary, and flies could be released at the larval stage, which is much simpler and which enables many more individuals to be carried in each release vehicle. Foster and Maddern (1978) have estimated that the use of genetic sterility could save up to half of the rearing and handling costs, and 80 to 90 per cent of the release costs of a conventional SIRM programme. Another substantial advantage of larval release is that released flies are likely to be much better acclimatized, having pupated naturally in the soil, and having then emerged naturally into the area where they are expected to perform.

Among the possible sources of genetic sterility in the blowfly, the use of *compound chromosomes*, CCs, shows some promise. The principle of compound chromosomes is illustrated in Fig. 10.12. It can be seen that they are equivalent to a pair of chromosomes with left arm joined to left arm, LL, and right arm joined to right arm, RR. As shown in

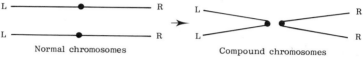

Normal chromosomes   Compound chromosomes

Fig. 10.12. Diagrammatic representation of the construction of compound chromosomes. L = left arm and R = right arm. The rearrangement is produced by exposing normal adults to irradiation just prior to mating. Some of their gametes then contain compound chromosomes. For further details, see Foster *et al.* (1972). (From Foster 1980.)

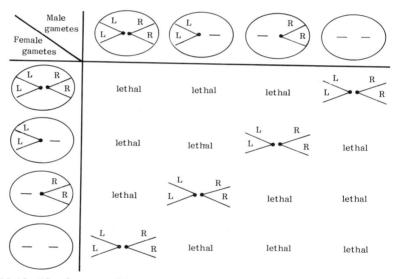

Fig. 10.13. The four possible gametes that can be produced by individuals with compound chromosomes, and the 16 possible zygotes that can result from the mating of two such individuals. Twelve of the zygote types are lethal, because of duplications or deficiencies of whole chromosomes. The remaining four zygotes are viable, having the same karyotype as their parents. If segregation of all four gametic types is random, then it follows that 25 per cent of zygotes will be viable. If centromeres always disjoin (producing only the second and third gamete types), then zygote viability will be 50 per cent. (From Foster *et al.* (1972) *Science* **176**, 875–80. Copyright 1972 by the AAAS.)

Fig. 10.13, there are four possible outcomes of meiosis in CC individuals, and, depending on the type of segregation, zygote viability can vary from 25 per cent to 50 per cent when two CC individuals are mated. In contrast, no viable offspring result from mating a CC fly to a wild-type fly, because all zygotes are unbalanced. This is the key to the use of compound chromosomes in genetic control.

The simplest use involves the release of one sex of a CC strain in sufficient numbers so that most matings are CC × wild-type. The release of just one sex, males, can be achieved by the use of a CC strain that also carries the Y-autosome female-killing translocation described above, or a similar translocation involving an autosomal recessive gene for white eyes located on chromosome 3. In the latter case, a laboratory strain is constructed that is homozygous for the white-eye gene and in which, therefore, all flies have white eyes. Into this strain is introduced a reciprocal translocation between a Y chromosome and a segment of chromosome 3 carrying the wild-type, normal, dominant eye colour allele. Thus, all males have normal eye colour but all females are white eyed. When released into the field, females are unable to see properly and so fail to survive, leaving males to do the mating with wild-type females.

Although the above use of compound chromosomes is likely to be effective, there is another way in which they can be used, to even greater effect. In order to understand this method, we have to recall from above that no viable offspring result from a CC × wild-type mating, and that the zygote viability from CC × CC matings varies between 25 per cent and 50 per cent. Noting that wild-type flies are completely fertile when they mate amongst themselves, and assuming there is random segregation in the CC strain, we then have the following relative fitnesses:

CC	CW	WW
0.25	0.00	1.00

where CC and WW represent offspring of the CC and wild-type strain respectively, and CW represents offspring of CC × wild-type matings.

It is evident that we have complete selection against heterozygotes, which, as we saw in Appendix A5.4.6, gives rise to an unstable equilibrium. If sufficient CC individuals can be introduced into a population for their frequency to be greater than the unstable equilibrium frequency, then the frequency of CC individuals will increase to 100 per cent, and a laboratory population will have completely replaced the wild-type population, as illustrated in Fig. 10.14. The relative number of flies that need to be released to ensure the desired result can be determined from the equilibrium frequency, which is given by $\hat{q} = 1/(1 + w)$, as derived in Appendix A5.4.6. In the example quoted

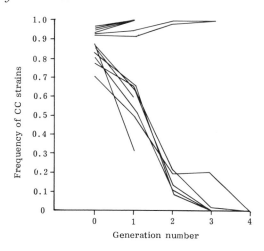

Fig. 10.14. Response to complete selection against heterozygotes in 13 populations of *Drosophila melanogaster*. The populations were formed by combining flies from a CC strain and a normal strain in various proportions such that the initial frequency of the CC strain varied from 0.71 to 0.96 at generation 0. In this particular study, the fitness of the CC strain was only 10 per cent of the normal strain, which means that the unstable equilibrium frequency was 0.9. It can be seen that in all populations in which the frequency of the CC strain was greater than 0.9, the CC strain completely displaced the normal strain, and vice versa. Notice that complete displacement is achieved very quickly, in four generations or less. (After Foster *et al.* (1972) *Science* **176**, 875–80. Copyright 1972 by the AAAS.)

above, $w = 0.25$, which gives $\hat{q} = 0.8$. This means that sufficient flies have to be released in order that slightly more than 80 per cent of the total population *after* release consists of CC flies. In other words, there must be a release of at least four times as many CC flies as the total size of the wild-type population. Obviously, this is a tall order. However, it may be possible to improve the fitness of the CC strain, in which case less CC flies would need to be released.

Suppose, for example, that the CC strain were made homozygous for a resistance gene that is absent or rare in the wild-type population. If the appropriate insecticide were used in the field during and after the release, then the resistant CC strain may have a considerable advantage over the susceptible wild-type strain. For example, if the resistant CC strain had twice the fitness of the susceptible wild-type strain in the presence of the appropriate insecticide, then $w = 2$, which gives $\hat{q} = \frac{1}{3}$. It would then be necessary to release only half as many CC flies as the total size of the wild-type population. This is an eight-fold reduction in the number of flies to be reared and released, and represents a very large reduction in the cost of a control programme.

One difficulty with this proposal is the public-relations problem created by the mass release of resistant flies: the general public is aware that resistant flies are a major problem, and yet here we have scientists breeding up resistant flies in the laboratory, and releasing them into the field in large numbers. The public's worst fears about scientists appear to be vindicated!

Fortunately, there is a simple solution to this dilemma. If the laboratory strain has also been made homozygous for a temperature-sensitive mutant gene that, for example, results in the death of all larvae if the temperature falls below 5 °C, then the laboratory population that has just taken over from the wild-type population could be completely eliminated in the winter. Alternatively, the CC population that has replaced the wild-type population could be eradicated by a conventional SIRM programme. This would be easier than using a conventional SIRM programme to eradicate the original wild-type population, if the CC population is less fit than the wild-type population.

In discussing Y-autosome reciprocal translocations, we overlooked the fact that in strains carrying such translocations, males are effectively heterozygous for the translocation, and will therefore show a reduction in fertility because of the production of unbalanced gametes (see Section 4.3.2.1). Bearing this in mind, and recalling that a Y-autosome eye colour translocation renders released females blind and hence effectively sterile, it is interesting to consider the effect of releasing large numbers of a translocation-eye colour (TE) strain. Because the released females are blind, the only mating of interest is TE male × wild-type female. In blowflies, about 30 per cent of fertilized eggs resulting from matings of this type will fail to develop because of unbalanced gametes coming from the male. If the TE strain has two Y-autosome translocations, then there is a larger proportion of male gametes that are unbalanced, and the consequent 'loss' of offspring resulting from the first generation matings is increased to around 70 per cent in blowflies. This infertility in the first generation is the *direct effect* of the release of a TE strain. For those matings that are fertile, it is shown in Fig. 10.15 that all female offspring have normal eyes but are heterozygous for the eye colour mutants, and that all male offspring are translocation carriers. In the second generation, matings will occur between the above two types of first-generation offspring, or between first-generation female offspring and TE males, if releases continue to be made. In either case, there will be the same 30 per cent or 70 per cent 'loss' of offspring as before. In addition, from the latter type of mating, a proportion of second-generation female offspring that do develop will be blind and hence will fail to survive, because of homozygosity for one or more eye colour mutants. This is the *indirect effect* of the release of a TE strain.

If there is only one eye colour mutant in the TE strain, then 50 per

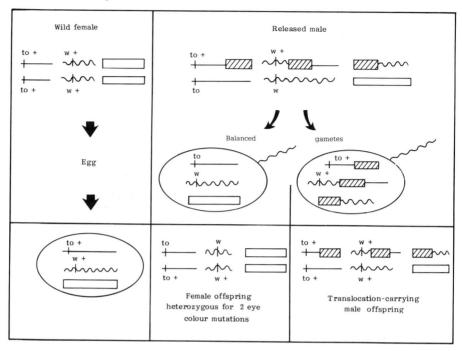

Fig. 10.15. The result of mating a released TE male with a wild-type female. In this particular example, the TE strain has two Y-autosome translocations, each with one eye colour mutant. The symbols *w* and *to* represent alleles for white and topaz eyes respectively, and *w+* and *to+* are normal (wild-type) alleles at the respective loci. The open box is the X chromosome and the striped box is the Y chromosome, of which there are three separate parts resulting from two translocations. Note that only two types of balanced gametes are produced by released males. (From Foster and Maddern 1978.)

cent of second-generation females will be lost. If a TE strain has two unlinked eye colour mutants attached to a part or parts of the Y chromosome, then 75 per cent of second-generation females will be effectively sterile, as shown in Fig. 10.16. And if three different eye colour mutants are used, the effective sterility expected in second-generation females increases to 87.5 per cent. Recalling that if the TE strain has just one translocation, only 70 per cent of second-generation offspring develop beyond the fertilized-egg stage, we can conclude that the total genetic death of second-generation females will be 65 per cent, 82 per cent, and 91 per cent for one, two, and three mutants respectively. Corresponding figures for a TE strain with two translocations are all greater than 90 per cent.

It is obvious that:

> *By combining direct and indirect effects of genetic sterility, a very high level of genetic death can be imposed on a population.*

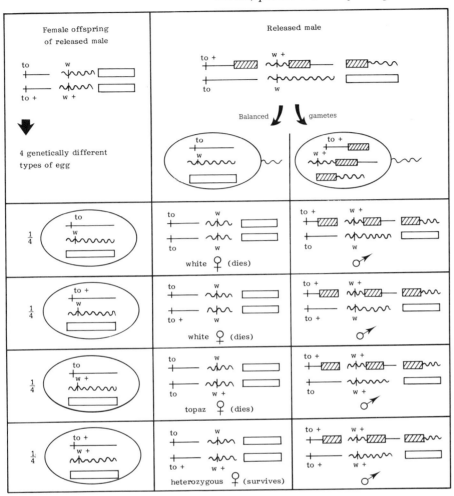

Fig. 10.16. The indirect effect of releasing a TE strain. With two unlinked eye colour mutants, only $\frac{1}{2} \times \frac{1}{2} = \frac{1}{4}$ of gametes from a first generation female contain both alleles for normal eye colour ($w+$ and $to+$). Thus 75 per cent of the female offspring of such a female are blind and hence fail to survive. The only surviving females are just like their mothers, and so will give rise to an equally high level of genetic death in the next generation. Note that only two types of balanced gametes are produced by released males. (From Foster and Maddern 1978.)

There are of course many technical problems to be overcome before any of the above techniques can be put into large-scale practice in the field. But many of these techniques have already been tried in experimental releases, and some of them have been found to work quite well. For an account of the results achieved with the sheep blowfly, and for a discussion of some of the problems encountered, see Whitten (1979) and Foster (1980).

While the above account has been concerned with the sheep blowfly, it should be evident that many of the techniques discussed are equally applicable to other species of insects. Readers requiring further information in relation to other species should consult Hoy and McKelvey (1979) and Curtis (1981).

10.5.3 Worms

In principle, the techniques described above for the biological control of insects are also applicable to worms. In practice, very little work has been done in this area. However, experiments are now in progress on the value of alternating releases of two closely related species that produce substantial hybrid sterility and thus give rise to selection against heterozygotes. The release of hybrid worms with substantial sterility is also being investigated (Le Jambre and Royal 1980).

Biological control techniques such as these may be of use in the longer term. In the short term, however, animal producers have to combat worms as best they can, by selecting for resistance in the host and by drenching. Following an extensive review of resistance to anthelmintics in worms, and bearing in mind the evidence on cross-resistance, Prichard *et al.* (1980) drew up recommendations for drenching strategies under various circumstances. In doing so, they classified available drenches into one of four groups as shown in Table 10.4. They then made the recommendations shown in Table 10.5.

The effectiveness of these recommendations depends on the extent of cross-resistance between groups, and, as mentioned above, the evidence on cross-resistance is not particularly straightforward. Much more work needs to be done before the effectiveness of these recommendations can be predicted with any certainty, but, in the meantime, they represent the best available set of guidelines.

10.5.4 Bacteria

Various attempts have been made to control resistant bacteria. In the United Kingdom, for example, the use of 'therapeutic' antibiotics as feed additives was banned in 1971. The aim of this legislation was to minimize the use as feed additives of any antibiotic that is likely to be used in the treatment of clinical cases in either animals or humans. Antibiotics not in this category were classified as 'feed' antibiotics and were allowed unrestricted use as feed additives. These decisions were taken on the advice of the Swann committee (Anon. 1969), which investigated the use of antibiotics in animal industries and concluded that their widespread non-therapeutic use certainly increased the selection pressure for resistance. And this in turn increased the number of resistant bacteria available for the transfer of resistance from an animal strain of bacteria to a human strain. Several European countries have enacted similar laws, but there is still much debate as to whether

Table 10.4. The four groups of anthelmintics for sheep roundworm control, based on apparent modes of action, structure, spectrum of activity, and efficacy against resistant populations

	Broad spectrum		Narrow spectrum	
	Benzimidazoles and pro-benzimidazoles		Salicylanilides and substituted nitrophenols	Organophosphates
Group number	1	2	3	4
	Thiabendazole Parbendazole Cambendazole Mebendazole Oxibendazole Fenbendazole Albendazole Oxfendazole Thiaphanate Febentel	Levamisole Morantel	Clioxanide Rafoxanide Bromosalans Nitroxynil	Napthalophos

(After Prichard *et al.* 1980.)

Table 10.5. Recommendations aimed at minimizing development of resistance or controlling an existing resistance problem in roundworms of sheep

1. Integrate pasture and animal management for parasite control so as to reduce dependence on anthelmintics.
2. When a narrow spectrum anthelmintic will suffice, for example for a *Haemonchus* problem, use anthelmintics from groups 3 and 4 regardless of whether there is resistance to one or other of the broad spectrum groups.
3. Where no resistance problem exists, use anthelmintics from groups 1 and 2 in a slow rotation, changing at approximately yearly intervals.
4. Where resistance is diagnosed, and if a broad spectrum anthelmintic is required, use the alternative group while the population remains resistant to the first group. Should tests show that the worms have reverted to susceptibility, reintroduce this group in a slow rotation. Its continued effectiveness will need to be monitored.

(From Prichard *et al.* 1980.)

such laws are necessary or effective. The debate is fuelled by evidence such as that presented by Threlfall *et al.* (1980), indicating not only that new strains of resistant bacteria are continuing to develop in animals in Great Britain, but that the plasmid-encoded resistance genes in these new strains have been transferred to human bacteria, creating new problems in human medicine. Some people cite these observations as evidence that the law is being abused, if not broken, by unscrupulous veterinarians and/or feed and drug companies. Other people see the same observations as a rare exception to the general rule that feeding antibiotics to animals does not present an increased risk to human health (Cherubin 1981).

In the USA, the addition of antibiotics to feeds has never been banned, despite threats from the Food and Drug Administration. Consequently, there has been a constant and large-scale application of selection pressure for resistance in bacteria in the USA, and it is unlikely to abate in the future. It seems likely, therefore, in the USA and else-where, and whether the use of various antibiotics in feed is banned or not, that we will have to learn to live with resistant bacteria.

Despite this rather pessimistic outlook, there is some hope for the future control of at least some types of pathogenic bacteria, as a result of research using recombinant DNA techniques. The bacteria that have received particular attention in this regard are the enterotoxigenic strains of *E. coli* that adhere to the epithelial cells of a host's small intestine, secreting enterotoxins and thus causing diarrhoea in humans, calves, and piglets. Adherence of the *E. coli* to the intestinal epithelial cells is usually due to specific antigens on the surface of the bacteria: colonizing-factor antigens in humans, K99 in calves, and K88 and 987P in piglets. We have already encountered the K88 antigen in Sections 5.6.7 and 10.3.2. As an example, we shall now consider briefly the results of recombinant DNA research into the K88 antigen, and shall then discuss how the results of this research may lead to more effective control of pathogenic K88 *E. coli* in the future.

It has been shown by Kehoe *et al.* (1981) that four structural genes are involved in the normal production of K88 antigen by pathogenic *E. coli*. All four genes are located on a large plasmid that is approxi-mately 70 kb long. Three of the structural genes (*A*, *B*, and *C*) are immediately adjacent to each other. The fourth gene (*D*), is separate from the others, as illustrated in Fig. 10.17. The actual K88 antigen, which is a 23 500-MW polypeptide, is coded by gene *D*. The 17 000-MW product of the *C* gene appears to act as a positive regulator of gene *D*. The 70 000-MW product of gene *A* (and possibly the 29 000-MW product of gene *B*) is involved in attachment of the K88 antigen to the surface of pathogenic *E coli*. How can this and other knowledge be applied to provide more effective control of pathogenic K88 *E. coli*?

It has been known for some time that the K88 antigen can be used as

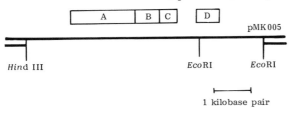

Fig. 10.17. The arrangement of the structural K88 genes of *E. coli*. The central dark line represents a 6.7 kb segment of plasmid from pathogenic K88 *E. coli*; it is bounded by a *Hind*III and an *Eco*RI cleavage site, these being the points at which this particular segment was joined to a vector plasmid (open boxes), to produce a recombinant plasmid pMK005. The boxes above the dark line represent the four structural genes whose polypeptide products are known. A fifth gene, *E* (not shown), has been identified to the left of *A*, but its polypeptide product is not yet known. (Adapted from Dougan *et al.* 1983.)

the basis of a useful vaccine against pathogenic *E. coli*, but commercial production of the vaccine has been hampered by the relatively low yield of K88 antigen obtained from pathogenic strains raised in the laboratory. One solution to this problem is to insert recombinant plasmids such as the one shown in Fig. 10.17 into non-pathogenic laboratory strains of *E. coli*, which produce a much higher yield of K88 antigen than pathogenic strains raised in the laboratory. In this way, recombinant DNA technology may be used to aid the production of an effective vaccine. Indeed, the first product of recombinant DNA technology to go on commercial sale was a K88 vaccine produced in this way.

A refinement of this procedure would be to use a plasmid lacking the structural gene *A*. As suggested by Kehoe *et al.* (1981), when such plasmids are inserted into non-pathogenic laboratory strains of *E. coli*, they would secrete large quantities of K88 antigen into the culture supernatant, because they lack the polypeptide required for attachment of the antigen to the bacterial cell wall. This would facilitate the harvest and purification of the antigen.

The possibility of a rather different approach arises from the fact that the enterotoxin that causes the clinical signs in the host is also a polypeptide and is hence the product of a gene. If the enterotoxin gene could be removed, or if it could be replaced by a non-toxin gene from a non-pathogenic strain of *E. coli*, then the resultant detoxified strain of K88 bacteria could be used as a vaccine, which may be more effective than K88 antigen alone.

Not all pathogenic bacteria are amenable to the methods of control outlined above. But as knowledge of the genetic structure of bacteria increases, we are likely to see the emergence of an increasing number of possibilities for novel methods of bacterial control.

10.6 Summary

In most situations in which hosts and parasites or pathogens interact with each other, a dynamic equilibrium results as each side attempts to get the upper hand. A good example of this is the interaction between rabbits and the myxoma virus that causes myxomatosis. A less straightforward example of host–pathogen interaction is scrapie in sheep, in which the incubation period of the pathogen may be longer than the life span of the host. This rather fascinating aspect of scrapie raises difficult questions for quarantine authorities. Some pathogens keep one step ahead of their hosts by the process of antigenic variation, which is the occurrence of a sequence of different antigenic variants all arising from the single population of pathogens that originally entered the host.

There is considerable genetic variation for resistance in hosts to parasites and pathogens, which implies that selection for resistance in the host may be worthwile. The many successful examples of selection for resistance indicate that this is so.

There is also, however, considerable genetic variation in parasites and pathogens for resistance to chemicals, which again implies that selection will be able to increase resistance, which in this case is undesirable. The difficulty is that any use of a chemical directed against a certain parasite or pathogen automatically imposes selection for resistance. In fact, the most likely consequence of the use of almost any chemical is that the organism against which it is directed will become resistant to it; and the more effective the chemical is initially, the faster will resistance develop.

An alternative to the use of chemicals is biological control. Those forms of biological control regarded as being genetic include the conventional sterile insect release method, and variations of this method, in which genetic sterility replaces radiation-induced sterility, and in which females can be genetically removed prior to release. Other genetic forms of control involve the use of Y-autosome translocations and/or compound chromosomes.

There are fewer alternatives available for the control of worms and bacteria. With the former, however, there are at least guidelines for the judicious use of different chemical compounds. In an attempt to control the spread of resistant bacteria, some countries have enacted bans on the use of certain antibiotics as additives in animal feed. There is considerable debate as to whether such bans are effective. It is possible that recombinant DNA techniques may lead to more effective vaccines against certain pathogenic bacteria.

10.7 Further reading

Abraham, E. P. (1981). The beta-lactam antibiotics. *Scientific American* **244**(6), 76–86. (A review of the history of the use of antibiotics, together with a look into the future.)

Bishop, J. A. and Cook, L. M. (Eds) (1981). *Genetic consequences of man made change*. Academic Press, London. (A thought-provoking collection of essays on the effect of human activities on other species. Chapters 3, 4, 7, and 8 are particularly relevant.)

Clowes, R. C. (1973). The molecule of infectious drug resistance. *Scientific American* 228(4), 19–27. (A review of the role of plasmids in the horizontal transmission of resistance to antibiotics in bacteria.)

Cohen, S. N. and Shapiro, J. A. (1980). Transposable genetic elements. *Scientific American* 242(2), 36–45. (A review of the role played by transposons in the rapid development of strains of bacteria with multiple resistance to antibiotics.)

Fenner, F. and Ratcliffe, F. N. (1965). *Myxomatosis*. Cambridge University Press, Cambridge. (An account of the dynamic interaction between rabbits and the myxoma virus.)

Foster, G. G. (1980). Genetic control of sheep blowfly (*Lucilia cuprina*) and the logistics of the CSIRO control programme. *Wool Technology and Sheep Breeding* 28(1), 5–10. (A review of the CSIRO's control programme, containing references to the various genetical projects involved.)

Henson, J. B. and Noel, J. C. (1979). Immunology and pathogenesis of African animal trypanomiasis. *Advances in Veterinary Science and Comparative Medicine* 23, 161–82. (A thorough review, including a good discussion of antigenic variation.)

Kimberlin, R. H. (1979b). An assessment of genetical methods in the control of scrapie. *Livestock Production Science* 6, 233–42. (A review of scrapie, with particular reference to the possibility of genetical control.)

La Chance, L. E. (1979). Genetic strategies affecting the success and economy of the sterile insect release method. In *Genetics in relation to insect management*. (Eds Hoy, M. A. and McKelvey, J. J. (Jr)) pp 8–18. The Rockefeller Foundation, New York. (A review of various SIRM control programmes, especially the screw-worm eradication programme in the USA.)

McKay, W. M. (1975). The use of antibiotics in animal feeds in the United Kingdom; the impact and importance of legislative controls. *World's Poultry Science Journal* 31, 116–28. (A rejoinder to the review of Smith (1975), see below, arguing strongly that there is no cause for concern.)

Murray, M., Morrison, W. I., Murray, P. K., Clifford, D. J., and Trail, J. C. M. (1979). Trypanotolerance: a review. *World Animal Review* 31, 2–12. (A thorough review of the genetic basis of tolerance to trypanosomes in cattle and in mice.)

Prichard, R. K., Hall, C. A., Kelly, J. D., Martin, I. C. A., and Donald, A. D. (1980). The problem of anthelmintic resistance in nematodes. *Australian Veterinary Journal* 56, 239–51. (A review of resistance in worms, with recommendations for appropriate drenching programmes.)

Prusiner, S. B. (1984). Prions. *Scientific American* 251(4), 48–57. (A review of available evidence on the nature of the scrapie agent.)

Richardson, R. H., Ellison, J. R., and Averhoff, W. W. (1982). Autocidal control of screwworms in North America. *Science* **215**, 361–70. (A review of the North American screw-worm project.)

Smith, H. W. (1975). Antibiotic-resistant bacteria in animals: the dangers to human health. *World's Poultry Science Journal* **31**, 104–15. (An account of the practical implications of the continual use of antibiotics in animal feed and also in therapy.)

Solomon, K. R. (1983). Acaracide resistance in ticks. *Advances in Veterinary Science and Comparative Medicine* **27**, 273–96. (A thorough review.)

Wakelin, D. (1978). Genetic control of susceptibility and resistance to parasitic infections. *Advances in Parasitology* **16**, 219–308. (An extensive review of the genetic basis of resistance in hosts.)

11

Genetic and environmental control of inherited disease

11.1 Introduction

In Chapter 10 we saw that there is genetic variation in hosts for resistance to parasites and pathogens, and we saw how this can be exploited by means of artificial selection for resistance to infectious disease. It remains for us now to turn our attention to non-infectious disease. As we saw in Chapter 6, it is unlikely that the heritability for liability to any disease is zero; by far the most common situation observed in practice is that there is at least some genetic variation in liability to any particular disease. This raises the possibility of artificial selection for decreased liability as a means of decreasing the incidence of the disease.

However, the existence of genetic variation in liability is associated with a popular misconception; many people firmly believe that if a disease shows any sign of being inherited, then there is nothing that can be done about it apart from selecting against it.

One purpose of this chapter is to show that this belief is wrong. By considering several examples, we will see that there are many potential non-genetic means of alleviating inherited diseases. Having proved this point, we shall then go on to consider some examples of genetic control of inherited diseases.

11.2 Environmental control of inherited disease

11.2.1 Hip dysplasia

> *The heritability of liability to hip dysplasia is certainly greater than zero, but there are several non-genetic methods of decreasing its frequency.*

The first method concerns the level of feeding after weaning, the effect of which was demonstrated by Lust et al. (1978). They took puppies from two types of matings, normal × dysplastic and dysplastic × dysplastic, and split the offspring from each mating type into two groups. Members of the first group were given food in normal quantities, and members of the other group were fed a restricted amount of the same diet. The results are shown in Table 11.1.

Table 11.1. The effect of the level of post-weaning feeding on the average age of onset and on the incidence of hip dysplasia

Parents	Diet	Average age of onset (weeks)	Incidence
Normal X dysplastic	Normal	17	5/5
	Restricted	68*	0/3
Dysplastic X dysplastic	Normal	17	3/4
	Restricted	35	4/4

* No signs of hip dysplasia had been observed by the 68th week, at which time the experiment ceased. (From Lust *et al.* 1978.)

Although the numbers in each class are small, it is possible to draw some conclusions about the effect of level of diet. Among progeny from dysplastic X dysplastic matings, the effect of restricting the level of feeding was to double the age at onset. For progeny from the normal X dysplastic matings, the effect of diet restriction was evident in both age of onset and incidence. The conclusion to be drawn from these results is that restricted feeding during the growing phase can reduce the average age of onset and the incidence of hip dysplasia. In other words:

> *Restricted feeding during the growing phase can reduce the liability to hip dysplasia.*

This indicates that food intake is a component (obviously a non-genetic component) of liability to hip dysplasia. There are other non-genetic factors that may also decrease liability to hip dysplasia. Administration of sodium ascorbate to pregnant bitches and to pups until they reach puberty, has been claimed to have a profound effect on the incidence of hip dysplasia (Belfield 1976). And restriction of the amount of exercise during the growing phase also seems to help.

The main point that must be emphasized in relation to these observations can be stated as follows.

> *The fact that liability to hip dysplasia can be reduced by non-genetic means does not mean that hip dysplasia is a non-genetic disease.*

It simply means that there are certain non-genetic factors, as well as many genetic factors, that contribute to the variation in liability to hip dysplasia.

11.2.2 Muscular dystrophy in chickens

The heritability of liability to muscular dystrophy, MD, is quite high in all species studied. But research into MD in chickens has shown that administration of penicillamine delays the onset of clinical signs (Chou *et al.* 1975), and that exercise plus injections of diphenylhydantoin improves the ability of MD chickens to right themselves from a lying position (Entrikin *et al.* 1977).

Once again, this evidence is not at all in conflict with the existence of genetic variation in liability to MD. It does, however, point the way to possible non-genetic methods of alleviating the effects of MD, that may be very useful in practice.

11.2.3 Haemolytic anaemia due to pyruvate kinase deficiency

The first example of inherited disease that we considered in this book, in Section 3.2, was haemolytic anaemia due to pyruvate kinase deficiency. As we saw, it is due to homozygosity for a recessive gene that codes for a deficient form of pyruvate kinase. The heritability of liability to this type of haemolytic anaemia is quite high.

But once again, there is a non-genetic form of treatment that has been used to overcome the anaemia. It involves whole body irradiation (to block the immune response) followed by transplantation of marrow cells from non-anaemic and preferably MHC-identical litter mates (Weiden *et al.* 1976).

11.2.4 Bleeding disorders

Inherited bleeding disorders were also discussed in Chapter 3. As with the case of haemolytic anaemia described above, transplantation of certain tissues has alleviated certain cases of bleeding disorders. For example, transplantation of normal liver into an affected pig alleviated von Willebrand's disease (Webster *et al.* 1976). And sometimes replacement therapy can help, using either whole blood or the patient's own red blood cells suspended in fresh serum.

11.2.5 Mannosidosis

Mannosidosis was also mentioned in Chapter 3, as an example of a lysosomal storage disease in which the specific biochemical defect, namely a deficiency of α-mannosidase, is known. Although there are no non-genetic treatments for mannosidosis currently available, there are several possibilities under consideration.

Enzyme replacement, for example, is a possibility for the future. This could involve either infusion of normal enzyme, or transplantation of an organ or tissue that could produce normal enzyme.

In relation to the second possibility, an interesting 'experiment of nature' was reported by Jolly *et al.* (1976). It concerned a male calf that developed the usual symptoms of mannosidosis at nine months of age, and which showed zero levels of α-mannosidase in its plasma, as expected in affected individuals. Its lymphocytes, however, showed an unexpectedly high level of α-mannosidase activity, similar to that normally seen in non-affected individuals. It was subsequently discovered that the affected calf was an XX/XY chimaera, with 77 per cent of 60,XX lymphocytes and 23 per cent of 60,XY lymphocytes. This indicated that the calf had had a female co-twin. The relatively normal enzyme levels in the affected calf's lymphocytes and lymph nodes were presumably due to the natural transplantation of lymphocyte-producing cells from the normal co-twin during foetal development.

The result of nature's experiment in transplantation was not particularly encouraging, as it did not prevent the calf from becoming affected. But this was largely because there are many sources of α-mannosidase apart from lymphocytes, and all these other sources were, of course, producing defective enzyme. If the calf had been chimeric in all tissues that produce α-mannosidase, then presumably the calf would not have been affected.

11.2.6 Phenylketonuria

The simplest and probably the best example of an environmental solution to a genetical problem comes from phenylketonuria, PKU, which is a recessive disease in humans caused by a deficiency of the enzyme phenylalanine hydroxylase. Affected individuals are unable to break down phenylalanine, which builds up in the body and leads to severe mental retardation. The importance of PKU is that it illustrates what might be called the paradox of inherited disease:

> *The more that is learnt about the genetic basis of a disease, the more it is likely that a non-genetic treatment will be developed.*

In the case of PKU, as soon as the biochemical basis of the genetic defect was discovered to be the lack of phenylalanine hydroxylase, a simple non-genetic treatment became evident. It followed logically from the well-known fact that humans are unable to synthesize phenylalanine; our only source of phenylalanine is the food we eat. Obviously, a simple solution to this inherited disease is to remove phenylalanine from the diet of individuals homozygous for the defective gene.

In this example, and in all the other examples discussed above, we have seen that it is possible to alleviate inherited diseases by non-genetic means, even if the heritability of liability is quite high. Indeed, in the case of PKU, broad-sense heritability (see Section 6.7.1) is 100 per cent, because under normal circumstances, there are no non-genetic factors contributing to variation in liability to PKU; everyone receives phenylalanine in their diet. However, this particular non-genetic factor can be changed easily, with dramatic effects; the removal of phenylalanine from the diet of homozygote recessives reduces their liability to such an extent that they now have a liability on the other side of the threshold.

11.2.7 Genetic effects of environmental control

The simple remedy described above has essentially eradicated the PKU syndrome. It has not, of course, removed the genes for PKU, and this raises one of the problems associated with environmental control of inherited diseases. If we enable individuals who would have otherwise died, to survive and reproduce and hence to pass on what were formerly harmful or even lethal genes, are we going to increase the frequency of these genes, giving rise to an ever-increasing number of individuals requiring treatment in the future?

The basic principles of population genetics tell us that if we remove selection pressure from a previously harmful gene so that all genotypes have equal relative fitness, then on average the frequency of each gene will remain constant, except for the effect of mutation, which will tend to increase the frequency of the previously undesirable gene, but at an imperceptibly low rate.

Thus, in principle there will be no extra problems created for the future by the use of environmental control of inherited disease, provided that all genotypes have an equal chance of contributing genes to the next generation. In practice, however, this proviso is often not met. For example, if an animal commands a large stud fee but has a certain genetic defect, it is tempting to correct or treat the defect and then to use that animal as extensively as possible. In effect, this increases the relative fitness of the undesirable genotype and consequently increases the frequency of the undesirable gene. In practice, therefore, environmental control of genetic diseases should be used with caution, and animals so treated should not be allowed to make large contributions to the next generation. They could, however, be used repeatedly in crossbred matings to produce commercial (non-pedigree) animals, such as those that will be slaughtered for meat. Or if they are companion animals, they could be neutered and sold as pets.

11.3 Genetic control of inherited disease

For the majority of this section we shall be concerned with single-locus diseases, because of the many different types of genetic control programmes that can be conducted for such diseases. We shall, however, discuss some examples of genetic control programmes for non-Mendelian familial diseases in Section 11.3.7.

In general, genetic control programmes involve preventing certain individuals from contributing their genes to subsequent generations of the population into which they were born. We shall refer to this type of selection as *culling*. It must be stressed that when used in this sense, culling does not mean that such individuals have to be killed. Indeed, as noted in the previous section, they can be neutered and sold as pets, or, if farm animals, they can be mated to animals from another breed with the aim of producing commercial animals that will not be used for breeding.

11.3.1 Clinical screening

There are many single-locus diseases that are considered to be serious problems but which do not prevent affected individuals from reproducing. Well-known examples include the inherited eye diseases in dogs, such as progressive retinal atrophy, retinal dysplasia, and various forms of cataract. If the disease can be detected by clinical examination prior to reproductive age, then it should be possible to select against the disease.

The simplest programme would involve culling affected individuals. If the disease is recessive, this would involve selection against homozygotes for the harmful recessive gene. If all pedigreed offspring in a breed were screened, and if all affected offspring were culled, we would have complete selection against recessive homozygotes, the results of which were discussed in Chapter 5. A more effective programme would involve culling all parents of affected individuals as well as affected individuals themselves. If the disease were recessive, this would amount to partial selection against heterozygotes in addition to selection against homozygote recessives.

11.3.2 Pedigree analysis and test matings

Irrespective of how high the frequency of an undesirable recessive gene becomes, the frequency of affected offspring can be reduced immediately to zero if, from a particular time onwards, all matings involve at least one parent that is homozygous for the normal gene. But how do breeders distinguish such parents from those that are carriers?

The cheapest and quickest method involves an analysis of available pedigree records, with the aim of estimating the *prior probability* that a prospective normal parent is homozygous. Of the many different methods of pedigree analysis available, the one proposed by Grant (1976) is probably the easiest to use. After drawing up a pedigree and following some simple rules, the probability of homozygosity can be read from tables presented in Grant's paper. Prospective parents for which this probability is high (say greater than 95 per cent) could be used with considerable confidence.

In many cases, however, the probability of homozygosity as calculated from pedigree analysis will be much less than 95 per cent for all prospective parents. When this is so, it is time to consider test mating the prospective parent with:

(1) homozygotes for the recessive gene; or
(2) known heterozygotes; or
(3) progeny of the prospective parent; or
(4) a random sample of the population.

There are two possible outcomes from a test mating. If an offspring from a test mating is affected and hence homozygous for the undesirable gene (*aa*), then that offspring must have received one of its *a* genes from the prospective parent. Obviously, in this case we conclude that the prospective parent is a carrier. The other possible outcome is that the offspring of a test mating is normal, in which case there are two alternative explanations: either the prospective parent is homozygous for the normal gene (*AA*), or it is really a carrier (*Aa*) but by chance remained undetected because it passed on its *A* gene rather than its *a* gene. The aim of a test mating programme is to distinguish as efficiently as possible between these two alternatives.

Consider matings with homozygote recessives. If the prospective parent is a carrier, then a test mating of this type can be written as *Aa* × *aa*, with the expected results being one-half *Aa* (normal) offspring and one-half *aa* (affected) offspring. Recalling that the carrier parent remains undetected if an *Aa* offspring is produced, we can conclude that the chance of a carrier parent remaining undetected after producing one offspring is 0.5. Since the genotype of each offspring is produced independently of all other offspring, the chance of a carrier parent remaining undetected after two offspring is $0.5 \times 0.5 = (0.5)^2 = 0.25$; after three offspring it is $(0.5)^3 = 0.125$, and after n offspring it is $(0.5)^n$. Thus, for each additional offspring, the chance of a carrier parent remaining undetected decreases.

The important question that has to be asked in practice is how many offspring do we need to observe in order for there to be only a very small chance of a carrier parent remaining undetected? Suppose, for example, that we wish there to be no greater than a 5 per cent chance that a carrier remains undetected. In this case, we require sufficient offspring per parent to ensure that $(0.5)^n$ is less than or equal to 5 per cent, which gives $n = 5$ as the required number of offspring. If we had wanted the chance to be 1 per cent, then $n = 7$ is the required number.

It should be obvious that:

We can never prove that a parent is homozygous normal in a programme such as this.

All we can do is to reduce the chance of the parent being undetected as a carrier to some acceptably low level. On the other hand, as soon as the first affected offspring is produced, then we have proven that the parent of that offspring is a carrier.

The principle is the same for the other types of test matings listed above; the main difference is in the chance of a carrier remaining undetected. For example, test matings to known heterozygotes can be written as *Aa* × *Aa* if the prospective parent is a carrier, and the expected offspring proportions are 0.75 $A-$ and 0.25 aa. Thus the chance of a carrier remaining undetected in matings to known carriers is $(0.75)^n$.

The third type of test mating is slightly different, because it requires that the prospective parent has already produced some offspring, and it tests for all undesirable genes rather than just one gene as in the previous types of test mating. If an undesirable gene is rare in the population, then the matings that gave rise to the prospective parent's progeny will have been $Aa \times AA$, from which one-half of all progeny will be Aa and the other half will be AA. It follows that the frequency of the A gene in these progeny is $(0.5 \times 0.5) + (0.5 \times 1.0) = 0.75$ and the frequency of the a gene is $(0.5 \times 0.5) + (0.5 \times 0.0) = 0.25$. Recalling from Chapter 5 that the frequencies of gametes produced by a group of individuals equal the gene frequencies in that group, and assuming that the prospective parent is a carrier and hence will produce 0.5 A gametes and 0.5 a gametes, we can see from Fig. 11.1 that the probability of such a test mating producing a normal offspring is 0.875. Thus the chance of not detecting the prospective parent as being a carrier is 0.875 for each mating with an offspring, and is $(0.875)^n$ for n such matings.

		Gametes from prospective parent	
		0.5 A	0.5 a
Gametes from progeny of prospective parent	0.75 A	0.375 AA	0.375 Aa
	0.25 a	0.125 Aa	0.125 aa

Fig. 11.1. The results of mating a prospective parent, who is a carrier, to its own progeny. Those genotypes not enclosed in the box have normal phenotypes, and their combined probability is $0.375 + 0.375 + 0.125 = 0.875$.

The fourth type of test mating, involving matings to a random sample of animals from the population, can be thought of as a generalization of the third type of test mating just described. Using a similar argument as used above, it can be seen from Fig. 11.2 that if the

		Gametes from prospective parent	
		0.5 A	0.5 a
Gametes from general population	p A	0.5p AA	0.5p Aa
	q a	0.5q Aa	0.5q aa

Fig. 11.2. The results of mating a prospective parent, who is a carrier, to a random sample of the population. The probability of a normal phenotype in the offspring is $0.5p + 0.5p + 0.5q = p + 0.5q = 1 - 0.5q$.

frequency of the A gene is p and the frequency of the a gene is q in the general population, and if the prospective parent is a carrier, then the chance of not detecting the prospective parent as being a carrier is $(1 - 0.5q)^n$ from n such matings.

The number of offspring required for various chances of failing to detect carriers is given in Table 11.2 for each type of test mating described above. It can be seen that matings to homozygote recessives are most efficient. If such individuals are not available, then the next best option is to use known carriers. Although the last two types of test matings appear to be much less efficient than the first two, they do offer certain advantages. For example, they do not involve the expense of establishing and maintaining a special group of tester animals. More importantly, they provide a test for undesirable recessive genes at all loci rather than just at one locus, as is the case for the first two types of test matings.

Table 11.2. The number of offspring required in test matings for a recessive gene. In the fourth type of test mating, q is the frequency of the recessive gene in the general population

Prospective parent mated to	Chance that a carrier remains undetected with n offspring	Number of offspring required to reduce the chance of non-detection to		
		0.05	0.01	0.001
1. Homozygote recessives	$(0.5)^n$	5	7	10
2. Known carriers	$(0.75)^n$	11	16	24
3. Offspring of prospective parent	$(0.875)^n$	23	35	52
4. Individuals chosen at random	$(1 - 0.5q)^n$, where q equals			
	0.2	29	44	66
	0.1	59	90	135
	0.01	598	919	1379

In comparing the last two types of test matings, it appears from Table 11.2 that it is more efficient to mate a prospective parent to its offspring than to a random sample of the population, unless the recessive gene is quite common in the population. Against this, however, is the fact that the use of offspring as testers greatly increases the time required to test a prospective parent.

It is evident from Table 11.2 that if a recessive gene is rare in the population, then a carrier could be used quite extensively without being detected with the fourth type of test mating. This is often cited as a major disadvantage of this type of test mating. If, for example, the carrier is a bull in an Artificial Insemination (A.I.) centre, then he could have produced many carrier offspring before being detected. However, if this does occur, then the frequency of the recessive gene will be much higher in the next generation of testers, with the consequence that carrier bulls will be detected much more efficiently and hence the recessive gene frequency will decrease again. Van Vleck (1967) has studied the implications of this cyclical change in gene frequency, and has shown that test mating prospective parents to randomly chosen testers provides quite effective selection against recessive genes in dairy cattle populations using A.I.

In the above consideration of the design of test matings, we have assumed that prospective parents are either homozygous normal or heterozygous, while in the pedigree analysis described previously, we were able to estimate the prior probability that prospective parents are either homozygous normal or heterozygous. The obvious step is to combine together the two approaches, such that the prior probabilities estimated from the pedigree analysis are used in the design and interpretation of test matings. This has been done in various situations, but the relevant theory is beyond the scope of this book. Readers requiring further information on this and related topics should consult Kidwell (1970) and Kidwell and Hagy (1973).

One final point should be mentioned in relation to test matings. It concerns the use of *multiple ovulation and embryo transfer* (MOET) and/or *early foetal recovery* to increase the effectiveness of test mating programmes. MOET can be used to increase the number of males tested with a given number of females, or to enable the test mating of females that would otherwise never produce sufficient offspring in a lifetime. Artificial termination of pregnancy at the earliest stage at which the relevant defect is detectable in the foetus also increases the number of test matings that can be achieved in a given time. Both these techniques have been exploited commercially in testing males (Johnson *et al.* 1980) and females (Leipold and Peeples 1981) for bovine syndactyly.

11.3.3 Insurance schemes

Many breed societies are not in a position to run a test mating programme; some are too small, while others are unable to obtain the necessary cooperation from a sufficient number of breeders. And even if a test mating programme can be organized, there is still the problem that prospective parents have to produce offspring before conclusions can be drawn as to their genotype.

Thus, for the many situations in which test mating programmes are not conducted, or for the common situation in which breeding animals are sold before they have produced any offspring, the buyer of breeding animals runs a certain risk of purchasing a carrier. In circumstances such as this, buyers would be much happier if they could insure against the future production of an affected offspring by the animal being purchased.

An example of such a scheme is provided by the Galloway Cattle Society in Great Britain, in relation to an autosomal recessive lethal defect called tibial hemimelia, which is characterized by a large abdominal hernia, shortened and twisted hind legs, and an opening in the head. Under the terms of the insurance scheme, a purchaser of a young bull at any official breed society sale is insured against that young bull throwing defective offspring within 21 months following the sale (9 months gestation plus 12 months use). It is compulsory for all young bulls sold at official sales to be insured, and the vendor is obliged to pay the initial premium. If purchasers wish to extend the period of cover beyond 21 months, then they have the option of paying further premiums.

The biggest advantage of this insurance scheme is that it provides a positive incentive for affected calves to be reported to the breed society, and thereby overcomes one of the greatest problems faced by breed societies in the past in relation to genetic defects: it was often in the interest of the breeder to pretend that the defect had not occurred. With the Galloway insurance scheme, however, purchasers know that they will receive the fair market value of any bull that throws an affected calf.

The idea of using an insurance scheme in this way is an attractive one and is equally applicable to all species. Indeed, it seems to be a very sensible approach that should be given serious consideration by any breed society faced with an unacceptably high frequency of an undesirable gene. Further details of the Galloway insurance scheme are given in Section 11.3.7.

11.3.4 Biochemical screening

> With inborn errors of metabolism, the phenomenon of gene-dosage (Section 3.2.1) provides a very simple and quick means of distinguishing homozygote normals from carriers at an early age.

A good example of this in animals is provided by mannosidosis, which as we have previously seen, is due to a deficiency of α-mannosidase. Because heterozygotes have one normal and one abnormal gene, they show a lower level of enzyme activity in blood plasma than homozygote normals. The only problem with this method of detecting carriers is that there are other sources of variation in enzyme level apart from the mannosidosis locus; genes at other loci, and various non-genetic factors such as age and season all cause variation in the level of α-mannosidase. The result of this variation is that there are two over-lapping distributions of enzyme level, as can be clearly seen in Fig. 11.3. Animals at either end of the scale are easy to categorize, but those with enzyme levels in the overlapping region could come from either distribution. In practice, these doubtful cases constitute less than 5 per cent of cattle tested, which means that the blood plasma test is a very efficient carrier test. For those few animals that cannot be classified on blood plasma levels alone, other tests can be conducted, the details of which need not concern us here.

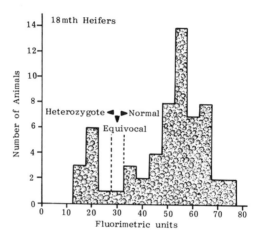

Fig. 11.3. Levels of α-mannosidase in plasma of Angus cattle. (From Jolly *et al.* 1974.)

Armed with an efficient blood plasma test, breed societies can readily establish programmes for genetic control of diseases such as mannosidosis. In New Zealand and Australia, for example, where mannosidosis was relatively frequent in the Angus breed, the relevant breed societies established genetic control programmes with the ultimate aim of eradicating the offending gene from the pedigreed section of the breed altogether. This was achieved by insisting that after a certain date, the only calves that could be registered with the breed society were those that had been tested and shown to be homozygote normal, or that were the progeny of two homozygote normal parents.

Of course, the gene may reappear from time to time because of mutation, and it may be present in commercial herds, but programmes such as these can very easily ensure, even with mutation, that affected calves do not appear. All that is required is for breeders to ensure that at least one parent has been shown to be homozygote normal.

11.3.5 *Voluntary programmes*

Voluntary programmes are those in which no obligation is placed upon the breeder to cull animals that should be culled because they are affected or because they are carriers. The obvious advantage of such programmes is that the rights of individual breeders are not limited in any way. The obvious disadvantage is that culling will most likely be much less extensive than it should be, and so the incidence of the disease or defect may decrease less slowly than expected, or may not decrease at all. The best examples of this latter situation come from certain attempts at selection against hip dysplasia in dogs, which are discussed in Section 11.3.7.7.

11.3.6 *Compulsory programmes*

In compulsory programmes, the breed society lays down certain rules that must be followed by all breeders. The main disadvantage of such programmes is that not all breeders are going to be happy with all of the rules all of the time; at best, the breed society executive will be involved in arguments with disgruntled breeders from time to time; at worst, the breed society could be taken to court. Despite these disadvantages, most programmes for the genetic control of inherited diseases involve at least some rules that are binding on all breeders. If the rules are based on sound genetic principles, then the genetic control programme is virtually guaranteed to succeed. Because of this, the majority of breeders in most breed societies that have instigated genetic control programmes, have been willing to abide by the rules.

11.3.7 *Examples of genetic control programmes*

11.3.7.1 *Tibial hemimelia in Galloway cattle* The tibial hemimelia control programme run by the Galloway Cattle Society in Great Britain provides an excellent example of what can be done to control an autosomal recessive defect for which there is no biochemical means of carrier detection.

The first part of the control programme involves maintaining a *register of carriers*, which is a list of all animals known to have produced an affected calf.

The second part of the Galloway programme is the *insurance scheme* described in Section 11.3.3. The insurance premium ranges from 2 per cent to 4 per cent of the purchase price of the young bull. If an insured bull produces an affected calf and thus shows himself to be a carrier, the breed society must be informed and the bull must be slaughtered. For the above premiums, the owner of the bull receives 'fair market value' not exceeding the purchase price. If the owner wishes to insure for a greater value, then a higher premium can be arranged at the time of purchase.

The third part of the control programme is the maintenance of a *test herd of known carrier cows*, which is run jointly by the Galloway Cattle Society and the East of Scotland College of Agriculture. Consisting of around 40 cows, this herd is mated to two or three of the best young bulls each year, with the aim of obtaining around 10 calves per bull.

If none of the calves of a certain bull is affected, then that bull is assumed to be homozygous for the normal gene (see Section 11.3.2), and is released for general use, often through an A.I. station. As soon as an affected calf is produced, however, then the sire of that calf is immediately known to be a carrier and is slaughtered.

Since maintaining the herd of tester cows is an expensive business, use is made of the fact that the defects associated with tibial hemimelia can be detected in a foetus as young as 90 days (Fig. 11.4). Thus cows can be aborted at three months and then re-mated as quickly as possible, thereby increasing the number of calves per cow per year.

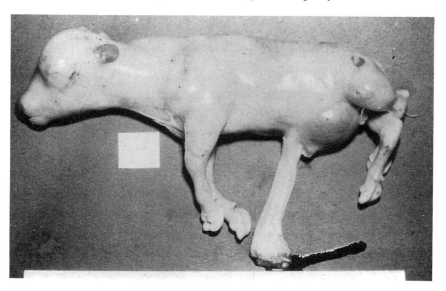

Fig. 11.4. A 90-day-old Galloway foetus showing the abdominal hernia and shortened and twisted hind legs that are characteristic of tibial hemimelia. (From Pollock *et al.* 1979.)

Considered as a whole, the approach adopted by the Galloway breed society in relation to tibial hemimelia is an excellent model for the genetic control of a simply inherited disease. Not all parts of the programme will be directly relevant to other situations, but there are important lessons in the Galloway programme for all breed societies in all species of domestic animals.

11.3.7.2 Mannosidosis   The New Zealand mannosidosis control programme in Angus cattle is an excellent example of a control programme where biochemical detection of carriers is possible at an early age. The aim of the programme was to remove gradually the recessive gene from all registered Angus cattle, so as eventually to reach the stage where those herds supplying the majority of commercial sires were free of the mannosidosis gene. The programme proceeded in four stages, according to rules laid down by the New Zealand Angus Association:

1975: a normal mannosidosis test was a pre-requisite for transfer of registration from one breeder to another
1978: a normal mannosidosis test was a pre-requisite to registration
1980: all known heterozygotes were de-registered, and the registration of any remaining untested cattle was suspended
1982: testing ceased, because all calves offered for registration were by now the progeny of two homozygote normal parents.

A cost–benefit study of this scheme was undertaken by Jolly and Townsley (1980). They showed that the ratio of benefits to costs ranged between 2.9 : 1 and 4.1 : 1, depending on assumptions. Part of the reason for these favourable ratios is that the frequency of carriers in New Zealand Angus cattle was quite high before the scheme commenced, being around 10 per cent. For other populations and/or other defects where the initial frequency is lower, the benefit : cost ratios might not be as favourable, and may even be less than unity. It is therefore important that:

> *The economic viability of a proposed genetic control programme should be investigated carefully before embarking on that programme.*

11.3.7.3 Inherited bleeding disorders in dogs   Genetic control programmes against inherited bleeding disorders in dogs in the USA have been remarkably successful (Dodds *et al.* 1981), especially when it is considered that they have been voluntary programmes. Their success has been largely due to good public relations on the part of those people doing the biochemical screening, and to *genetic counselling*, which is the provision of genetic advice in relation to inherited defects or diseases. In one such successful programme, the incidence of factor X deficiency (see Table 3.3) was reduced from about 20 per cent in 1972 to less than 3 per cent in 1981, as a result of the screening of more than 2500 dogs for factor X activity in a nation-wide heterozygote detection programme. Another autosomal defect, von Willebrand's disease, has also been subjected to a similar screening programme (for von Willebrand factor protein, VWF: see Table 3.3 and Section 3.4) in seven different breeds. In the two-year period 1979 to 1981, incidence decreased significantly in three of these breeds, and showed a trend towards reduction in two others. The most likely explanation for only partial success in some breeds is that the control programmes are not compulsory.

Biochemical screening programmes have also been conducted in dogs for the two X-linked bleeding defects, haemophilia A and haemophilia B (see Section 3.4). Because these defects are X-linked and recessive, the screening programmes have concentrated on detecting carrier females. This is done by testing for coagulation factor IX in the case of haemophilia B, and for factor VIII procoagulant activity protein, VIIIC, and VWF in the case of haemophilia A (see Table 3.3 and Section 3.4). Despite the difficulties caused by random X-inactivation in relation to carrier detection of X-linked defects in females, as described in Section 3.4, these control programmes have successfully eliminated haemophilia A and B from many families.

11.3.7.4 Inherited eye diseases in dogs   Currently, there are no biochemical screening techniques available for the detection of carriers of inherited eye diseases in dogs. Nevertheless, some control programmes against inherited eye diseases have been very successful. The best example is the programme for control of early-onset progressive retinal atrophy (PRA) in Irish Setters in England, which effectively eradicated the disease over a period of 13 years, commencing in 1947. In this programme, breeders were obliged by the Irish Setter Association to sign a 'form of undertaking' not to breed from affected dogs or from proven carriers. In addition, the Kennel Club required that an animal could be registered only if:

(1)   its owner signed a declaration that the animal was neither affected nor known to have produced an affected offspring, and
(2)   the owners of the sire and dam signed similar declarations.

Finally, because PRA in Irish Setters is not lethal, it was possible to test-mate prospective parents by mating them to known homozygotes, and because early-onset PRA can be detected in animals less than one year old, the results of the test matings could be obtained in reasonable time.

Several similar genetic control programmes for inherited eye diseases in dogs are being conducted in Great Britain and in the USA. For descriptions of these schemes, see Dodds *et al.* (1981) and Jolly *et al.* (1981).

11.3.7.5 K88 Escherichia coli scours in pigs   In Section 5.6.7 we saw that pigs that are homozygous for a recessive gene *s* are resistant to

K88 *E. coli* scours, because they lack a receptor to which the *E. coli* need to attach in order to be pathogenic. Susceptible pigs are either *SS* or *Ss*. Although this disease is caused by a pathogen, it is included here because there is a very simple and yet unusual genetic means of immediately controlling the disease; K88 *E. coli* scours can be prevented by using only resistant (*ss*) boars to breed offspring.

At first sight this might not seem to be a sensible proposal, because if any sows are susceptible (*SS* or *Ss*), then there is either a 100 per cent chance or a 50 per cent chance that their offspring will be susceptible if these sows are mated to only resistant (*ss*) boars. However, as explained in Section 5.6.7, susceptible offspring born to susceptible sows receive passive immunity from their sows and consequently do not suffer from K88 *E. coli* scours. In fact, the only offspring that suffer from the disease are susceptible offspring from the mating of susceptible boars to resistant sows. It follows that:

> *The problem of K88* E. coli *scours can be immediately controlled by using only resistant boars.*

This conclusion provides an interesting contrast to the situations described above, where the favourable gene is dominant and hence the disease can be controlled by ensuring that at least one parent is homozygous for the dominant gene.

Initially, a major drawback to the practical application of the above proposal was in identifying resistant boars; the only possibilities were test matings which were expensive and time-consuming, or an *in vitro* test using a section of small intestine which could be obtained only by slaughtering the pig. Fortunately, Snodgrass *et al.* (1981) successfully modified the *in vitro* test so that it could be carried out with a small sample of intestinal epithelial cells obtained by biopsy. The test is simple and effective, and enables easy and rapid identification of resistant pigs prior to breeding age.

11.3.7.6 Stress-susceptibility in pigs   There is a recognized syndrome in pigs known as *porcine stress syndrome* (PSS). It is characterized by sudden death in response to stress, and by pale, soft exudative (PSE) meat in carcases of affected pigs. The rapid death is the result of the affected pig exhibiting malignant hyperthermia syndrome (MHS), which, as we saw in Section 6.3, can be triggered by stress or by exposure to halothane. The key to the genetic control of PSS in pigs is that reaction to halothane is a single-locus, autosomal recessive characteristic. Thus pigs susceptible to PSS can be detected by a halothane test, as described in Section 6.3.

In terms of genetic control programmes, the halothane test is a form of clinical screening. It could therefore be used, possibly in conjunction with pedigree analysis and test matings, to reduce the frequency of PSS within a population. However, reaction to halothane is positively correlated with certain economically important characteristics including eye muscle area, killing-out percentage and percentage of lean in the carcase. Because of these correlations, there may be situations in which breeders wish to maintain the 'halothane' allele in a population of pigs that is used as a source of parents for commercial, slaughter-generation pigs. This proposal is quite feasible, because so long as the other parent of slaughter-generation pigs is homozygous for the normal (non-reactor) gene, then all slaughter-generation pigs will be either heterozygous or homozygous normal, and hence will not show PSS. Since the halothane gene appears to be associated also with reduced litter size, it would be best to have the female parent homozygous normal. The male parent could then come from a line of pigs selected strongly for desirable carcase characteristics, which may well have the halothane gene at a relatively high frequency. Obviously, PSS is likely to be a problem in this particular line, but breeders are often willing to cope with PSS in a parent-generation line, so long as they can avoid it in the slaughter-generation.

Although it is used extensively throughout the world, the halothane test has several disadvantages, including the need for experienced operators to conduct the test successfully, and the potential danger to these operators of over-exposure to halothane. It would be far simpler if reactors could be detected with a blood test, and there are indications that this may be possible in the future, because reactors show increased activity of the enzyme creatine phosphokinase.

Another possibility for the future involves exploiting the fact that the halothane locus is very closely linked to three loci whose products are easily detectable in a blood test. These loci are the H blood group locus (see Table 8.2), the locus for phosphohexose isomerase (Phi), and the locus for 6-phosphogluconate dehydrogenase (6-PGD). The closeness of the linkage is indicated by the observation that H and Phi are about 2 centimorgans apart, and that the halothane locus lies in between these two loci, but closer to Phi than to H (Andresen *et al.* 1980). Because of this close linkage, genes at these four loci tend to be inherited together in the same haplotype for many generations, in a manner similar to that observed with genes of the major histocompatibility complex, which was discussed in Section 8.4. It follows that a particular combination of easily detectable genes at the H, Phi and 6-PGD loci could be an accurate marker for the halothane gene, and another set of genes could be an equally accurate marker for the normal gene. These combinations would probably differ from population to population, but within a population they could enable heterozygote detection and hence a much greater degree of genetic control over PSS.

11.3.7.7 Hip dysplasia in dogs Hip dysplasia (HD) is probably the best-known of all animal disorders that are familial but not due to a single gene; it has been the subject of more research, and more controversy than any other similar disorder. One of the causes of controversy is that HD is traditionally diagnosed not on clinical signs but by subjective evaluation of a radiograph, as shown in Fig. 6.2. The problem with this is that in most populations, the incidence of abnormal hips as diagnosed by radiography is much higher than the incidence of clinical hip dysplasia (CHD). This immediately creates a credibility gap between veterinarians and breeders, because all too often a dog that can jump a fence six feet high is diagnosed as being dysplastic according to its radiograph. The solution to this problem is for breeders and veterinarians to realize that there is a difference between *selection criteria* and *selection objectives*. As explained in Section 16.9, the latter are the characters that we wish to improve by selection. They are chosen irrespective of whether they are cheap or expensive or even impossible to measure. In the present context, there is just one selection objective, namely CHD. Having decided upon the selection objective, the next step in designing a breeding programme is to decide upon the selection criteria, which are the characters that will actually be measured in the programme. Selection criteria are often the same as selection objectives, but they do not have to be the same. Since CHD is very difficult to measure, and may not be expressed at a young age, the selection criterion in most HD control programmes is 'radiographic' hip dysplasia (RHD), which is readily assessable at an early age. The fact that RHD is assessed subjectively does not detract from its usefulness as a selection criterion. All that is required is that RHD be measured on an arbitrary scale of, say, 1 to 5, and that this measurement has a positive genetic correlation (see Section 16.8) with CHD. Although no estimates of this correlation have been made, the available evidence indicates that it is positive and sufficiently high to justify the use of RHD as the selection criterion in control programmes.

The failure to distinguish between selection objectives and selection criteria has led to much confusion in relation to HD control programmes. The problem is that most articles and papers on HD discuss radiographic diagnosis as if it were clinical diagnosis. Thus, when people talk about the incidence of HD being altered as a result of a selection programme, they are almost always referring to the incidence of RHD and not CHD. Similarly, when people estimate the heritability of HD, they always estimate it from data on RHD rather than CHD. Because of the dearth of information on CHD, we have no alternative in this chapter but to discuss HD control programmes in terms of RHD. Readers should appreciate, however, that it would be far more satisfactory if the following summary of control programmes could be presented in terms of CHD rather than RHD.

The heritability of liability to RHD has been estimated on many occasions, with estimates usually falling in the range 0.25 to 0.50. This is sufficiently high to justify a selection programme based on simple *mass selection*, which is selection of individuals according to their own phenotype. As mentioned above, phenotype for RHD is usually measured on an arbitrary scale from, say, 1 to 5.

When we discussed familial disorders in Chapter 6, we concluded that:

(1)   the more severely an individual is affected, the more frequent and severe will be the disorder in the offspring of that affected individual, and

(2)   among normal individuals, the lower their genetic relationship with affected individuals and the larger the proportion of their relatives that are normal, the less frequent and severe will be the disorder in their offspring.

In the early stages of an HD control programme, when the incidence is still quite high, the first of these conclusions is particularly relevant; all potential parents should be ranked according to severity, and as many as possible should be culled according to severity. As the incidence decreases over time, a stage will be reached when the least affected category of potential parents contains more animals than are needed as actual parents. This is the stage at which the second conclusion becomes relevant; the aim then is to select those animals having the lowest genetic relationship with affected individuals and the largest proportion of normal relatives. As the programme progresses, affected animals will gradually disappear from the pedigrees of selected animals.

When selection has been applied along these lines in well-controlled breeding programmes over several years, the incidence of RHD has fallen substantially. Examples of successful programmes in relatively large populations are given in Table 11.3, and there are many undocumented examples of successful selection programmes in individual kennels.

However, not all HD control programmes have been successful; some have produced no change in the incidence of RHD despite the radiographing of thousands of dogs and despite the existence of soundly-based guidelines for breeders. In most cases, these failures can be attributed to the fact that only a small proportion of breeders in these programmes actually followed the guidelines; the remainder either bred from dogs that had not been radiographed because it was thought that those dogs had a high chance of failing the radiographic test, or bred from dogs that had failed the radiographic test. If the guidelines were compulsory, then these practices would not be possible. But in most cases the guidelines are voluntary, and it is therefore not surprising that some HD control programmes have been unsuccessful.

Table 11.3. The results of successful genetic control programmes against hip dysplasia in dogs

Change in incidence		Number of years	Breed	Country	Reference
From	To				
0.45	0.29	7	Swiss sheepdogs	Switzerland	Freudiger *et al.* (1973)
0.40	0.20	4	German Shepherds	USA	Riser (1973)
0.44	0.12	7	German Shepherds	German Democratic Republic	Böhme *et al.* (1978)
0.41	0.28	14	German Shepherds	Finland	Paatsama (1979)
0.43	0.10	14	Boxers	Finland	Paatsama (1979)
0.50	0.27	5	German Shepherds	Sweden	Hedhammar *et al.* (1979)

However, particular breeders who have been willing and able to follow the guidelines have achieved success.

11.3.8 DNA screening

Although biochemical screening is a powerful technique for heterozygote detection, it has one major limitation: it can be used only for those loci whose polypeptide product is present in readily available body fluids or tissues such as blood plasma or blood cells. Fortunately, recent advances in molecular genetics have now opened the way for overcoming this limitation, by enabling screening of DNA itself rather than screening of the polypeptide product of DNA.

The technique relies on the existence of *DNA polymorphisms*, which are differences between individuals in relation to the presence or absence of particular restriction enzyme sites (see Section 2.2). Suppose, for example, that one individual is homozygous for the nucleotide sequence TCGA at one particular site in its genome, that another individual is homozygous for the sequence TCAA at the same site, and that a third individual is heterozygous at this site. Suppose further, that this particular site of four nucleotides occurs within a section of DNA for which a probe (see Section 2.7) exists. If the DNA from any nucleated cell, e.g. lymphocytes, of these three individuals is digested with the relevant restriction enzyme, which in this case is *Alu*I, and then treated with the appropriate probe after Southern blotting (see Section 2.7), a different pattern of electrophoretic bands will be revealed for each of the three genotypes, as shown in Fig. 11.5.

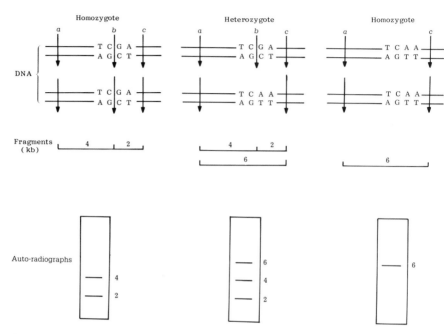

Fig. 11.5. A hypothetical example of detection of a DNA polymorphism. Cleavage sites for the enzyme *Alu*I are indicated by arrows. Since sites *a* and *c* are the same in all individuals, their nucleotide sequence $\left(\begin{matrix} T C G A \\ A G C T \end{matrix}\right.$; see Table 2.1$\left.\right)$ is not shown. The nucleotide sequence at site *b* exists in two different forms, only one of which can be cut by the restriction enzyme *Alu*I. The end result is that the three different genotypes can be distinguished according to the different lengths (shown in kb) of DNA fragments produced by digestion with *Alu*I. The patterns shown in this figure will result if the probe (see Section 2.7) used for detection includes the *b* cleavage site as well as regions immediately adjacent to it on both sides. If the probe detects a region only between sites *a* and *b* but not including site *b*, then the 2 kb band will not be visible. Similarly, if the probe detects a region only between sites *b* and *c*, then the 4 kb band will not be visible. But in all three cases, the three genotypes are distinguishable according to band patterns.

DNA polymorphisms can be used in screening for single-locus diseases either directly or indirectly. The direct method is applicable if the disease is due to a change in nucleotide sequence that coincidentally gives rise to a DNA polymorphism that is detectable because it corresponds to a restriction enzyme site. The sickle cell gene in humans (see Section 3.5) is now detectable in this way, because the normal sixth codon, CTC, of the β chain is part of a restriction site, whereas the sixth codon in the sickle cell β chain, CAC, is not. Many more single-locus diseases are likely to become detectable in this manner in both humans and animals in the future.

The indirect method is relevant to situations in which a DNA polymorphism is closely linked to the disease locus. This method is analogous to the use of closely linked marker loci in the detection of PSS in pigs, as described in Section 11.3.7.6. But it is much more powerful because DNA polymorphisms occur throughout the genome. In addition, whereas a blood group or enzyme locus that is 'closely linked' to a disease locus may be one or two centimorgans (equivalent to one or two thousand kb) from the disease locus, a 'closely linked' DNA polymorphism may be only, say, 0.005 centimorgans (five kb) from the disease locus. Recalling that 0.005 centimorgans represents a recombination fraction of 0.005 per cent (see caption to Fig. 1.22), it is obvious that appropriate DNA polymorphisms are likely to be much more reliable markers for disease loci, because crossing-over is much less likely to destroy the association between the marker and the disease locus.

Since the techniques of DNA screening are in their infancy, there are no good animal examples available at present. But the techniques are so much more powerful than other techniques, that it must be only a matter of time before they come into general use.

11.3.9 Gene therapy

Gene therapy is a logical extension of enzyme replacement therapy, which, as we saw in Section 11.2, involves supplying the patient with regular doses of a particular enzyme, or with cells or an organ that will provide a regular supply of that enzyme. One of the major disadvantages of the second approach, namely the problem of rejection, could be overcome if the patient's own cells could be removed, 'repaired', and then replaced. This is a form of gene therapy, because the 'repairing' phase involves the addition of normal genes to cells that were previously defective. Considerable research effort has been devoted to this form of gene therapy, and although some progress has been made, there are still many technical problems to be solved. The steps involved are:

1. Identify and isolate the normal gene at the relevant locus, or construct it (Fig. 2.7) if the appropriate amino acid sequence is known.
2. Produce large numbers of copies of the gene, by cloning.
3. Insert this 'foreign' DNA into appropriate cells previously removed from the patient.
4. Replace the 'repaired' cells in the patient.

Even if it can be made to work regularly and effectively, this form of gene therapy is applicable only to those diseases in which the relevant gene is expressed in only one tissue. In any case, it is doubtful whether such therapy will be of much practical use in domestic animals. The reason for this is that the technology of heterozygote detection using DNA polymorphisms is likely to progress at least as fast as that of gene therapy, so that by the time gene therapy becomes a practical reality, heterozygote detection will most likely be available for nearly all single-locus diseases. Once this stage is reached, it would be far easier to remove harmful genes from populations by conventional selection. Only in species such as humans, where this type of selection is unacceptable, is the addition of normal genes to previously defective cells likely to have much impact.

The one area in which gene manipulation at the molecular level may be of use in domestic animals is where it may be advantageous to introduce additional copies of relevant genes, or to introduce a gene from a different species. It is possible, for example, that sheep with extra keratin genes would produce more wool, or that a particular gene from goats may confer strong resistance to a certain disease in sheep. In situations such as this, gene 'therapy' applied to fertilized eggs or to embryos may be of considerable use in domestic animals.

The most promising approach involves adding the relevant foreign DNA to a fertilized egg or to an embryo at the blastocyst stage. In the former case, the DNA is injected into one of the pronuclei of the fertilized egg, before the two pronuclei fuse (Fig. 11.6). In the latter case, the DNA is added to laboratory cultures of teratocarcinoma cells, which have the unusual property of being able to participate in normal embryogenesis when injected into a blastocyst, contributing to all tissue types found in the resultant adult. Thus teratocarcinoma cells act as vectors for the transfer of foreign DNA into an embryo. Once again, there are considerable technical difficulties to be overcome before either of these techniques provides regularly reproducible results, but some progress has been made. For example, foreign DNA injected into the male pronucleus of mouse eggs has subsequently been expressed in the appropriate tissues of the resultant adults, and has been passed on, via natural mating, to the offspring of these adults (e.g. Wagner *et al.* 1981). This last observation highlights the major difference between gene therapy applied to single tissues and gene therapy applied to fertilized eggs or embryos. In the former case, the germ line is not involved and thus surviving 'patients' will pass on their own genes but not their foreign genes to future generations. In contrast, gene therapy applied to fertilized eggs or to embryos involves inserting foreign DNA into all or most of the cells, including the germ cells, of the resultant individuals, who then pass on their own genes and their foreign genes to future generations.

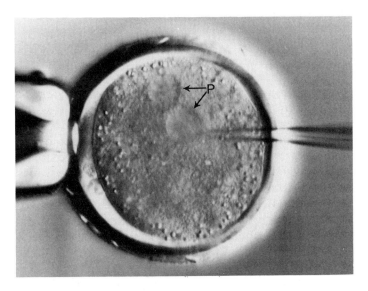

Fig. 11.6. Injecting DNA into a fertilized sheep egg. The egg is kept in position by a holding pipette (left), while the DNA is inserted into the egg by means of an injection pipette (right). The DNA is injected into either of the two pronuclei (P), one of which is the haploid nucleus of the sperm cell (called the male pronucleus), while the other is the haploid nucleus of the egg cell (called the female pronucleus). (Courtesy of J. Murray.)

11.4 Summary

It is commonly believed that if a disease is inherited, then there is nothing that can be done about that disease except to select against it. An equally common belief is that if a non-genetic treatment is found to be effective against a certain disease, then that disease cannot have a genetic basis. Both these beliefs are without foundation. In fact, the paradox of inherited diseases is that as knowledge about the genetic basis of the disease increases, so too does the likelihood of finding a non-genetic treatment for it. There are many examples of non-genetic treatments reducing the liability to inherited diseases.

There are several different types of programmes for the genetic control of inherited disease. If the disease can be detected by clinical examination prior to reproductive age, then it can be selected against by culling affected individuals and, more effectively, by culling parents of affected individuals. A more effective approach to genetic control of single-locus recessive diseases involves ensuring that in all matings at least one parent is homozygous normal. Pedigree analyses and test matings can be used to help distinguish homozygous normal individuals from carriers. Insurance schemes have the double virtue of providing financial protection for buyers of young breeding stock, and encouraging breeders to report the birth of all affected offspring. With inborn errors of metabolism, the phenomenon of gene–dosage provides a very quick and simple means of distinguishing homozygous normals from carriers at an early age. Although voluntary control programmes are easier to run, they are usually much less effective than compulsory programmes. Examples of inherited diseases that have been or could be controlled by genetic means include tibial hemimelia and mannosidosis in cattle, bleeding and eye disorders in dogs, K88 *E. coli* scours and porcine stress syndrome in pigs, and hip dysplasia in dogs.

DNA screening relies on the existence of DNA polymorphisms, which are differences between individuals in relation to the presence or absence of particular restriction enzyme sites. This type of screening can be used either directly or indirectly to distinguish all genotypes at particular disease loci.

Gene therapy involves the addition of foreign DNA to cells or to a fertilized egg or to an embryo. The only type of gene therapy likely to be of use in domestic animals is where it may be advantageous to introduce additional copies of relevant genes, or to introduce genes from different species.

11.5 Further reading

Anderson, W. F. and Diacumakos, E. G. (1981). Genetic engineering in mammalian cells. *Scientific American* 245(1), 60–93. (A review of the injection of 'foreign' genes into mammalian cells cultured in the laboratory.)

Dodds, W. J., Moynihan, A. C., Fisher, T. M., and Trauner, D. B. (1981). The frequencies of inherited blood and eye diseases as determined by genetic screening programs. *Journal of the American Animal Hospital Association* 17, 697–704. (A review of screening and genetic control programmes in dogs.)

Done, J. T. (1981). Hereditary disease in cattle: detection of carrier bulls. *Veterinary Annual* 21, 100–5. (A review of methods of heterozygote detection, with particular emphasis on the economic aspects.)

Jolly, R. D. (1978). Mannosidosis and its control in Angus and Murray Grey cattle. *New Zealand Veterinary Journal* 26, 194–8. (An account of the mannosidosis control programme in New Zealand.)

Jolly, R. D., Dodds, W. J., Ruth, G. R. and Trauner, D. B. (1981). Screening for genetic diseases: principles and practice. *Advances in Veterinary Science and Comparative Medicine* 25, 245–76. (A detailed review of screening and genetic control programmes in domestic animals.)

Pollock, D. L., Fitzsimons, J., Deas, W. D., and Fraser, J. A. (1979). Pregnancy termination in the control of the tibial hemimelia syndrome in Galloway cattle. *Veterinary Record* 104, 258–60. (A summary of the tibial hemimelia genetic control programme, with particular emphasis on early foetal recovery as a means of increasing the effectiveness of the herd of carrier cows.)

Smith, C. and Webb, A. J. (1981). Effects of major genes on animal breeding strategies. *Zeitschrift für Tierzüchtung und Züchtungsbiologie* **98**, 161–9. (This paper includes a brief discussion of prior and posterior probabilities, and also considers the ways in which the halothane gene can be exploited in breeding programmes.)

Trueman, J. W. H. (1978). A program to reduce the incidence of combined immuno-deficiency. *Theriogenology* **10**, 365–70. (A useful illustration of the calculation prior and posterior probabilities for particular genotypes, and the application of these calculations to a genetic control programme.)

Part III
Genetics and animal breeding

Introduction

Part III of this book covers genetic aspects of animal breeding, with an emphasis on the application of genetic knowledge to improvement of animal populations. We start by considering the use of single genes in animal breeding (Chapter 12). We then move on to consider some basic concepts that are relevant to animal improvement programmes, namely relationship and inbreeding (Chapter 13) and variation and heritability (Chapter 14). The remaining chapters are concerned with the achievement of animal improvement by means of selection and crossing.

Many books contain material that is directly relevant to the topics mentioned above. For the benefit of readers who may require another point of view or further information on a particular topic, a list of relevant books is given below.

Text and reference books

Becker, W. A. (1984). *Manual of quantitative genetics* (4th edn). Academic Enterprises, P.O. Box 666, Pullman, Washington.

Bowman, J. C. (1984). *An introduction to animal breeding* (2nd edn). Edward Arnold, London.

Bulmer, M. G. (1980). *The mathematical theory of quantitative genetics.* Clarendon Press, Oxford.

Chapman, A. B. (Ed.) (1985). *General and quantitative genetics.* (Vol. A4, *World Animal Science* series.) Elsevier, Amsterdam.

Cunningham, E. P. (1969). *Animal breeding theory.* Landbruksbokhandelen, Universitetsforlaget Vollebekk, Oslo. (Available from the author at The Agricultural Institute, Dunsinea Research Centre, Castleknock, Co. Dublin, Ireland.)

Dalton, D. C. (1976). *Animal breeding: first principles for farmers.* Aster Books, Wellington.

Dalton, D. C. (1980). *An introduction to practical animal breeding.* Granada, London.

Falconer, D. S. (1981). *Introduction to quantitative genetics* (2nd edn). Longman, London.

Falconer, D. S. (1983). *Problems in quantitative genetics.* Longman, London.

Johansson, I. and Rendel, J. (1968). *Genetics and animal breeding.* W. H. Freeman, San Francisco.

Lasley, J. F. (1978). *Genetics of livestock improvement* (3rd edn). Prentice Hall, Englewood Cliffs, New Jersey.

Lerner, I. (1958). *The genetic basis of selection.* Wiley, New York.

Lerner, I. and Donald, H. P. (1966). *Modern developments in animal breeding.* Academic Press, London.

Lush, J. L. (1945). *Animal breeding plans* (3rd edn). Iowa State University Press, Ames, Iowa.

Maciejowski, J. and Zieba, J. (1983). *Genetics and animal breeding.* Elsevier, Amsterdam.

Ollivier, L. (1981). *Éléments de génétique quantitative.* Masson, Paris.

Pirchner, F. (1983). *Population genetics in animal breeding* (2nd edn). Plenum Press, New York.

Rice, V. A., Andrews, F. N., Warwick, E. J., and Legates, J. E. (1970). *Breeding and improvement of farm animals* (6th edn). McGraw-Hill, New York.

Turner, H. N. and Young, S. S. Y. (1969). *Quantitative genetics in sheep breeding.* Macmillan, Melbourne.

Van Vleck, L. D. (1979). *Notes on the theory and application of selection principles for the genetic improvement of animals.* Department of Animal Science, Cornell University, Ithaca, New York.

Single genes in animal breeding

12.1 Introduction

In previous chapters we discussed many characters that are determined by alleles at single loci, including blood-cell antigens and a large variety of defects and diseases. In addition to those described previously, there are many other characters determined by single loci, some of which are of particular importance to animal breeding. The aim of this chapter is to discuss such characters, and then to consider some of the practical uses to which single-gene characters can be put.

12.2 Coat colour

Coat colour in mammals is due to the presence in hair and wool of pigment granules consisting of melanins in a protein framework. The melanins in these granules are formed by a series of metabolic pathways that convert tyrosine into either *eumelanins* (dark colour) or *phaeomelanins* (light colour). Eumelanins are often described as black, although they include brown and the derivatives of black and brown. Phaeomelanins, which are often called yellow, can range from a rather bright yellow through to red. The cells in which melanin production occurs are called *melanocytes*, and the resultant pigment granules are known as *eumelanosomes* or *phaeomelanosomes*.

Although much remains to be discovered about the genetic control of pigment formation, it is now generally accepted that at least six autosomal loci, each with multiple alleles, influence the production and distribution of pigment. The most interesting aspect of these six loci is that they appear to exist in all mammals. Although not all species of mammals have all known alleles at any locus, there are enough examples of similar coat colours and/or patterns inherited in a similar way in several different species to present a convincing picture of homology of coat colour and coat pattern alleles among species (Searle 1968).

Because there is a series of more than two alleles at most of the coat-colour loci, each locus is often referred to as a *series*. A summary of characteristics of the six main series (loci) is given in Table 12.1. Further details are given below.

Table 12.1. Characteristics of the six main allelic series (loci) determining coat colour in mammals. Each locus is autosomal

Series	Symbol	Main alleles	Effects and mode of action	Site of action
Agouti	A	$A^y, A^w, A,$ a^t, a, a^e	Controls regional distribution of black and yellow pigments in body and in individual hairs, from all yellow to all black	Outside melanocyte
Brown	B	B^{lt}, B, b, b^l	Affects eumelanins, changing black to brown (bb). May lighten eyes (bb) and under-fur ($B^{lt}-$)	Within melanocyte
Albino	C	$C, c^{ch}, c^b,$ c^s, c^a, c	Reduces intensity of pigmentation, first yellow and then black, until none is left in cc (albino)	Within melanocyte
Dilute	D	D, d, d^l	Dilutes both black and yellow colours by clumping pigment granules. Dilute lethals (d^l) have convulsions	Within melanocyte, changing shape
Extension	E	$E^d, E,$ e^{br}, e	Extends black (dominant) or yellow (recessive) pigment in body as a whole, with e^{br} giving a black/yellow variegation	Within melanocyte
Pink-eyed	P	P, p, p^s	Main effect on eumelanosomes, with dark colours much more diluted than light ones. Retinal pigmentation removed in p and p^s, with the latter also causing male sterility	Within melanocyte

Adapted from Prota and Searle (1978).

12.2.1 Agouti (A) series

The term agouti refers to the 'salt and pepper' appearance of coats of wild mice and rabbits, and also of the South American rodent from which the series derives its name. The distinctive coat colour of the wild-type, A, allele is due to the presence of a terminal or subterminal yellow band in otherwise black hairs. Some of the other alleles at this locus increase or decrease the width of the yellow band within each hair, giving rise to the two extremes of sable or yellow (where the whole hair is yellow) and black or non-agouti (where the yellow band is absent). Other alleles, such as 'saddle' in German Shepherd dogs, and bicolour (black and tan) in Collies, affect the distribution of pigment over the body.

12.2.2 Brown (B) series

Alleles at this locus are concerned with the protein portion of pigment granules, which affects the shape of granules. The B allele gives rise to normal, elongated granules which are black. The b allele, which is usually recessive to B, produces ovoid or spherical granules which are brown or chocolate or (in the case of horses) chestnut.

12.2.3 Albino (C) series

It seems likely that this locus is the structural locus for the enzyme tyrosinase which catalyses the first step in the conversion of tyrosine into melanin. The C allele produces normal tyrosinase and hence normal pigmentation. Other alleles produce less efficient forms of the enzyme which result in fewer granules and hence less pigment. Seven different alleles have been recognized at this locus, including those that have become the 'trademarks' of Burmese (c^b) and Siamese (c^s) cats, i.e. all Burmese cats are homozygous for c^b and all Siamese cats are homozygous for c^s. It is thought that these two alleles give rise to different temperature-sensitive forms of tyrosinase which function more efficiently in the body extremities, because such extremities are colder than other parts of the body.

12.2.4 Dilution (D) series

Alleles at this locus affect the intensity of pigmentation, but by controlling the clumping together of granules rather than by reducing the number of granules. In this way, the allele d dilutes black to blue and yellow to cream. Although d is recessive in most species, it appears to be codominant with D in horses, giving rise to chestnuts (DD), palominos (Dd) and cremellos (dd). However, as often happens in discussions of coat-colour genetics, not everyone agrees that palominos and cremellos are due to the d allele. An alternative explanation is that they are due to heterozygosity and homozygosity for a codominant allele in the Albino series.

12.2.5 Extension (E) series

Alleles at this locus are concerned with the extent of yellow and black pigment in the coat as a whole, as distinct from within each hair. The allele e^{br}, for example, gives a variegation of black and yellow, producing brindle coats in dogs and cattle.

12.2.6 Pink-eyed dilution (P) series

The p allele in this series affects the shape and structure of black pigment granules only, leaving yellow melanins untouched. Its name is derived from another effect of some of the alleles at this locus, namely the complete failure of pigment production in the retina.

12.2.7 Other loci

Other loci are known to be important in determining coat colour and pattern in certain species. In cats, for example, the autosomal Tabby locus has three alleles, giving rise to Abyssinian tabbies (T^a), mackeral tabbies (T) and blotched tabbies (t^b). Another easily recognizable locus in cats is the X-linked orange locus, which gives rise to the well-known tortoiseshell coat colour which was discussed in Section 1.8. The orange allele, O, gives full extension of yellow pigment in the same way as the yellow allele, A^y, does at the autosomal agouti locus. The non-orange allele, o, allows normal expression of agouti.

Another locus of some importance is the autosomal merle locus in dogs, with alleles m (normal) and M (merle). The M allele is codominant with respect to coat colour, but is also a recessive lethal. Thus MM dogs are unable to survive, which means that breeding merle (Mm) dogs can lead to problems. It should be possible to avoid these problems by mating merle (Mm) to non-merle (mm), which will produce approximately equal numbers of merle and non-merle offspring.

12.2.8 Complications

Although the inheritance of coat colours is quite straightforward in principle, there are several complications that arise in practice.

One of these is *epistasis*, or interaction between loci. An example of epistasis in relation to coat colour is the complete masking of the various tabby alleles at the tabby locus by the non-agouti allele; it is usually impossible to tell which tabby allele or alleles are present in cats that are homozygous for non-agouti, because such cats are uniformly black.

Another complication is that alleles at different loci can sometimes be used with more or less equal validity to explain a certain colour. In German Shepherd dogs, for example, black coat colour has been attributed to an allele at the A locus by some writers, and to an allele at the E locus by others. Extensive data from many matings are often required before a firm decision can be made in favour of one or other of the possibilities.

One of the most perplexing complications arises from the ambiguity with which coat colours are so often named. In horses, for example, even experienced breeders sometimes disagree as to what to call the coat colour of a particular horse. This problem is not helped by the tendency for some coats to change colour at different times of the year, and at different stages of a horse's life. It is very important, therefore, that any discussion of the inheritance of particular colours, especially in horses, should contain precise definitions of each relevant colour, supported if possible by colour photographs.

12.2.9 Coat colour rules in horses

Despite the problem just outlined, there are two rules for coat colour inheritance in horses that are valid. The *Chestnut Rule* states that the mating of chestnut by chestnut will not produce a black, brown, bay, or grey. And the *Grey Rule* states that a grey horse must have at least one grey parent.

Both these rules have now been sufficiently tested for them to be of considerable value in checking the validity of pedigrees. If, for example, the results of a certain mating appear to break one or other of these rules, then the progeny of that mating should not be registered until a thorough check of the pedigreee has been made, because it is most likely that an error has been made in recording the pedigree.

12.2.10 The proliferation of new 'varieties'

With segregation at so many coat colour loci in various domestic species, it has been possible for enthusiasts to develop a large number of different genotypes, many of which are recognized as distinct varieties or even as distinct breeds. One of the best examples of this is in cats, the breeds and varieties of which are listed in Table 12.2. While there may now be other genetic differences between varieties and between breeds, it is important to realize that, in many cases, what are now called different breeds originally differed by alleles at only one or a few coat-colour loci. These differences in coat colour have in many cases become the trademark of the breed. Problems arise when, as in Palomino horses for example, the coat-colour trademark is due to heterozygosity at a coat-colour locus; it is then impossible to obtain a separate population that will breed true for the trademark.

12.2.11 Coloured wool

Wool colour can also be explained in terms of alleles at several of the loci described above. Although knowledge in this area is far from complete, the following tentative conclusions can be drawn. At the A locus, a dominant allele A^{wh} inhibits all eumelanin production but

Table 12.2. Genotypes of breeds and varieties of cats

Short hairs	Genotype	Long hairs	Genotype
Abyssinian, blue	ddT^a-	Balinese	llT^a-
Abyssinian, chocolate	bbT^a-	Bicoloured	$aallS-$
Abyssinian, sorrel	$b^lb^lT^a-$	Birman, blue	$aac^sc^sddggll$
Abyssinian, lilac	$bbddT^a-$	Birman, chocolate	$aabbc^sc^sggll$
Abyssinian, normal	T^a-	Birman, cream	$c^sc^sddggllO$
Abyssinian, red	OT^a-	Birman, lilac	$aabbc^sc^sddggll$
Abyssinian, silver	IT^a-	Birman, red	c^sc^sggllO
Albino, blue eyed	c^ac^a	Birman, seal	aac^sc^sggll
Bicoloured, black	$aaS-$	Black	$aall$
Bicoloured, blue	$aaddS-$	Blue	$aaddll$
Bicoloured, tabby	$aaS-t^bt^b$	Blue-cream	$aaddllO$
Black	aa	Cameo, cream	$ddI-llO$
Blue	$aadd$	Cameo, shaded	$I-llO$
British Blue	$aadd$	Cameo, shell	$I-llO$
Burmese, blue	aac^bc^bdd	Cameo, smoke	$I-llO$
Burmese, blue-cream	aac^bc^bddOo	Chocolate	$aabbll$
Burmese, brown	aac^bc^b	Cream	$ddllO$
Burmese, chocolate	$aabbc^bc^b$	Chinchilla	$I-ll$
Burmese, cream	c^bc^bddO	Colourpoint, blue	aac^sc^sddll
Burmese, lilac	$aabbc^bc^bdd$	Colourpoint, chocolate	$aabbc^sc^sll$
Burmese, red	c^bc^bO	Colourpoint, cream	c^sc^sddllO
Burmese, tortie	aac^bc^bOo	Colourpoint, lilac	$aabbc^sc^sddll$
Chartreuse	$aadd$	Colourpoint, red	c^sc^sllO
Chinese Harlequin, black	$aaSS$	Colourpoint, seal	aac^sc^sll
Chinese Harlequin, blue	$aaddSS$	Colourpoint, tortie	aac^sc^sllOo
Chocolate	$aabb$	Cymric	$llM-$
Cinnamon	aab^lb^l	Lilac	$aabbddll$
Egyptian Mau, bronze	bb	Maine Coon, black	$aall$
Egyptian Mau, cinnamon	b^lb^l	Tabby	llt^bt^b
Egyptian Mau, silver	$I-$	Red Self	Ot^bt^b
European Albino	c^ac^a	Shaded silver	$I-llt^bt^b$
Foreign white	c^sc^sW-	Silver	$I-llt^bt^b$
Havana	$aabb$	Smoke	$aaI-ll$
Japanese Bob-tail, black	$aaSS$	Somali, blue	$ddllT^a-$
Japanese Bob-tail, red	OSS	Somali, chocolate	$bbllT^a-$
Japanese Bob-tail, tortie	$aaOoSS$	Somali, lilac	$bbddllT^a-$
Korat	$aadd$	Somali, normal	llT^a-
Lilac	$aabbdd$	Tabby, brown	llt^bt^b
Lilac, light	aab^lb^ldd	Tabby, cream	$ddllO$
Russian Blue	$aadd$	Tabby, red	Ot^bt^b
Siamese, blue	aac^sc^sdd	Tabby, silver	$I-llt^bt^b$
Siamese, chocolate	$aabbc^sc^s$	Tortoiseshell	$aallOo$
Siamese, cream	c^sc^sddO	Tortoiseshell & White	$aallOoS-$
Siamese, lilac	$aabbc^sc^sdd$	Turkish, cream	$ddllOSS$
Siamese, red	c^sc^sO	Turkish, red	$llOSS$
Siamese, seal	aac^sc^s	White, blue eyed	$llW-$
Siamese, tabby, seal	c^sc^s	White, odd eyed	$llW-$
Siamese, tabby, blue	c^sc^sdd	White, orange eyed	$llW-$

Short hairs	Genotype
Siamese, tabby, choc.	bbc^sc^s
Siamese, tabby, lilac	bbc^sc^sdd
Siamese, tortie, seal	aac^sc^sOo
Siamese, tortie, blue	aac^sc^sddOo
Siamese, tortie, choc.	$aabbc^sc^sOo$
Siamese, tortie, lilac	$aabbc^sc^sddOo$
Silver	$I{-}t^bt^b$
Tabby, blue	ddt^bt^b
Tabby, brown	t^bt^b
Tabby, choc.	bbt^bt^b
Tabby, cinnamon	$b^lb^lt^bt^b$
Tabby, cream	$ddOt^bt^b$
Tabby, lilac	$bbddt^bt^b$
Tabby, red	Ot^bt^b
Tabby, silver	$I{-}t^bt^b$
Tonkanese	aac^bc^s
Tortoiseshell	$aaOo$
Tortoiseshell & White	$aaOoS{-}$
White	$W{-}$

A locus: a = non-agouti
B locus: b = brown, b^l = light brown
C locus: c^b = Burmese, c^s = Siamese, c^a = blue-eyed albino
D locus: d = dilution
G locus: g = gloving
I locus: I = melanin inhibitor
L locus: l = long hair
O locus: O = orange, o = non-orange (black)
S locus: S = dominant white spotting
T locus: T^a = abyssinian tabby, t^b = blotched tabby
W locus: W = dominant white

If a locus is not shown for a particular genotype, then the genotype at that locus is $A{-}$, $B{-}$, $C{-}$, $D{-}$, $G{-}$, ii, $L{-}$, oo or o, ss, $--$, and ww respectively, for the loci listed above (Reprinted with permission from R. Robinson *Genetics for cat breeders* (2nd edn), Copyright 1977, Pergamon Press. Corrections supplied by R. Robinson have been included.)

allows formation of phaeomelanin, giving rise to white or tan wool. Among the other alleles at this locus are A^b (badger face) and A^w (reverse badger face, or black mouflon), which inhibit eumelanin production on the upper and lower parts of the body respectively. The non-agouti allele, a, is recessive to all others at this locus in sheep and results in an entirely coloured fleece. Two alleles are recognized at the B locus, namely B (black) and b (brown or moorit). Among other important loci, the S (white spotting) locus has two alleles, S and s. In contrast to the situation in other mammals, white spotting in sheep appears to be due to the recessive allele, s.

In most modern domestic sheep, the dominant allele A^{wh} is at quite a high frequency, with the result that most sheep in these breeds are white. The recessive allele, a, is often at a low frequency, and alleles

B, *b*, *S*, and *s* may well be at intermediate frequencies. Since A^{wh} prevents the formation of eumelanin, it masks the effect of *B*, *b*, *S*, and *s*. It is only, therefore, in non-agouti (*aa*) sheep that the presence of these other alleles becomes evident.

With the upsurge in interest in the use of naturally coloured wools for the home-spinning industry, and with many people wishing to breed their own coloured wool, the genetic basis of wool colour has developed a new-found importance. Detailed studies of the inheritance of the many different colours and patterns that do occur around the world are now being undertaken. For a detailed review of knowledge in this area, readers are referred to Ryder (1980) and Adalsteinsson (1983).

12.3 Carpet wool

The Romney breed has been the predominant sheep breed in New Zealand since the 1920s. A typical New Zealand Romney fleece consists of a majority of unpigmented wool fibres, which are solid, and a minority of hair fibres, which have a core or medulla down the centre, filled with air. Traditionally, there has been strong selection against medullated (hairy) fibres in the New Zealand Romney.

In 1931, a purebred Romney ram lamb with an exceptionally hairy fleece was born on a property belonging to a Mr Nielson. It was given to Professor Dry at what is now Massey University, where it was mated to normal Romney ewes. Approximately equal numbers of hairy and normal lambs of both sexes were produced. Further matings confirmed that the extreme hairiness was due to an autosomal, incompletely-dominant allele which is now designated N^d (*N* for Nielson, and *d* for Dry), and which must have resulted from a mutation of the normal, non-hairy allele, *n*. Fleeces from sheep that are homozygous for N^d have around 65 per cent by weight of medullated fibres, which is ideal for carpet manufacture. Heterozygotes ($N^d n$) have a level of medullation intermediate between the two homozygotes. They can be easily distinguished as lambs, because they also have an intermediate distribution of 'halo' hairs, which are long, coarse fibres projecting above other fibres in the birth coat.

Being virtually free of pigmentation, and having good spinning qualities, wool resulting from the N^d allele soon became popular with carpet manufacturers, and the allele became known as the carpet wool gene. The numbers of $N^d N^d$ and $N^d n$ sheep were increased as rapidly as possible, giving rise to a new breed called the Drysdale, which is now the basis of a firmly established carpet wool industry.

Other carpet wool alleles have appeared in New Zealand from time to time, presumably due to other mutations at the same locus. The N^t allele, for example, arose in a purebred Romney flock belonging to a Mr Coop on his farm called Tuki Tuki. When homozygous, its effects are the same as that of N^d, but it differs from N^d in being completely dominant to n. Thus heterozygotes ($N^t n$) are indistinguishable from homozygotes ($N^t N^t$). Despite the difficulties that this creates for breeders wishing to establish a line of sheep homozygous for N^t, this allele has been used quite extensively for carpet wool production, and has been the basis for the formation of another carpet wool breed, known as the Tukidale. Yet another carpet wool allele, called N^j, arose in a part Border-Leicester/part Romney flock belonging to a Mr Johnson, and has given rise to yet another carpet wool breed: the Carpetmaster.

The creation of new breeds that differ from others by just a single allele is not new; it has happened in the past, for example, with Siamese and Burmese cats, and with Red Angus cattle. What is fascinating about the carpet wool alleles is that simple mutations at a particular locus have given rise to a new industry of considerable economic importance, not only in New Zealand, but also in several other parts of the world to which the various carpet wool alleles have been exported.

12.4 Prolificacy in sheep

Another sheep gene that has created considerable interest is the *fecundity* gene, F, that appears to be the major cause of the prolificacy seen in the Booroola strain of Australian Merino. Developed initially by two commercial sheep breeders (the Seears brothers) on a property called 'Booroola' in New South Wales, and further developed by the CSIRO, Booroola Merinos are one of the most prolific strains of sheep in the world, and are the only prolific strain having unpigmented wool of Merino quality.

The effects of the F gene on ovulation rate and on litter size in Merinos are illustrated in Fig. 12.1. It can be seen that FF homozygotes have an ovulation rate approximately three times that of homozygotes for the normal (+) gene, and that heterozygotes ($F+$) have an intermediate ovulation rate. The increases in ovulation rate are reflected in FF homozygotes having a litter size approximately double that of ++ homozygotes, with heterozygotes again being intermediate.

Although there is still much research to be done in relation to the F gene, it is already being exploited commercially in several countries, because it appears to offer a relatively easy means of significantly increasing prolificacy in the Merino and in other breeds of sheep.

12.5 Polledness

In general, the presence or absence of horns can be attributed to the action of two alleles at an autosomal locus, with polled (P) being dominant to horned (p). Many breeds of domestic livestock are horned and are therefore homozygous for p. With increasing concern over the

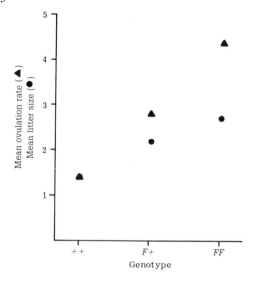

Fig. 12.1. Mean ovulation rate (▲) and litter size (●) for ewes that are homozygous (*FF*) or heterozygous (*F+*) or that lack (*++*) the fecundity gene. The data were obtained from ewes born in the CSIRO Booroola flock between 1973 and 1980. (Redrawn from Piper *et al.* 1984.)

practical problems of carcase damage and difficult handling that are often associated with horned animals, many practical breeders now prefer their animals to be polled. There are two ways this can be achieved: environmentally, by dehorning; and genetically, by the use of homozygous polled animals as parents. As in the case of carpet wool alleles, polled alleles arise from time to time in horned flocks or herds, as a result of mutation. Where there has been sufficient demand, these alleles have been used in, or introduced into, established breeds to create polled versions of these breeds. The main difficulty inherent in creating a polled version of any breed is the usual difficulty of attempting to select in favour of a dominant allele, as discussed in Section 5.6.4: it is very difficult to achieve complete homozygosity, because the recessive (horned) allele remains 'hidden' in heterozygotes. We shall return to this problem in Section 18.4.1, when we discuss grading-up.

A rather unusual problem arises with the occurrence of the polled gene in goats. In normal circumstances in other species of mammals, individuals that have two X chromosomes develop into normal females. And even in goats, XX individuals that are horned (*pp*) or heterozygous for polled (*Pp*) are normal females. But all XX goats that are homozygous for the polled allele are intersexes. In addition, a proportion of XY goats that are homozygous for the polled allele are sterile. The association between intersexuality and polledness in goats is summarized in Table 12.3. Although the reason for the association is not yet

Table 12.3. The association between intersexuality and polledness in goats. Notice that the *P* allele is dominant for polledness and recessive for intersexuality

Sex chromosomes	Genotype at the polled locus		
	PP	*Pp*	*pp*
XX	Polled intersex	Polled female	Horned female
XY	Polled male (some sterile)	Polled male	Horned male

known, its existence enables researchers to produce intersex goats whenever they need them, by mating, for example, a homozygous polled male that is fertile with a heterozygous polled female. Because the inheritance of sex chromosomes and of alleles at the polled locus follows the normal Mendelian rules, approximately one-half of the offspring have two X chromosomes and of these, approximately one-half are homozygous for *P* and are therefore intersexes. By investigating the nature of this 'ready-made' intersexuality in goats, reproductive biologists such as Shalev *et al.* (1980) are obtaining valuable insights into the basis of sex determination.

Although the above account of the genetic basis of polledness provides an adequate explanation for most cases of polledness in domestic animals, some occurrences require a more complicated explanation. For example, two extra loci, called African horn (Ha) and scurred (Sc), have been proposed in order to explain the inheritance pattern in some cattle (White and Ibsen 1936; Long and Gregory 1978). However, Frisch *et al.* (1980) have shown that not even these two extra loci can account satisfactorily for all observations.

Despite these difficulties, the single gene explanation for polledness is still a very useful guide to the inheritance of polledness in many domestic animals.

12.6 Dwarf poultry

There are alleles at several different loci in the domestic chicken that give rise to birds with a markedly lower mature body size. Most of these so-called *dwarfing genes* seriously affect viability or hatchability and are therefore of no commercial interest. There is, however, a Z-linked dwarfing gene, *dw*, that does appear to have some commercial value.

For practical purposes the *dw* gene can be considered as recessive.

Thus dwarf males are $Z^{dw}Z^{dw}$ and dwarf females are $Z^{dw}W$. At hatching they have the same weight as normal chickens, but their adult weights are reduced by 30 per cent in females and by 40 per cent in males. While there are still some doubts about the effect of the *dw* gene on reproductive ability, it appears that laying rate is reduced by 10 per cent or more in light strains but may not be affected in heavier strains. Egg weight is certainly reduced in absolute terms, but is probably increased relative to body weight. The effect on sexual maturity is variable, but hatchability is generally increased. The two most important effects of the *dw* gene are that daily food consumption is markedly decreased because of a reduced basal metabolism per unit body weight, and more birds can be housed in a shed of a given size, thus enabling housing costs to be spread over more birds.

Despite its several advantages, the *dw* gene has been used very little, if at all, in commercial layer strains, mainly because of problems with reduced laying rate and smaller egg size. In broiler breeding, however, it is creating considerable interest and several commercial broiler lines homozygous for *dw* have been developed. These *dw* lines are used as a source of female parents which are mated to males from a normal broiler strain. Mainly because of reduced feed consumption and a smaller space requirement, these *dw* broiler breeder lines are more economical than normal broiler lines. The resultant broilers have the same feed conversion efficiency and yield of meat as broilers from normal breeder lines. The only difference is that body weight is slightly reduced (by 3 or 4 per cent) because the *dw* gene is not completely recessive. This is, however, a small price to pay for the benefits outlined above.

The *dw* gene has been the subject of many studies in anatomy, nutrition, physiology, and behaviour. Readers requiring further information should consult the review by Guillaume (1976).

12.7 Genes for sexing chickens

Poultry breeders are continually faced with the need to separate day-old chickens into groups of males and females. The conventional method by which sexes are distinguished is vent sexing which requires skilled operators and which is therefore relatively expensive.

A much cheaper alternative, now widely used commercially, is to exploit genetically-determined differences between the sexes which can be distinguished easily by unskilled workers. There are two main systems of genetical sexing available in chickens, and we shall now discuss each one in turn.

12.7.1 Z-linked sexing

The most widely used system exploits differences in day-old chicks produced by alleles at one or other of two Z-linked loci, namely the feathering locus and the gold/silver locus.

At the feathering locus, slow feathering, K^s, is dominant to rapid feathering, k^+. In the presence of K^s, the primary flight feathers in day-old chicks are much shorter than another set of feathers called coverts, while with k^+ they are much longer. At the other locus, the silver allele, S, which inhibits phaeomelanin production, is dominant to the gold allele, s^+, which allows production of both eumelanins and phaeomelanins. Since all four alleles exert discernible effects in day-old chicks, either pair of alleles can be used to enable sexing to be carried out by unskilled workers.

Either of these loci can be used for sexing chicks produced by the crossing of two distinct lines, strains, or breeds. An illustration of their use is given in Fig. 12.2. Since most commercial layer and broiler breeding programmes involve the crossing of three or four separate lines (as we shall see in Section 19.4), Z-linked sexing can be very useful, especially if it can be exploited at those stages of the breeding programme where large numbers of one sex are required (as, for example, in the line that provides the female parent for the final cross).

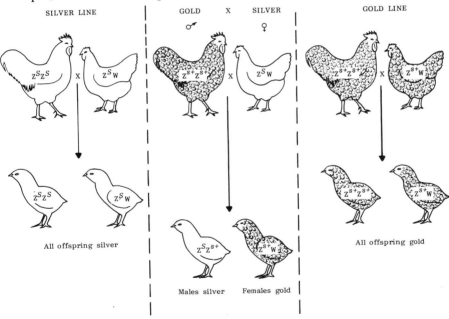

Fig. 12.2. The use of gold/silver Z-linked sexing in the offspring of a cross between two lines of chickens, one being homozygous for gold (Z^{s+}) and the other being homozygous for silver (Z^S). The cross shown is between a silver female and a gold male, giving all females gold and all males silver. It is left to the reader to verify that the reciprocal cross (gold female by silver male) would produce a different result that would not enable day-old chickens to be sexed according to colour.

Its major disadvantage is the time and effort required to incorporate the appropriate allele into the appropriate line.

While Z-linked sexing is useful for sexing the offspring from a cross between two lines, it is of no use in distinguishing the sex of day-old chicks *within* a line. This is because if there is segregation within a line at either of the above Z-linked loci, then males and females of both types will be produced within that line: Z-linked sexing works only if one line is homozygous for one allele and the other line is homozygous for the other allele. In order to sex chicks within a line, we have to use a different form of genetic sexing.

12.7.2 Automatic sexing within lines

As discussed in Section 1.8, the absence of dosage compensation in chickens results in barred males, $Z^B Z^B$ (with two doses of the barred gene) being noticeably lighter than barred females, $Z^B W$ (with only one dose of Z^B). If this effect were noticeable in day-old chicks, then a method of genetic sexing within a line would be readily available. Unfortunately, the effect of the barred gene is not visible in day-old chicks. It is, however, possible to exaggerate the effect of the barred gene by exploiting the effect of autosomal recessive genes that reduce pigmentation, such as *mo* (mottling). In lines homozygous for this gene and the barred gene, the difference between barred males and barred females is evident in day-old chicks, as shown in Fig. 12.3.

Fig. 12.3. Automatic sexing of day-old chickens in a line that is homozygous for barred, Z^B, and for mottling, *mo*. The *mo* gene enhances the difference between males and females caused by the barred gene, producing a discernible difference in day-old chicks. The female has broad stripes, while the male is essentially white. (From Silverudd 1978.)

In principle, therefore, the effect of *mo* and the barred gene could be exploited jointly to replace vent sexing within a line. In practice, there are several problems to overcome, not the least of which is the difficulty of incorporating these genes into any particular line. Although not used very extensively, automatic sexing within lines has been championed by Silverudd (1978), to whom readers are referred for further information.

12.8 Pedigree verification

Many breeding animals are bought on the strength of their pedigree. In some situations such as Thoroughbred yearling sales, it is the major source of information available on the animal being sold. It is obviously important for the buyer to be certain that the available pedigree is in fact correct.

It is easy, however, to think of reasons why the pedigree may not be correct. An honest mistake could have been made, for example, in recording the identity of the sire. This possibility is greatly enhanced in the case of artificial insemination, where straws or pellets could be incorrectly labelled, or where the wrong straw or pellet could be used in an insemination. Dishonest 'mistakes' are also possible. The most common examples are those that occur when the breeder wants the pedigree to look better than it actually is; a strong temptation indeed when large sums of money are at stake.

Because of the high risk of incorrect pedigrees occurring from time to time, breed societies are often keen to conduct random checks on pedigrees, and to verify the pedigree of each animal sold at official sales. The easiest and most effective method of checking pedigrees is to see if they are compatible with Mendelian inheritance at as many single loci as can be easily identified. Coat-colour loci and some of the other loci discussed in this chapter are sometimes useful. Even more useful, however, are blood-cell antigens and blood proteins.

In order to understand why blood-cell antigens and blood proteins are useful in checking pedigrees, we need to understand certain implications of their inheritance.

12.8.1 *Blood proteins*

Different forms of each blood protein are detected by the different speeds with which they travel in an electrophoretic gel; more highly charged molecules travel faster and therefore further in a given time. Because this technique can distinguish only between variants of a particular protein that differ in electric charge, it does not usually enable identification of all variants. Instead, it effectively groups together all variants with the same overall electric charge. However, irrespective of how many groups of variants exist for a particular protein, each group is transmitted from generation to generation as if it were the product of a codominant allele at the locus for that protein; individuals that have only one band on the gel are homozygous, and individuals with two bands are heterozygous. By observing an individual's phenotype (the number and position of bands on the gel), we can immediately write down its genotype; there is a one-to-one correspondence between phenotype and genotype.

The main complication that can occur with blood proteins is if an allele produces no protein, as happens in some cases, for example, with enzymes. If one of these so-called *null alleles* is present at a locus in a population, then there is no longer a one-to-one correspondence between phenotype and genotype, and the null allele is recessive to all other alleles at that locus; individuals with one band could be homozygous for the allele producing that band, or could be heterozygous for the null allele. All individuals that have two bands must still be heterozygous for alleles corresponding to the two bands, and hence all alleles that do produce protein will still be codominant with each other.

12.8.2 *Blood-cell antigens*

The situation is much the same for blood-cell antigens, which were discussed in Section 8.3. All individuals in which two antigens of a particular system are detected must be heterozygous for the two corresponding alleles. But in many situations, and especially in the early stages of research into a particular system, not all of the antigens actually present in a population will be detectable, because suitable typing sera will not yet have been developed. Thus individuals in which only one antigen is detected may be homozygous, or may be heterozygous for an allele that produces an antigen that cannot yet be detected. Such an allele is called 'blank' and is recessive to all alleles whose antigens can be detected. Among all alleles whose antigens can be detected, gene action is codominant.

12.8.3 *An example of parentage testing*

An example is shown in Table 12.4 of a Thoroughbred foal that was typed at six different red-cell antigen loci, and at two blood protein loci. For technical reasons, the foal's dam was typed at only three of the six antigen loci, but was typed at the same two protein loci. The alleged sire was typed at all six antigen loci and at the two protein loci. It is known that blank alleles exist at all six of the antigen loci, but that there are no null alleles at either of the protein loci. Using this knowledge, the genotypes corresponding to each phenotype for each animal can be written down, as shown in Table 12.4. Each locus is then checked for compatibility with Mendelian inheritance.

Starting with the A locus, it can be seen that since the foal lacks the a^H allele of its dam, it must have inherited the other allele, a^{A_1}, from

Table 12.4. Parentage test of a Thoroughbred foal

		Red-cell antigens						Proteins	
	Locus	A	C	D	K	P	Q	Albumin	Transferrin
Phenotype	Foal	A_1	C	DJ	K	P_1	Q	A	D
	Dam	A_1H	C	J	–	–	–	A	DH
	Sire?	A_1	C	J	K	P_1	Q	B	D
Genotype	Foal	a^{A_1}–	c^C–	$d^D d^J$	k^K–	p^{P_1}–	q^Q–	$Alb^A Alb^A$	$Tf^D Tf^D$
	Dam	$a^{A_1} a^H$	c^C–	d^J–	––	––	––	$Alb^A Alb^A$	$Tf^D Tf^H$
	Sire?	a^{A_1}–	c^C–	d^J–	k^K–	p^{P_1}–	q^Q–	$Alb^B Alb^B$	$Tf^D Tf^D$
Exclude Sire?		NO	NO	YES	NO	NO	NO	YES	NO

Adapted from Van Vleck (1977) in *The horse* by J. W. Evans, A. Borton, H. F. Hintz, and L. D. Van Vleck. Copyright 1977 by W. H. Freeman and Company. All rights reserved.

that parent. Thus the dash (representing either a^{A_1} or blank) must have come from the sire, which is quite possible. Thus, there is no evidence against the sire from the A locus. Neither is there from the C locus. For the next locus, D, the foal's genotype is known exactly, and it must have inherited the d^J allele from its dam. This in turn means that the foal's other allele, d^D, must have come from its sire, which is not compatible with the alleged sire's genotype. Providing that no mistakes have been made in blood typing, the conclusion is drawn that the alleged sire cannot be the true sire of the foal. The albumin locus provides supporting evidence.

12.8.4 The exclusion principle

It is important to realize that even though we have now proved that the alleged sire is not the true sire, we would have accepted him as the true sire if only the first two loci had been used in the test. This illustrates an important rule that applies to all cases of parentage checking. Known as the *exclusion principle*, it states that:

> *Although you may be able to prove from a parentage test that a particular animal is not the true parent, you can never prove that a particular animal is the true parent.*

12.8.5 Efficiency of parentage testing

The efficiency of parentage testing is measured in terms of the *exclusion probability*, which is the probability of excluding all but the true parent.

Consider some extreme examples. If only one locus is tested and if all individuals in the population are homozygous for the same allele at that locus (the gene frequency equals one), then the exclusion probability is zero because every male has the same phenotype. If there are two alleles at this locus in the population, then there is a chance of excluding the wrong sire. If one allele has a very high frequency, then this chance will be very low because nearly all individuals will still have the same phenotype. But as the allele frequencies tend to intermediate values, the exclusion probability increases. It is also increased if both alleles are codominant because then there are three different phenotypes rather than the two that would be seen with a recessive allele. In addition, the exclusion probability will be increased by increases in the number of alleles per locus, and in the number of loci, both of which give rise to more phenotypes.

In summary, the efficiency of parentage testing increases with the number of different phenotypes that are likely to occur, and this depends on: the number of loci; the number of alleles at each locus; the frequency of alleles at each locus; and the type of gene action.

It is now evident why coat colour loci are not as helpful as blood proteins or antigens in relation to parentage testing; there are usually only a small number of informative loci in any one breed, the number of alleles is generally small, and dominance is fairly common. Epistatic interactions between coat colour loci and ambiguity in definition of different colours make them even less useful for parentage testing.

Exact formulae have been developed for calculating the efficiency of parentage testing in uniparous and multiparous species. These are described and illustrated by Gundel and Reetz (1981).

12.9 Summary

Apart from the large number of defects and diseases that are inherited in a simple, Mendelian manner, there are many other single-locus characters that are relevant to animal breeding. Some of these, such as blood groups, have already been discussed. Others include coat colours, wool type, polledness, dwarfism in poultry, and characters used for sexing chickens.

Coat colour in most species of mammals can be interpreted in terms of multiple alleles at each of several loci (A: agouti series, B: brown series, C: albino series, D: dilution series, E: extension series), together with other loci whose effects are distinguishable only in certain species (including tabby, white spotting, and merle). Acting in many different ways, and showing considerable interactions between loci (epistasis), the alleles at these loci control the amount and distribution of eumelanin (brown or black) and phaeomelanin (yellow or red). In several cases, coat colours and patterns determined by particular alleles have become the trademark of a particular breed. When this trademark is associated with heterozygosity, as in Palomino horses, the breed concerned cannot 'breed true'.

Mutations occurring in various sheep flocks in New Zealand have produced hairy fleeces well suited to carpet manufacture. These so-called carpet-wool genes are dominant or incompletely dominant, and have led to the establishment of specialty carpet-wool breeds such as the Drysdale ($N^d N^d$), the Tukidale ($N^t N^t$) and the Carpetmaster ($N^j N^j$).

Polledness has considerable economic advantage in certain circumstances, and from a practical point of view can be said to be due to an autosomal dominant allele. A certain type of dwarfism in poultry, due to a Z-linked recessive allele, also appears to have economic merit in some circumstances.

Easily detectable differences between male and female day-old chickens can be exploited through Z-linked sexing (which utilizes segregation at the Z-linked loci determining gold/silver feather colour, or fast/slow feathering rate) or automatic sexing (which utilizes auto-somal alleles to enhance the effect of the Z-linked barring allele).

The exclusion principle in parentage testing states that you cannot prove that a particular animal *is* the true parent, but you may, under certain circumstances, prove that a particular animal is *not* the true parent. The efficiency of parentage testing depends on the number of loci that can be used, the number and frequency of alleles at each locus, and the type of gene action at each locus.

12.10 Further reading

Adalsteinsson, S. (1983). Inheritance of colours, fur characteristics and skin quality traits in North European sheep breeds: a review. *Livestock Production Science* **10**, 555–67. (A review of the genes affecting coat colour in sheep, and of the heritability of many fur and skin characteristics.)

Bindon, B. M. (1984). Reproductive biology of the Booroola Merino sheep. *Australian Journal of Biological Sciences* **37**, 163–89. (A review of the genetic and physiological characteristics of the Booroola Merino.)

Guillaume, J. (1976). The dwarfing gene *dw*: its effects on anatomy, physiology, nutrition and management; its application in the poultry industry. *World's Poultry Science Journal* **32**, 285–304. (A review of the dwarfing gene in chickens.)

King, R. C. (Ed.) (1975). *Handbook of genetics. Vol. 4. Vertebrates of genetic interest.* Plenum, New York. (This is a valuable reference book, containing review chapters describing many single-locus characters in amphibia, fish, birds, and mammals. It also contains a review of blood-group alleles in domesticated mammals, and a series of tables illustrating the distribution and probable homologies of coat colour genes in mammals.)

Ryder, M. L. (1980). Fleece colour in sheep and its inheritance. *Animal Breeding Abstracts* **48**, 305–24. (A review of colour inheritance in sheep.)

Searle, A. G. (1968). *Comparative genetics of coat colour in mammals.* Logos, London. (A detailed review of coat colour genetics in mammals, highlighting the probable homologies of coat colour genes in different species.)

Silverudd, M. (1978). Genetic basis of sexing automation in the fowl. *Acta Agri-culturae Scandinavica* **28**, 169–95. (A review of genetic methods of chicken sexing.)

Todd, N. B. (1977). Cats and commerce. *Scientific American* **237**(5), 100–7. (A popular account of how the distribution of different coat colour alleles around the world appears to correlate with human migration.)

Trommershausen-Smith, A., Suzuki, Y., and Stormont, C. (1976). Use of blood typing to confirm principles of coat-colour genetics in horses. *Journal of Heredity* **67**, 6–10. (A practical example of parentage testing, in this case being used to check the rules of coat-colour inheritance.)

Wickham, G. A. (1978). Development of breeds for carpet wool production in New Zealand. *World Review of Animal Production* **14**, 33–40. (A review of carpet wool genes.)

13
Relationship and inbreeding

13.1 Introduction

The general idea of relationship is well understood by most people. Parents and their offspring, for example, are commonly and correctly regarded as being more closely related than grandparents and their grandchildren. One aim of this chapter is to provide a specific understanding of the concept of relationship, so as to enable calculation of the relationship between any two individuals. As we shall see in Chapters 14 and 16, such calculations have important practical applications in heritability estimation, and in artificial selection programmes.

An understanding of relationship goes hand in hand with an understanding of inbreeding. An equally important aim of this chapter, therefore, is to explain the concept of inbreeding and to discuss its practical implications.

13.2 The inbreeding coefficient

We shall start by considering the simplest situation, which involves two offspring, B and C, having one parent, A, in common. Using arrows to indicate the direction of inheritance, this situation can be represented as follows:

A is called the *common ancestor* of B and C. If B and C each receive a copy of the same gene, i.e. the same segment of DNA at any locus in A, then we can say that B and C have genes that are *identical by descent* from A. Thus, genes are identical by descent if they are copies of the same segment of DNA.

Suppose that B and C are mated together to produce an offspring, O:

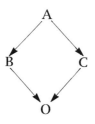

If B and C have genes that are identical by descent, then for any locus, there is a chance that O will inherit two genes that are identical by descent, one from A through B, and the other from A through C. We are now in a position to define the inbreeding coefficient.

> *The inbreeding coefficient of an individual is the probability that the two genes present at a locus in that individual are identical by descent.*

The inbreeding coefficient is usually given the symbol F. If an individual has two genes that are identical by descent at a certain locus, then that individual must be homozygous at that locus. Thus the inbreeding coefficient is a reflection of homozygosity. However, it is important to note that it does *not* measure homozygosity in an absolute sense. The reason for this is explained in Section 13.4.

Having now been introduced to inbreeding, we shall consider relationship in some detail, and shall then return to inbreeding.

13.3 Relationship

The most common situation in practice, and the one most easily understood, is where neither of the related individuals is inbred. In this section, therefore, unless otherwise stated we shall assume that relatives are not inbred, i.e. that $F = 0$ for each relative.

13.3.1 Direct relationship

The most basic relationship is that between a parent (P) and its offspring (O). At any locus an offspring has two genes, one of which (the paternal gene) is a copy of one of its father's genes, and the other of which (the maternal gene) is a copy of one of its mother's genes. Since the same is true for all loci, we can say that an offspring has exactly $\frac{1}{2}$ of its genes in common with each of its parents. In other words, a parent and an offspring have exactly $\frac{1}{2}$ of their genes in common.

> *The expected proportion of genes in common between two non-inbred individuals is the relationship between them.*

The relationship between offspring and parent, therefore, is $\frac{1}{2}$.

What is the relationship between offspring (O) and grandparent (GP)?

In order to answer this question we need to follow the transmission of genes from a particular grandparent (GP_1) to a parent (P) to an offspring (O). It is obvious that GP_1 transmits exactly $\frac{1}{2}$ of its genes to P who in turn transmits exactly $\frac{1}{2}$ of its genes to O. *On average* therefore, O will receive $\frac{1}{2} \times \frac{1}{2} = \frac{1}{4}$ of its genes from GP_1. Thus the relationship between an offspring and any one of its grandparents is $\frac{1}{4}$.

It is important to notice that whereas the actual proportion of genes in common between offspring and parent is always exactly $\frac{1}{2}$, the actual proportion of genes in common between an offspring and its grandparent, and between all other relatives, may not be the same as the expected proportion. For example, the actual proportion of genes in common between a particular offspring and a particular grandparent could be greater or less than $\frac{1}{4}$. It could be as high as $\frac{1}{2}$ (if a gamete from P happens to contain all of the genes that P inherited from GP_1) and it could be as low as zero (if a gamete from P happens to contain all of the genes that P inherited from its other parent, and none from GP_1). For any particular O and GP, we have no way of knowing the actual proportion of genes in common. All we can say is that the expected proportion, and hence the relationship, is $\frac{1}{4}$.

By an argument similar to that used above, the relationship between an offspring and one of its great-grandparents is

$$\frac{1}{2} \times \frac{1}{2} \times \frac{1}{2} = \left(\frac{1}{2}\right)^3 = \frac{1}{8}.$$

> *In general, the relationship between an individual and an ancestor decreases by $\frac{1}{2}$ for each generation that separates the individual from that ancestor.*

Thus, the relationship between an individual and an ancestor is

$$\left(\frac{1}{2}\right)^n,$$

where n is the number of generations between the individual and the ancestor in the pathway of direct descent from the ancestor to the individual.

It follows that irrespective of how superior a particular ancestor may have been, if it occurs only once in an individual's pedigree and several generations back in the pedigree, then there is only a small chance that the individual in question (whom we shall call O) will have inherited any of that ancestor's genes. Thus it is pointless to select an individual solely because its pedigree contains a distant ancestor that happened to be very famous.

The situation is a little different if a particular ancestor occurs more than once in a pedigree. In this case, each occurrence of the ancestor in the pedigree provides an independent opportunity for the individual, O, to have inherited the ancestor's genes. The total relationship is the sum of the independent contributions from each pathway of direct descent. Thus, if an ancestor occurs several times in a pedigree, and not too far back in the pedigree, then the relationship between O and the ancestor could be quite high. This is exactly what happens with *line breeding*, which is the name given to mating schemes in which the aim is to have a particular ancestor occur very frequently in the one pedigree. Since the end result of line breeding is that O has a high proportion of its genes in common with the ancestor, we can say that the end result of line breeding is almost to reproduce the entire set of genes of that particular ancestor. If the ancestor was very famous then this may seem to be a sensible thing to do. But aiming to reproduce a particular ancestor is equivalent to admitting that no improvement has been made in the population or breed since that ancestor was born. And not many breeders would be willing to make such an admission. Line breeding, therefore, is more concerned with resurrecting the past than with breeding better animals for the future and as such, it is not a particularly progressive form of breeding.

The relationships considered above are called *direct relationships* because they refer to relationships that can be traced in a direct line or pathway of descent from an ancestor to a descendent. The other type of relationship is where two individuals have an ancestor in common, and so may each have inherited the same gene from that ancestor. This is called *collateral relationship.*

13.3.2 Collateral relationship

We shall return to the simplest situation described in Section 13.2, which involves two offspring, B and C, having one parent, A, in common. In this case, B and C are called *half-sibs* and, as noted before, A is their common ancestor. As we also saw before, if B and C each receive a copy of the same gene at any locus in A, then we can say that B and C have genes that are identical by descent from A. We can now consider a more general definition of relationship.

> *The relationship between any two individuals is the expected number of genes at a locus in one individual that are identical by descent with a randomly chosen gene at the same locus in the other individual.*

This definition is correct in all situations, whether the related individuals are inbred or not. If the related individuals are not inbred, then this definition is equivalent to the one given in Section 13.3.1, namely that the relationship between two individuals is the expected proportion of genes in common between them.

We are now in a position to use a very common method of calculating coefficients of relationship and of inbreeding. The derivation of relevant formulae and illustrations of their use are given in Appendix

13.1. The most important conclusion to be drawn from Appendix 13.1 is that:

> The inbreeding coefficient of an individual is $\frac{1}{2}$ the relationship between its parents.

13.4 The base population

It was stated in Section 13.2 that the inbreeding coefficient does not measure homozygosity in an absolute sense. In fact, it measures the decrease in heterozygosity relative to a base population in which all individuals are assumed to be unrelated and to have zero inbreeding. In other words, it measures the extent to which an individual is less heterozygous than individuals that are assumed to have zero inbreeding. In practice, we usually assume zero inbreeding ($F = 0$) for all individuals on which we have no information.

For example, consider a population with gene frequencies of 0.4 and 0.6 and with Hardy–Weinberg genotype frequencies of 0.16 A_1A_1, 0.48 A_1A_2, and 0.36 A_2A_2. If we know nothing about the breeding history of this population, then we assume that all its members have zero inbreeding. The fact that the probability of an individual being homozygous is $0.16 + 0.36 = 0.52$ does not matter; homozygotes in this population have genes that are regarded as being *alike in state* rather than identical by descent. That is, they are homozygous for copies of different ancestral genes rather than for two copies of the one ancestral gene.

If an individual, O, has an inbreeding coefficient of $F = 0.3$ relative to this base population, then the probability of O being heterozygous at that locus is $1 - 0.3 = 0.7$ times the probability of a randomly-chosen member of the base population being heterozygous. Now, the chance of a randomly-chosen member of the base population being heterozygous is simply the frequency of heterozygotes in the base population, which is $2 \times 0.6 \times 0.4 = 0.48$. It follows that the chance of O being heterozygous is $0.7 \times 0.48 = 0.34$. Looking at this from the opposite point of view, it can be seen that there is a chance of $1 - 0.34 = 0.66$ that O will be homozygous at that locus.

We can summarize the meaning of the inbreeding coefficient by considering the two extreme cases of $F = 0$ and $F = 1$. If an individual has an inbreeding coefficient of $F = 0$ relative to a particular base population, then the chance of that individual being heterozygous at any locus equals the frequency of heterozygotes at that locus in the base population. If an individual has an inbreeding coefficient of $F = 1$, then the chance of that individual being heterozygous at any locus equals zero; it has genes identical by descent at all loci, in which case it must be homozygous at all loci.

Some readers may be asking how measures of relationship and of inbreeding can be of any use when their actual magnitudes are dependent on the arbitrary choice of a base population, the members of which are conveniently assumed to have no ancestors in common. At first sight, it is difficult to reconcile this with the fact that if we go back a sufficient number of generations, then any two individuals must have ancestors in common and hence must be related. In practice, this does not present a problem, so long as we remember that coefficients of relationship and inbreeding are relative, rather than absolute, quantities. Thus, if an individual has a zero inbreeding coefficient relative to a particular base population, then we know that this individual has the same level of heterozygosity as members of the base population. If the actual level of heterozygosity in the base population is quite low, due, say, to substantial mating of close relatives amongst the ancestors of the base population, then the individual with an inbreeding coefficient of zero will actually have quite a low level of heterozygosity. The important point is, however, that if a second individual has an inbreeding coefficient of 0.3 relative to the same base population, then we immediately know that this second individual is 30 per cent less heterozygous than the individual whose inbreeding coefficient is zero. This highlights the fact that inbreeding coefficients enable the *relative* levels of heterozygosity to be compared among individuals whose pedigrees can be traced back to the same base population. Examples of the practical application of such comparisons are given in Section 13.6.

13.5 Inbreeding in populations

Inbreeding can be defined as the mating of related individuals.

In previous sections, we have seen that the greater the relationship between two individuals, the greater will be the inbreeding coefficient of their offspring. We have also seen that the greater the inbreeding coefficient of an individual, the lower is the chance of that individual being heterozygous. It follows that the greater the extent of inbreeding within a population, the lower the frequency of heterozygotes within that population. In the extreme case where all members of a population are completely inbred ($F = 1$), all individuals are homozygous at all loci, and the frequency of heterozygotes is zero.

It is apparent that inbreeding changes genotype frequencies, because the frequency of heterozygotes decreases, and the frequency of homozygotes increases. However, it is important to realize that inbreeding does not change gene frequencies. In the population described in

Section 13.4, for example, the effect of the decrease in heterozygosity is simply to relocate an equal number of A_1 and A_2 genes from heterozygotes into homozygotes: genotype frequencies change but, on average, gene frequencies do not change. In the extreme case where heterozygosity has been reduced to zero ($F = 1$), the population will consist, on average, of 0.4 A_1 homozygotes and 0.6 A_2 homozygotes.

We conclude that:

> *Inbreeding changes genotype frequencies but does not, on average, change gene frequencies.*

13.6 Inbreeding depression

We have seen that inbreeding decreases heterozygosity and hence increases homozygosity. We have also seen, in Chapter 5, that quite a few harmful recessive genes are maintained in populations by a balance between selection and mutation, and that the majority of these genes are hidden in heterozygotes. Since inbreeding decreases the frequency of heterozygotes, it tends to bring these harmful recessive genes out into the open, in the form of homozygotes. Consider, for example, a recessive gene with a frequency of 0.05. In a non-inbred, random mating population, the frequency of homozygotes for this gene, and hence the frequency of the recessive phenotype, will be $(0.05)^2 = 2.5$ per thousand. The frequency of heterozygotes will be $2 \times 0.05 \times 0.95 = 95$ per thousand. An individual that has an inbreeding coefficient of $F = 0.4$ relative to this base population will be 40 per cent less likely to be heterozygous; the loss in heterozygosity will be 40 per cent of 95 per thousand, which equals 38 per thousand. Since this loss in heterozygosity is apportioned equally between the two types of homozygotes, the frequency of homozygote recessives will have risen from 2.5 per thousand to $(2.5 + 19.0) = 21.5$ per thousand, which is a very large increase.

Bearing in mind that inbreeding affects all loci in the same way, it is evident that inbreeding will increase the frequency of all genetic defects and abnormalities that are due to recessive genes. Since many of these defects and abnormalities decrease the productive and/or reproductive performance of animals, inbreeding usually leads to a decrease in performance.

> *The decrease in performance resulting from inbreeding is called inbreeding depression.*

The role of harmful recessive genes in leading to inbreeding depression, as discussed above, is just one example taken from the overall set of causes of inbreeding depression.

> *In general, inbreeding depression occurs at any locus at which the perform-*
> *ance of the heterozygote is greater than the midpoint between the two*
> *homozygotes.*

In other words, inbreeding depression occurs whenever genes that increase performance show any degree of dominance at a locus. The greater the departure of the performance of heterozygotes from a level intermediate between the homozygotes, the greater the inbreeding depression. Also, the closer gene frequencies are to intermediate values, the greater the inbreeding depression.

Inbreeding depression is generally greatest for characters associated with natural fitness such as viability and reproductive ability, because there is more dominance at the loci affecting these characters than at loci affecting other characters.

13.6.1 *Production of inbred lines*

A lot of medical and veterinary research is conducted using inbred lines of laboratory animals such as mice and rats. And several attempts have been made to develop inbred lines of larger animals such as chickens and pigs. In order to learn about the effects of inbreeding in general, we shall investigate the production of such inbred lines. The example that we shall consider is the development of inbred lines of mice by Bowman and Falconer (1960). So as to illustrate the main principles of inbreeding without being concerned with details of experimental design, the following description of the experiment of Bowman and Falconer will sometimes involve a simplified version of what actually happened.

Bowman and Falconer started with 20 pairs of double-first cousins, which are individuals with both pairs of grandparents in common. Each pair consisted of one male and one female which were mated together to produce a litter of *full-sibs*, which are individuals having both parents in common. Since the relationship between double-first cousins is 25 per cent (as calculated in Appendix 13.1.3), the average inbreeding coefficient (F) of their offspring is 12.5 per cent, which in this experiment corresponded to generation zero. Each litter constituted the foundation for a different inbred line. From within each litter, one male and one female were chosen as parents of the next generation. For that and all subsequent generations, the only mating in each line was between one male and one female chosen from within each litter. Thus, after the first generation, each of the 20 inbred lines remained completely separate, with each subsequent generation in each line resulting from a mating between a pair of full-sibs bred within that line. The inbreeding coefficient of the members of each litter was calculated by the method described in Appendix 13.1. An illustration of the design of the inbreeding programme is given in Fig. 13.1, and the most important results are presented in Table 13.1 and Fig. 13.2.

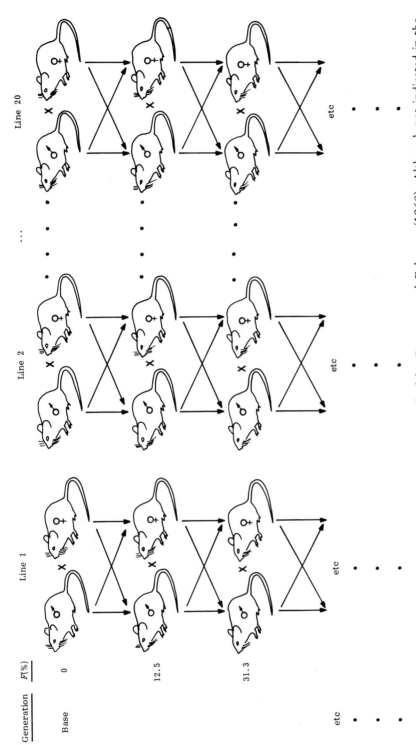

Fig. 13.1. The production of inbred lines of mice, as described by Bowman and Falconer (1960). Although not indicated in the diagram, individuals in the base generation were double-first cousins, which have a relationship of 25 per cent and hence give rise to offspring with an inbreeding coefficient of 12.5 per cent.

Table 13.1. The production of inbred lines of mice

Generation number	F (%) of litters	Number of lines	
		Lost	Remaining
Base	0	0	20
0	12.5†	0	20
1	31.3†	0	20
2	43.8	0	20
3	54.7	1	19
4	63.3	10	10
5	70.3	15	5
6	76.0	17	3
7	80.6	17	3
8	84.3	17	3
9	87.3	17	3
10	89.7	17	3
11	91.7	18	2
12	93.5	19	1
.			
.			
.			
20	98.8	19	1

† Derived in Appendix A13.1.3. Adapted from Bowman and Falconer (1960) 'Inbreeding depression and heterosis of litter size in mice'. *Genetical Research* 1, 262–74. Cambridge University Press, Cambridge.

The most dramatic result of inbreeding was the reduction in mean litter size and consequent loss of lines. Starting at 7.8, mean litter size fell to 5.4 at generation two, by which time the inbreeding coefficient had reached 44 per cent. Mean litter size fell again in the next generation, with one of the lines failing to produce at least one male and one female and therefore becoming extinct. In the next generation, nine more lines became extinct for the same reason. A further five became extinct in the next generation, and two in the next after that, leaving only three of the original 20 lines still present after only six generations of full-sib mating. Amongst these three remaining lines, litter size was quite reasonable, with a mean of 6.6. Averaged over all 20 original lines, however, litter size was now only 1.0, compared with 7.8 at the beginning. The three remaining lines survived for a few more generations, but by the twelfth generation, when average inbreeding was 94 per cent, only one line remained out of the original 20; the other 19 had become extinct because they had failed to produce one male and one female replacement. This one remaining line appeared

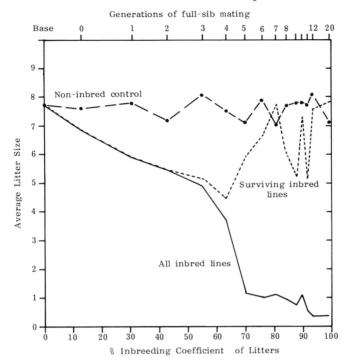

Fig. 13.2. Mean litter size of all 20 inbred lines (———) and of surviving inbred lines (-----) during inbreeding, compared with mean litter size of a non-inbred control line (— — —). Starting at generation zero with the progeny of double-first cousins, inbreeding was conducted for 20 generations in the most intense form possible in mammals, namely by mating full-sibs in every generation. Drawn from the data of Bowman and Falconer (1960) 'Inbreeding depression and heterosis of litter size in mice'. *Genetical Research* **1**, 262–74. Cambridge University Press, Cambridge.

to be quite normal and had a litter size as large as that in the base population. Being thus not in danger of extinction, it continued in existence until generation 20, when the experiment ceased. Although the inbreeding coefficient was then as high as 99 per cent, this one remaining line gave every indication that it could have continued in existence indefinitely.

The 19 lines that became extinct did so because they became homozygous for harmful genes at a sufficient number of loci for viability and reproductive ability to be seriously impaired. The one remaining line must have been homozygous for favourable genes at all the relevant loci. Being almost completely homozygous, it bred true for all inherited characteristics; all individuals in that line had essentially the same genotype.

If other characters had been measured in this experiment, a decline in their performance would also have been observed, with the magnitude of decline depending on the type of character; those closely associated with natural fitness would have shown more inbreeding depression than those of only minor importance to natural fitness.

In most animal breeding programmes, inbreeding is not as rapid as in the above experiment, although quite high values of F can be reached. Since inbreeding depression is such a general phenomenon, and since the aim of most animal breeding programmes is to improve performance, it is obvious that:

> *One of the aims of breeding programmes should be to minimize inbreeding by avoiding the mating of close relatives.*

13.6.2 Endangered species and zoo animals

From the data available in a survey of 939 offspring belonging to a total of 16 different species of captive ungulates, Ralls *et al.* (1979) determined that at least 380 or 40 per cent had related parents and hence were inbred. Amongst these inbred offspring, 49 per cent died before they were six months old. In contrast, only 23 per cent of non-inbred offspring died in the same period. These figures provide a further illustration of the importance of inbreeding depression, and highlight the problems encountered in breeding animals in captivity. However, these figures should not be taken as meaning that inbreeding always results in decreased viability and reproductive performance. The current populations of Chillingham white cattle and Père David's deer, for example, are surviving and reproducing quite well despite being derived from very small numbers of ancestors and having very high levels of inbreeding. The point to understand is that these cases, and others like them, correspond to the one line out of 20 that survived inbreeding in Bowman and Falconer's experiment. Thus, for every one case of Chillingham white cattle or Père David's deer, there will be many otherwise similar populations that are now extinct; there is no doubt that inbreeding on average decreases viability and reproductive ability. This general conclusion provides a warning to those well-meaning conservationists who believe that their job is done once a species that is endangered in the wild, can be bred in captivity.

The California condor, for example, is an endangered species. Its total numbers have dwindled to such an extent that the species now consists of only 20 to 30 birds, and shows signs of decreasing even further in the future. A Condor Recovery Plan has been launched, with the aim of capturing some of the remaining birds, with a view to getting them to breed in captivity, free from the perils in the wild that threaten to lead the species into extinction. If the Plan is successful, so the argument goes, then at least the species will not become extinct, even if it does disappear from the wild.

But what are the chances of getting the California condor to reproduce in captivity, and if it does reproduce, what are the longer-term consequences of such action? Unfortunately, the chances of getting the California condor to reproduce in captivity are not particularly high, and even if this were to be successful on a limited scale, the longer-term consequences are not particularly favourable. Suppose, for example, that out of the California condors captured, only one pair produces offspring. When the original birds die, and the only surviving offspring are ready to reproduce, there will be no alternative to the mating of full-sibs, which results in offspring with 25 per cent inbreeding. Although this is an extreme example, it does indicate that if California condors do reproduce in captivity, but only on a limited scale, then the level of inbreeding may increase quite rapidly and reproductive performance will consequently decline, as illustrated by the above survey results. No matter what is done, therefore, the California condor may be doomed. That is not to say that attempts should not be made to save the California condor; it is simply to say that saving the California condor may not be as simple as just getting it to reproduce in captivity.

Problems similar to that of the condor are familiar to conservationists all over the world. There are many species and breeds whose total population consists of a handful of animals. Valiant attempts are made to conserve these species and breeds in zoos and in other centres. With such small numbers, however, the problems of inbreeding depression in characters such as viability and reproductive ability become very important. Often there is little that can be done; at best, matings can be planned so as to minimize the relationship between mates, and to this end, exchanges of breeding animals between different zoos can often be helpful. But in many situations such exchanges are either impossible or too expensive.

There are, unfortunately, no easy solutions to the problems inherent in the conservation of rare species or breeds. The problems themselves have been raised in this chapter simply to illustrate the relevance of an understanding of relationship and inbreeding to the breeding of animals in captivity and to animal conservation in general.

13.7 Summary

The inbreeding coefficient of an individual is the probability that the two genes present at a locus in that individual are identical by descent. It is a measure of the extent to which an individual is less heterozygous and hence more homozygous than individuals that are assumed to have zero inbreeding.

The most basic relationship is that between parent and offspring. Since each offspring receives copies of exactly $\frac{1}{2}$ of its parent's genes, parent and offspring have exactly $\frac{1}{2}$ of their genes in common, and are said to have a relationship of $\frac{1}{2}$.

The relationship between any two individuals that are in a direct line or pathway of descent is called a direct relationship, and equals $(\frac{1}{2})^n$ where n is the number of generations separating the two individuals. Individuals that share a common ancestor are called collateral relatives.

The relationship between any two individuals is the expected number of genes at a locus in one individual that are identical by descent with a randomly chosen gene at the same locus in the other individual. If any two individuals are not inbred, then the relationship between them equals the expected proportion of genes that they share in common.

The inbreeding coefficient of an offspring is $\frac{1}{2}$ the relationship between its parents.

The mating of related individuals is called inbreeding. The result of inbreeding is a decrease in the frequency of heterozygous genotypes and a corresponding increase in the frequency of homozygous genotypes. Thus, inbreeding changes genotype frequencies. On average, however, it does not change gene frequencies. Inbreeding is often associated with decreases in average performance (known as inbreeding depression) for many economically important characters, and especially for viability and reproductive ability. In animal breeding, therefore, the general aim is to minimize inbreeding.

In the case of zoo animals, and particularly when dealing with endangered species and rare breeds, it is often very difficult to achieve this aim, because the population size has already been reduced to a low level.

13.8 Further reading

Frankel, O. H. and Soulé, M. E. (1981). *Conservation and evolution*. Cambridge University Press, Cambridge. (A thorough review of the genetic and evolutionary aspects of plant and animal conservation. The accounts of inbreeding (in Chapter 3) and of genetics of captive propagation (Chapter 6) are directly relevant to Section 13.6.2 of this chapter.)

Lamberson, W. R. and Thomas, D. L. (1984). Effects of inbreeding in sheep: a review. *Animal Breeding Abstracts* **52**, 287–97. (A valuable summary of all available data.)

Olney, P. J. S. (Ed.) (1977). *1977 International Zoo Yearbook* (Vol. 17). Zoological Society of London, London. (Section 1 of this volume contains the proceedings of the Second World Conference on Breeding Endangered Species in Captivity. Many of the papers are directly relevant to Section 13.6.2 of this chapter.)

Soulé, M. E. and Wilcox, B. (Eds) (1980). *Conservation biology*. Sinauer Associates, Sunderland, Massachusetts. (A valuable collection of papers, including five particularly concerned with captive propagation and conservation.)

Van Vleck, L. D. (1977). Genetics of the horse. In *The horse*. Eds Evans, J. W., Borton A., Hintz, H. F., and Van Vleck, L. D. pp. 427–552. Freeman,

San Francisco. (Chapter 16 of *The horse* is an easy-to-read account of relationship and inbreeding, using examples from horses. Among other things, it gives a very clear guide to the tabular or covariance method of calculating relationships.)

Appendix 13.1

Calculation of relationship and inbreeding

A13.1.1   Relationship

We saw in Section 13.3.1 that the relationship between an individual and its ancestor decreases by $\frac{1}{2}$ for each generation that separates the individual from that ancestor. We also saw that the overall relationship is the sum of the independent contributions of each pathway of descent from the ancestor to the individual.

As with direct relatives, relationship between two collateral relatives decreases by $\frac{1}{2}$ for each generation that separates the two relatives along a certain pathway. The only difference is that in the case of collateral relatives, the pathway between them consists of two lines of descent; from the common ancestor to one relative and from the common ancestor to the other relative. To emphasize this, the total number of generations separating two collateral relatives is written as $n + n'$, where n is the number of generations between the common ancestor and the first relative, and n' is the number of generations between the common ancestor and the second relative. Thus for any pathway, the relationship between two collateral relatives is

$$\left(\frac{1}{2}\right)^{n+n'}.$$

Sometimes there may be more than one pathway through a certain common ancestor, and sometimes there may be more than one common ancestor. In all these cases, the overall relationship is the sum of the relationships contributed by each pathway.

The common ancestor is obviously of critical importance in determining collateral relationships. What if this ancestor is inbred? We saw in Section 13.2 that the inbreeding coefficient of an individual is the chance that two genes at a locus in that individual are identical by descent. It follows that the higher the inbreeding coefficient of the common ancestor, the greater is the chance that it will transmit the same gene to each of the collateral relatives descended from it, and hence the greater the chance that the two relatives will each have genes that are identical by descent. In fact, if the common ancestor is inbred, the relationship between two collateral relatives descended from the common ancestor is increased by a proportion equal to the inbreeding coefficient of the common ancestor. Algebraically, this is equivalent to multiplying

$$\left(\frac{1}{2}\right)^{n+n'}$$

for each pathway by $1 + F_A$, where F_A is the inbreeding coefficient of the common ancestor in that pathway.

Taking all this into account and using the symbol a for relationship, we can now express the relationship between any two individuals as

$$a = \sum_{i=1}^{p} \left(\frac{1}{2}\right)^{n_i+n_i'} (1 + F_{A_i}), \tag{1}$$

where summation is over p pathways. This expression can also be used to calculate direct relationships, by noting that in such cases there is only one line of descent for each pathway, and so $n_i' = 0$ for each pathway.

Although derived from the simpler definition, the relationship calculated by the above method corresponds to the exact definition of relationship, namely the expected number of genes at a locus in one individual that are identical by descent with a randomly chosen gene at the same locus in the other individual. For reasons that are explained in Chapter 14, the relationship calculated in equation (1) is often called the *additive relationship*, which is why it is given the symbol 'a'. It is also sometimes called the *numerator relationship*, because it is the numerator of another measure of relationship that was introduced by Wright (1922). If the relatives in question are not inbred, then the additive relationship has a minimum of zero and a maximum of one, and corresponds to the simpler definition of relationship, namely the proportion of genes in common. If one or both of the relatives in question is inbred, then the simpler definition is no longer useful, because in this case the additive relationship can be greater than one and in fact has a maximum value of two. In order to understand why this is so, consider two individuals, B and C, whose possible genotypes are shown in Table A13.1, where A_1, A_2, A_3, and A_4 are four genes (segments of DNA) that are *not* identical by descent, i.e. they have each arisen from a different ancestral gene.

It can be seen that the limitation of the simple definition is that it cannot distinguish between the last two situations shown in Table A13.1, i.e. it cannot distinguish between two individuals each having the same pair of genes in common, and two individuals each being homozygous for the same gene. In both cases, the two individuals have 100 per cent of their genes in common. However, if B and C are mated together, then the two different cases will give rise to very different results in terms of the inbreeding coefficient of the resultant offspring. In the

Table A13.1

Genotype		Inbreeding coefficient		Additive relationship between B and C	
B	C	F_B	F_C	Simple definition	Exact definition
$A_1 A_2$	$A_3 A_4$	0	0	0	0
$A_1 A_2$	$A_3 A_2$	0	0	$\frac{1}{2}$	$\frac{1}{2}$
$A_1 A_2$	$A_1 A_2$	0	0	1	1
$A_1 A_1$	$A_1 A_1$	1	1	1	2

latter case, the offspring of B and C are certain to have two copies of the same gene (A_1) and therefore to have an inbreeding coefficient of $F = 1$, whereas in the case where both B and C are $A_1 A_2$, the chance that an offspring will have two copies of the same gene is $\frac{1}{4}$ (for $A_1 A_1$) and $\frac{1}{4}$ (for $A_2 A_2$), giving an inbreeding coefficient of $F = \frac{1}{4} + \frac{1}{4} = \frac{1}{2}$.

A13.1.2   Inbreeding

Continuing with our example of B mating with C to produce off-spring O, consider the randomly chosen gene that B passes on to O. In order to calculate the inbreeding coefficient of O, we need to calculate the probability that the other gene at this locus in O (the gene from C) is identical by descent with the gene from B. We can do this by noting that the additive relationship, a, between B and C is the expected number of genes in C that are identical by descent with a randomly chosen gene at the same locus in B. But the randomly chosen gene in B is the gene that was passed from B to O. Since there is a chance of one-half that any gene in C will be passed on to O, it follows that the probability that O will receive a gene from C that is identical by descent with the gene that O received from B, is equal to the additive relationship between B and C, multiplied by one-half.

> In general, the inbreeding coefficient of any individual is $\frac{1}{2}$ the additive relationship between the parents of that individual.

From equation (1), this means that the inbreeding coefficient of any individual is:

$$
\begin{aligned}
F &= \frac{1}{2} a \\
&= \frac{1}{2} \sum_{i=1}^{p} \left(\frac{1}{2}\right)^{n_i + n_i'} (1 + F_{A_i}) \\
&= \sum_{i=1}^{p} \left(\frac{1}{2}\right)^{n_i + n_i' + 1} (1 + F_{A_i}).
\end{aligned}
\tag{2}
$$

A13.1.3 Some practical examples

We shall illustrate the calculation of additive relationship and of inbreeding coefficients by using examples that were encountered in Chapter 13. In so doing, we will illustrate a method that involves working from first principles in a manner that clearly indicates what the above formulae actually mean. If you understand the method, then you will have no need to memorize the formulae.

The first example involves half-sibs, which are individuals with one parent in common. Using letters to indicate identification of animals, the traditional pedigree of the offspring from two half-sibs would be drawn as follows:

$$
O
\begin{cases}
B \begin{cases} A \\ D \end{cases} \\
C \begin{cases} A \\ E \end{cases}
\end{cases}
$$

It is far easier if we redraw the traditional pedigree as a *path diagram*, in which each individual appears only once and in which arrows indicate the direction of inheritance, as shown below.

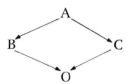

Notice that not all individuals appear in the path diagram; those that are not common ancestors and who do not occur in any pathway between a common ancestor and the individual of interest have been excluded (D and E in this case).

We shall now calculate the relationship between the two half-sibs, B and C, and the inbreeding coefficient of their offspring, O. We proceed by drawing up a table as follows:

Pathway	n	n'	Contribution to relationship
B $\underline{A}$ C	1	1	$\left(\dfrac{1}{2}\right)^{1+1}(1 + F_A) = \dfrac{1}{4}$

The purpose of underlining the A is to indicate that it is the common ancestor for that pathway. With only one pathway in the path diagram, there is only one entry in the table, which gives the relationship between B and C as

$$a_{BC} = \left(\frac{1}{2}\right)^{1+1}(1 + 0)$$

$$= \frac{1}{4}$$

It then follows that the inbreeding coefficient of O is given by

$$F_O = \frac{1}{2}a_{BC}$$

$$= \frac{1}{8}$$

For our second example, we shall consider the matings that were used in Bowman and Falconer's inbreeding experiment, as described in Section 13.6.1. The full-sib descendants of double-first cousins will have a traditional pedigree as follows:

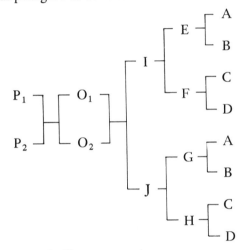

Redrawing this as a path diagram, we have

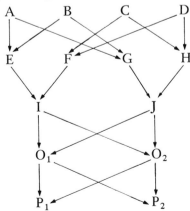

In order to calculate the additive relationship between I and J, we set up the following table:

Pathway	n	n'	Contribution to relationship
I E <u>A</u> G J	2	2	$\left(\dfrac{1}{2}\right)^{2+2}(1 + F_A) = \left(\dfrac{1}{16}\right)(1 + 0)$
I E <u>B</u> G J	2	2	$\left(\dfrac{1}{2}\right)^{2+2}(1 + F_B) = \left(\dfrac{1}{16}\right)(1 + 0)$
I F <u>C</u> H J	2	2	$\left(\dfrac{1}{2}\right)^{2+2}(1 + F_C) = \left(\dfrac{1}{16}\right)(1 + 0)$
I F <u>D</u> H J	2	2	$\left(\dfrac{1}{2}\right)^{2+2}(1 + F_D) = \left(\dfrac{1}{16}\right)(1 + 0)$

Summing over pathways we get

$$a_{IJ} = 4 \times \frac{1}{16} = \frac{1}{4},$$

and hence

$$F_{O_1} = F_{O_2} = \frac{1}{2}a_{IJ} = \frac{1}{8},$$

which is the inbreeding coefficient of individuals at generation zero in Table 13.1 (see p. 374).

We now proceed to calculate the relationship between O_1 and O_2.

Pathway	n	n'	Contribution to relationship
O_1 I E <u>A</u> G J O_2	3	3	$\left(\dfrac{1}{2}\right)^{3+3}(1 + F_A) = \left(\dfrac{1}{64}\right)(1 + 0)$
O_1 J G <u>A</u> E I O_2	3	3	$\left(\dfrac{1}{2}\right)^{3+3}(1 + F_A) = \left(\dfrac{1}{64}\right)(1 + 0)$
O_1 I E <u>B</u> G J O_2	3	3	$\left(\dfrac{1}{2}\right)^{3+3}(1 + F_B) = \left(\dfrac{1}{64}\right)(1 + 0)$
O_1 J G <u>B</u> E I O_2	3	3	$\left(\dfrac{1}{2}\right)^{3+3}(1 + F_B) = \left(\dfrac{1}{64}\right)(1 + 0)$
O_1 I F <u>C</u> H J O_2	3	3	$\left(\dfrac{1}{2}\right)^{3+3}(1 + F_C) = \left(\dfrac{1}{64}\right)(1 + 0)$

O_1 J H $\underline{C}$ F I O_2 3 3 $\left(\dfrac{1}{2}\right)^{3+3}(1+F_C)=\left(\dfrac{1}{64}\right)(1+0)$

O_1 I F $\underline{D}$ H J O_2 3 3 $\left(\dfrac{1}{2}\right)^{3+3}(1+F_D)=\left(\dfrac{1}{64}\right)(1+0)$

O_1 J H $\underline{D}$ F I O_2 3 3 $\left(\dfrac{1}{2}\right)^{3+3}(1+F_D)=\left(\dfrac{1}{64}\right)(1+0)$

O_1 $\underline{I}$ O_2 1 1 $\left(\dfrac{1}{2}\right)^{1+1}(1+F_I)=\left(\dfrac{1}{4}\right)(1+0)$

O_1 $\underline{J}$ O_2 1 1 $\left(\dfrac{1}{2}\right)^{1+1}(1+F_J)=\left(\dfrac{1}{4}\right)(1+0)$

Summing over pathways we get

$$a_{O_1O_2}=\left(8\times\dfrac{1}{64}\right)+\left(2\times\dfrac{1}{4}\right)$$

$$=\dfrac{5}{8},$$

and hence

$$F_{P_1}=F_{P_2}=\dfrac{1}{2}a_{O_1O_2}$$

$$=\dfrac{5}{16}$$

$$=31.3\%,$$

which is the inbreeding coefficient of individuals at generation one in Table 13.1.

It will be obvious to any reader who has worked through the above examples that this method of calculating relationships could become very complex and tedious for large pedigrees. In fact, a stage is soon reached at which it is virtually impossible to draw up a readable path diagram. If a large number of relationships or inbreeding coefficients have to be calculated on a regular basis in the one population, then another method called the tabular or covariance method is often preferable. Some very clear examples of its use are given by Van Vleck (1977).

14
Variation and heritability

14.1 Introduction

The majority of characters of interest in animal breeding are continuously varying in the sense that individuals cannot be readily classified into distinct classes. Milk production, fleece weight, body weight, and speed over a certain distance are just a few examples of continuously varying characters. Even a character like egg production, which is not strictly continuous, can be considered as continuously varying because there are a large number of classes into which hens can be fitted, according to the number of eggs laid in a certain period. Continuously varying characters are called *quantitative characters* or *metric characters*, and variation in them is called *quantitative variation* or *continuous variation*. Because they are measured on a continuous scale, it is not immediately obvious how variation in quantitative characters can result from the action of genes, which of course are discrete entities.

The aim of this chapter is to describe the basis of quantitative variation, and to show how it can be partitioned using the concept of heritability.

14.2 Quantitative variation

Since quantitative variation is a statistical concept, we have to use statistical methods to investigate it.

We shall start with a frequency distribution of first lactation milk yields in Friesians, as illustrated in Fig. 14.1. Even though milk yield is measured on a continuous scale, heifers have to be divided into an arbitrary number of small classes in order to construct the distribution. Once this is done, it becomes apparent that the distribution is more or less symmetrical and that it corresponds quite closely to a Normal distribution. To illustrate this, the mean and the variance of the actual distribution in Fig. 14.1 have been calculated, and a Normal distribution with the same mean and the same variance has been superimposed on the data. It can be seen that the Normal distribution is quite a good description of the data, and the same conclusion applies to most quantitative characters. This is fortunate, because many of the

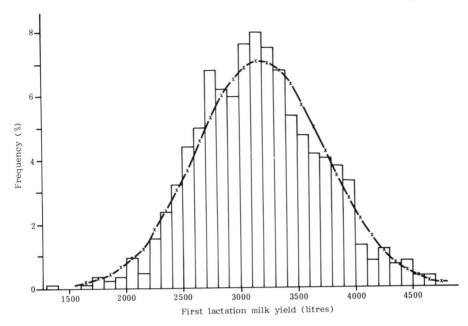

Fig. 14.1. Frequency distribution of first lactation milk yield in a population of 840 Friesian cows. The two statistical concepts used to describe these observations are the mean and the variance. The mean is estimated as $\Sigma P_i/n$, where P_i is the performance (milk yield) of the ith heifer, n is the total number of heifers, and Σ indicates summation from $i = 1$ to $i = n$. The variance is estimated as $\{\Sigma P_i^2 - (\Sigma P_i)^2/n\}/(n - 1)$, and is a measure of the extent of the differences in performance between individuals. If P_i is the performance of the ith heifer expressed as a deviation from the population mean, then the variance is estimated as $\Sigma P_i^2/(n - 1)$. In this population, the mean is 3180 litres and the variance is 315 352 (litres)^2. A Normal distribution with this mean and variance has been superimposed on the frequency distribution.

applications of quantitative genetics involve the assumption that the data follow a Normal distribution. For those characters that are not Normally distributed, e.g. faecal worm-egg counts, it is usually possible to find a simple transformation, such as square-root or logarithmic, that will enable the useful properties of the Normal distribution to be exploited.

How does continuous variation arise from the action of discrete genes? In order to answer this question, we shall start by considering just one locus. Imagine that there are two alleles at this locus, with allele *A* increasing milk yield by one litre and allele *a* giving no increase in milk yield. Imagine further that there is no dominance at this locus, so that the three genotypes *aa, Aa,* and *AA* contribute 0, 1, and 2 litres to milk yield. If the frequency of each allele in a herd of cows is 0.5, then the frequency distribution of milk yield resulting from variation at this one locus will be symmetrical and will consist of three classes, as shown in Fig. 14.2a. Suppose now that exactly the same situation exists at a second locus, with the three genotypes *bb*, *Bb*, and *BB* also contributing 0, 1, and 2 litres to milk yield. Considering these two loci together, the resultant frequency distribution now has five classes and is shown in Fig. 14.2b. As we increase the number of loci contributing to the quantitative character, we increase the number of classes.

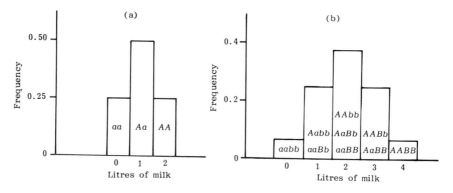

Fig. 14.2. Frequency distribution of milk yield resulting from variation at (a) a single locus, and (b) two loci. At each locus, the upper-case allele contributes one litre of milk, and the frequency of each allele is 0.5.

Not all alleles will have the same effect on milk yield. If there are several with a very small effect, then the difference between some of the classes will be as small as the error of measurement of milk yield, and it will no longer be possible always to place heifers in their correct class; the edges of the different class will be blurred. Besides the effect of alleles at a number of loci, there are many non-genetic or environmental factors that affect milk yield, and these will also tend to blur the edges of the discontinuous classes. If the effect of a particular allele is very large relative to the total variation, then discrete classes will still be evident even in the presence of environmental factors. But the majority of alleles affecting quantitative characters have an effect that is sufficiently small so as to be not readily detectable.

> We can conclude that quantitative characters are determined by the combined action of alleles at many loci, most of which have a relatively small effect on the character, and a number of non-genetic or environmental factors.

14.3 The performance of an individual

The performance of an individual in relation to a particular character is called its *phenotypic value, P,* for that character. The two factors that together determine phenotypic value are *genotypic value, G,* and *environmental deviation, E.* The genotypic value of an individual refers to the combined effect of all that individual's genes at all loci that affect the character in question, taking account of the way in which the genes are paired together into a genotype. The environmental deviation represents the combined effect of all non-genetic factors that have influenced the phenotypic value. For any individual, G is determined at conception, and E represents the combined effect of all factors that have influenced the particular character in that individual between conception and the time when P is measured.

Although both P and G have been defined above in terms of many loci, we can more easily understand their meaning, and the meaning of E, by thinking in terms of just one locus. However, as described in Section 14.2, most loci that affect a quantitative character are not able to be identified, because of the small effect that they exert on that character. One solution to this problem is to find a locus whose alleles have a large and easily identifiable effect on some other character, and then to examine the effect of this locus on the quantitative character in question. As an example, we shall use the transferrin locus in cattle, the genotypes of which can be identified using starch-gel electrophoresis. For the quantitative character on which these alleles exert an influence, we shall choose milk yield in Jersey cattle. The data used in this example are based on the results of research conducted by Ashton *et al.* (1964).

There were only two transferrin alleles, Tf^A and Tf^D, in the population of Jerseys reported in this study, giving three identifiable genotypes. So as to simplify notation in the following discussion, we shall refer to these two alleles as T and t respectively. Thus the three identifiable genotypes are TT, Tt, and tt. Knowing the milk yield (phenotypic value, P) of each cow and its transferrin genotype, we can calculate the average milk yield for each of the three different genotypes, and then compare the average yield of different genotypes.

> The average performance of a particular genotype is by definition the genotypic value, G, of that genotype.

In other words, genotypic value equals mean phenotypic value for that genotype.

In one herd, for example, the average milk yield of all TT cows was 1882 litres. This is the genotypic value, G, for this particular genotype in that herd. Of course, not all TT cows in that herd produced 1882 litres of milk; there was considerable variation about this average value. Suppose that the best *TT* cow yielded 1978 litres and the worst *TT* cow produced only 1773 litres of milk. If the transferrin locus were the only locus affecting milk yield, we could then say that the best *TT* cow had a phenotypic value, *P*, of 1978, a genotypic value, *G*, of 1882 and hence an environmental deviation, *E*, of 1978 − 1882 = +96 litres. In other words, the combined effect of all non-genetic factors on milk yield in this particular cow was to increase milk yield by 96 litres. Using a similar argument, the worst *TT* cow had *P* = 1773, *G* = 1882, and *E* = 1773 − 1882 = −109; the combined effect of all non-genetic factors affecting milk production in this cow was to decrease milk yield by 109 litres.

We can summarize the above discussion of phenotypic value, genotypic value, and environmental deviation by writing phenotypic value as

$$P = G + E, \tag{14.1}$$

where *E* can be either positive or negative, depending on the combined effect of all non-genetic factors that have influenced the character in that individual.

Equation (14.1) provides a useful description of an individual's own performance for a particular character. However, we need to know more about an animal than its own performance; we also need to know its value as a breeding animal, as judged by the average performance of its offspring, when compared with the average performance of all offspring. In other words, we need to know how much better or worse is the performance of the offspring of that individual, when compared with the population mean. The difference in performance between a particular offspring and the population mean is called the *deviation* of that offspring from the population mean. Using this term, we can define breeding value as follows.

If an animal is mated with a random sample of animals from a population, then that animal's *breeding value* is twice the average deviation of its offspring from the population mean. The average performance of offspring is doubled in this definition so as to take account of the fact that only half the genes of any offspring come from the animal in question, the other half coming from the random sample of animals to which it was mated.

We shall now illustrate the meaning of breeding value by continuing with the transferrin example that was used earlier. The overall results of Ashton *et al.* (1964) showed that *tt* cows gave on average 200 litres more milk than did either *Tt* or *TT* cows, which were not significantly different from each other. Since the genotypic value of *TT* was 1882 litres, as quoted earlier, it follows that the genotypic values for *tt* and *Tt* were 2082 litres and 1882 litres, respectively. Since the aim of most dairy improvement programmes is to increase milk yield, we can call

t the favourable allele and T the unfavourable allele. From the results given above, we can also say that T is completely dominant to t, with respect to milk yield.

Now, in the herd mentioned above, the frequencies of the T and t alleles were 0.67 and 0.33, respectively, and the frequencies of genotypes corresponded to Hardy–Weinberg frequencies. It follows that the overall mean of this population of Jersey cows was

$$(0.67)^2 (1882) + 2(0.67)(0.33)(1882) + (0.33)^2 (2082) = 1904 \text{ litres.}$$

Notice how the population mean is determined solely by gene frequencies and by genotypic values. Since we have no control over the latter, it follows that the only way we can change the population mean is to change gene frequencies. For example, the average milk production of this herd will increase if we increase the frequency of the favourable allele, t. In fact,

> When we select the top producers and cull the poor producers in any selection programme, we are doing nothing more than trying to increase the frequency of favourable genes.

Returning to our example, we shall now express the three genotypic values in a manner similar to that of breeding values, namely as deviations from the overall mean. This gives genotypic values of -22 litres for TT and Tt, and $+178$ litres for tt.

Next, we shall imagine that we had one bull of each transferrin genotype, and that each bull was mated to a random sample of cows from the Jersey population being studied. Since the frequencies of T and t in this Jersey population are 0.67 and 0.33 respectively, it follows that in the random sample of cows mated to each bull, the probability of a gamete containing a T allele equals 0.67, and the probability of a gamete containing a t allele equals 0.33. Thus on average, we expect each random sample of cows to produce two-thirds T gametes and one-third t gametes.

The results of mating each of the three bulls to a random sample of cows are illustrated in Fig. 14.3. All the gametes from the TT bull will contain a T allele, which, when combined with two-thirds of T and one-third of t gametes from the random sample of cows, will produce two-thirds TT and one-third Tt offspring. But since T is dominant to t with respect to milk yield, all these offspring will have the same genotypic value, namely -22 litres.

In contrast, the Tt bull is expected to produce equal numbers of T and t gametes, which, when combined with two-thirds T gametes and one-third t gametes from the random sample of cows to which he was mated, will result in offspring genotypes in the proportions $\frac{1}{3} TT$, $\frac{1}{2} Tt$, and $\frac{1}{6} tt$. Thus, $\frac{1}{3} + \frac{1}{2} = \frac{5}{6}$ of this bull's offspring will have

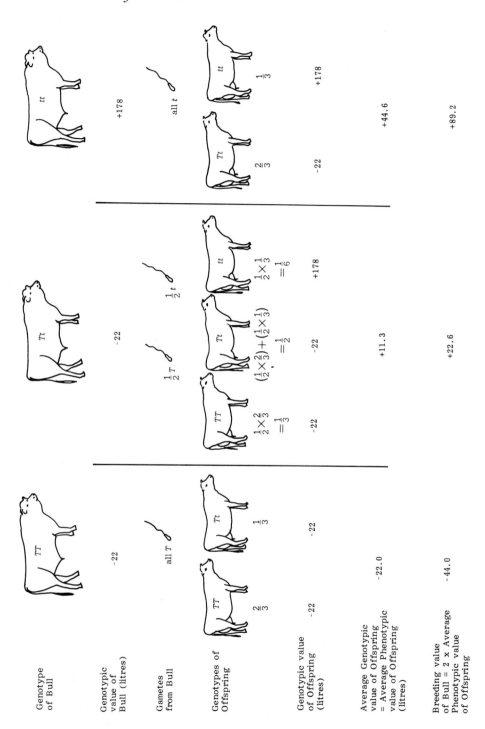

a G of -22, and the other $\frac{1}{6}$ (the t homozygotes) will have a G of $+178$ litres, giving an average G of $\{\frac{5}{6} \times (-22)\} + (\frac{1}{6} \times 178) = +11.3$ litres.

The third bull, tt, produces only t gametes and so is expected to result in two-thirds of Tt and one-third of tt offspring, which will have an average G of $\{\frac{2}{3} \times (-22)\} + (\frac{1}{3} \times 178) = +44.6$ litres.

Now, when we defined genotypic value as the average phenotypic value for that genotype, we were effectively saying that the environmental deviations have a mean of zero. By continuing to think of environmental deviations in this way, we can also say that the expected average performance (average phenotypic value) of any group of animals equals the average of their genotypic values. It follows that the results of mating each of the above three bulls to a random sample of cows will be offspring averages of -22.0, $+11.3$, and $+44.6$ for bulls with genotypes TT, Tt, and tt, respectively. According to the definition given above, these results mean that the breeding values of the three bulls are -44.0, $+22.6$, and $+89.2$, respectively. Notice that the breeding value of tt is $89.2 - 22.6 = 66.6$ litres greater than the breeding value of Tt, which in turn is $22.6 - (-44.0) = 66.6$ litres greater than the breeding value of TT. Expressed graphically, we now have the situation as illustrated in Fig. 14.4.

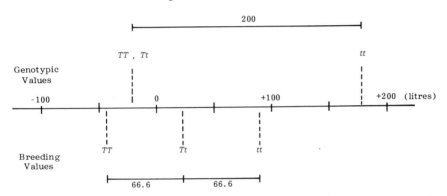

Fig. 14.4. A comparison of genotypic values and breeding values, expressed as deviations from the population mean. In this particular example, bulls of the three different genotypes have been mated to random samples of cows from a population in which T (frequency of $\frac{2}{3} = 0.67$) is completely dominant to t (frequency of 0.33), with a difference of 200 litres between the genotypic values of the two homozygotes. Notice that the breeding value of the heterozygote is exactly midway between the breeding values of the two homozygotes.

Fig. 14.3. The results of mating three different Jersey bulls each to a random sample of Jersey cows, in which the frequency of the T gene is $\frac{2}{3}$, and the t gene is $\frac{1}{3}$. This means that each random sample of cows is expected to produce $\frac{2}{3}$ T gametes and $\frac{1}{3}$ t gametes. Genotypic values and breeding values are expressed as deviations from the population mean, which is 1904 litres.

There are two points to notice about these results:

> (1) *an animal's breeding value does not necessarily equal its genotypic value, and*
>
> (2) *the breeding value of a heterozygote is exactly midway between the breeding value of the two relevant homozygotes.*

What if the transferrin alleles had been at different frequencies in the available population of Jersey cows? Would the same conclusions still apply? We shall answer this question by considering two extreme situations.

If the cows were all homozygous for T, then the only genotypes resulting from any of the three bulls would have been TT or Tt, in which case all offspring would have had the same performance: the breeding value of each bull would have been -44 litres. Using the same approach as described above, readers can verify that if the cows were all homozygous for t, then the breeding values of bulls TT, Tt, and tt would have been -44, $+156$, and $+356$, respectively. Notice that the heterozygote's breeding value is exactly midway between that of the two homozygotes. In fact, even though the actual magnitudes of breeding values change with changes in the frequency of allele t, the breeding value of the heterozygote is always exactly intermediate between the breeding value of each homozygote, for any frequency of t greater than zero.

In addition to gene frequency, there are two other factors that affect breeding value. They are the type of gene action, and the difference in genotypic value between the two homozygotes. It is left to the reader to verify, by working through appropriate examples, that even though the actual magnitudes of breeding values change with changes in these two factors, the breeding value of the heterozygote is always exactly intermediate between the breeding value of each homozygote.

We can conclude that the breeding value of a heterozygote is always exactly midway between the breeding value of the two relevant homozygotes, irrespective of the gene frequency, the type of gene action, and the difference in genotypic value between the two homozygotes. This is a reflection of the fact that:

> *An animal's breeding value is directly proportional to the number of favourable genes that it carries.*

Thus, changing from no t genes to one t gene increases breeding value by exactly the same amount as changing from one t gene to two. In other words, an animal's breeding value is determined by adding together the effect of each gene that it possesses. In order to emphasize this *additive* nature of breeding value, it is usually given the symbol A.

It should be evident from the above discussion that an animal's breeding value is not a constant; the actual magnitude of an animal's breeding value can change from one population to the next, depending on gene frequencies, type of gene action, and the difference in genotypic value between homozygotes. Of the three factors that affect the actual magnitude of breeding value, only the first (gene frequency) can be altered by humans. An important practical implication of this is that an animal's breeding value in a particular population can change as a result of selection changing gene frequencies in that population.

We are now able to distinguish clearly between genotypic value, G, and breeding value, A. Genotypic value is determined by genes combined in pairs to form a genotype. However, individuals pass on their genes and not their genotypes. Thus, the average performance of the offspring of any individual depends on the number of favourable genes possessed by that individual and not on how those genes were combined together in that individual's genotype.

How then can we describe the relationship between genotypic value and breeding value? The most convenient way is to consider that genotypic value is determined by the number of favourable genes present (the breeding value) plus a deviation due to the way in which those genes are combined in pairs into a genotype. Expressed symbolically, we have

$$G = A + D, \tag{14.2}$$

where D is the deviation due to dominance, which can be positive or negative. If there is no dominance, then by definition the heterozygote is exactly intermediate between the two homozygotes. In this case gene action is entirely additive, and genotypic value equals breeding value. Another way to think of dominance is as an interaction *within* a locus. If there is not interaction within a locus, then gene action is entirely additive and there is no dominance.

We are now ready to expand our consideration to the more realistic situation in which more than one locus affects the quantitative character. The concepts discussed above apply equally to this situation. Thus:

> *Breeding value is determined by the sum of the effects of all favourable genes at all loci that affect the character.*

At each locus, there may be a deviation from additive gene action due to dominance. The sum, over all loci, of these deviations is the overall dominance deviation, D. With more than one locus, however, we also have to take into account the possibility of interaction between loci, which is called *epistasis* (see Section 12.2.8). If there is no epistasis, then the genotypic value is entirely determined by breeding value and dominance deviation. If there is epistasis between any loci that affect the character, however, then this is an independent source of deviation from the additive gene action that determines the breeding value. We can cater for epistatic interactions by adding another term to our expression for genotypic value, to give

$$G = A + D + I, \tag{14.3}$$

where I is the overall deviation due to interactions between loci. The combined deviations due to dominance and epistasis are sometimes referred to as *non-additive* deviations, because they both arise from non-additive gene action.

If gene action is entirely additive within and between loci, then $D = I = 0$, and $G = A$. If, on the other hand, there is complete dominance at every locus that contributes to a character, and if interactions between loci are very large, then A still has a certain value, because every gene still has an additive effect. The only difference is that D and I also have some non-zero value (positive or negative), in order to describe the deviations from additive gene action due to dominance and to epistasis.

Combining equations (14.1) and (14.3), we can now write

$$\begin{aligned} P &= G + E \\ &= A + D + I + E \end{aligned} \tag{14.4}$$

as an adequate description of the components of an individual's phenotypic value.

How can we use this equation to say something useful about observations in the real world? Firstly, we can use it as an aid to understanding why some high performers leave only mediocre offspring and why sometimes an ordinary performer will produce exceptional offspring. In the first case the high performer obviously has a very high phenotypic value but, assuming that it was mated to reasonable performers, its breeding value must be relatively low. In other words, it does not have many favourable genes, but the sum of its dominance and interaction deviations and its non-genetic influences, i.e. the sum of $D + I + E$, is large and positive (Fig. 14.5a). The alternative case, of an ordinary performer leaving exceptional offspring when mated to a random sample of individuals, indicates a large number of favourable genes, but a combined effect of dominance and interaction deviations and non-genetic influences such that the sum of $D + I + E$ is large and negative (Fig. 14.5b).

The second point that can be made from equation (14.4) concerns selection programmes, the aim of which is to improve the average performance of a population of animals, i.e. to produce offspring with higher phenotypic values, P. Humans have no control over dominance deviations, D, and interaction deviations, I. And they also have no

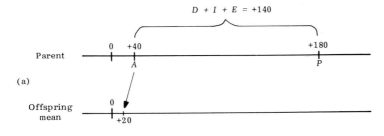

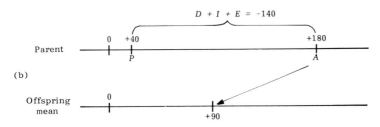

Fig. 14.5. A comparison of average offspring performance from an individual with (a) a high performance (phenotypic value, P) but relatively low breeding value, A, and (b) a high breeding value, A, but much lower performance (phenotypic value, P). In the first case, the sum of $D + I + E$ equals $+140$, and the average performance of the individual's offspring is much lower than the individual's own performance. In the second case, the sum of $D + I + E$ equals -140, and the average performance of the individual's offspring is much greater than the individual's own performance.

control over many of the non-genetic influences, E, that affect phenotypic values. However, we have seen that an individual's breeding value, A, is the average performance of that individual's offspring, when compared with the overall population mean. It follows that the way to produce offspring with the highest phenotypic values is to select as parents those individuals with the highest breeding values. Unfortunately, we cannot determine an animal's breeding value unless we mate it with a very large number of randomly chosen individuals, and compare the average performance of the resultant offspring with the overall population mean. Obviously, this is impractical in most circumstances. However, at the time when we wish to select the parents of the next generation, we have access to various pieces of information that provide us with clues to the breeding value of each animal being considered for selection. The way in which we make use of these clues in a selection programme, to help us decide which animals are most likely to have the highest breeding values, is described in Section 16.6.

14.4   The differences between individuals

The most popular parameter used for measuring the extent to which individuals differ from each other is the variance, which we first encountered in Section 14.2. The extent to which individuals differ in their phenotypic values is measured by the *phenotypic variance*, V_P. This is a quantity that can be estimated easily (as shown in the caption to Fig. 14.1) for any character on which a reasonable number of observations (values of P) are available. If phenotypic values are expressed as deviations from the population mean, then V_P is simply the average of the squared phenotypic values. Given a value of V_P so calculated, we can make use of our knowledge about the components of P (as shown in equation 14.4) to learn something about the components of V_P. In fact, for every component of P there is a corresponding component of V_P. Thus, we can write

$$V_P = V_A + V_D + V_I + V_E, \tag{14.5}$$

where V_A is variance in breeding values, V_D is variance in dominance deviations, V_I is variance in deviations due to epistatic interactions, and V_E is variance in non-genetic deviations.

Equation (14.5) represents the partitioning of V_P into a set of components. Because breeding values represent the additive effect of genes, V_A is often called the *additive genetic variance*. In contrast, V_D and V_I are components of *non-additive genetic variance*. The sum of the first three components is the *total genetic variance*, or the *genotypic variance*, V_G, which measures the extent to which different individuals have different genotypes. Since genotypes are determined at conception, V_G measures the extent to which individuals differ in factors that are fixed at the moment of conception. The other component, V_E, measures the extent to which individuals differ in all non-genetic factors that have influenced the character from the moment of conception to the time when phenotypic value was measured.

There is one important difference to note between equations (14.1) to (14.4), and equation (14.5). In the first four equations, each component can have a positive or a negative sign, so that we cannot think in terms of a phenotypic value being divided into proportions attributable to A or D or I or E. The components of V_P, on the other hand, are always positive and so do represent proportions of V_P. We can see this most easily by dividing through equation (14.5) by V_P to give

$$1 = \frac{V_A}{V_P} + \frac{V_D}{V_P} + \frac{V_I}{V_P} + \frac{V_E}{V_P}. \tag{14.6}$$

The relative contribution of any of the components to V_P is simply the proportion of V_P attributable to that component.

14.4.1 Heritability

We mentioned in Section 14.3 that breeding values are of prime importance in selection programmes. We are likely, therefore, to be interested in the proportion of phenotypic variance attributable to variance in breeding values. This is represented by the fraction V_A/V_P, which is called *heritability* and which is given the symbol h^2. Strictly speaking, it is called heritability in the *narrow sense*, in contrast to the ratio V_G/V_P, which is heritability in the *broad sense*, and which describes the relative contribution of genotypic variance to V_P. Since V_A is never greater than V_G, it is evident that narrow-sense heritability is never greater than broad-sense heritability. (This point was stated without explanation in Section 6.7.1.) Because the concept of heritability is most commonly used in animals in relation to selection programmes, and because it is breeding values that are of prime importance in selection programmes, the term heritability is normally used in the narrow sense unless otherwise specified.

14.4.2 Response to selection

It has been stated above that the aim of selection programmes is to select those individuals with the highest breeding values. It has also been stated that heritability indicates the relative contribution of variation in breeding values to phenotypic variance. We can link these two observations together and explain why heritability is such an important concept by considering two extreme situations in relation to response to selection.

Consider a population of animals at breeding age. Suppose that a measurement of phenotypic value for each animal has been made and hence V_P has been calculated. The top 20 per cent of individuals are to be selected as parents of the next generation, and the remaining 80 per cent are to be culled.

We shall start by supposing that heritability is zero, which means that V_A is zero. If there is no variation in breeding values, then all individuals have exactly the same breeding value. Thus, any of the top 20 per cent of animals must have exactly the same breeding value as any other animals in the population. Consequently, when members of the top 20 per cent are mated together, the average breeding value of their offspring will be exactly the same as the breeding value of any member of the parental population before selection. The end result is that the offspring distribution will have the same mean as the unselected parental distribution, as shown in Fig. 14.6. None of the superiority of the selected parents has been passed on to their offspring.

At the other extreme, we could assume that heritability was one, which means that the whole of V_P is due to variation in breeding values. Since the only differences between individuals are now differences in breeding values, the top 20 per cent of individuals will have

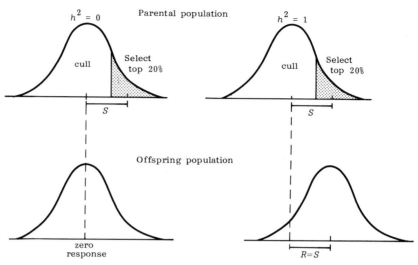

Fig. 14.6. Response to selection for the two extreme values of heritability, namely zero and one. S is the selection differential, or superiority of selected parents, and equals the difference between the mean of selected parents and the mean of the unselected parental population. R is response to selection, and equals the difference between the offspring mean and the mean of the unselected parental population. When heritability is zero, $R = 0$; when heritability is one, $R = S$.

the top 20 per cent of breeding values; ranking in order of phenotypic value will be equivalent to ranking on breeding value. In this case, the superiority of selected parents is due solely to superiority in breeding value and hence we expect the offspring mean to equal the mean of selected parents, as also shown in Fig. 14.6. All of the superiority of selected parents has been passed on to their offspring.

In order to summarize the conclusions to be drawn from Fig. 14.6, we need to introduce two new terms. The first is *selection differential*, S, which is the phenotypic superiority of selected parents, and which is equal to the mean of selected parents minus the mean of the whole parental population from which they were selected. The other term is *response to selection*, R. Providing there has been neither deterioration nor improvement in the environment between generations, then response to selection equals the mean of the offspring population minus the mean of the whole parental population.

It can be seen from the above discussion and from Fig. 14.6, that if $h^2 = 0$ then $R = 0$, and if $h^2 = 1$ then $R = S$. In fact, these two extreme situations are just special cases of the general prediction

$$R = h^2 S, \qquad (14.7)$$

which is true for all values of heritability. From this statement we can obtain another definition of heritability that indicates its importance in determining response to selection.

> *Heritability is the proportion of phenotypic superiority of parents that is seen in their offspring.*

It is now obvious that we need to know the heritability of a character if we are to predict how rapidly it will respond to selection. How do we determine the heritability of a character?

14.4.3 Estimation of heritability

The basis of methods for estimating heritability is resemblance between relatives. In order to understand why this is so, we need to consider why relatives tend to resemble each other more than unrelated individuals.

> *The extent of resemblance between relatives depends upon the extent to which relatives have:*
>
> (1) *genes in common;*
> (2) *genotypes in common;*
> (3) *environments in common.*

The extent to which individuals have genes in common is really the additive relationship between them, which was discussed in Chapter 13. In fact, the reason for putting 'additive' in front of 'relationship' and for giving it the symbol a should now be evident: the extent to which individuals have genes in common is the same as the extent to which they have breeding values in common. In other words, additive relationship refers specifically to the additive effects of genes; it refers to the extent to which individuals have genes in common as distinct from having genotypes in common.

We must now consider another factor that affects resemblance between relatives. Starting with an extreme situation, suppose that all members of a population, whether relatives or not, have exactly the same set of genes, i.e. they all have the same breeding value, which means that $V_A = 0$. If this is so, then the fact that relatives within that population have genes in common, i.e. have genes identical by descent from a common ancestor, contributes nothing to the resemblance between those relatives. However, the greater the tendency for unrelated individuals within a population to have different genes, i.e. the larger the variation in breeding values (the greater the value of V_A), then the more will individuals having genes in common tend to resemble each other, when compared with unrelated members of the population. Since the maximum value of V_A is V_P, we can say that, as far as resemblance between relatives is concerned, the importance of having genes in common ranges from zero if $V_A = 0$, to a maximum when $V_A = V_P$. In order to cover all possibilities, including the two extreme situations just described, we can say that the relative importance of having genes in common equals V_A/V_P. And this ratio is, of course, heritability, h^2.

We have now established that resemblance between relatives increases with increases in additive relationship, a, and with increases in heritability, h^2. In fact, the contribution to resemblance between relatives that is attributable to relatives having genes in common equals the product of relationship and heritability, ah^2. (Strictly speaking, this applies only when the relatives are not inbred, which is the usual situation for domestic animals whose performance is being measured for heritability estimation. Consideration of the use of inbred relatives in the estimation of heritability is beyond the scope of this book.) From this we can conclude that if having genes in common were the *only* cause of resemblance between relatives, then we would have

$$\text{resemblance} = ah^2. \tag{14.8}$$

It follows from this equation that if we can measure resemblance between relatives whose additive relationship is known, then we can estimate heritability as

$$\hat{h}^2 = \frac{\text{resemblance}}{\text{additive relationship}} \tag{14.9}$$

where $\hat{\ }$ indicates 'estimate of'.

This simple expression is the basis of most estimates of heritability. Although the principle is very simple, the actual mechanics of heritability estimation are often quite complex, because the measuring of resemblance between relatives is often complex. We shall not go into the details of methods in this book. We shall instead simply point out that they usually involve an analysis of variance and/or a regression analysis, the details of which are given by Becker (1984).

In estimating heritability as described above, we assumed that the only reason for individuals resembling each other was that they have genes (breeding values) in common. However, at the beginning of this section it was stated that resemblance can also be due to the sharing of genotypes and to the sharing of environments.

We saw above that the contribution of having genes (breeding values) in common depends upon the value of V_A/V_P. In a similar manner, the contribution of having genotypes in common depends upon the values of V_D/V_P and V_I/V_P, and the contribution of having environments in common depends upon V_{Ec}/V_P, where V_{Ec} is the *common environmental variance*. By analogy with h^2 being equal to V_A/V_P, the ratio V_{Ec}/V_P is sometimes written c^2.

In order to understand the meaning of V_{Ec}, we shall consider the following example. When individuals are raised in the same litter, they experience a certain set of non-genetic factors in common with all other individuals in the same litter. In the case of pigs, for example, this so-called *common environment* or *C* effect is largely a reflection of milk production by the sow, which often differs from litter to litter. The result is that piglets raised in the same litter are more likely to resemble each other than piglets raised in different litters by the same sow, because of the difference in common environment (milk production) from one litter to the next. This variation in common environment is part of the non-genetic variance, and is given the symbol V_{Ec}. If there is no variation in common environment, i.e. if $V_{Ec} = 0$, then having environments in common will make no contribution to resemblance between relatives. For example, if $V_{Ec} = 0$, then the resemblance between full-sib piglets raised in different litters will be the same as the resemblance between full-sib piglets raised in the same litter. On the other hand, the greater the differences in common environment from one litter to the next, i.e. the greater the value of V_{Ec}, then the greater will be the resemblance between full-sib piglets raised in the same litter, when compared with full-sibs raised in different litters. Also, since some sows produce more milk on average during each lactation than other sows, it follows that full-sib pigs reared in different litters (but obviously from the same sow) will also have environmental factors in common, and thus will tend to resemble each other more closely than one would expect on the basis of the genes and genotypes that they share.

Another good example of common environmental effects is provided by the effect of the lymphoid leucosis virus (LLV) in domestic chickens (Gavora *et al.* 1983). The most important method by which LLV spreads is vertically, via the egg, from the hen to her offspring. This means that the members of any particular full-sib group, i.e. the off-spring of a particular hen, tend to be either all carrying LLV (if that particular hen carried the virus), or to be all free of LLV (if their dam was not carrying the virus). Apart from spreading vertically, LLV can also spread horizontally among chickens being raised together. Since full-sibs are most likely to be raised together, this increases the tendency for groups of full-sibs to be either all carrying LLV or all free of LLV. Apart from its contribution to mortality, the presence of the virus causes considerable reduction in performance, in terms of decreased egg production and egg size, lower shell quality, and delayed onset of lay. To the extent that members of any particular full-sib family tend to be either all infected or all free of the virus, it follows that within any full-sib family, performance will tend to be either uniformly decreased, or uniformly unaffected. Once again, therefore, full-sibs will tend to resemble each other in performance more than would be expected on the basis of the genes and genotypes that they share.

It should now be evident that equation (14.8) is an oversimplification of the real situation. In order to take account of the fact that resemblance can be due to individuals having genotypes and environments in common, as well as genes in common, equation (14.8) needs to be rewritten. In doing so, we shall neglect the role of interaction variance,

V_I, because it is relatively unimportant, and because it cannot be incorporated easily into the following account. The expanded version of equation (14.8) is then

$$\text{resemblance} = ah^2 + d\frac{V_D}{V_P} + \frac{V_{Ec}}{V_P} \qquad (14.10)$$

where d = probability of relatives having the same genotype (called the *dominance relationship*).

Now, we saw in equation (14.9) that heritability is usually estimated by dividing resemblance by additive relationship. From equation (14.10) this gives

$$\frac{\text{resemblance}}{\text{additive relationship}} = h^2 + \frac{d}{a}\frac{V_D}{V_P} + \frac{1}{a}\frac{V_{Ec}}{V_P}. \qquad (14.11)$$

It follows from this equation that if relatives have genotypes or environments in common, then the resultant estimate of heritability may be biased upwards. In many situations it is possible to eliminate sources of bias. Half-sibs, for example, have no genotypes in common which means that d is zero. And half-sibs are less likely to have environments in common than full-sibs. Half-sib estimates of heritability, therefore, are usually unbiased. Full-sibs on the other hand, have $d = \frac{1}{4}$, and often have environments in common. Thus, it is very likely that full-sib estimates of heritability will be biased upwards. The other relatives commonly used in heritability estimation are offspring and parents, which have $d = 0$ but which may share common environments, depending on the experimental design.

Since V_A and V_P can vary from population to population, and from time to time within a population, and since estimates of heritability are subject to sampling variation, it is not surprising that heritability estimates for the same character vary from one estimate to the next. However, for any particular character there is a certain degree of similarity between estimates. Table 14.1 lists representative heritability values of various characters in domestic animals. The main conclusion that can be drawn is that:

> In general, those characters most closely associated with viability and reproductive ability have lower heritability than other characters.

This has important practical implications, as we shall see in Chapter 19.

Table 14.1. Representative values of the heritability of various characters in domestic animals. Since there is considerable variation between estimates of the heritability of any particular character, values in the table have been rounded to the nearest 5 per cent

	%	Reference
Horses		
Log of earnings—jumping	20	(1)
—3 day event	20	(1)
Best time—Thoroughbred	25	(1)
—Trotter	25	(1)
Pulling ability	25	(1)
Log of earnings—Trotter	40	(1)
—Thoroughbred	50	(1)
Handicap weight—Thoroughbred	50	(1)
Performance rate—Thoroughbred	55	(1)
Dairy cattle		
Protein yield	20	(2)
Milk yield	25	(2)
Butterfat yield	25	(2)
Percent protein	40	(2)
Percent butterfat	50	(2)
Beef cattle		
Survival to weaning—trait of dam	5	(3)
—trait of offspring	5	(3)
Weaning weight	25	(4)
Postweaning gain (feed lot)	35	(4)
Efficiency of gain (feed lot)	35	(4)
Final weight (feed lot)	40	(4)
External fat thickness	40	(4)
Marbling	40	(4)
Percentage bone	55	(4)
Percentage retail product	65	(4)
Sheep		
Survival to weaning—trait of offspring	5	(3)
Probability of conceiving—fertility	5	(5)
Survival to weaning—trait of dam	10	(3)
Number of lambs per pregnancy—fecundity	15	(5)
Clean fleece weight	40	(6)
Fibre diameter	45	(6)
Staple length	50	(6)
Greasy fleece weight	60	(6)

	%	Reference
Pigs		
Litter size	10	(7)
Killing-out proportion	20	(8)
Daily gain	25	(8)
Food conversion ratio	35	(8)
Eye muscle area	50	(8)
Carcase lean proportion	50	(8)
Domestic chickens		
Fertility	5	(9)
Hatchability	5	(9)
Liveability	5	(9)
Hen-housed egg production to 273 days	10	(9)
Egg weight	55	(9)
Body weight	50	(9)

(1) Average of many estimates. Hintz (1980). (2) Average of many estimates. White *et al.* (1981). (3) Average of many estimates. Cundiff *et al.* (1982*a*). (4) Best-bet estimates from the literature and from the Germ Plasm Evaluation Program of the USDA. Koch *et al.* (1982). (5) Median of several estimates. Bindon and Piper (1976). (6) South Australian Merino sheep. Gregory (1982*a*). (7) Average of many estimates. Hill (1982). (8) Values used by the Meat and Livestock Corporation in Great Britain. Mitchell *et al.* (1982). (9) Best-bet estimates from the literature and from several Canadian layer populations. Gowe and Fairfull (1980).

14.5   Summary

Quantitative characters are those that vary on a scale that is essentially continuous. They are determined by the action of genes at many loci and also by many non-genetic (environmental) factors.

The phenotypic value, P, of an animal is its measured performance for a particular character. Phenotypic value can be thought of as being determined by a genotypic value, G, and a non-genetic (environmental) deviation, E. An animal's genotypic value is determined by its breeding value, A, which depends on the number and additive effect of favourable genes possessed by that animal, relative to other animals in the population, and by deviations due to dominance, D, and epistasis, I, which reflect the way in which the genes are combined together into a genotype. Since individuals pass on their genes and not their genotypes, it is an individual's breeding value that determines the relative performance of the offspring of that individual. In fact, the breeding value of an animal is defined as twice the average performance of offspring produced by mating that animal with a random sample of animals, expressed as a deviation from the population mean.

Different animals have different phenotypic values, and the extent of these differences in a population is expressed as the variation in phenotypic values or phenotypic variance, V_P, which is composed of variation in breeding values, V_A, dominance and epistatic variance, $V_D + V_I$, and non-genetic (environmental) variance, V_E. The relative contribution of V_A to V_P is expressed as V_A/V_P, which is called heritability (in the narrow sense). It measures the proportion of phenotypic superiority in parents that is seen in their offspring, and hence is very important in predicting the extent of response to selection. Because of this importance, estimates of heritability are required for all economically important characters.

These estimates are obtained by estimating resemblance between relatives, on the assumption that the only reason that relatives resemble each other is that they have genes in common. In practice, relatives may also resemble each other because they may have genotypes in common and/or environments in common. If either of these factors contributes to the resemblance between relatives used to obtain a particular heritability estimate, then that estimate will be biased upwards.

In general, those characters most closely associated with viability and reproductive ability have lower heritability than other characters.

14.6 Further reading

Becker, W. A. (1984). *Manual of quantitative genetics* (4th edn). Academic Enterprises, P.O. Box 666, Pullman, Washington, USA. (This book provides a description of the statistical methodology used in the analysis of quantitative variation.)

Bulmer, M. G. (1980). *The mathematical theory of quantitative genetics*. Clarendon Press, Oxford. (A thorough review of the theory of quantitative genetics.)

Falconer, D. S. (1981). *Introduction to quantitative genetics* (2nd edn). Longman, London. (This is the classic textbook on quantitative genetics. It is essential reading for anyone wishing to understand this field of genetics. Chapters 6 to 10 in Falconer's book provide a more detailed account of the subject matter of the present chapter.)

15

Selection between populations

15.1 Introduction

Artificial selection occurs whenever humans choose to breed from certain animals and not from others. The choice can be made between populations and/or within populations. In the case of the former, it involves a decision as to the most appropriate source of breeding stock or commercial progeny. In the latter case, it refers to continual efforts to alter a population by breeding from only those animals which by some criteria are the most desirable within that population.

In both cases, selection is exploiting genetic variation. In the former case, it is genetic variation between populations that is being exploited, and in the latter case, selection is utilizing genetic variation within populations.

The aim of this chapter is to discuss the main points involved in selection between populations. It is followed, in Chapter 16, by a discussion of selection within populations.

15.2 Comparisons between populations

Most breeders have their favourite breed or strain, and most have their own reasons for their particular preference. It is obvious that individual preferences differ a great deal among different breeders, even among breeders engaged in very similar enterprises in the same area; it is not difficult to think of many situations in which completely different breeds or strains are used by different breeders for the same purpose.

In deciding which breeds or strains to use, breeders are carrying out selection between populations. For many years, this type of selection was in most cases a very subjective process, because breeders had no really useful information on the relative merits of various populations under any particular set of conditions. The only way to obtain such information is to conduct fairly large and well-designed comparisons of different populations, a job that no single breeder can ever hope to carry out, because of the relatively long time required, and because of the large numbers of animals needed in order to produce useful results. However, governments, universities, breed societies and some other large institutions sometimes have the resources necessary for comparisons, and have become actively involved in this type of work, both within countries and between countries.

15.2.1 Design of comparisons

Deciding on the best design for comparisons is quite a complicated business, and should be undertaken only after consultation with people experienced in this area. The factors that must be taken into account include randomization, equality of treatment, the type of comparison, the number of sires and the number of repeat measurements.

The question of randomization arises at several levels. In many situations, it is best if each population is represented by a random sample of animals. If this is not possible, then great care must be taken to see that the method and extent of selection of animals should be as similar as possible for each population. Also, if the animals are to be evaluated according to the performance of their progeny when mated to a so-called *tester* population, then the mates must be chosen at random from the tester population. Furthermore, if the comparison is to be conducted at more than one site, then there must be a random allocation of stock from each population to each site.

The requirement that all stock be treated equally is so obvious that it should not need to be mentioned. However, because there are sometimes quite large profits or losses to be made depending on the outcome of a comparison, all possible precautions should be taken to ensure that the comparison is conducted in a manner that will minimize the chances of any population receiving favoured treatment.

Another factor that affects the efficiency of distinguishing between populations is the type of comparison that is conducted, which was discussed by Smith (1976). For example, a *direct comparison* (all populations under test are compared at the same location) is more efficient than an *indirect comparison* (each population measured in a different location, with comparisons being made indirectly through a tester population measured in all locations). Also, a comparison between pure bred offspring (where each offspring in the comparison is the progeny of a male and a female belonging to one of the populations under test) is more efficient than a comparison between crossbred offspring (where the offspring being compared are obtained from mating just one sex, usually male, from the populations under test, to a random sample of members of a tester population).

Another factor that plays a major role in determining the efficiency of a comparison is the number of measurements made on each population under test. There are two important generalizations that can be made in regard to this factor. Firstly, the efficiency of a comparison is determined primarily by the number of sires represented in each population under test, rather than the number of offspring per sire.

Indeed, comparisons involving only, say, two or three sires per population are virtually worthless, even if hundreds of offspring from each sire have been measured, because the average performance of all these offspring will be much more a reflection of the average breeding value of those particular sires than of the population from which the sires came. Thus, when planning a comparison, it is vitally important to ensure that each population under test is represented by as many sires as possible. Another point to note is that unless the characters being measured have a very low repeatability, it is usually better to measure a different set of offspring each season or year than to take repeated measurements on the same set of offspring.

There are several ways in which the number of observations required from each population can be determined. First, there is the traditional approach, which involves asking how many observations are required in order to detect a certain difference between populations, given certain probabilities of concluding that there is no difference when in fact there is (Type II error), and of concluding that there is a difference when in fact there is not (Type I error). Another approach involves asking how many observations are required in order to have a certain probability of the best population actually being ranked first (Hill 1978). Yet another alternative is to decide on the numbers required by using economic criteria such as costs and likely benefits (Hill 1974).

Readers requiring further information in relation to design of comparisons should consult the references cited in this section, and at the end of this chapter.

15.2.2 Genotype-environment interaction

Comparisons generally involve testing the performance of groups of animals under one or a limited number of environments. The Food and Agricultural Organization, F.A.O., for example, has conducted a large-scale comparison of various Friesian strains from around the world, by using appropriate semen on a large number of cows on collective farms in Poland (Stolzman *et al.* 1981; Jasiorowski *et al.* 1983). Two questions immediately come to mind. Do the results obtained in Poland apply to other countries as well? Would the bulls within countries of origin, and the countries of origin themselves, rank in the same order if the trial had been conducted in, say, Canada or New Zealand?

If the results of comparisons vary depending on the environments in which they are conducted, we say there is *genotype-environment interaction*, G × E. The variation in results may involve just a change in relative performance, or a change in ranking (Fig. 15.1). The problem with G × E is that in many cases it is difficult to predict in advance whether or not such an interaction is likely to be important. All that can be done in practice is to conduct the comparison in all relevant environments, which is usually impracticable, or to interpret the results from a different environment cautiously, in the light of what is known in general about G × E in relation to the characters concerned.

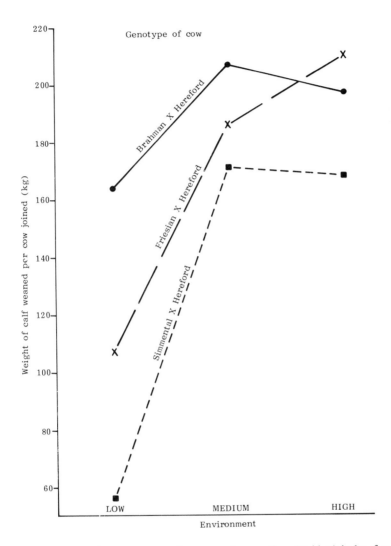

Fig. 15.1. Examples of genotype–environment interaction (G × E) in beef production. Three different cow genotypes were evaluated under low, medium and high levels of nutrition on a research station at Grafton, N.S.W., Australia. All cows were mated to Hereford bulls. Both types of G × E are illustrated: a change in the differences between genotypes (low versus medium environments), and a change in the rankings of genotypes (medium versus high environments). If there were no G × E, the lines joining different environments would be parallel. (Data from Barlow *et al.* 1978.)

15.3 Examples of comparisons

We shall now consider some examples of comparisons, so as to illustrate the type of results that are available to breeders.

15.3.1 Germ plasm evaluation programmes

In 1969, the United States Department of Agriculture, in conjunction with Kansas State University and the University of Nebraska, embarked upon a large-scale, long-term programme with the aim of characterizing a large number of breeds and crossbreds in cattle, in relation to economic beef production. Large numbers of Hereford, Angus, Jersey, South Devon, Red Poll, Limousin, Simmental, Charolais, Brown Swiss, Gelbvieh, Maine Anjou, Chianina, Pinzgauer, Tarentaise, Brangus, Santa Gertrudis, Brahman and Sahiwal bulls were mated to Hereford and Angus cows, and reproductive and productive data were collected from all offspring. Brown Swiss and Red Poll cows were also mated to bulls from some of the above breeds. The daughters of many of the above matings were retained and were in turn mated to bulls from various breeds.

It is not the aim of this chapter to even summarize the results of this or any other comparison. For those readers who do wish to study the results of this particular programme, progress reports can be obtained as described at the end of the chapter. This programme was cited solely as an example of the type of comparison that is being conducted in many different countries.

The end result of all this activity is comparative information on the relative performance of various breeds and various crossbreds, under certain well-defined environmental conditions in various countries. It is the type of information that, if properly interpreted, can provide breeders with useful guidance when they select between breeds.

15.3.2 Random sample tests

Although the comparisons described above often require many years for completion, they are usually based on one or a small number of samples from various strains or breeds.

A somewhat different type of comparison is the *random sample test*, in which samples are taken from various strains at regular (usually annual) intervals. If the comparisons are conducted in the same environment each year, then the results of random sample tests provide not only a regular comparison of various strains, but also an indication of genetic progress over time.

Random sample tests have been conducted in poultry for many years. They involve the supposedly random sampling of eggs from the hatchery of each of a number of breeding organizations, and the subsequent raising of the resultant chickens in a standard and well-defined set of conditions. The tests are conducted by an independent body, which measures performance and publishes the results. Despite a some-what checkered history in both broiler and layer chickens, the idea of random sample testing has now spread to pigs, especially in Europe, and to sheep.

Despite the recent resurgence in interest, there are still now, and always have been, certain objections to the concept of random sample tests.

One of these objections concerns the different aims of different breeding organizations; one strain of broilers, for example, may have been developed for one particular feeding regime such as *ad libitum*, while another strain may have been developed for a restricted feeding regime. Even with all the best intentions, an independent authority will be hard-pressed to decide on a feeding regime that would satisfy all entrants, if the above situation exists.

Another problem is that very large numbers of observations are often required in order to detect economically-important differences between strains, especially for reproductive characters which are particularly variable. For example, if a comparison involved two strains of pigs with a true difference in litter size of 0.5 pigs per litter (which is quite a large and economically-important difference), and if it were desired to have the probabilities of Type I and Type II errors (see Section 15.2.1) both equal to five per cent, then it would be necessary to measure 758 litters in each strain (Hill 1978). This is much greater than the total testing capacity of most pig testing stations. Because the total facilities avail-able are usually quite limited, and because it is usually required that stock be tested from a reasonable number of entrants, the number of observations per entrant is often very limited. The result is that the rankings of entrants often vary considerably from test to test, with luck or chance playing an alarmingly important role in determining the ranking of entrants. Not surprisingly, this tends to undermine the public's and the entrants' faith in the results. Some tests have attempted to overcome this problem by concentrating on three-year rolling averages, rather than on the results of any one year, and this does seem to be a step in the right direction. However, it is quite likely that one entrant may be improving more rapidly than another, so that real changes in relative merit may occur over a short time. If this is happen-ing, then the use of three-year rolling averages may be misleading.

The final problem concerns the supposedly random sampling of animals to enter the test. Despite careful precautions by the testing body, it is often possible for certain entrants to provide better than average stock, which, of course, defeats the whole purpose of random sample testing.

Despite these problems, random sample tests continue to be con-ducted in poultry and in pigs, in many different countries.

There is a tendency for all entrants to view the results as a sort of league table, and they are all keen to be at or near the top. Indeed, there are often considerable financial advantages to be gained for those at the top of the table, who take every opportunity to inform the public that their stock are better than anyone else's. They conveniently forget about the problems inherent in the tests, as described above. In contrast, those at the bottom of the table are more inclined to point out these problems. Occasionally, breeding companies become disenchanted with the test and refuse to participate. Given the problems outlined above, this is often the wisest course of action available. However, because of the financial benefits for the winners, and despite the problems inherent in the tests, there is usually quite strong competition among breeding organizations for a place in the test.

15.4   Summary

Most breeders have their favourite breed or strain, and most have their own reasons for their particular preference. In deciding which breed or strain to use, breeders have traditionally relied on mainly subjective assessments of the relative economic merits of various breeds and strains. But this situation is now gradually changing, with many large-scale comparative trials being conducted in various countries. Because of the scale of operations usually required for these comparisons, they are most commonly conducted by governments, universities, breed associations or other large institutions.

It is not unusual for the relative performance of breeds or strains to alter from one environment to the next if they are evaluated in more than one environment, and sometimes rankings may even change. Such alterations indicate genotype–environment interaction (G × E), the potential importance of which must be taken into account when evaluating the results of comparative trials.

Random sample tests have been conducted for many years among strains of broiler and layer chickens, and have recently been applied to different strains of pigs as well. While there are several theoretical and practical objections to the way in which these tests are usually conducted, the potential financial rewards for the company whose stock comes out on top of the 'league table' have in many cases been sufficient to maintain the demand for tests to be conducted.

15.5   Further reading

Anon. (1974*a*). *Proceedings of the Working Symposium on Breed Evaluation and Crossing Experiments with Farm Animals.* Research Institute for Animal Husbandry 'Schoonoord', Zeist, The Netherlands. (The papers presented at this symposium provide an excellent coverage of the requirements for efficient breed evaluation, and also provide many examples of evaluations in progress at that time.)

Anon. (1974*b*). *Germ plasm evaluation program.* Report No. 1. U.S. Meat Animal Research Center, Nebraska. (The first report of results from the cattle comparison described in Section 15.3.1. This report and subsequent reports are available from the U.S. Meat Animal Research Center, Clay Center, Nebraska, 68933, USA.)

Baker, H. K., Bech Andersen, B., Colleau, J., Langholz, H., Legoshin, G., Minkema, D., and Southgate, J. (1976). Cattle breed comparison and crossbreeding trials in Europe: a survey prepared by a working party of the European Association of Animal Production. *Livestock Production Science* 3, 1–11. (A review of cattle comparisons in progress in Europe in the early 1970s, with recommendations on the methodology for future comparisons.)

Dickerson, G. E. (1962). Implications of genetic-environmental interaction in animal breeding. *Animal Production* 4, 47–63. (A standard reference on genotype-environment interaction.)

Dickerson, G. E. (1965). Random sample performance testing of poultry in the U.S.A. *World's Poultry Science Journal* 21, 345–57. (A review of random sample testing in USA, including a useful discussion of the principles involved.)

Hartmann, W. (1974). Random sample poultry tests; underlying principles, achievements and future prospects. *World Animal Review* (11), 44–9. (Another review of random sample testing.)

Hill, W. G. (1978). How reliable is CPE? *Pig Farming* 26(2), 40–3. (A provocative analysis of the problems inherent in random sample testing of pigs. For enlightening responses to this article, see articles and correspondence in the same journal by Smith, D. 26(2), 43–4; Hill, W. G. 26(4), 93; and Smith, D. 26(5), 129.)

King, J. W. B., Curran, M. K., Standal, N., Power, P., Heaney, I. H., Kallweit, E., Schröder, J., Maijala, K., Kangasniemi, R., and Wallstra, P. (1975). An international comparison of pig breeds using a common control stock. *Livestock Production Science* 2, 367–79. (An account of a large-scale indirect comparison of breeds in pigs.)

Sutherland, R. A., Webb, A. J., and King, J. W. B. (1985). A survey of world pig breeds and comparisons. *Animal Breeding Abstracts* 53, 1–22. (A summary of comparisons between pig breeds.)

16
Selection within populations

16.1 Introduction

Ever since animals were first domesticated, humans have been attempting to alter animal populations, by means of artificial selection within those populations. The vast array of different breeds and strains of domestic animals in existence today is testimony to the effectiveness of such selection. Despite the many changes that have already occurred as a result of artificial selection, there are still many improvements needed, especially in terms of production of milk, eggs, wool, meat, and other animal products. And there is also a continual demand for faster horses, better showjumpers and more intelligent guidedogs. In most situations, the characters that need to be improved are quantitative characters.

The aim of this chapter is to discuss the factors that affect response to selection for a single quantitative character, to describe how response can be predicted, and to discuss how predictions for different breeding programmes can be compared so as to enable an optimum programme to be chosen. We shall then consider selection for more than one character, and shall finish with a discussion of economic evaluation of selection programmes.

Response to selection for a single character depends on five factors:

    (1) *variation in breeding values;*
    (2) *generation interval;*
    (3) *intensity of selection;*
    (4) *effective population size;*
    (5) *accuracy of selection.*

In order to investigate the effect of each factor, we shall first see what happens to response if we change just one factor at a time. We shall then consider all five factors at the same time, and hence obtain a prediction of response to selection, and determine the size of population required for selection.

16.2 Variation in breeding values

The basic material on which artificial selection acts is variation in breeding values, V_A. If all animals have exactly the same breeding value ($V_A = 0$), then there can be no response to selection. And the greater the differences between animals in breeding value (the larger V_A), the greater will be the response to selection.

For example, if V_A for milk yield in a group of dairy bulls is small, then the total range of breeding values may extend from, say, only −150 litres to +150 litres. If we select the top 20 per cent as shown in Fig. 16.1a, then the selected bulls will have an average breeding value of around +70 litres. In contrast, if V_A is larger, then breeding values may range from, say, −300 litres to +300 litres. The top 20 per cent of bulls in this case will have an average breeding value of around +140 litres (Fig. 16.1b). It follows that response to selection within the second lot of bulls, where V_A is larger, will be much larger than response to selection within the first lot, where V_A is small.

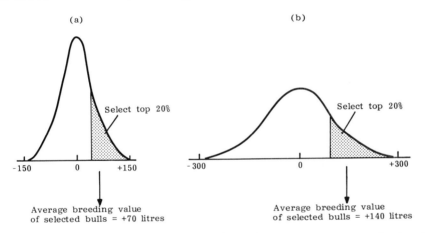

Fig. 16.1. The effect of variation in breeding values on response to selection in dairy bulls. Population (b) has more variation in breeding values and hence a greater response to selection.

> In general, response to selection increases as variation in breeding values increases.

16.3 Generation interval

Consider two herds, A and B. Suppose that bulls and cows are not joined until they are $2\frac{1}{4}$ years old in herd A, whereas those in B are joined a year earlier, at $1\frac{1}{4}$ years. If the parents in each herd are the result of selecting the same proportion for the same characters, then by the time the offspring from the first matings have been produced in each herd, the amount of genetic improvement will be the same in each herd. But it has taken one year longer to achieve the same amount of improvement in herd A compared with herd B. Response to selection *per year* is obviously greater in the herd that breeds from younger animals.

This indicates that the rate of response to selection depends on the age of parents when their offspring are born. If parents produce more than one offspring or set of offspring, then the rate of response depends on the average age of parents when their offspring are born, which is called the *generation interval*. The symbol for generation interval is L.

16.4   Selection intensity

It is clear that more genetic progress will be made if we select and breed from only the top 10 per cent of animals, i.e. cull the worst 90 per cent, than if we select and breed from, say, the top 30 per cent, i.e. cull the worst 70 per cent. So, for any character, the smaller the proportion selected, the greater the response to selection. Now, as the proportion selected becomes smaller, selection becomes more intense. In fact, the most convenient measure of the amount of selection applied is *selection intensity, i*, which is the superiority of selected parents, standardized according to the amount of variation in the character concerned. The superiority of selected parents is the selection differential, S, as defined in Section 14.4.2, and the usual method of standardizing the selection differential is to divide S by the square root of V_P. Thus, if we know the selection differential and if we know V_P for the character being selected, then selection intensity can be calculated from the formula

$$i = S/\sigma_P, \tag{16.1}$$

where σ_P is called the phenotypic standard deviation, which is the square root of V_P.

In many situations, however, and especially when predictions from different possible selection programmes are being compared, we need to know the selection intensity without actually measuring the superiority of parents. Fortunately, there is a simple relationship between selection intensity and proportion selected, so that the appropriate value of i can be obtained straight from the proportion selected, irrespective of which actual character is being selected. The usual method of doing this is to consult a table of values like those shown in Table 16.1.

At the beginning of this section we saw that the smaller the proportion selected, the greater the response to selection. It is now evident that this is equivalent to saying that selection intensity should be as large as possible. This can be achieved by increasing the number of

Table 16.1. Selection intensity, i, as a function of proportion selected, p

Proportion selected p	Selection intensity, i				
	Number of individuals from which parents are to be selected*				
	$n = \infty$	$n = 100$	$n = 50$	$n = 20$	$n = 10$
1.0	0	0	0	0	0
0.9	0.20	0.19	0.19	0.18	0.17
0.8	0.35	0.35	0.34	0.33	0.32
0.7	0.50	0.49	0.49	0.48	0.46
0.6	0.64	0.64	0.63	0.62	0.60
0.5	0.80	0.79	0.79	0.77	0.74
0.4	0.97	0.96	0.95	0.93	0.89
0.3	1.16	1.15	1.14	1.11	1.07
0.2	1.40	1.39	1.37	1.33	1.27
0.1	1.76	1.73	1.71	1.64	1.54
0.05	2.06	2.02	—	1.87	—
0.01	2.67	2.51	—	—	—

* The simple relationship between i and p that is mentioned in the text strictly applies only when $n = \infty$. However, it can be seen in the table that for a given p, the size of n has relatively little effect on i, especially at higher values of p. For more extensive tables, see Becker (1984), Tables 2 and 3, and Falconer (1981), Appendix Tables A and B. (From Becker 1984).

offspring raised from a fixed number of parents, or by selecting a smaller number of replacements from a fixed total number of offspring.

In practice, limitations on selection intensity are usually imposed by natural reproductive ability. If reproductive ability in a herd or flock is low, then selection intensity will be low. And any technique that improves reproductive rate can, if properly used, also increase selection intensity and hence lead to an increase in response to selection. Reproductive rate can be greatly increased in males by the use of artificial insemination and in females by the use of multiple ovulation and embryo transfer. Unfortunately, both techniques have sometimes been used to increase the reproductive rate of only average and even below-average performers. If used in this way, artificial insemination and multiple ovulation and embryo transfer obviously do not lead to an increase in response to selection. However, if used to increase the reproductive rate of the most superior animals in a well-designed selection programme, then either or both of these techniques can lead to considerable increases in response to selection (Robertson and Rendel 1950; Land and Hill 1975; Nicholas and Smith 1983).

Even if these two techniques for increasing reproductive rate are used, however, there are still practical limitations on the extent to which selection intensity can be increased. For example, if the total number of parents is reduced too far, then inbreeding problems and other problems associated with having small numbers of parents will be encountered. We shall discuss these problems in Section 16.5.

Another limitation is that the number of offspring raised from a fixed number of parents is usually limited by the total facilities available for raising offspring. However, it is possible to increase the number of offspring raised from a fixed number of parents, by breeding from the parents on more than one occasion. This will certainly increase the selection intensity, but it also increases the generation interval, because the longer parents are kept for breeding, the greater will be their average age when their offspring are born.

This highlights the fact that generation interval and selection intensity are interrelated factors; if we vary one, we also often vary the other. Ollivier (1974) has considered the interrelationship of these two factors in detail, and has shown how to obtain the optimum combination of i and L in various species of domestic animals. We shall return to this issue in Section 16.7.

16.5   Effective population size

The concept of effective population size is relevant to selection programmes in two different ways. We shall now discuss each of these in turn.

16.5.1   Inbreeding

It is obvious that as the number of parents in a selection programme decreases, then it becomes more and more likely that relatives will be mated together. Since inbreeding is defined as the mating of relatives, it follows that the smaller the number of parents, the greater will be the rate of inbreeding.

However, it is not simply the actual number of parents that is important. For example, the mating of relatives and hence inbreeding will be much less likely if there are 50 parents of each sex than if there are 99 female parents and just one male parent. And yet in each case the total number of parents is 100. Clearly we need another measure of population size; we need a measure that is a truer reflection of the amount of inbreeding expected. As shown in Appendix 16.1, an appropriate measure is *effective population size, N_e*, which can be calculated as

$$N_e = \frac{4sdL}{s + d}$$

(16.2)

where s = number of sires entering the population per year, d = number of dams entering the population per year, and L = average of male and female generation intervals, in years.

As also shown in Appendix 16.1, the increase in inbreeding per year, ΔF, is approximately

$$\Delta F = 1/(2N_eL) \qquad (16.3)$$

which for situations in which many more females are selected than males, is approximately

$$\Delta F = 1/(8sL^2). \qquad (16.4)$$

Although these formulae are only approximations, they are sufficiently accurate to be very useful in many practical circumstances for providing predictions of how much inbreeding should be expected in a particular selection programme.

There are two reasons for keeping a watchful eye on ΔF when designing breeding programmes. The first is inbreeding depression which, as explained in Chapter 13, can result in considerable decrease in performance, especially for characters associated with viability and/ or reproductive ability. The second reason is to do with the fact that inbreeding tends to decrease the amount of V_A in a population, with individuals becoming more and more homozygous for the same alleles and hence more and more genetically alike. And as we saw in Section 16.2, any decrease in V_A causes a decrease in response to selection.

How, then, do we try to ensure that ΔF is at an acceptably low level? In practice, we generally follow two rather arbitrary rules, which are:

(1) avoid mating very close relatives such as full-sibs, or parents and offspring, and
(2) ensure that effective population size is sufficient for ΔF to be less than an arbitrarily chosen value, such as 1 per cent per year. From equation (16.3), this amounts to ensuring that N_e is larger than $50/L$.

16.5.2 Chance and luck: genetic drift

As we saw in Chapter 5, genetic drift refers to random changes in gene frequency due entirely to chance. It results from the sampling of small numbers of genes that is inevitable in small populations or, to be more specific, in populations with small N_e. There are two ways in which genetic drift affects response to selection.

16.5.2.1 Decreased response The aim of artificial selection is to increase the frequency of favourable genes until the population is homozygous or *fixed* for each favourable gene. If this stage is ever reached, then all members of the population will be homozygous for the same set of favourable genes; there will be no variation in breeding values and there will be no more response to selection; the population will have reached its maximum possible *selection limit* or *selection plateau*.

As effective population size decreases, however, the more likely it is that changes in the frequency of favourable genes will be random rather than in the desired direction. And the smaller the effective population size, the greater is the chance that random changes in gene frequency will result in the actual loss of favourable genes from the population. As soon as a favourable gene is lost from the population, the average performance of that population will be less than if the gene had not been lost, and hence the response to selection at that stage will be less, and the ultimate limit to selection will also be less and will be reached more quickly. These three results of genetic drift are clearly illustrated in Fig. 16.2.

We can conclude that as effective population size decreases:

   (1) *selection response decreases;*
   (2) *selection limit decreases;*
   (3) *selection limit is reached more quickly.*

16.5.2.2 Variability of response   Besides affecting the average response to selection as described above, genetic drift also affects the variation in response. In order to understand the importance of variation in response, we have to imagine a situation in which the same selection programme is conducted in many identical populations at the same time. If we were to draw a separate line for response in each population, we would obtain a graph like that shown in Fig. 16.3a or in Fig. 16.3b. The only difference between these two sets of selection responses is that $N_e = 2$ in the former and $N_e = 86$ in the latter. Notice that there is much more variation in response when N_e is relatively small. At any generation, for example, there is a larger difference between the best and the worst line when $N_e = 2$ compared with when $N_e = 86$.

In practice, of course, there is usually just one selection line. The problem is that we have no way of knowing in advance which of all possible lines it will be; if we have bad luck, then our actual line may be one of the worst possible, and if we have good luck we may have a line that is amongst the best possible. If N_e is relatively large, then this will not be particularly important, as there is not much difference between the best and worst possible lines. But if N_e is small, then the difference in response between a good line and a bad line could determine the difference between success and failure of a selection programme. And remember that if N_e is small in an otherwise well-designed programme, the difference between large response and small response is determined

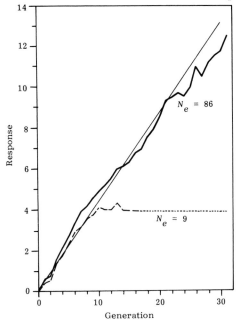

Fig. 16.2. The effect of N_e (effective population size) on response to selection. The lines for $N_e = 86$ and $N_e = 9$ represent the average response of at least two replicate selection programmes. The thin, straight line indicates the expected response to selection, obtained from $R = i r_{AC} \sigma_A$ (equation (2) in Appendix 16.4). Notice that with large N_e, observed response was very near to expected response for the first 20 generations. After that, response seems to have decreased slightly, but there is no sign of a selection limit. With small N_e, however, response to selection was always lower than with large N_e, and a limit (indicated by the dotted line) was soon reached at around generation 10. Because there was no further response in these lines, they were discarded after generation 13. (Drawn from the results of an artificial selection experiment with abdominal bristle number in *Drosophila melanogaster*, reported by Rathie and Nicholas 1980.)

mainly by chance; the breeder in charge has little control over what the observed response will be.

> *Thus, no matter how well-designed a selection programme is in terms of the other factors that affect response to selection, the only way for a breeder to be confident that expected response will actually be achieved in practice is to ensure that the effective population size is not too small.*

This raises the obvious question as to how large does a population have to be in order to reduce variation in response to an acceptably low level? For most situations likely to be encountered in practice, Nicholas (1980) has shown how to calculate the size of population required to

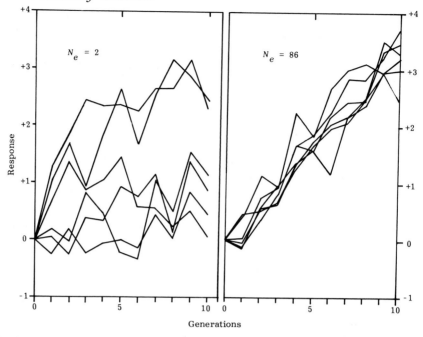

Fig. 16.3. Variation in response to selection, for replicate populations in which $N_e = 2$ or $N_e = 86$. Each line represents a separate population. In all twelve populations, the top 33 per cent of males and females were selected as parents each generation. It is evident that response is much more variable with small effective population size. (Drawn from results of an artificial selection experiment with abdominal bristle number in *Drosophila melanogaster*, conducted by Hammond 1973.)

satisfy particular requirements. Suppose it is decided, for example, that the coefficient of variation of response (the standard deviation of response divided by mean response) should be no greater than 20 per cent. In order to satisfy this requirement if a programme is to be evaluated after t years, Nicholas has shown that the effective population size should be at least

$$50 L/(i^2 h^2 t),$$

where L = average generation interval, i = average selection intensity, and h^2 = heritability.

For example, consider a beef cattle selection programme in which the top 5 per cent of bulls and all cows are selected as parents. From Table 16.1, we see that this corresponds to respective values of 2.06 and zero for i, giving the average i as $(2.06 + 0)/2 = 1.03$. If the heritability is $h^2 = 0.3$, if the average generation interval is $L = 4$ years, and if the programme is to be evaluated after $t = 10$ years, then the effective population size should be at least $50(4)/\{(1.03)^2 (0.3)(10)\} = 63$.

Notice that this is a more stringent requirement than the value of $50/L = 50/4 = 12.5$ required to keep the rate of inbreeding less than 1 per cent per year (see Section 16.5.1).

We can conclude that, when designing selection programmes, it is far easier to avoid problems with inbreeding than it is to avoid problems associated with genetic drift.

16.6   Accuracy of selection

When selection is carried out, the usual aim is to select as parents those animals that have the highest breeding values out of all the animals available for selection, so as to achieve the highest possible average performance in the offspring of selected parents. If we knew exactly the true breeding value of each animal, we could achieve this aim with maximum efficiency by ranking animals according to true breeding value, and selecting those at the top of the list.

In practice, however, we do not know an animal's true breeding value. Instead, all that we have is one or more clues to the animal's true breeding value. These clues consist of one or more measurements of performance (phenotypic values) taken on the animal itself and/or on one or more of its relatives. By means that will be described below, we can use these clues to estimate the true breeding value of each animal, and the animals can then be ranked according to *estimated breeding value*, which is often abbreviated to EBV. By ranking animals according to their EBV, we will tend to rank them according to true breeding value. The more accurate our estimate of true breeding value, the more accurate will be the ranking.

It is convenient to think of accuracy on a scale from 0 per cent, where the clue provides no information on breeding value, to 100 per cent, where the clue is the true breeding value itself. If accuracy is 0 per cent, then animals will, in effect, be ranked at random, and so selecting those animals at the top of the list will produce no response to selection, on average. At the other extreme, if accuracy is 100 per cent, then animals will be ranked according to true breeding value, and response to selection will be maximized.

Strictly speaking, *accuracy of selection* is the correlation between the available clues and the true breeding value of an animal. Use is made of this definition in the comparison of accuracy of different methods of selection, given in Appendix 16.2. The symbol for accuracy is r_{AC}, where r is the usual symbol for correlation, A represents true breeding value, and C stands for clues.

16.6.1   Clues to a candidate's breeding value

At regular intervals during a selection programme, groups of animals become available for selection and therefore have to be ranked in terms of estimated breeding value, EBV. These animals are called *candidates*. The potential clues to a candidate's true breeding value are:

(1) a single measurement or the average of repeated measurements on a candidate or on any relative, and
(2) the average performance of a group of relatives (for example, full-sibs, half-sibs, or progeny), where the records available on each relative are either single measurements or the average of repeated measurements.

We shall now consider each of these clues in turn.

16.6.1.1 Single measurement on a candidate   The simplest and most commonly available clue to a candidate's breeding value is a single measurement of its own performance. Measurement of such performance is called *performance testing*, and selection on the results of performance testing is called *individual* or *mass selection*. As shown in Appendix 16.2.1,

> *The accuracy of selecting on a single measurement of a candidate's own performance is the square root of heritability.*

Thus, accuracy increases as heritability increases, as shown in Fig. 16.4.

16.6.1.2 Single measurement on a relative of a candidate   In Section 14.4.3 we saw that the additive relationship between two individuals can be thought of as the extent to which they have breeding values in common. Bearing this in mind, we will now consider two extreme situations, both of which involve a candidate and one other animal.

At one extreme, suppose that the candidate and the other animal are identical twins, in which case they have identical breeding values and an additive relationship of one. Because their breeding values are identical, a measurement of performance on the other animal will be just as accurate a clue to the candidate's true breeding value as a measurement of the candidate's own performance.

On the other hand, if the additive relationship between the candidate and the other animal is zero, then their breeding values are completely independent of each other, and a measurement of performance on the other animal will provide no clue to the candidate's true breeding value.

In the first case, the accuracy of selecting on a measurement of performance on the other animal equals the accuracy of selecting on the candidate's own performance, which as we saw in Section 16.6.1.1, equals the square root of heritability. In the second case, accuracy of selecting on a measurement of performance on the other animal is zero.

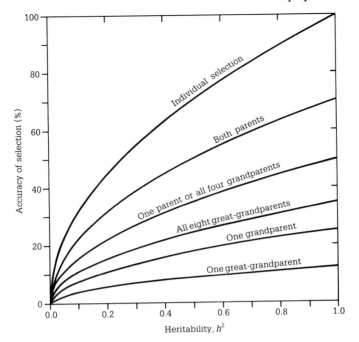

Fig. 16.4. Accuracy of individual selection, and of various types of pedigree selection, for all values of heritability. Values of accuracy were obtained as h for individual selection (Appendix 16.2.1), $0.5h$ for selection on one parent (Table 16.2), $h\sqrt{\frac{1}{2}} = 0.71h$ for selection on both parents (Appendix 16.2.2), $0.25h$ for selection on one grandparent (Table 16.2), $h\sqrt{\frac{1}{4}} = 0.5h$ for selection on all four grandparents (Appendix 16.2.2), $0.125h$ for selection on one great-grandparent (Table 16.2), and $h\sqrt{\frac{1}{8}} = 0.35h$ for selection on all eight great-grandparents (Appendix 16.2.2).

In order to cover all possibilities, including the two extreme situations just described, we can say that:

$$
\left\{
\begin{array}{l}
\text{Accuracy of selection} \\
\text{based on a single} \\
\text{measurement of per-} \\
\text{formance on a relative}
\end{array}
\right\}
=
\left\{
\begin{array}{l}
\text{Extent to which the} \\
\text{candidate and the} \\
\text{relative have breeding} \\
\text{values in common}
\end{array}
\right\}
\times
\left\{
\begin{array}{l}
\text{Accuracy of relative's} \\
\text{performance as a clue} \\
\text{to the relative's true} \\
\text{breeding value}
\end{array}
\right\}
$$

$$
=
\left\{
\begin{array}{l}
\text{additive relationship} \\
\text{between the candidate} \\
\text{and the relative}
\end{array}
\right\}
\times
\left\{
\begin{array}{l}
\text{square root of} \\
\text{heritability}
\end{array}
\right\}
$$

This can be written more briefly as

$$\text{Accuracy} = ah. \tag{16.5}$$

By drawing upon our knowledge of additive relationships, as detailed in Chapter 13, we can now write down the accuracy of selection based on a single measurement of performance on any relative. The accuracy of single measurements on various relatives is compared with the accuracy of selecting on a candidate's own performance in Table 16.2.

Table 16.2. Accuracy of ranking candidates or accuracy of selection based on a single measurement of performance on the candidate itself or on a relative

Single measurement of performance on	Accuracy
Candidate itself	h
One parent	$\frac{1}{2}h$
One grandparent	$\frac{1}{4}h$
One great-grandparent	$\frac{1}{8}h$
One identical twin	h
One full-sib	$\frac{1}{2}h$
One half-sib	$\frac{1}{4}h$
One progeny	$\frac{1}{2}h$

We can conclude that, with the exception of identical twins, a single measurement on one relative is always less accurate than a single measurement on the candidate itself.

However, the accuracy of selection based on any type of relative can be increased by taking measurements on more than one of them. We shall now consider this situation for each type of relative that is commonly available. The relevant expressions for accuracy of selection are derived in Appendix 16.2.2.

16.6.1.3 Average performance of ancestors Selection on the basis of ancestor performance is called *pedigree selection*. The accuracy of various types of pedigree selection is compared with the accuracy of individual selection in Fig. 16.4. It can be seen that except at low values of heritability, accuracy of pedigree selection is much lower than accuracy of individual selection. We can conclude that:

Pedigree selection alone is really only useful when selection has to be made before the candidate's own performance can be measured.

16.6.1.4 Average performance of sibs Selection on sib performance is called *sib selection*. We have already seen in Section 16.6.1.2 that the accuracy of sib selection based on one full-sib is 0.5h and on one half-sib is 0.25h. If we select on the basis of the average of more than one sib, we would expect the accuracy of sib selection to increase above these values as the number of sibs increases. This trend is illustrated in Fig. 16.5. It can be seen that the curves tend to flatten off as the number of sibs increases; for example, the gain in accuracy is much greater in going from 5 to 10 sibs than from 45 to 50 sibs. Notice also that even with a very large number of sibs, accuracy of sib selection is always less than one. In fact, even with an infinite number of full-sibs, the accuracy is never greater than $\sqrt{0.5} = 0.71$, and for half-sibs the maximum accuracy is $\sqrt{0.25} = 0.5$. Also indicated on the curves are the points at which the accuracy of sib selection just equals that of individual selection. If heritability is 0.1, for example, at least five full-sibs or 26 half-sibs are needed before sib selection is as accurate as selection on a single measurement of the candidate's own performance. However, it follows from the maximum possible values of accuracy given above, that selection on any number of full-sibs is never as accurate as individual selection if heritability is greater than 0.5, and similarly, that selection on any number of half-sibs is never as accurate as individual selection if heritability is greater than 0.25. This is the reason why the half-sib curves for $h^2 = 0.3$ and 0.7 and the full-sib curve for $h^2 = 0.7$ in Fig. 16.5 do not have asterisks.

16.6.1.5 Average performance of progeny The evaluation of candidates on the basis of their progeny's performance is called *progeny testing*. The accuracy of selection based on just one progeny is the same as that based on one parent or on one full-sib, namely 0.5h. As the number of progeny increases, so does the accuracy of selection increase, and, as shown in Fig. 16.6, the increase in accuracy is much larger than for sib selection. Consequently, the accuracy of selection based on average progeny performance quickly reaches a level greater than that of individual selection, and tends towards 100 per cent, for any value of heritability. This is exactly what we would expect, because the true breeding value of a candidate is defined in terms of the average performance of that candidate's progeny (see Section 14.3).

It is evident that progeny testing can give a higher accuracy of selection than clues obtained from any other type of relative. Balanced against this is the major disadvantage that it takes a much longer time to evaluate candidates on their progeny's performance than on their own performance. Indeed, in many situations, the resultant increase in generation interval is sufficient to offset any increase in accuracy, with the result that response to selection is actually slower with progeny testing than with, for example, individual selection.

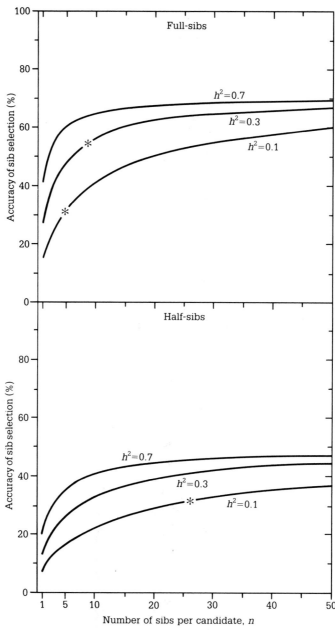

Fig. 16.5. Accuracy of sib selection, as a function of the number, n, of sibs measured per candidate. Asterisks indicate the number of sibs required to give an accuracy equivalent to that of individual selection. Curves are drawn from the equation $r_{AC} = h \sqrt{[na^2/\{1 + (n-1)ah^2\}]}$, where a is the additive relationship between sibs, being equal to $\frac{1}{2}$ for full-sibs and $\frac{1}{4}$ for half-sibs. This expression is derived as equation (17) in Appendix 16.2.2.

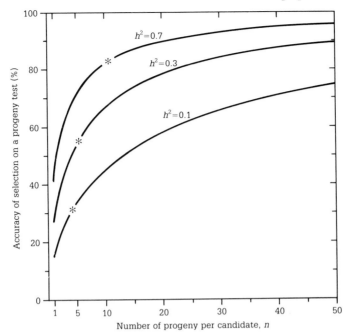

Fig. 16.6. Accuracy of selection based on progeny testing, as a function of the number, n, of progeny measured per candidate. Asterisks indicate the number of progeny required to give an accuracy equivalent to that of individual selection. Curves are drawn from the equation $r_{AC} = h\sqrt{[n/\{4 + (n-1)h^2\}]}$, which is derived as equation (18) in Appendix 16.2.2. This expression applies to the most common situation in animal breeding, in which progeny are related as half-sibs.

16.6.1.6 The effect of common environment   In Section 14.4.3 we saw that under certain circumstances, common environmental or C effects can be an important source of resemblance between relatives. It is shown in Appendix 16.2.2.2 that when selection is based on the average performance of a group of relatives, the increase in resemblance caused by C effects results in a reduction in the accuracy of selection. This means that when selection is based on any group of relatives, the accuracies discussed in previous sections are really maximum values: if C effects are important in any particular situation, then the accuracy of selection will be less than that described above.

16.6.1.7 Average of repeated measurements   For some characters, it is possible to measure performance more than once on the same animal. Fleece weight, for example, can be measured each year on each sheep, and maternal ability of beef cows, as judged by the weaning weight of their calves, can be measured on the same cow each time she weans a calf. Such characters are said to be *repeatable,* and the correlation between repeated measurements on the same animal is called the *repeatability*, which ranges from 0 per cent to 100 per cent.

The meaning of repeatability becomes clearer when it is compared with heritability. Repeatability is an indication of the extent to which an animal's superiority in one measurement will be seen in subsequent measurements of that same animal, i.e. *within its own lifetime*, whereas heritability indicates the extent to which the superiority of parents will be seen *in their offspring*.

Repeatability is concerned with the extent to which differences between individuals are permanent, i.e. remain throughout their lives. These permanent differences must be due to factors that remain permanent throughout the individual's lifetime. These include the individual's genotype ($G = A + D + I$) and a deviation due to any *permanent environmental factors*, E_p, i.e. a deviation due to environmental factors that exert the same effect on all measures of performance throughout that individual's lifetime. This is in contrast to the deviation due to *temporary environmental factors*, which affect one measurement but not the next. This deviation is written as E_t. Using this distinction between E_p and E_t, we can say that permanent differences between the performance of individuals are due to differences in $G + E_p$. We know from Section 14.4.1 that heritability indicates the relative contribution of V_A to V_P (as shown in equation 14.6). In a similar manner, repeatability indicates the relative contribution of $V_G + V_{E_p}$ to V_P, where V_{E_p} is the *permanent environmental variance*, being that part of V_E due to permanent differences in environmental factors. Since $V_G = V_A + V_D + V_I$ (Section 14.4), it follows that repeatability equals $(V_A + V_D + V_I + V_{E_p})/V_P$. Because of this, we can conclude that for repeatable characters, heritability is never greater than repeatability.

For animals in which repeated measurements can be made, breeding value can be estimated from the average of repeated measurements, which is often called *average lifetime performance*. Since each additional measurement provides extra information on that animal, accuracy of selection increases with the number of measurements included in the average, as shown in Fig. 16.7. Notice that accuracy increases less markedly with extra measurements for characters of high repeatability. This is because in a highly repeatable character, one measurement of performance gives a very good indication of future performance.

In previous sections, we have discussed the accuracy of selection based on a single measurement of performance on the candidate itself, or on one or several of its relatives. We can now extend this list of clues to include the average of repeated measurements on the candidate or on one or several of its relatives. In each case the conclusions in relation to increased accuracy resulting from the use of repeated measurements, as described above, will apply.

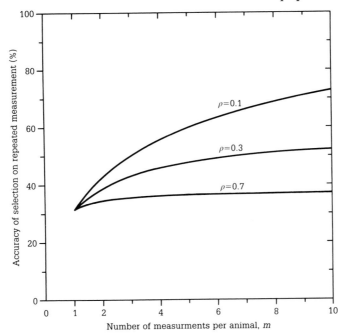

Fig. 16.7. Accuracy of selection based on the average of m measurements per animal, for various values of repeatability, ρ, and for a heritability, h^2, of 10%. The curves are drawn from the equation $r_{AC} = h \sqrt{[m/\{1 + (m - 1)\rho\}]}$, which is derived as equation (23) in Appendix 16.2.3. Notice that when there is only one measurement ($m = 1$), this equation tells us what we first encountered in Fig. 16.4, namely that accuracy of selection on a single measurement is the square root of heritability.

16.6.2 *Combining clues from more than one source*

In practical selection programmes, clues to a candidate's true breeding value for a particular character are often available from more than one source. In selecting hens for egg production, for example, data are usually available on the hen's own performance, on the average performance of the full-sib family of which she is a member (called a *dam-family average*) and on the average performance of the half-sib family of which she is a member (called a *sire-family average*). Corresponding male candidates can be ranked according to the same sire-family and dam-family averages, but obviously without any measurement of their own performance being included. In pigs, both male and female candidates are often ranked in terms of their own performance and that of their sibs.

It seems obvious that the accuracy of selection will increase as more clues are used, but how should the clues be combined together into one

overall estimate of a candidate's breeding value, and how can the resultant accuracy be determined?

Imagine that we have k different clues,

$$C_1, C_2, \ldots, C_k,$$

where a clue may be (1) a single measurement or the average of repeated
measurements on a candidate or on any relative, or (2) the average
performance of a group of the candidate's relatives (for example,
full-sibs, half-sibs or progeny), where the record available on each
relative is either a single measurement or the average of repeated
measurements.

The single overall estimate of a candidate's true breeding value that
is usually obtained from these different clues is called a *selection index*.
It is the sum of all clues, with each clue being weighted in such a way as
to maximize the accuracy of the index as an estimate of the candidate's
true breeding value. The index is written as

$$I = b_1 C_1 + \ldots + b_k C_k, \tag{16.6}$$

where I is the estimated breeding value, EBV, and where the b's are
weighting factors chosen so as to maximize the correlation between the
index, I, and the candidate's true breeding value, A. As derived in
Appendix 16.2.4, the accuracy of the index, r_{AI}, is given by

$$r_{AI} = \sqrt{(b_1 a_{1\alpha} + b_2 a_{2\alpha} + \ldots + b_k a_{k\alpha})}, \tag{16.7}$$

where $a_{1\alpha}$ is the additive relationship between the candidate, called α,
and any one of the relatives from which the 1st clue was obtained, and
similarly for the other a terms.

The general principles involved in obtaining appropriate weighting
factors and the resultant accuracy are described in Appendix 16.2.4.
Algebraic expressions for combinations of clues that are commonly
encountered have been listed by Van Vleck (1977, Table 17.5) and by
Becker (1984).

In writing down the index and deriving the appropriate b values,
care must be taken in specifying whether the same measurement occurs
in more than one clue. In the egg selection index cited above, for
example, a female candidate's measurement is usually included in the
dam-family average and in the sire-family average, and all measurements on members of the dam-family are included in the sire-family as
well. With male candidates, however, there is no measure of the candidate's own performance and so the dam-family average and sire-family
average obviously cannot include the candidate's own performance,
although the sire-family average still incorporates all of the dam-family
measurements. The reason why we must be careful in specifying whether
the same measurement occurs in more than one clue is that the values
of b and hence of accuracy differ depending on whether certain
observations occur in more than one clue.

16.6.3 *Prediction of a candidate's own future performance*

All of the discussion in Section 16.6 to date has been concerned with using clues to estimate a candidate's breeding value. In other words, we have been thinking solely in terms of predicting the performance of a candidate's progeny. There are, however, several situations in which we wish to predict the future performance of the candidate itself rather than the performance of its progeny.

In beef cattle herds, for example, cows have to be ranked each year in terms of their own predicted future maternal ability. And thoroughbred and standardbred yearlings have to be ranked in terms of their own predicted future racing ability.

The beef cattle example just mentioned above, involves predicting the future performance of a candidate from one or more measurements of its past performance. This is usually done by calculating a production index that is called *Most Probable Producing Ability*, MPPA. This index is constructed in the same manner as for the selection index shown above. Thus if the only clue to the candidate's future performance is the average of m measurements of its previous performance, $\bar{P}$, then

$$I = \text{MPPA} = b\bar{P}. \tag{16.8}$$

The only difference is that b is now chosen so as to maximize the correlation between I and future performance, rather than between I and true breeding value as in the case of a normal selection index. The derivation of b required for the calculation of MPPA is given in Appendix 16.2.5.

In the case of thoroughbred and standardbred yearlings, no previous measurements of the candidate's own performance are available, but there are records from various ancestors, each of which may have repeated measurements. In the same general manner as before, a linear combination of clues can be assembled into an index of the form

$$I = b_1 C_1 + \ldots + b_k C_k, \tag{16.9}$$

where the C's are clues and the b's are determined so as to maximize the correlation between I and future performance.

It turns out that if there are no previous records on the candidate, the b values for predicting the candidate's future performance are the same as those for predicting the candidate's breeding value, and the accuracy of predicting future performance is h times the accuracy of predicting the performance of the candidate's breeding value, where h is the square root of heritability.

16.6.4 *Adjustment factors*

It is common knowledge that the age of a cow when she calves exerts a strong influence on her total milk production in that lactation. Thus, age of the cow at calving is an 'environmental' bias that leads to some cows producing more milk than others.

If we make no allowance for age of the cow at calving, and simply select cows according to their actual total lactation production, then we will be tending to select cows that are older rather than cows that have the highest breeding value for milk production. If we could somehow allow for the fact that different cows have calved at different ages, then our selection would be more accurate; there would be a greater chance of selecting those cows with the highest true breeding values.

In some selection programmes, it is possible to allow for the effect of some of the more serious environmental biases by selecting within groups having the same environmental biases. For example, it may be possible to select within groups of cows that are all approximately the same age. In many situations, however, this is impractical.

In such cases, we can make allowance for the identifiable sources of environmental bias by adjusting the actual performance of certain cows so as to give all cows the production figures that they would have achieved had they all, for example, calved at the same age. This is done by means of adjustment factors, which are usually either *additive* or *multiplicative*. Additive adjustment factors are numbers of units of the character that are added to, or subtracted from a candidate's actual performance. Multiplicative factors are proportions by which a candidate's actual performance is multiplied.

Adjustment factors are obtained from the performance records of large numbers of animals. If properly used, they can substantially increase accuracy of selection. In the simplest cases, they do this by reducing V_E and hence reducing V_P, which in turn increases heritability, and, as we saw in the introduction of Section 16.6, the higher the heritability the greater the accuracy of selection.

16.6.5 BLUP

When performance records are used as clues in a selection index, it is automatically assumed that the records have been adjusted previously for all known sources of environmental bias, using adjustment factors as described in the previous section. However, there are many situations in which adjustment factors are not available from prior data. Consider, for example, the ranking of dairy sires using a set of production records obtained from 1000 herds over a three year period. Since each season in each year in each herd usually represents a unique combination of environmental factors, adjustment factors are required for each of the $1000 \times 3 \times 4 = 12000$ different combinations of herds and years and seasons (called herd–year–seasons). Now, since each herd–year–season is unique, it is obvious that herd–year–season adjustment factors can be estimated only from the data that we wish to adjust. With so many different herd–year–seasons, and with the number of records varying considerably between herd–year–seasons, the estimation of adjustment factors from such data is a major computing task, fraught with statistical difficulties. Assuming that the adjustment factors can be estimated, there are still two more sets of calculations required before bulls can be ranked. The first of these involves applying the appropriate adjustment factor to each original record. The second step involves calculating the estimated breeding value (EBV) of each sire, using $I = b_1 C_1$, where C_1 is the average adjusted performance of all daughters of a given sire, and b_1 is the appropriate index weight.

Fortunately, it is now possible to combine all these steps into one set of calculations, using the so-called BLUP method of estimation developed by Henderson (1949, 1973). BLUP stands for best linear unbiased prediction, which describes the statistical properties of the estimates obtained using the method. In practical terms, EBVs obtained using the BLUP method have all the desirable properties of EBVs obtained from a conventional selection index. Indeed, the conventional selection index is a special case of BLUP. However, the real merit of the BLUP method is that it is more powerful than the conventional selection index approach. For example, the BLUP method can be used to provide directly-comparable estimates of the average breeding value of groups of animals born in different years, thus providing an estimate of response to selection. In addition, BLUP can account for complications such as non-random mating, sires coming from more than one distinct group or population, environmental trends over time, herd differences in the average breeding value of dams, and bias due to selection and culling.

Because of the widespread occurrence of the above phenomena in cattle populations, and because no other method of estimating breeding values can account for them, BLUP has become the method of choice for calculating EBVs of cattle under many different circumstances throughout the world. Applications of BLUP in other species, particularly sheep, are also now beginning to emerge.

A brief account of the BLUP method is given in Appendix 16.3.

16.7   The five factors considered together

In Section 14.4.2 we saw that a very general prediction of response to selection for a single character was

$$R = h^2 S, \tag{16.10}$$

where R is response per generation, and S is selection differential.

In the present chapter we have been thinking in terms of response per year, which equals response per generation divided by generation interval. By making this alteration to equation (16.10), and by doing a little re-arranging as shown in Appendix 16.4, we can obtain the prediction

$$R_y = ir_{AC}\sigma_A/L, \tag{16.11}$$

where R_y is response per year, i is selection intensity, r_{AC} is accuracy of selection, σ_A is the square root of variation in breeding values, i.e. $\sigma_A = \sqrt{V_A}$, and L is generation interval.

We can summarize the meaning of equation (16.11) by saying that:

> *The rate of response to selection per year increases as:*
> *intensity of selection increases;*
> *accuracy of selection increases;*
> *variation in breeding values increases; and*
> *generation interval decreases.*

And we have seen earlier in this chapter that:

> *The rate of response to selection per year decreases as effective population size decreases.*

> *The first four of these factors together determine the predicted rate of response to selection. The fifth factor, effective population size, determines the extent to which the predicted response is actually achieved.*

In practice, the problem is that not all five factors are independent; by altering one factor in a desirable direction, another factor is often altered in an undesirable direction. For example, progeny testing increases the accuracy of selection, which is desirable, but this usually requires that parents be kept for longer, which is undesirable because it increases the generation interval. And, as we saw in Section 16.4, increasing selection intensity often involves increasing generation interval.

Designing a selection programme, therefore, involves striking several compromises among the above five requirements. The best selection programme for a particular situation is the one that has the optimum combination of these factors. This is generally decided by predicting response per year and calculating effective population size for various alternative selection programmes, and then choosing the programme with the highest response per year from amongst those for which effective population size is sufficiently large.

In other words,

> *The aims in designing a selection programme are:*
> (1) *decide on a minimum effective population size in relation to inbreeding and genetic drift, and*
> (2) *maximize $R_y = ir_{AC}\sigma_A/L$.*

Since σ_A is essentially fixed once the population has been established, it follows that maximizing R_y really amounts to maximizing ir_{AC}/L. In some circumstances, all three of these factors are interrelated, while in other circumstances, r_{AC} may be independent of the other two, and may even be constant in all the possible alternative breeding programme designs being compared at a particular time. As an example of the latter situation, suppose that a beef breeder has decided to select male and female replacements on a performance test of yearling weight, which means that r_{AC} is known, and can be regarded as fixed. However, a decision still has to be made in relation to male and female replacement rates, i.e. in relation to the length of time that males and females should remain in the herd. In this situation, the replacement rates that maximize response per year are those that maximize the ratio i/L, because both σ_A and r_{AC} are the same for all the possible alternative replacement rates being considered.

Whether it is ir_{AC}/L or i/L that is being maximized, the main point to understand is that it is the value of the ratio that counts, rather than the individual values of i or r_{AC} or L. Indeed, in many cases, the selection programme with the highest i or the lowest L may not give the maximum value of ir_{AC}/L or i/L. For example, Ollivier (1974) has shown that in order to maximize i/L, and hence response per year, sows should be kept for two farrowings, ewes for three lambings, and beef cows for seven calvings. In each of these situations, i is less than its possible maximum, and L is greater than its possible minimum. Readers requiring further information on maximizing response to selection should consult Ollivier (1974).

16.8 Correlated responses

Selection for one character almost always produces changes in other characters. For example, selection for increased growth rate in broilers leads to a correlated improvement in food conversion efficiency; selection for increased egg number in chickens results in a correlated decrease in egg size; and selection for increased fleece weight in sheep leads to a correlated increase in fibre diameter.

A change in an unselected character resulting from selection of another character is called a *correlated response*. The size of the correlated response depends, among other things, on the *genetic correlation* between the two characters, which is the extent to which the two characters are determined by the same genes. A genetic correlation can also be thought of as the correlation between breeding values for the two characters concerned. We shall use the symbol r_G for genetic correlation.

The genetic correlation between two characters can be determined from statistical analyses very similar to those used for estimating heritabilities. Becker (1984) gives all the necessary details on methodology.

There is, however, another measure of association between two characters that is important. It is called the *phenotypic correlation*, r_P, which is the correlation between the measurements of two characters on the same animal. It reflects the extent to which performance in one character is associated with performance in the other character.

Values for genetic and phenotypic correlations amongst some characters commonly measured in domestic animals are given in Table 16.3.

A phenotypic correlation arises from the correlation between breeding values for the two characters, r_G, and from the so-called *environmental correlation*, r_E, which results from correlations amongst non-additive genetic factors and environmental factors influencing the two characters. A description of the exact relationship amongst these three correlations is beyond the scope of this book. Readers requiring further information should consult Falconer (1981), Chapter 19. For our purposes, it is sufficient for us to be aware of the concepts of genetic and phenotypic correlation so as to be able to appreciate the following implications for practical selection programmes.

Firstly, the direction and magnitude of correlated response in an unselected character depends on the sign and the magnitude, respectively, of the genetic correlation between the unselected character and the selected character. If, for example, the genetic correlation between two characters is negative, then an increase in the character being selected will produce a decrease in the correlated character.

Secondly, the existence of genetic correlations raises the possibility of improving a character that is difficult or expensive to measure, by selecting on a character that is easy and cheap to measure. The practice of improving one character by selecting on another is called *indirect selection*. It is used, for example, in broiler chickens, where growth rate has a high positive genetic correlation with food conversion efficiency. This means that we can improve food conversion efficiency by selecting solely on growth rate, without going to the considerable extra expense of actually measuring food conversion efficiency.

Thirdly, the existence of genetic correlations provides us with an important new source of clues to a candidate's breeding value for a particular character; the performance of a candidate and/or of any relative in any character that is genetically correlated with the character to be improved, provides additional clues to the candidate's breeding value for the character to be improved. Thus, even if the character to be improved can be measured easily and cheaply on candidates themselves, it may be useful to measure the performance of the candidate and/or of relatives for other genetically correlated characters, and to make use of all these measurements when estimating the breeding value of candidates.

The way in which these various measurements can be used in estimating breeding values is described in the next section.

Table 16.3. Examples of genetic (r_G) and phenotypic (r_P) correlations. Because there is considerable variation between estimates, values in the table have been rounded to the nearest five per cent. Estimates obtained at other times or from other populations may be quite different from those shown below

	r_G	r_P	Reference
Dairy			
Milk yield: fat yield	+0.80	+0.90	(1)
Milk yield: protein yield	+0.90	+0.95	(1)
Milk yield: percent fat	−0.30	−0.20	(1)
Fat yield: protein yield	+0.90	+0.95	(1)
Pigs			
Growth rate: food conversion ratio	−0.80	−0.75	(2)
Growth rate: food intake	+0.40	+0.45	(2)
Growth rate: backfat thickness	−0.25	−0.10	(2)
Food conversion ratio: backfat thickness	+0.30	+0.20	(2)
Sheep			
Clean fleece weight: yield	+0.50	+0.40	(3)
Clean fleece weight: staple length	+0.40	+0.30	(3)
Clean fleece weight: greasy fleece weight	+0.60	+0.80	(3)
Clean fleece weight: crimps per inch	−0.35	−0.30	(3)
Domestic chickens			
Egg number: age at first egg	−0.35	−0.25	(4)
Egg number: body weight	−0.20	0.00	(4)
Egg number: egg weight	−0.30	−0.05	(4)
Age at first egg: body weight	+0.10	0.00	(4)
Age at first egg: egg weight	+0.15	+0.10	(4)
Body weight: egg weight	+0.40	+0.33	(4)

References: (1) Maijala and Hanna (1974), from various sources. (2) McPhee *et al.* (1979), from various European sources. (3) Gregory (1982*b*), from South Australian Merino sheep. (4) Emsley *et al.* (1977), from White Leghorn strain-crosses.

16.9 Taking account of more than one character

Almost always in practice, the purpose of a selection programme is to improve the performance of more than one character. Some characters, such as growth rate, are easily measurable on candidates of each sex; others are measurable in only one sex, e.g. egg production, or are expensive to measure, e.g. food conversion efficiency, or are not yet able to be measured on live animals, e.g. yield of saleable meat.

16.9.1 Selection objectives and net economic values

> The first step in designing a selection programme is to determine which characters should be improved, irrespective of whether they are cheap or expensive or even impossible to measure on the candidate.

These characters are called *selection objectives*.

> The next step involves determining the net economic value, v, of improvement in each selection objective.

For each objective, this is defined as the additional profit expected from a unit improvement in that objective.

As an example, we shall consider a pig selection programme in which the aim is to improve growth rate, G, food conversion ratio, F, which is g of food consumed per g of body weight gain, and percentage lean in the carcase, L. Thus G, F and L are the three selection objectives. In the particular example being considered here, we shall assume that every increase of 1 g per day in G is worth an extra 1 pence profit; every unit decrease in F is worth an additional 864 pence profit, and every percentage-point increase in L is worth an extra 55 pence profit. Thus, the net economic values are $v_1 = +1$ pence, $v_2 = -864$ pence, and $v_3 = +55$ pence. These values correspond to those used by the Meat and Livestock Commission in Great Britain, as quoted by Mitchell *et al.* (1982). Notice that the sign of the net economic value indicates whether the aim is to increase (+) or decrease (−) that particular selection objective.

The next step involves writing down an expression for the *true breeding value of overall merit*, which is the sum of the true breeding values for each selection objective, with each true breeding value being multiplied by the relevant net economic value. The true breeding value for overall merit is sometimes called *aggregate breeding value*, and is given the symbol *H*. For the example considered above, the aggregate breeding value of any candidate can be written as

$$H = v_1 A_1 + v_2 A_2 + v_3 A_3, \tag{16.12}$$

where v_1 is the net economic value of the first selection objective, A_1 is the true breeding value of the candidate for the first selection objective and similarly for the other objectives, and where objective 1 is growth rate (G), objective 2 is food conversion ratio (F), and objective 3 is percentage lean in the carcase (L).

> The overall aim of a selection programme is to maximize the improvement in aggregate breeding value.

16.9.2 *Selection criteria*

Having obtained an expression for aggregate breeding value, the next step in designing a selection programme is to decide which characters will actually be measured, and on which animals measurements will be taken. In other words, we must decide which measurements will be used as clues to the aggregate breeding value. These clues are called *selection criteria* and will be given the symbol C.

A selection criterion may be

(1) a single measurement or the average of repeated measurements for any selection objective or for any character that is correlated with a selection objective, taken on the candidate or on any relative, or

(2) the mean performance of a group of relatives, where the record available on each relative is either a single measurement or the average of repeated measurements for any selection objective or for any character that is correlated with a selection objective.

It is important to realize that there is essentially no limitation on what can be used as selection objectives or selection criteria, provided that the latter are measurable on candidates for selection or on their relatives, and provided that estimates of the relevant genetic, phenotypic, and economic parameters, as listed below, are available. At one extreme, the set of selection criteria used in any particular selection programme may be the same as the set of selection objectives for that programme. At the other extreme, the criteria may be completely different from the objectives. In many cases in practice, the criteria and the objectives have some but not all characters in common. Another point to note is that the number of criteria may be greater than, equal to, or less than the number of objectives.

To continue with our pig example, we note that the two most easily measured characters on pigs are growth rate, G, and backfat thickness, B. We shall therefore use these two measurements as clues to the aggregate breeding value. Thus, the selection criteria are G and B measured on each candidate. So as to keep the example as simple as possible, we shall not consider information from relatives.

16.9.3 *Relative importance of selection criteria*

Having now decided on our selection criteria, we next have to decide on the relative importance of each criterion, bearing in mind that the overall aim of our selection programme is to maximize the improvement in aggregate breeding value. This is one of the more difficult aspects of designing a selection programme, because in order to determine the correct values for relative importance of each selection criterion, we need to know:

(1) V_P and h^2 for each selection objective and for each selection criterion;
(2) r_P between each pair of selection criteria;
(3) r_G between each pair-wise combination of selection objective and selection criterion; and
(4) the net economic value for each selection objective.

If estimates for each of the above parameters are available, then the correct values of relative importance for each selection criterion can be obtained by a relatively complicated arithmetic procedure which is described in Appendix 16.5. Briefly, it involves defining an aggregate selection criterion, K, as the sum of all selection criteria, with each criterion being multiplied by its relative importance, b, the values of which are to be determined. In terms of our pig example,

$$K = b_1 C_1 + b_2 C_2, \qquad (16.13)$$

where K is the aggregate selection criterion, b_1 is the relative importance of the first selection criterion, C_1 is the first selection criterion, and similarly for the second criterion; and where criterion 1 is growth rate (G) and criterion 2 is backfat thickness (B).

Using available estimates of all relevant genetic, phenotypic and economic parameters, a set of calculations as described in Appendix 16.5 produces values of relative importance (b) that maximize the improvement in aggregate breeding value. In many situations it is really only feasible to do the necessary calculations on a computer.

It is all very well to say that you cannot calculate the correct values for the relative importance of each selection criterion without having estimates of the relevant genetic, phenotypic, and economic parameters and without access to a computer. What about all those breeders who have been conducting selection programmes for years with neither estimates of parameters nor access to a computer?

The answer is that, consciously or unconsciously, they have been guessing the values of each relevant genetic, phenotypic, and economic parameter, and have been using those guessed values to guess the relative importance of each selection criterion. Because so much guesswork has been involved, the values of relative importance that they have used in their selection programmes may have been far from the correct values. But the important point to realize is that:

> *Whenever more than one selection criterion is used in a selection programme, the relative importance of each criterion has to be determined in one way or another, either consciously or unconsciously, before selection can proceed.*

This statement is as true of selection programmes conducted by prehistoric humans during domestication of livestock as it is of selection programmes conducted today.

To return to our pig example, it is shown in Appendix 16.5 that the values of relative importance for growth rate and backfat are $b_1 = +1.269$ and $b_2 = -15.028$, respectively. The signs of the b's indicate that maximum improvement in aggregate breeding value will be achieved by selecting for high (+) growth rate and for low (−) backfat thickness.

When interpreting the magnitude of the b's, it must be noted that their actual value depends on the units of measurements. For example, if growth rate were measured as kg per day rather than as g per day, then b_1 would be 1269 instead of 1.269. If measurements were standardized (divided by the relevant phenotypic standard deviation, which is 54 for G and 2.8 for B, as derived from p. 481), then the b values would have to be multiplied by the relevant phenotypic standard deviation, giving rise to 'standardized' b values of $b_1' = 1.269 \times 54 \doteq +69$ and $b_2' = -15.028 \times 2.8 \doteq -42$.

Having either guessed or calculated the relative importance of each selection criterion, we can now proceed with selection, of which there are three main methods.

16.9.4 Tandem selection

Tandem selection involves ranking candidates in any one year in terms of only one selection criterion, with the relative number of years devoted to each criterion depending on the relative importance of each criterion. Since calculation of the actual number of years to devote to each criterion is not straightforward, and since tandem selection is rarely, if ever, used in practice, we shall not discuss the method further in this book.

16.9.5 Independent culling levels

The method of independent culling levels involves selecting animals which exceed a particular level of performance (the culling level) for *each* criterion. In other words, an animal will not be selected, i.e. will be culled, if its performance is worse than the culling level for any one criterion, irrespective of how much its performance may be better than the culling level for each of the other criteria. The actual culling levels are determined from the relative importance of each criterion and are chosen so as to give, overall, a certain proportion of candidates selected. The method of calculating the correct culling levels is rather complicated (see Turner and Young 1969, Section 12.2), but the necessary calculations can now be done with a computer, using software such as that developed by Saxton (1982). Once calculated, the culling levels can be used in any selection programme for which they are relevant, without any further use of a computer.

The actual position of each culling level, i.e. whether it is above or below the mean, and how far it is from the mean, depends primarily on the relative size of the standardized b values. In the case of our pig example, these values were shown in Section 16.9.3 to be $b_1' = +69$ and $b_2' = -42$, giving a ratio of 1.6 : 1 in favour of growth rate. This indicates that more selection pressure should be placed on growth rate than on backfat. For this particular example, it turns out that, if the overall proportion to be selected is 20 per cent, and if the mean performance of all candidates is 750 g/day for G and 20.0 mm for B, then the culling levels should be approximately 780 g/day for G and 21.4 mm for B. Bearing in mind that we wish to select for high growth rate and low backfat, this means that any pig having a growth rate greater than 780 g/day *as well as* having less than 21.4 mm of backfat will be selected as a replacement (Fig. 16.8). Selection using these culling levels is approximately equivalent to selecting pigs which are in the top 27 per cent for growth rate and also in the bottom 69 per cent for backfat.

16.9.6 Index selection

The third method involves ranking and selecting candidates according to their aggregate selection criterion (equation 16.13), which in the present context is often called a *selection index, I*, being written as

$$I = b_1 C_1 + b_2 C_2. \tag{16.14}$$

In the case of the pig example, the index is

$$I = 1.269\,(G) - 15.028\,(B) \tag{16.15}$$

which can be simplified (by dividing by 1.269, and rounding to the nearest whole number) to give

$$I = G - 12\,(B).$$

The effect of selecting replacements using this index is also illustrated in Fig. 16.8. In contrast to the method of independent culling levels, it can be seen that selection on an index enables very good performance in one criterion to compensate for only mediocre performance in another criterion; the emphasis is on overall performance rather than on meeting a minimum requirement for each criterion.

16.9.7 A comparison of the three methods

> *In terms of response to selection, index selection is better than or equal to independent culling levels, which in turn is better than or equal to tandem selection.*

In some situations, independent culling levels and tandem selection may be cheaper and easier to apply in practice than a selection index, although this is becoming less likely as techniques of index selection improve. In the past, for example, the use of a selection index required

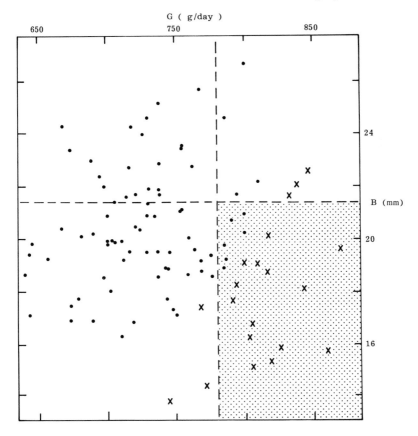

Fig. 16.8. A comparison between the use of independent culling levels and the use of a selection index in the selection of 20 replacement pigs out of 100 candidates. In this example, as described in Section 16.9, there are two criteria, namely growth rate, G, and backfat, B, measured on each candidate. The phenotypic correlation between G and B is $r_P = -0.1$. Each dot and each cross represents a candidate. The mean growth rate of all candidates is 750 g/day and the mean backfat is 20 mm. As explained in the text, the standardized b values for the criteria are $b'_1 = +69$ for G, and $b'_2 = -42$ for B. The fact that b'_1 is positive and b'_2 is negative indicates that selection is for increased growth rate and decreased backfat. The relative magnitude of the b' values indicates that selection is more intense on growth rate than on backfat. With the method of independent culling levels, the actual culling levels (dashed lines) corresponding to the above b' values are 780 g/day and 21.4 mm (see Section 16.9.5); with this method, all 20 pigs in the lower right-hand quadrant (shaded area) are selected as replacements. The selection index corresponding to the above b' values is $I = G - 12(B)$, as described in Section 16.9.6. The 20 replacements selected using this index are indicated by crosses. Notice that three of the replacements selected using the index are actually worse than the culling level for backfat, and another three are worse than the culling level for growth rate. In each case however, performance is very good for the other criterion.

every candidate to be measured for each criterion whereas independent culling levels could be applied in stages, with only those candidates actually selected in terms of the first criterion being measured for the criterion to be considered next, and so on. But now exactly the same multi-stage method can be used with a selection index (Cunningham 1975). One of the remaining disadvantages of the selection index method is that candidates usually have to be handled twice; once to be measured and the second time to be ranked in terms of an index value. Exceptions to this are situations in which none of the selection criteria is measured on the candidates themselves but rather on their relatives, or where an index can be calculated at the time of measurement. Another disadvantage of the use of a selection index is that candidates and relatives usually have to be identified, and production records have to be kept. Again, the exception to this is where an index can be calculated for each candidate at the time of measurement. For this situation, and for the other two methods of selection, it is possible to establish fixed culling levels either for the index, or for each criterion that has to be measured on candidates. In this case, measurements still have to be made, but do not have to be recorded because the selection decision on each candidate can be made immediately. If, however, some of the selection criteria involve measurements on relatives, then records have to be kept and candiates have to be identified for all three methods.

One final point must be noted. In the previous sections, we have seen that before any selection involving more than one criterion occurs for any of the three methods, the relative importance of each criterion must be determined. In addition, before the method of independent culling levels can be used, the actual culling levels must be determined from the relative importance of each criterion.

Traditionally, there has been a tendency to overlook these facts when comparing the methods. It has been commonly believed, for example, that there is no need to determine the relative importance of each criterion, i.e. the b values, if independent culling levels are used. It should be obvious from the above discussion that this belief is incorrect.

16.10 Economic evaluation of selection programmes

In Section 16.7 we discussed the use of two criteria by which alternative selection programmes can be compared. These criteria were response per year, R_y, and effective population size, N_e. In many situations, they will provide a quite adequate comparison of different programmes. But in some situations, we need to take other factors into account as well.

Consider, for example, two alternative pig selection programmes, each with the same effective population size and each with the same selection objectives, namely growth rate, G, percentage lean in the carcase, L, and food conversion ratio, F. Furthermore, suppose that the selection criteria in females, namely G and B, are the same in both programmes, and that the only difference between these two programmes is that programme A has the same selection criteria in males as in females, while programme B includes provision for measuring individual food intake, and hence food conversion ratio, in males. Let us suppose that the addition of F to the selection criteria in males increases accuracy of selection and hence response per year in terms of aggregate breeding value by around 30 per cent. In terms of response per year, therefore, programme B is obviously superior. But how does it compare in economic terms?

Let us suppose that the increase of 30 per cent in response to selection per year in programme B compared with programme A amounts to an extra £5 net profit per sow per year. Since genetic gains are cumulative, this means that additional profit will be £10 per sow in the second year, £15 in the third year, etc. Balanced against this additional profit, we have the additional cost in the first year for construction of housing facilities for individual feeding of boars.

In summary, we have two alternative programmes, in one of which the initial establishment cost and the subsequent net profit per sow per year are higher than in the other. The obvious question to ask is then: which of the programmes will show the greater overall profitability at the end of, say, 10 years?

In order to answer this question, we need to think in terms of *discounted cash flow*, which can most easily be explained through the following illustration. If we have £100 available now, we can invest it at, say, 10 per cent interest, and thus end up with £100(1 + 0.10) = £110 after one year. If we invest this amount for a further year at 10 per cent, we shall have £110(1 + 0.10) = £100(1 + 0.10)^2 = £121 at the end of two years. In general, if we start with £100 and invest it at $100d$ per cent compound interest, we will end up with £100(1 + d)^t after t years.

If we now look at this from the other point of view, it becomes clear that £121 earned in two years' time is equivalent to only £121/(1 + 0.10)^2 = £100 earned now. Likewise a cost of £121 in two years' time is equivalent to a cost of only £100 now.

It follows that if we are to compare programmes in which costs and/or income occur differentially into the future, we should discount all future income and costs back to their present value, and compare the programmes in relation to overall profitability in terms of net present value. We can do this by multiplying all future costs and returns by a discount factor $1/(1 + d)^t$ for each year of the programme, and then summing the resultant present values to get one figure for net present value of overall profitability. In terms of our pig example, we can simply look at the additional profit resulting each year from programme B, and discount that back to the present. We will then have a net present value of additional profitability with which to compare the net present value of additional establishment costs, in terms of facilities for individual feeding of boars. Since these establishment costs are incurred at the commencement of the project, they are equivalent to present costs and do not have to be discounted.

The necessary calculations are presented in Table 16.4, in which it can be seen that the net present value of additional profit during the first 10 years of the programme is £22 894. In terms of this particular example, this means that if the programme is to break even within 10 years, i.e. if any additional costs are to be recouped within 10 years, then the net present value of additional cost must not exceed £22 894. In other words, the additional cost of providing individual feeding facilities for boars must not exceed £22 894.

Table 16.4. Calculation of net present value of additional profit expected from measuring food intake in a selection programme conducted in a herd of 200 sows

Year of programme, t	Discount factor $= 1/(1 + d)^t$	Undiscounted additional profit £	Net present value of additional profit £	Cumulative net present value of additional profit £
	(X)	(Y)	(X × Y)	
0	1.000			
1	0.909			
2	0.826	1000	826	
3	0.751	2000	1502	2328
4	0.683	3000	2049	4377
5	0.621	4000	2484	6861
6	0.564	5000	2820	9681
7	0.513	6000	3078	12759
8	0.467	7000	3269	16028
9	0.424	8000	3392	19420
10	0.386	9000	3474	22894

In the above calculations, the discount rate is taken as 10%, and it is assumed that the financial benefit of measuring food intake is an additional profit of £5 per sow per year, which is first seen in the second year after building the shed that is required for measuring food intake.

It must be emphasized that the figures used in this example have been chosen simply to illustrate the principles of economic evaluation of selection programmes. The actual figures applicable to any specific pig selection programme could be quite different from those used above.

16.11 Summary

The rate of response to selection within a population depends on five factors, namely variation in breeding values (V_A), generation interval (L), intensity of selection (i), effective population size (N_e), and accuracy of selection (r_{AC}).

Candidates for selection are ranked according to estimated breeding value, EBV, which is determined from available clues to their true breeding value. Some clues are more accurate than others.

The potential clues to a candidate's true breeding value are (1) a single measurement or the average of repeated measurements on the candidate or on any relative of the candidate, and (2) the average performance of a group of the candidate's relatives, e.g. full-sibs, half-sibs or progeny, where the record available on each relative is either a single measurement or the average of repeated measurements.

If more than one of these clues are to be used to estimate the breeding value of each candidate, the clues are combined together into a selection index.

Response to selection per year (R_y) equals $ir_{AC}\sigma_A/L$, where σ_A is the square root of V_A. In designing a breeding programme, the aim is to (1) maximize $ir_{AC}\sigma_A/L$ and (2) ensure that N_e is not too small.

A correlated response is the change in an unselected character resulting from selection of another character. The magnitude of correlated response depends on the genetic correlation (r_G) between the two characters. The correlation between measurements of two characters on an individual is the phenotypic correlation (r_P).

When determining how animals should be selected in a breeding programme, the following steps must be taken: (1) decide on the selection objectives; (2) determine the net economic value of each selection objective; (3) define the aggregate breeding value; (4) decide on the selection criteria; and (5) determine the relative importance of each selection criterion.

Values for the relative importance of selection criteria are usually guessed by breeders, but they can be calculated so as to maximize improvement in profitability. Once the relative importance of each criterion has been determined, there are three methods of selection from which to choose: (1) tandem selection; (2) independent culling levels; and (3) index selection.

In terms of improvement in profitability, index selection is better than or equal to independent culling levels which is better than or equal to tandem selection. In practice, the method of independent culling levels is sometimes the preferred method, because in certain circumstances it is substantially easier to use than a selection index.

The aim of most selection programmes is to improve profitability. But different programmes may incur different costs and may generate different incomes at different times in the future. Valid assessments of the profitability of any selection programme, and comparisons between different programmes, can be made by discounting all future costs and incomes back to a common base year, using standard methods of discounted cash flow.

16.12 Further reading

Bird, P. J. W. N. and Mitchell, G. (1980). The choice of discount rate in animal breeding investment appraisal. *Animal Breeding Abstracts* **48**, 499-505. (A review of economic evaluation of animal breeding programmes, with particular emphasis on choice of discount rate.)

Cunningham, E. P. (1969). *Animal breeding theory*. Landbruksbokhandelen, Universitetsforlaget Vollebekk, Oslo. (Chapters 10 and 11 of this book provide an excellent introduction to selection in general, and to selection indices in particular, at a more sophisticated level than in the present chapter.)

Falconer, D. S. (1981). *Introduction to quantitative genetics* (2nd edn). Longman, London. (Chapters 11, 12, 13, and 19 of this classic textbook provide a detailed discussion of artificial selection.)

Hazel, L. N. (1943). The genetic basis for constructing selection indices. *Genetics* **28**, 476-90. (The first application of selection indices to animal improvement.)

Hill, W. G. (1971). Investment appraisal for national breeding programmes. *Animal Production* **13**, 37-50. (An example of economic evaluation of alternative selection programmes.)

James, J. W. (1979). Selection theory with overlapping generations. *Livestock Production Science* **6**, 215-22. (A review of recent developments in selection theory, describing several important issues not mentioned in this chapter.)

James, J. W. (1982). Construction, uses, and problems of multitrait selection indices. *Proceedings of the Second World Congress on Genetics Applied to Livestock Production* **5**, 130-9. (A brief review of selection indices, with a discussion of some of the problems associated with their use.)

Miller, R. H. and Pearson, R. E. (1979). Economic aspects of selection. *Animal Breeding Abstracts* **47**, 281-90. (A comprehensive review of economic evaluation of selection programmes.)

Ollivier, L. (1974). Optimum replacement rates in animal breeding. *Animal Production* **19**, 257-71. (A good example of how to maximize response to selection per year, with particular reference to how rapidly parents should be replaced by offspring.)

Smith, C. (1978). The effect of inflation and form of investment on the estimated value of genetic improvement in farm livestock. *Animal Production* **26**, 101-10. (A discussion of some of the factors that determine what interest rate should be chosen in economic evaluation of selection programmes.)

Turner, H. N. and Young, S. S. Y. (1969). *Quantitative genetics in sheep breeding*. Macmillan, Melbourne. (Chapters 9, 10, 11, and 12 in this book present a more detailed account of some of the material covered in the present chapter. Chapter 12 contains a particularly good discussion of the various methods of selecting for more than one character. Readers should note, however, that in Turner and Young's discussion of selection indices, the only examples considered are those in which the selection objectives are the same as the selection criteria. Thus, in comparing different indices, they are also comparing different objectives.)

Van Vleck, L. D. (1977). Genetics of the horse. In *The horse*. (Eds Evans, J. W., Borton, A., Hintz, H. F., and Van Vleck, L. D.) pp. 427–552. W. H. Freeman, San Francisco. (Van Vleck's Chapter 17 contains a useful account of most of the material covered in the present chapter. It contains particularly useful accounts of selection indices.)

Van Vleck, L. D. (1979). *Notes on the theory and application of selection principles for the genetic improvement of animals*. Department of Animal Science, Cornell University, Ithaca, New York. (Like Cunningham's book, these notes provide an excellent introduction to selection in general, and to selection indices in particular.)

Appendix 16.1

Rate of inbreeding, and effective population size

A16.1.1   Rate of inbreeding

Consider a typical three-generation pedigree:

Generation
number

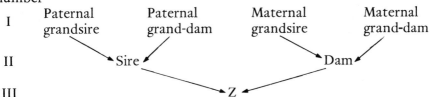

If none of the individuals in generations I and II is inbred (i.e., $F_I = F_{II} = 0$), then the only way for the offspring Z in generation III to be inbred is either: (1) its paternal grandsire and its maternal grandsire are the same individual; or (2) its paternal grand-dam and its maternal grand-dam are the same individual.

Consider alternative (1) first. Assuming that there are S male parents in generation I, we have:

$$
\begin{Bmatrix} \text{Chance that both} \\ \text{grandsires are the} \\ \text{same individual} \end{Bmatrix} = \begin{Bmatrix} \text{Chance that the} \\ \text{paternal grandsire} \\ \text{is a male from} \\ \text{generation I} \end{Bmatrix} \times \begin{Bmatrix} \text{Chance that a} \\ \text{randomly chosen} \\ \text{male from genera-} \\ \text{tion I will be the} \\ \text{same male as the} \\ \text{paternal grandsire} \end{Bmatrix}
$$

$$= \quad 1 \times \text{(Chance of randomly choosing one particular male out of } S \text{ males available)}$$

$$= \quad 1 \times 1/S$$

$$= \quad 1/S \tag{1}$$

If both grandsires are the same individual, then this grandsire is a common ancestor of Z, and we have the following path diagram:

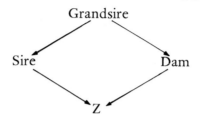

From first principles, it is obvious that the inbreeding coefficient of Z is given by

$$\frac{1}{2}\left(\frac{1}{2}\right)^{1+1} = \frac{1}{8} \tag{2}$$

Combining equations (1) and (2) together, we have:

$$F_Z = \left\{\begin{array}{l}\text{Inbreeding coefficient} \\ \text{of Z, assuming that} \\ \text{both grandsires are} \\ \text{the same individual}\end{array}\right\} \times \left\{\begin{array}{l}\text{Chance that both} \\ \text{grandsires are the} \\ \text{same individual}\end{array}\right\}$$

$$= \frac{1}{8} \times \frac{1}{S}$$

$$= \frac{1}{8S} .$$

Applying the same argument in the case of females, and assuming there are D female parents in generation I, we have:

$$F_Z = \frac{1}{8D} .$$

Combining these two possible pathways together, we have:

$$F_Z = \frac{1}{8S} + \frac{1}{8D} .$$

We already have said that $F_I = F_{II} = 0$, and noting that Z is any offspring in generation III, we have

$$F_{III} = \frac{1}{8S} + \frac{1}{8D}$$

so that the *increase in inbreeding* per generation is given by

$$\Delta F = F_{III} - F_{II}$$

$$= \frac{1}{8S} + \frac{1}{8D} \tag{3}$$

This expression was first derived by Wright (1931), using a different approach.

In the above derivation, S and D refer respectively to the number of male and female parents entering the population each generation. If s sires and d dams enter each year, and if the average generation interval is L years, then

$$S = sL,$$

and

$$D = dL,$$

in which case

$$\Delta F = \left(\frac{1}{s} + \frac{1}{d} \right) / 8L \text{ per generation}$$

$$= \left(\frac{1}{s} + \frac{1}{d} \right) / 8L^2 \text{ per year.}$$

Since d is usually much larger than s, $1/d$ is often negligible compared with $1/s$, in which case

$$\Delta F = 1/(8sL^2) \text{ per year, approximately.}$$

Although it is approximate, this equation works quite well in practice.

It is obvious from equation (3) that ΔF depends on the number of sires and the number of dams. Can we replace these with just one variable, which gives a useful measure of population size in terms of ΔF? An obvious reference point is the case where the number of male parents equals the number of female parents.

A16.1.2 Effective population size

If there are equal numbers of male and female parents, and if the total number of parents is N, then $S = D = N/2$. Substitution of these values into equation (3) gives

$$\Delta F = \frac{2}{8N} + \frac{2}{8N}$$

$$= \frac{1}{2N} \tag{4}$$

as the increase in inbreeding if there are equal numbers of parents of each sex, and if the total number of parents is N. But if S and D are not the same, then, as given previously by equation (3)

$$\Delta F = \frac{1}{8S} + \frac{1}{8D} \tag{3}$$

We can now ask the question: how many parents in total are needed in a population in which there are equal numbers of male and female parents, in order to give the same ΔF as a population with S male and D female parents? The answer to this question is *effective population size*, N_e, which is the value of N for which ΔF in (4) equals ΔF in (3). By equating (3) and (4) we get

$$\frac{1}{2N} = \frac{1}{8S} + \frac{1}{8D}$$

where N is now the effective population size, N_e. Rearranging, we get

$$N_e = \frac{4SD}{S + D}$$

Recalling that $S = sL$ and $D = dL$, we obtain

$$N_e = \frac{4sdL}{s + d}$$

In summary,

$$N_e = \frac{4SD}{S + D}$$

$$= \frac{4sdL}{s + d},$$

and

$$\Delta F = 1/(2N_e) \text{ per generation}$$

$$= 1/(2N_e L) \text{ per year}$$

$$= 1 \left/ \left\{ 2\left(\frac{4sdL}{s + d}\right)L \right\} \right. \text{ per year}$$

$$= \left(\frac{1}{s} + \frac{1}{d} \right) / 8L^2 \text{ per year,}$$

which gives, if $d \gg s$,

$$\Delta F = 1/(8sL^2) \text{ per year, approximately.}$$

It should be noted that the above formulae strictly apply only when replacements are chosen at random and are then mated at random. If there is directional selection of replacements, then in most cases the values of N_e will be less than that shown above; and the greater the intensity of selection, the smaller will be the value of N_e.

Appendix 16.2

Accuracy of selection

Candidates are ranked according to clues. Accuracy is defined as the correlation between the clue (C) and true breeding value (A), and is written as r_{AC}. Like any other correlation,

$$r_{AC} = \text{cov}(A, C)/\sqrt{(V_A V_C)},$$

where $\text{cov}(A, C)$ is the covariance between A and C. We can obtain expressions for the accuracy of each of the clues commonly used for ranking candidates, by obtaining values for $\text{cov}(A, C)$, V_A, and V_C.

In order to do this, we need to make use of some rules concerning variances and covariances. If X, Y, and Z are variables, and if a is a constant, then these rules can be written as follows:

$$\text{cov}(X, aY) = a \cdot \text{cov}(X, Y)$$

$$\text{cov}(X, Y + Z) = \text{cov}(X, Y) + \text{cov}(X, Z)$$

$$\text{cov}(X, X) = V(X)$$

$$V(aX) = a^2 V(X)$$

$$V(X + Y) = V(X) + V(Y) + 2 \cdot \text{cov}(X, Y).$$

If X and Y are independent, i.e. uncorrelated, then

$$\text{cov}(X, Y) = 0$$

and consequently

$$V(X + Y) = V(X) + V(Y).$$

A16.2.1   A candidate's own performance

In this case, the clue is phenotypic value, $P = A + D + I + E$, and so

$$\mathrm{cov}\,(A,\ C) = \mathrm{cov}\,(A,\ P)$$

$$= \mathrm{cov}\,(A,\ A + D + I + E)$$

$$= \mathrm{cov}\,(A,\ A) + \text{all other terms zero}$$

(because A, D, I, and E are independent of each other)

$$= V_A.$$

Thus,

$$\mathrm{cov}\,(A,\ P) = V_A, \tag{1}$$

and since

$$C = P,$$

we have

$$r_{AP} = V_A/\sqrt{(V_A\,V_P)}$$

$$= \sqrt{(V_A/V_P)}$$

$$= \sqrt{(h^2)}$$

$$= h. \tag{2}$$

A16.2.2 *Performance of relatives*

If the clue to the candidate's breeding value is a single measurement of the performance of a relative, then

$$r_{AC} = \left\{\begin{array}{l}\text{extent to which}\\ \text{the candidate and}\\ \text{the relative have}\\ \text{breeding values}\\ \text{in common}\end{array}\right\} \times \left\{\begin{array}{l}\text{accuracy of the relative's}\\ \text{phenotype as a clue to its}\\ \text{breeding value}\end{array}\right\} \tag{3}$$

$$= \quad \text{relationship} \qquad \times \quad h.$$

Since relationship is less than one for all relatives except identical twins, we conclude that a single observation on one relative is always less accurate than a single observation on the candidate itself. We can, however, increase the accuracy by selecting on the average performance of more than one relative. The algebra involved in obtaining expressions for the accuracy in these circumstances soon becomes rather complex, but the principle remains simple.

Let the clue to a candidate's breeding value be the average performance, $\bar{P}$, of a certain number of a particular type of relative, such as parents, grandparents, great-grandparents, sibs, or progeny. In general, the accuracy of selection will then be

$$r_{AC} = r_{A\bar{P}}$$

$$= \text{cov}\,(A,\ \bar{P})/\sqrt{(V_A\,V_{\bar{P}})}. \tag{4}$$

We shall now evaluate (4) for each of the types of relatives listed above.

A16.2.2.1 Average performance of ancestors　　If we have a single observation on the sire, P_S, and on the dam, P_D, then

$$\bar{P} = (P_S + P_D)/2,$$

and hence

$$\begin{aligned}
\text{cov}\,(A,\ \bar{P}) &= \text{cov}\,\{A,\ (P_S + P_D)/2\} \\
&= \tfrac{1}{2}\,\text{cov}\,(A,\ P_S + P_D) \\
&= \tfrac{1}{2}\,\{\text{cov}\,(A,\ P_S) + \text{cov}\,(A,\ P_D)\} \\
&= \tfrac{1}{2}\,\{\text{cov}\,(A,\ A_S + D_S + I_S + E_S) \\
&\qquad + \text{cov}\,(A,\ A_D + D_D + I_D + E_D)\} \\
&= \tfrac{1}{2}\,\{\text{cov}\,(A,\ A_S) + \text{cov}\,(A,\ A_D) + \text{all} \\
&\qquad \text{other terms zero}\}.
\end{aligned}$$

If there is random mating, the breeding value of an offspring (in this case A) on average equals one-half of the breeding value of each parent. Thus

$$\begin{aligned}
\text{cov}\,(A,\ A_S) &= \text{cov}\,(\tfrac{1}{2}A_S,\ A_S) \\
&= \tfrac{1}{2}V_A,
\end{aligned}$$

and

$$\begin{aligned}
\text{cov}\,(A,\ A_D) &= \text{cov}\,(\tfrac{1}{2}A_D,\ A_D) \\
&= \tfrac{1}{2}V_A.
\end{aligned}$$

Therefore,

$$\begin{aligned}
\text{cov}\,(A,\ \bar{P}) &= \tfrac{1}{2}(\tfrac{1}{2}V_A + \tfrac{1}{2}V_A) \\
&= \tfrac{1}{2}V_A. \tag{5}
\end{aligned}$$

We also need an expression for

$$\begin{aligned}
V_{\bar{P}} &= V\{(P_S + P_D)/2\} \\
&= \tfrac{1}{4}V(P_S + P_D) \\
&= \tfrac{1}{4}(V_{P_S} + V_{P_D} + \text{zero}),\ \text{if mating} \\
&\qquad \text{is at random} \\
&= \tfrac{1}{2}V_P. \tag{6}
\end{aligned}$$

Substituting (5) and (6) into (4), we have the accuracy of ranking candidates on the average performance of both parents equal to

$$r_{A\bar{P}} = \text{cov}\,(A, \bar{P})/\sqrt{(V_A V_{\bar{P}})}$$
$$= \tfrac{1}{2} V_A/\sqrt{(V_A \tfrac{1}{2} V_P)}$$
$$= \sqrt{(\tfrac{1}{4} V_A /\tfrac{1}{2} V_P)}$$
$$= h\sqrt{\tfrac{1}{2}}.$$

Notice that this is greater than the accuracy of ranking on just one parent ($\tfrac{1}{2}h$) but is still less than the accuracy of ranking on the candidate's own performance (h).

For the case of selection on the average performance of the candidate's four grandparents, it is left to the reader to verify, using an approach similar to that given above, that

$$\text{cov}\,(A, \bar{P}) = \tfrac{1}{4} V_A$$

and that

$$V(\bar{P}) = \tfrac{1}{4} V_P,$$

which gives the accuracy of selecting on all four grandparents as

$$r_{A\bar{P}} = \text{cov}\,(A, \bar{P})/\sqrt{(V_A V_{\bar{P}})}$$
$$= \tfrac{1}{4} V_A/\sqrt{(V_A \tfrac{1}{4} V_P)}$$
$$= \sqrt{(\tfrac{1}{16} V_A /\tfrac{1}{4} V_P)}$$
$$= h\sqrt{\tfrac{1}{4}}.$$

Using a similar argument, readers can verify that the accuracy of selecting on the average performance of all eight great-grandparents is $h\sqrt{\tfrac{1}{8}}$.

A16.2.2.2 Average performance of sibs   To determine the accuracy of sib selection (see Section 16.6.1.4), we set $\bar{P}$ equal to the average performance of n sibs. We then have

$$\text{cov}\,(A, \bar{P}) = \text{cov}\,\{A, (P_1 + \ldots + P_n)/n\}$$
$$= [\text{cov}\,\{A, (P_1 + \ldots + P_n)\}]\,/n$$
$$= \{\text{cov}\,(A, P_1) + \ldots + \text{cov}\,(A, P_n)\}/n. \qquad (7)$$

Now, $\text{cov}\,(A, P_j)$ is the covariance between the breeding value of one individual (the candidate) and the phenotypic value of the jth sib. Noting that in general, $P = A + D + I + E$, and assuming that the A of the candidate is independent of the D, I, and E of each sib, we have

$$\text{cov}\,(A,\,P_j) = \text{cov}\,(A,\,A_j) + \text{all other terms zero}$$

$$= a V_A, \tag{8}$$

where a is the additive relationship between the sibs. Substituting (8) into (7), we have

$$\text{cov}\,(A,\,\bar{P}) = n\{\text{cov}\,(A,\,P_j)\}/n$$

$$= a V_A. \tag{9}$$

Next, we have

$$V_{\bar{P}} = V\{(P_1 + \ldots + P_n)/n\}$$

$$= \{V(P_1 + \ldots + P_n)\}/n^2$$

$$= \{n V_P + n(n-1)\,\text{cov}\,(P_i,\,P_j)\}/n^2, \quad i \neq j$$

$$= \{V_P + (n-1)\,\text{cov}\,(P_i,\,P_j)\}/n. \tag{10}$$

Now, $\text{cov}\,(P_i,\,P_j)$ is the covariance between the ith and jth sibs. If we let t equal the correlation among sibs, then we have

$$t = \text{cov}\,(P_i,\,P_j)/\sqrt{(V_P V_P)},$$

$$= \text{cov}\,(P_i,\,P_j)/V_P, \tag{11}$$

which gives

$$\text{cov}\,(P_i,\,P_j) = t V_P. \tag{12}$$

Substituting (12) into (10), and rearranging, we have

$$V_{\bar{P}} = \{V_P + (n-1)t V_P\}/n$$

$$= V_P\{1 + (n-1)t\}/n. \tag{13}$$

Consequently, the accuracy of sib selection is

$$r_{A\bar{P}} = \text{cov}\,(A,\,\bar{P})/\sqrt{(V_A V_{\bar{P}})}$$

$$= a V_A/\sqrt{[V_A V_P\{1 + (n-1)t\}/n]}$$

$$= h\sqrt{[na^2/\{1 + (n-1)t\}]}. \tag{14}$$

Now, if all gene action is additive, i.e. if $D = I = 0$, and if the environmental deviations, E, are independent for each sib, then we have

$$\text{cov}\,(P_i,\,P_j) = \text{cov}\,(A_i + E_i,\,A_j + E_j)$$

$$= \text{cov}\,(A_i,\,A_j) + \text{other terms zero},$$

$$= a V_A, \tag{15}$$

in which case the correlation among sibs is

$$t = \text{cov}\,(P_i,\, P_j)/V_P$$
$$= aV_A/V_P$$
$$= ah^2. \tag{16}$$

The accuracy of sib selection can then be written as

$$r_{A\bar{P}} = h\sqrt{[na^2/\{1 + (n-1)ah^2\}]}. \tag{17}$$

If there are any common environmental effects (C effects, as described in Section 16.6.1.6), then the correlation among sibs is increased from

$$t = ah^2$$

to

$$t = ah^2 + c^2,$$

where $c^2 = V_{Ec}/V_P$, as described in Section 14.4.3. Since the correlation among sibs occurs in the denominator of the expression for the accuracy of sib selection, and since C effects increase the correlation, it follows that the presence of C effects will decrease the accuracy of selection.

A16.2.2.3 Average performance of progeny   With selection based on progeny testing, we now let $\bar{P} = (P_1 + \ldots + P_n)/n$ be the average performance of n progeny of the candidate. Using arguments similar to those given above, readers should be able to verify that for the case of selection based on progeny testing,

$$\text{cov}\,(A,\, \bar{P}) = \tfrac{1}{2}V_A,$$

and

$$V_{\bar{P}} = V_P\{1 + (n-1)t\}/n,$$

where t is the correlation amongst the progeny. Thus, the accuracy of selection based on progeny testing is

$$r_{A\bar{P}} = \text{cov}\,(A,\, \bar{P})/\sqrt{(V_A V_{\bar{P}})}$$
$$= \tfrac{1}{2}V_A/\sqrt{[V_A V_P\{1 + (n-1)t\}/n]}$$
$$= h\sqrt{[n/4\{1 + (n-1)t\}]}.$$

The most common situation in progeny testing is for the progeny to be half-sibs to each other. If there are no common environmental effects, then $t = h^2/4$, and

$$r_{A\bar{P}} = h\sqrt{[n/\{4 + (n-1)h^2\}]}. \tag{18}$$

If there are some common environmental effects, then $t = (h^2/4) + c^2$, and hence the accuracy of selection will be decreased.

The importance of C effects can be indicated by considering the accuracy of selection based on progeny testing in the limiting case of $n \to \infty$. If there are no C effects, i.e. if $c^2 = 0$, then by letting $n \to \infty$ in equation (18), it can be seen that the limiting accuracy is 100%, as expected from the definition of breeding value. If there are some C effects, then the accuracy is

$$r_{A\bar{P}} = h\sqrt{[n/\{4 + (n-1)(h^2 + 4c^2)\}]},$$

and the limiting accuracy as $n \to \infty$ is $h\sqrt{[1/(h^2 + 4c^2)]}$. If $h^2 = 0.25$, and if C effects contribute to resemblance as much as the sharing of genes, i.e. if $c^2 = h^2/4$, then the limiting accuracy is 71% rather than 100%.

A16.2.3 Average of several measurements of a candidate's own performance

Now the clue is the average lifetime performance of the animal to date, which can be written as

$$\bar{P} = (P_1 + P_2 + \ldots + P_m)/m,$$

where P_1 is the 1st measurement of the candidate's performance, etc., and m is the total number of measurements. In this case,

$$
\begin{aligned}
\text{cov}(A, C) &= \text{cov}(A, \bar{P}) \\
&= \text{cov}\{A, (P_1 + P_2 + \ldots + P_m)/m\} \\
&= \{\text{cov}(A, P_1 + P_2 + \ldots P_m)\}/m \\
&= m\{\text{cov}(A, P)\}/m,
\end{aligned}
$$

which from (1) can be written as

$$\text{cov}(A, \bar{P}) = V_A. \tag{19}$$

We also need

$$
\begin{aligned}
V_C &= V_{\bar{P}} \\
&= V\{(P_1 + P_2 + \ldots + P_m)/m\} \\
&= \{V(P_1 + P_2 + \ldots + P_m)\}/m^2 \\
&= \{m V_P + m(m-1)\,\text{cov}(P_i, P_j)\}/m^2 \\
&= \{V_P + (m-1)\,\text{cov}(P_i, P_j)\}/m, \tag{20}
\end{aligned}
$$

where $i \neq j$.

Now we define *repeatability* (ρ) as the correlation between any two measurements of the same character on one individual.

Thus for $i \neq j$,

$$\rho = \text{cov}\,(P_i, P_j)/\sqrt{(V_{P_i} V_{P_j})}$$
$$= \text{cov}\,(P_i, P_j)/V_P,$$

since $V_{P_i} = V_{P_j} = V_P$. After rearrangement we have

$$\text{cov}\,(P_i, P_j) = \rho V_P, \tag{21}$$

which when substituted into (20) gives

$$V_{\bar{P}} = \{V_P + (m-1)\rho V_P\}/m$$
$$= V_P\{1 + (m-1)\rho\}/m. \tag{22}$$

The accuracy of ranking candidates on lifetime performance is then

$$r_{A\bar{P}} = \text{cov}\,(A, \bar{P})/\sqrt{(V_A V_{\bar{P}})},$$

which after substitution from (19) and (22) gives

$$r_{A\bar{P}} = h\sqrt{[m/\{1 + (m-1)\rho\}]}. \tag{23}$$

A16.2.4 *Various combinations of clues combined into a selection index*

We can now describe a completely general situation that will enable us to rank candidates on any combination of clues, including any of the examples given above.

Imagine that we have k different clues

$$C_1, C_2, \ldots, C_k,$$

where a clue may be: (a) a single measurement or average lifetime performance of a candidate or of any relative; (b) the mean performance of a group of relatives (for example, full-sibs, half-sibs, or progeny), where the record available on each relative is either a single measurement or average lifetime performance.

Two questions now arise:

(1) How do we combine all these clues into a single number with which we will rank candidates?
(2) What is the accuracy of that single number as an overall guide to the candidate's breeding value?

The single number with which we rank candidates is called a *selection index*, which is a linear combination of all clues, with each clue being weighted in such a way as to maximize the accuracy of the index as an estimate of the candidate's breeding value.

The index is written as

$$I = b_1 C_1 + b_2 C_2 + \ldots + b_k C_k,$$

where the b's are the weighting factors chosen so as to maximize the correlation between I and the candidate's breeding value, A. This correlation (or accuracy) can be written in the usual manner as

$$r_{AI} = \text{cov}(A, I)/\sqrt{(V_A V_I)} \qquad (24)$$

and it is this expression which has to be maximized. In fact, it is algebraically more convenient to maximize the logarithm of r_{AI}, which has the same effect as maximizing r_{AI} itself. Thus, we wish to maximize

$$\log r_{AI} = \log \text{cov}(A, I) - \tfrac{1}{2}\log V_A - \tfrac{1}{2}\log V_I. \qquad (24a)$$

The method of maximizing $\log r_{AI}$ is the usual method of maximizing any function, namely to differentiate the expression, set the differential equal to zero, and solve for the parameter in question. In the case of a selection index, we need to obtain a value for each weighting factor, b_i, $i = 1, \ldots, k$, and so we take partial derivatives with respect to each b_i, and set all the partial differentials equal to zero. Recalling from high-school calculus that

$$\frac{\partial}{\partial x}\log f(x) = \frac{1}{f(x)}\frac{\partial}{\partial x}f(x),$$

we have, from (24a), and using b_1 as an example,

$$\frac{\partial}{\partial b_1}\log r_{AI} = \frac{1}{\text{cov}(A, I)}\frac{\partial}{\partial b_1}\text{cov}(A, I) - 0 - \frac{1}{2V_I}\frac{\partial}{\partial b_1}V_I. \qquad (25)$$

Now, since

$$\text{cov}(A, I) = \text{cov}(A, b_1 C_1 + \ldots + b_k C_k)$$
$$= b_1 \text{cov}(A, C_1) + \ldots + b_k \text{cov}(A, C_k),$$

we have

$$\frac{\partial}{\partial b_1}\text{cov}(A, I) = \text{cov}(A, C_1) + 0. \qquad (26)$$

And since

$$V_I = V(b_1 C_1 + \ldots + b_k C_k)$$
$$= b_1^2 V(C_1) + 2b_1 b_2 \text{cov}(C_1, C_2)$$
$$+ \ldots + 2b_1 b_k \text{cov}(C_1, C_k)$$
$$+ \text{other terms not including } b_1,$$

we have

$$\frac{\partial}{\partial b_1}V_I = 2b_1 V(C_1) + 2b_2 \text{cov}(C_1, C_2)$$
$$+ \ldots + 2b_k \text{cov}(C_1, C_k) + 0. \qquad (27)$$

Substituting (26) and (27) into (25), and setting the partial differential equal to zero, gives

$$0 = \frac{\text{cov}(A, C_1)}{\text{cov}(A, I)} - \frac{b_1 V(C_1) + b_2 \text{cov}(C_1, C_2) + \ldots + b_k \text{cov}(C_1, C_k)}{V_I}$$

After rearrangement, we have

$$b_1 V(C_1) + b_2 \text{cov}(C_1, C_2) + \ldots + b_k \text{cov}(C_1, C_k) = \text{cov}(A, C_1) \frac{V_I}{\text{cov}(A, I)}$$

(28)

Similar equations can be obtained by differentiating $\log r_{AI}$ with respect to each of the other b's.

When this is done, it is found that the ratio $V_I/\text{cov}(A, I)$ occurs on the right-hand side of all equations. Because it is a constant quantity common to the right-hand side of all equations, it can be given any value we like, without altering the relative value of the b's. Usually, the ratio $V_I/\text{cov}(A, I)$ is set equal to one, which is equivalent to setting $V_I = \text{cov}(A, I)$. In addition to simplifying the equations, this has the added advantage that it sets one unit of the index equivalent to one unit of predicted or expected breeding value. In other words, differences in index values between animals are equal to differences in expected breeding values. A further advantage in setting $V_I = \text{cov}(A, I)$ is that it gives a relatively simple expression for accuracy of selection, as shown in (34) below.

The equations can be further simplified by noting that, for any clue from any relative, R,

$$\text{cov}(A, C_R) = \text{cov}(A, P_R)$$

$$= \text{cov}(A, A_R + D_R + I_R + E_R)$$

$$= \text{cov}(A, A_R) + \text{all other terms zero}$$

$$= \left\{ \begin{array}{l} \text{relationship between} \\ \text{candidate and the} \\ \text{relative, R} \end{array} \right\} \times V_A$$

$$= a_{i\alpha} V_A,$$

(29)

where $a_{i\alpha}$ is the relationship between the ith relative and the candidate, whom we shall designate as α. In a similar manner, it can be shown that for any two clues C_i and C_j,

$$\text{cov}(C_i, C_j) = a_{ij} V_A,$$

(30)

where a_{ij} is the relationship between the ith relative and the jth relative. Finally, if the first clue is a single measurement on a single animal, then

$$V(C_1) = V_P.$$

Equation (28) can now be written as

$$b_1 V_P + b_2 a_{12} V_A + \ldots + b_k a_{1k} V_A = a_{1\alpha} V_A,$$

which can be rewritten as

$$b_1/h^2 + a_{12} b_2 + \ldots + a_{1k} b_k = a_{1\alpha}.$$

If the first clue involves more than one measurement and/or more than one member of a group of relatives, then $1/h^2$ in the first term is replaced by a more complex function which we shall call d_1.

Writing all k equations in the manner just described, the result is a set of 'normal' equations of the form:

$$
\begin{aligned}
d_1 b_1 + a_{12} b_2 + a_{13} b_3 + \ldots + a_{1k} b_k &= a_{1\alpha} \\
a_{21} b_1 + d_2 b_2 + a_{23} b_3 + \ldots + a_{2k} b_k &= a_{2\alpha} \\
a_{31} b_1 + a_{32} b_2 + d_3 b_3 + \ldots + a_{3k} b_k &= a_{3\alpha} \\
&\vdots \\
a_{k1} b_1 + a_{k2} b_2 + a_{k3} b_3 + \ldots + d_k b_k &= a_{k\alpha},
\end{aligned}
\tag{31}
$$

where a_{ij} = the relationship between an ith relative and a jth relative, $a_{i\alpha}$ = the relationship between an ith relative and the candidate, α, and

$$d_i = [\{(1 + (m_i - 1)\rho)/m_i h^2\} + (n_i - 1)a_{ii'}]/n_i, \tag{32}$$

where the ith clue is the average of m_i observations on each of n_i relatives in the groups, where all relatives within the group have a relationship of $a_{ii'}$ with each other, and where h^2 is heritability and ρ is repeatability.

We now have k normal equations with k unknowns, namely $b_1, \ldots, b_k$. The solution of these equations gives a set of b values which, when used in the index, will maximize the accuracy of selection.

Since we have set $V_I = \mathrm{cov}(A, I)$, it follows from (24) that the accuracy of selection can be written as

$$
\begin{aligned}
r_{AI} &= \mathrm{cov}(A, I)/\sqrt{[V_A \, \mathrm{cov}(A, I)]} \\
&= \sqrt{[\mathrm{cov}(A, I)/V_A]}.
\end{aligned}
\tag{33}
$$

Now

$$
\begin{aligned}
\mathrm{cov}(A, I) &= \mathrm{cov}(A, b_1 C_1 + \ldots + b_k C_k) \\
&= b_1 \mathrm{cov}(A, C_1) + \ldots + b_k \mathrm{cov}(A, C_k).
\end{aligned}
$$

Using equation (29), this can be written as

$$
\begin{aligned}
\mathrm{cov}(A, I) &= b_1 a_{1\alpha} V_A + \ldots + b_k a_{k\alpha} V_A, \\
&= (b_1 a_{1\alpha} + \ldots + b_k a_{k\alpha}) V_A,
\end{aligned}
$$

which gives accuracy of selection as

$$r_{AI} = \sqrt{(b_1 a_{1\alpha} + \ldots + b_k a_{k\alpha})}, \tag{34}$$

which is easy to evaluate for any index.

To illustrate the use of the normal equations, we shall derive the selection index for a progeny test based on half-sib progeny (as in cattle), and obtain the expression for the accuracy of a progeny test. Using the above notation, we have

$$k = 1 \quad \text{(only one clue)},$$

$$m_1 = 1 \quad \text{(only one record on each progeny)},$$

$$n_1 = n \quad \text{(the number of progeny)},$$

$$a_{11'} = \tfrac{1}{4} \quad \text{(relationship of half-sibs)},$$

which gives

$$d_1 = \{(1/h^2) + (n-1)/4\}/n,$$

and

$$a_{1\alpha} = \tfrac{1}{2} \quad \text{(relationship between progeny and the candidate)}.$$

With only one clue, we have only one normal equation, which is

$$d_1 b_1 = a_{1\alpha},$$

which gives

$$b_1 = a_{1\alpha}/d_1$$
$$= n/2\{(1/h^2) + (n-1)/4\}$$
$$= 2nh^2/\{4 + (n-1)h^2\}.$$

Then, in order to rank candidates according to a progeny test, they are ranked according to the value of

$$I = b_1 C_1, \tag{35}$$

where C_1 is the average performance of n progeny of the candidate, expressed as a deviation from the mean of progeny of all candidates, and b_1 is the weighting factor given above. The accuracy of a progeny test is then

$$r_{AI} = \sqrt{(b_1 a_{1\alpha})}$$
$$= h\sqrt{[n/\{4 + (n-1)h^2\}]}. \tag{36}$$

Readers should check that (36) is the same as (18), which is the accuracy of a progeny test derived from first principles.

As a second illustration of the use of the normal equations, consider the case where there are two clues, e.g. the performance of a candidate and one of its parents. Using the same notation, we have

$k = 2$	(two clues)
$m_1 = m_2 = 1$	(only one record on each individual)
$n_1 = 1$	(the candidate)
$n_2 = 1$	(one parent)
$a_{11'} = 1$	(relationship of candidate with itself)
$a_{22'} = 1$	(relationship of parent with itself)
$a_{12} = a_{21} = a_{2\alpha} = \frac{1}{2}$	(relationship between candidate and its parent)
$a_{1\alpha} = 1$	(relationship of the candidate with itself)

This gives

$$d_1 = 1/h^2,$$

and

$$d_2 = 1/h^2.$$

With two clues, we have two normal equations, which are

$$d_1 b_1 + a_{12} b_2 = a_{1\alpha}$$
$$a_{21} b_1 + d_2 b_2 = a_{2\alpha}.$$

After substituting from above, these become

$$b_1/h^2 + \tfrac{1}{2} b_2 = 1$$
$$\tfrac{1}{2} b_1 + b_2/h^2 = \tfrac{1}{2}.$$

From the first equation, we have

$$b_1 = h^2(1 - \tfrac{1}{2} b_2).$$

Substituting for b_1 in the second equation, and solving for b_2 eventually gives

$$b_2 = 2h^2(1 - h^2)/(4 - h^4)$$

and then

$$b_1 = h^2(4 - h^2)/(4 - h^4).$$

Candidates can now be ranked according to the value of

$$I = b_1 C_1 + b_2 C_2,$$

where C_1 is the performance of the candidate, C_2 is the performance of the candidate's parent, and b_1 and b_2 are given above.

The accuracy of selection is then

$$r_{AI} = \sqrt{(b_1 a_{1\alpha} + b_2 a_{2\alpha})}$$
$$= h\sqrt{[(5 - 2h^2)/(4 - h^4)]}\,.$$

For more detailed information, and for values of b for many different combinations of relatives, readers should refer to Becker (1984), Cunningham (1969), and Van Vleck (1977, 1979).

A16.2.5 Prediction of a candidate's future performance

In the previous section, we used a selection index to rank candidates according to their estimated breeding value. Now we wish to use an index to rank candidates according to their predicted future performance (their most probable producing ability or MPPA), using just one clue, namely $\bar{P}$, which is the average of m measurements of the candidate's previous performance, where each measurement is expressed as a deviation from its respective average.

The appropriate index is

$$I = \text{MPPA} = b\bar{P},$$

where the value of b is chosen so as to maximize

$$r_{P'I} = \text{cov}\,(P', I)/\sqrt{(V_{P'} V_I)},$$

which is the correlation between I and the future performance, P', of the candidate.

Since there is only one clue, there is only one normal equation when $r_{P'I}$ is maximized. It turns out that the equation is the same as the first of the normal equations (31) obtained by maximizing r_{AI}, except that the right-hand side is changed from $a_{1\alpha}$ to ρ/h^2, where ρ is repeatability (Van Vleck 1977, p. 541). Thus the relevant normal equation is

$$d_1 b_1 = \rho/h^2$$

where, from equation (32),

$$d_1 = \{1 + (m - 1)\rho\}/mh^2.$$

This gives

$$b_1 = \frac{m\rho}{1 + (m - 1)\rho}\,,$$

and hence

$$I = \text{MPPA} = \left(\frac{m\rho}{1 + (m - 1)\rho}\right)\bar{P}.$$

Appendix 16.3

BLUP

A16.3.1 General principles

In using BLUP, the aim is to obtain estimates of the breeding value of various animals, taking account of so-called fixed effects such as herd–year–season. This is done by using an adaptation of the least-squares method of estimation. In this section, the basic method of obtaining BLUP estimates of breeding value is described, and in the following section, an example is presented. Some readers may find it easier to work through the example before continuing with this section.

The basic steps in obtaining BLUP estimates of breeding value are:

1. Write down an expression (called a *model*) that describes an individual's performance in terms of all of the factors that need to be taken into account or to be estimated.

    For example, if we wish to estimate the effect of sires while allowing for the effect of herd–year–season, then an appropriate model would be

    $$Y_{ijk} = \mu + f_i + s_j + e_{ijk}, \qquad (1)$$

    where Y_{ijk} is a measurement taken on the kth offspring of the jth sire in the ith herd–year–season, f_i is the effect of the ith herd–year–season, s_j is the effect of the jth sire, and e_{ijk} is the residual error term.

2. Write down the set of so-called *least-squares equations*, corresponding to the model.

    In principle, this involves the following steps:

    (a) from the model, obtain an expression for the sum of squares of the residual error;

    (b) differentiate this expression with respect to each f_i, each s_j and with respect to μ;

    (c) set the differentials equal to zero, thereby creating a set of least-squares equations, consisting of one equation for each f_i, each s_j and for μ.

In practice, it is possible to write down the least-squares equations directly, as illustrated in the example given below, in Section A16.3.2.

3. Add σ_e^2/σ_s^2 to the diagonal coefficient of the left-hand side of each equation representing a sire effect, where σ_s^2 is the variance of sire effects, being equal to $\frac{1}{4}V_A = \frac{1}{4}h^2 V_P$ (see Falconer 1981, p. 155), and σ_e^2 is the residual error variance, being equal to

    $$V_P - \sigma_s^2 = (1 - \tfrac{1}{4}h^2)V_P,$$

where V_P is the phenotypic variance.

The equations are now called *mixed-model equations*.

4. Obtain an estimate of each effect by solving the mixed-model equations after imposing any necessary constraints, such as $\mu = 0$.

Steps 1 and 2 are the same as for the least-squares method. Step 3 is the adaptation that gives rise to BLUP estimates. In principle, the effect of this adaptation is to change the estimate of a sire effect from a simple mean of daughter performance (analogous to setting $I = C_1$ in equation (35) of Appendix 16.2.4) to a 'regressed' mean, as in a selection index (analogous to $I = b_1 C_1$, where b_1 is the appropriate regression or weighting factor). In other words, the addition of σ_e^2/σ_s^2 to the appropriate diagonal coefficients has an effect analogous to multiplying the daughter mean by the appropriate selection index weighting factor.

If two of the sires were related, and if a selection index were being used to estimate the breeding value of each sire, then the index for each sire would be of the form $I = b_1 C_1 + b_2 C_2$, where the b's are the appropriate weighting factors for each of the two available clues, and where C_1 and C_2 are the mean daughter performance for sires 1 and 2 respectively. Analogous weighting factors can be obtained automatically from mixed-model equations by setting up a matrix of additive relationships (called the *relationship matrix*). For the particular model being considered here (see equation (1) above), the *ij*th element of this matrix is the additive relationship between the *i*th sire and the *j*th sire. If animals other than sires are included in the model, then these other animals are also included in the relationship matrix. An illustration of the way in which the relationship matrix is utilized in the BLUP method is given in the example below.

It is important to note that a relationship matrix of some form is an important component of most BLUP estimation procedures. This highlights the practical importance of the concepts discussed in Chapter 13.

The estimate of each sire effect obtained in Step 4 is really an estimate of one half of the breeding value of that sire, because it corresponds to the expected performance of future offspring of that sire. If there are no fixed effects, e.g. if sire effects are the only effects in the model, then the estimated breeding values (EBVs) obtained from the BLUP method are exactly the same as those obtained by using a conventional selection index of the form EBV $= I = b_1 C_1$, as given in equation (35) of Appendix 16.2.4. If there are fixed effects and other complicating factors, then the EBVs obtained from the BLUP method are in effect equivalent to the EBVs that would be obtained from the conventional selection index after adjusting C_1 for all relevant fixed effects and after allowing for the other complicating factors.

An important point to emerge from the above outline is that in the BLUP method, there is a separate equation for each animal whose breeding value is to be estimated, and for each fixed effect being taken into account. In many practical situations, where hundreds or even thousands of animals are being evaluated, and where there are thousands of individual fixed effects, the BLUP method is obviously going to require the simultaneous solution of a very large number of equations. Although there are various ways of reducing the number of equations, the fact still remains that the BLUP method usually involves a very large amount of computation. As advances continue to be made in computing techniques and in computer design, the computational problems involved in solving these large numbers of equations are gradually being overcome.

A16.3.2 An example (based on an example given by Van Vleck and Hintz 1976).

Suppose that we have the following milk yields from 14 heifers:

Herd–year–season (i)	Sire (j)	Heifer identification (ijk)	Heifer record (litres of milk) (Y_{ijk})
1	1	111	3677
1	1	112	4161
1	1	113	3506
1	1	114	3904
1	3	131	3957
1	3	132	3447
2	1	211	3534
2	2	221	2941
2	2	222	3366
2	2	223	3755
2	3	231	2318
2	3	232	2730
2	3	233	3629
2	3	234	3158

These data can be summarized as follows:

Sire	Number in herd–year–season		Number of records for each sire	Sum of records for each sire
	$i = 1$	$i = 2$		
$j = 1$	4	1	5	18 782
$j = 2$	0	3	3	10 062
$j = 3$	2	4	6	19 239
Number of records in each herd–year–season	6	8	14	
Sum of records for each herd–year–season	22 652	25 431		48 083

The steps in writing mixed-model equations are:

1. Write down the least-squares equations, by equating each sum to its components determined from the model. Starting with μ, we have:

$$14\hat{\mu} + 6\hat{f_1} + 8\hat{f_2} + 5\hat{s}_1 + 3\hat{s}_2 + 6\hat{s}_3 = 48\,083,$$

where $\hat{\ }$ indicates that the effect is to be estimated. For the other effects, we have

$$
\begin{aligned}
6\hat{\mu} + 6\hat{f_1} \qquad\quad + 4\hat{s}_1 \qquad\quad + 2\hat{s}_3 &= 22\,652, \quad \text{for } f_1 \\
8\hat{\mu} \qquad + 8\hat{f_2} + 1\hat{s}_1 + 3\hat{s}_2 + 4\hat{s}_3 &= 25\,431, \quad \text{for } f_2 \\
5\hat{\mu} + 4\hat{f_1} + 1\hat{f_2} + 5\hat{s}_1 \qquad\qquad\qquad\; &= 18\,782, \quad \text{for } s_1 \\
3\hat{\mu} \qquad + 3\hat{f_2} \qquad + 3\hat{s}_2 \qquad\quad &= 10\,062, \quad \text{for } s_2 \\
6\hat{\mu} + 2\hat{f_1} + 4\hat{f_2} \qquad\qquad\qquad + 6\hat{s}_3 &= 19\,239, \quad \text{for } s_3
\end{aligned}
$$

2. Convert these least-squares equations to mixed-model equations by adding $\sigma_e^2/\sigma_s^2 = (1 - \frac{1}{4}h^2)V_P/\frac{1}{4}h^2 V_P = (4 - h^2)/h^2 = 15$ (if $h^2 = 0.25$) to the diagonal coefficients of the 3 × 3 submatrix representing sire effects. We then have the following set of mixed-model equations:

$$
\begin{aligned}
14\hat{\mu} + 6\hat{f_1} + 8\hat{f_2} + \qquad\quad 5\hat{s}_1 + \qquad 3\hat{s}_2 + \qquad 6\hat{s}_3 &= 48\,083 \\
6\hat{\mu} + 6\hat{f_1} \qquad\quad + \qquad 4\hat{s}_1 \qquad\qquad\quad + \quad 2\hat{s}_3 &= 22\,652 \\
8\hat{\mu} \qquad + 8\hat{f_2} + \qquad 1\hat{s}_1 + \qquad 3\hat{s}_2 + \quad 4\hat{s}_3 &= 25\,431 \\
5\hat{\mu} + 4\hat{f_1} + 1\hat{f_2} + (5 + 15)\hat{s}_1 \qquad\qquad\qquad\qquad &= 18\,782 \\
3\hat{\mu} \qquad + 3\hat{f_2} \qquad\qquad + (3 + 15)\hat{s}_2 \qquad\qquad &= 10\,062 \\
6\hat{\mu} + 2\hat{f_1} + 4\hat{f_2} \qquad\qquad\qquad\qquad + (6 + 15)\hat{s}_3 &= 19\,239
\end{aligned}
$$

3. Impose necessary constraints.

In the above example, even though we have six equations with six unknowns, the six equations are not independent. In fact, the first equation is the sum of the second and third equations, which means that we really have only five independent equations, but we still have six unknowns. A common way to overcome this problem is to set $\mu = 0$, which leaves five unknowns for which a unique set of estimates can be obtained from the five independent equations. After setting $\mu = 0$ and after removing the first equation because it is the sum of the second and third equations, the mixed-model equations become:

$$\begin{aligned}
6\hat{f}_1 + \quad 4\hat{s}_1 \quad\quad + \ 2\hat{s}_3 &= 22\,652 \\
8\hat{f}_2 + \ 1\hat{s}_1 + \ 3\hat{s}_2 + \ 4\hat{s}_3 &= 25\,431 \\
4\hat{f}_1 + 1\hat{f}_2 + 20\hat{s}_1 \quad\quad\quad\quad &= 18\,782 \\
3\hat{f}_2 \quad\quad + 18\hat{s}_2 \quad\quad &= 10\,062 \\
2\hat{f}_1 + 4\hat{f}_2 \quad\quad\quad + 21\hat{s}_3 &= 19\,239
\end{aligned}$$

In matrix notation, these equations can be written in the form $\mathbf{Bc} = \mathbf{d}$ as follows:

$$\begin{bmatrix} 6 & 0 & 4 & 0 & 2 \\ 0 & 8 & 1 & 3 & 4 \\ 4 & 1 & 20 & 0 & 0 \\ 0 & 3 & 0 & 18 & 0 \\ 2 & 4 & 0 & 0 & 21 \end{bmatrix} \begin{bmatrix} \hat{f}_1 \\ \hat{f}_2 \\ \hat{s}_1 \\ \hat{s}_2 \\ \hat{s}_3 \end{bmatrix} = \begin{bmatrix} 22\,652 \\ 25\,431 \\ 18\,782 \\ 10\,062 \\ 19\,239 \end{bmatrix}$$

The solutions of these equations can be obtained as $\mathbf{c} = \mathbf{B}^{-1}\mathbf{d}$, giving

$$\begin{aligned} \hat{f}_1 &= 3777 & \hat{s}_1 &= +24 \\ \hat{f}_2 &= 3191 & \hat{s}_2 &= +27 \\ & & \hat{s}_3 &= -51 \end{aligned}$$

Thus, the mean performance of heifers with an average sire is expected to be 3777 litres in herd 1 and 3191 litres in herd 2. Also, the mean performance of daughters of sire 1 in an average herd is expected to be 24 litres above average, which is 3 litres less than, and 75 litres better than, the performance of daughters of sires 2 and 3, respectively, in an average herd.

Recalling the definition of breeding value as given in Section 14.3, the estimated breeding values (EBVs) of the three sires are then as follows:

Sire	EBV
1	+48
2	+54
3	−102

We shall now suppose that two of the three sires were related, with sire 1 and sire 2 being half-sibs, and hence having an additive relationship of $a_{12} = a_{21} = 0.25$. The relationship matrix between all three sires is:

$$\begin{bmatrix} a_{11} & a_{12} & a_{13} \\ a_{21} & a_{22} & a_{23} \\ a_{31} & a_{32} & a_{33} \end{bmatrix} = \begin{bmatrix} 1 & 0.25 & 0 \\ 0.25 & 1 & 0 \\ 0 & 0 & 1 \end{bmatrix}$$

Previously, when the sires were unrelated, we converted the least-squares equations into mixed-model equations by adding $\sigma_e^2/\sigma_s^2 = (4 - h^2)/h^2 = 15$ to the diagonal coefficients of the 3×3 submatrix corresponding to the random effects. When the sires are related, we multiply σ_e^2/σ_s^2 by the inverse of the relationship matrix, and add the resultant matrix to the submatrix corresponding to the random effects. The inverse of the above relationship matrix is

$$\begin{bmatrix} 1 & 0.25 & 0 \\ 0.25 & 1 & 0 \\ 0 & 0 & 1 \end{bmatrix}^{-1} = \begin{bmatrix} 1.067 & -0.267 & 0.000 \\ -0.267 & 1.067 & 0.000 \\ 0.000 & 0.000 & 1.000 \end{bmatrix}$$

In practice, inverting the relationship matrix is usually avoided by following a set of rules (Henderson 1976) that enable the elements of the inverse of the relationship matrix to be written down directly.

Returning to the example, we multiply the inverse by $\sigma_e^2/\sigma_s^2 = 15$, giving

$$\begin{bmatrix} 16 & -4 & 0 \\ -4 & 16 & 0 \\ 0 & 0 & 15 \end{bmatrix},$$

which is added to the submatrix of random effects to give the mixed-model equations as:

$$\begin{bmatrix} 6 & 0 & 4 & 0 & 2 \\ 0 & 8 & 1 & 3 & 4 \\ 4 & 1 & (5+16) & (0-4) & (0+0) \\ 0 & 3 & (0-4) & (3+16) & (0+0) \\ 2 & 4 & (0+0) & (0+0) & (6+15) \end{bmatrix} \begin{bmatrix} \hat{f}_1 \\ \hat{f}_2 \\ \hat{s}_1 \\ \hat{s}_2 \\ \hat{s}_3 \end{bmatrix} = \begin{bmatrix} 22\,652 \\ 25\,431 \\ 18\,782 \\ 10\,062 \\ 19\,239 \end{bmatrix},$$

and the solutions are

$$\hat{f}_1 = 3772 \qquad \hat{s}_1 = +30$$
$$\hat{f}_2 = 3188 \qquad \hat{s}_2 = +33$$
$$\hat{s}_3 = -50$$

We now have:

EBV

Sire	Sires unrelated	Sires 1 and 2 half-sibs
1	+48	+60
2	+54	+66
3	−102	−100

The EBV of sire 3 has changed very little, because no new clues to its breeding value have been provided. In contrast, the fact that sires 1 and 2 are related means that sire 1's daughters now provide additional clues to sire 2's breeding value, and vice versa. Since both sires have above-average daughters, it is not surprising that both EBVs are higher when the relationship between these two sires is taken into account.

A16.3.3 Extensions of the model

The simple model used in the above example was chosen in order to illustrate the main principles of BLUP. Readers requiring examples of more complex models, incorporating cow evaluation as well as sire evaluation, and taking account of selection, common environment, multiple characters, and all lactations, should consult Van Vleck (1979) initially, and should then refer to Henderson (1973, 1974).

Appendix 16.4

Derivation of formulae for predicting response to selection

In Section 14.4.2, we saw that if selection is based on individual performance, then response to selection per generation is given by

$$R = h^2 S, \qquad (1)$$

where h^2 is heritability and S is selection differential.

But we know from equation (16.1) that $S = i\sigma_P$, where i is intensity of selection and σ_P is phenotypic standard deviation. We also know from Section 14.4.1 that $h^2 = V_A/V_P$, which can be re-written as $h\sigma_A/\sigma_P$, where h is the square root of heritability and σ_A is the standard deviation of breeding values. Finally, we saw in Section 16.6.1.1 that for selection on individual performance, h is the accuracy of selection, r_{AC}. Thus we can rewrite h^2 as $r_{AC}\sigma_A/\sigma_P$.

Substituting for S and for h^2 in equation (1) above, we see that response per generation is

$$R = i r_{AC} \sigma_A. \qquad (2)$$

Dividing by the generation interval, L, we obtain the response per year as

$$R_y = ir_{AC}\sigma_A/L, \tag{3}$$

as given in equation (16.11).

Now, since S is the phenotypic superiority of selected parents, it follows that h^2S, which equals $ir_{AC}\sigma_A$, can be thought of as the genetic superiority of the selected parents, i.e. the superiority that selected parents pass on, via their genes, to the next generation. It follows that R_y can be described as the genetic superiority of selected parents, divided by generation interval.

In many practical situations, i and/or r_{AC} differ between males and females. When this occurs, the improvement gained from selection of sires must be considered separately from the improvement gained from selection of dams. This can be done by calculating the genetic superiority $(ir_{AC}\sigma_A)$ and the generation interval (L) separately for each so-called *pathway* by which improvement can occur. Omitting the usual subscripts for accuracy, and using the subscripts S and D to refer to the sire and dam pathways respectively, response per year is then

$$R_y = \frac{\text{average genetic superiority}}{\text{average generation interval}}$$

$$= \frac{(i_S r_S \sigma_A + i_D r_D \sigma_A)/2}{(L_S + L_D)/2}$$

$$= \frac{i_S r_S + i_D r_D}{L_S + L_D}\sigma_A, \tag{4}$$

as first derived by Dickerson and Hazel (1944).

In progeny testing programmes, and in some other practical situations, selection of grandparents to produce male parents is more intense and may be more accurate than selection of grandparents to produce female parents. In cases such as these, we have to calculate the average genetic superiority and the average generation interval over four pathways which correspond to the selection of (1) sires of sires (SS); (2) sires of dams (SD); (3) dams of sires (DS); (4) dams of dams (DD). The appropriate prediction formula is then

$$R_y = \frac{\text{average genetic superiority}}{\text{average generation interval}}$$

$$= \frac{i_{SS} r_{SS} + i_{SD} r_{SD} + i_{DS} r_{DS} + i_{DD} r_{DD}}{L_{SS} + L_{SD} + L_{DS} + L_{DD}}\sigma_A, \tag{5}$$

which was first derived by Rendel and Robertson (1950). If i, r and L are the same in grandsires as in granddams, then equation (5) reduces to equation (4), and if *i*, *r* and *L* are the same in sires and dams, then equation (4) reduces to equation (3). Thus equations (3), (4) and (5) are different forms of the same basic formula for predicting response to selection.

Appendix 16.5

Calculation of the relative importance of selection criteria

Define an aggregate selection criterion, *K*, such that

$$K = b_1 C_1 + \ldots + b_n C_n,$$

where C_1 is the 1st selection criterion or clue and b_1 is the relative importance of the 1st selection criterion, etc.

Note that *C* is a single measurement or average lifetime performance for a single character measured on the candidate or on any relative of the candidate or on any group of relatives of the candidate such as half-sibs or progeny.

Using the same approach as that described when combining information from various relatives for a single character, we will choose values of *b* such that the correlation between *H* and *K*, written as r_{HK}, is maximized, where *H* is aggregate breeding value, as defined in equation (16.12).

By differentiating $\log r_{HK}$ with respect to each *b* and setting the differentials equal to zero, we obtain the set of normal equations shown in Table A16.5.

These normal equations can be written conveniently in matrix notation as:

$$\mathbf{Pb} = \mathbf{Gv},$$

where **P** is an $n \times n$ matrix of covariances among the *n* selection criteria, **b** is an $n \times 1$ vector of weighting factors to be determined, **G** is an $n \times m$ matrix of covariances between breeding values for the *n* selection criteria and the *m* selection objectives, and v is an $m \times 1$ vector of economic values. The weighting factors are then obtained as

$$\mathbf{b} = \mathbf{P}^{-1}\mathbf{Gv}.$$

A16.5.1 *An example*

We shall illustrate the calculation of *b* values using the pig example described in Section 16.9. The selection objectives are (1) growth rate, G, (2) food conversion ratio, F, and (3) percentage lean in carcass, L, and the selection criteria are (1) growth rate, G, and (2) backfat, B, both objectives being measured on each candidate.

Table A16.5. Normal equations for calculating the relative importance of selection criteria. This set of equations is obtained by differentiating $\log r_{HK}$ with respect to each b, and setting the differentials equal to zero.

$$b_1 \operatorname{cov}(C_1 C_1) + b_2 \operatorname{cov}(C_1 C_2) + \ldots + b_n \operatorname{cov}(C_1 C_n) = v_1 \operatorname{cov}(A_1 A_1') + v_2 \operatorname{cov}(A_1 A_2') + \ldots + v_m \operatorname{cov}(A_1 A_m')$$

$$b_1 \operatorname{cov}(C_2 C_1) + b_2 \operatorname{cov}(C_2 C_2) + \ldots + b_n \operatorname{cov}(C_2 C_n) = v_1 \operatorname{cov}(A_2 A_1') + v_2 \operatorname{cov}(A_2 A_2') + \ldots + v_m \operatorname{cov}(A_2 A_m')$$

$$\cdot$$
$$\cdot$$

$$b_1 \operatorname{cov}(C_n C_1) + b_2 \operatorname{cov}(C_n C_2) + \ldots + b_n \operatorname{cov}(C_n C_n) = v_1 \operatorname{cov}(A_n A_1') + v_2 \operatorname{cov}(A_n A_2') + \ldots + v_m \operatorname{cov}(A_n A_m')$$

In these equations, n = number of selection criteria, m = number of selection objectives, v_k = net economic value for kth selection objective, $\operatorname{cov}(C_i C_j)$ = covariance between ith and jth selection criteria, and $\operatorname{cov}(A_i A_k')$ = covariance between the breeding value for the ith selection criterion and the kth selection objective.

If the selection criteria are single measurements on the candidates themselves, then

1. $\operatorname{cov}(C_i C_j) = r_{P_{ij}} \sqrt{(V_{P_i} V_{P_j})}$, where $r_{P_{ij}}$ = phenotypic correlation between the ith and jth selection criteria, and V_{P_i} = phenotypic variance of the ith selection criterion.

2. $\operatorname{cov}(A_i A_k') = r_{G_{ik}} \sqrt{(h_i^2 V_{P_i} h_k^2 V_{P_k})}$, where $r_{G_{ik}}$ = genetic correlation between the ith selection criterion and the kth selection objective, and h_i^2 = heritability of the ith character.

3. If $i = j$, then $r_{P_{ii}} = 1$ and $\operatorname{cov}(C_i C_j) = V_{P_i}$.

4. If the ith selection criterion is the same character as the kth selection objective, then $r_{G_{ik}} = 1$, and $\operatorname{cov}(A_i A_k') = h_i^2 V_{P_i}$.

The first step is to obtain values for the necessary parameters, as listed in Section 16.9.3. The following values have been obtained primarily from Mitchell *et al.* (1982) and from McPhee *et al.* (1979).

1. V_P and h^2 for each objective and criterion

	Objectives			Criteria	
	Growth rate (G) g/day	Food conversion ratio (F) g food/ g gain	Percentage lean in carcass (L) percent	Growth rate (G) g/day	Backfat thickness (B) mm
V_P	2916	0.0256	15.21	2916	7.84
h^2	0.25	0.35	0.50	0.25	0.50

2. r_P between each pair of selection criteria

	G	B
G	1	−0.10
B		1

3. r_G between each pair-wise combination of selection objective and selection criterion

		Objectives		
		G	F	L
Criteria	G	+1.00	−0.80	+0.35
	B	−0.25	+0.30	−0.25

4. Net economic value for each selection objective

	G	F	L
v (pence)	+1	−864	+55

In order to write down the normal equations, we need values for $\text{cov}(C_iC_j)$ and $\text{cov}(A_iA'_k)$, which can be obtained as follows:

$$\text{cov}(C_1C_1) = V_{PG} = 2916.0$$

$$\text{cov}(C_2C_2) = V_{PF} = 7.84$$

$$\text{cov}(C_1C_2) = \text{cov}(C_2C_1) = r_{PG,B}\sqrt{(V_{PG}V_{PB})} = -0.10\sqrt{[(2916.0)(7.84)]}$$
$$= -15.12$$

$$\text{cov}(A_1A'_1) = h^2_G V_{PG} = (0.25)(2916.0) = 729.0$$

$$\text{cov}(A_1A'_2) = r_{GG,F}\sqrt{(h^2_G V_{PG} h^2_F V_{PF})}$$
$$= -0.80\sqrt{[(0.25)(2916.0)(0.35)(0.0256)]} = -2.04460$$

$$\text{cov}(A_1A'_3) = r_{GG,L}\sqrt{(h^2_G V_{PG} h^2_L V_{PL})}$$
$$= +0.35\sqrt{[(0.25)(2916.0)(0.50)(15.21)]} = +26.06042$$

$$\text{cov}(A_2A'_1) = r_{GB,G}\sqrt{(h^2_B V_{PB} h^2_G V_{PG})}$$
$$= -0.25\sqrt{[(0.50)(7.84)(0.25)(2916.0)]} = -13.36432$$

$$\text{cov}(A_2A'_2) = r_{GB,F}\sqrt{(h^2_B V_{PB} h^2_F V_{PF})}$$
$$= +0.30\sqrt{[(0.50)(7.84)(0.35)(0.0256)]} = +0.05622$$

$$\text{cov}(A_2A'_3) = r_{GB,L}\sqrt{(h^2_B V_{PB} h^2_L V_{PL})}$$
$$= -0.25\sqrt{[(0.50)(7.84)(0.50)(15.21)]} = -1.36500$$

Using these values, the normal equations are

$$b_1(2916.0) + b_2(-15.12) = (1)(729.0) \quad + (-864)(-2.04460)$$
$$+ (55)(26.06042)$$

$$b_1(-15.12) + b_2(7.8400) = (1)(-13.36432) + (-864)(0.05622)$$
$$+ (55)(-1.36500).$$

These equations can be written in matrix form as

$$\begin{bmatrix} 2916.00 & -15.12 \\ -15.12 & 7.84 \end{bmatrix} \begin{bmatrix} b_1 \\ b_2 \end{bmatrix} = \begin{bmatrix} 729.00000 & -2.04460 & 26.06042 \\ -13.36420 & 0.05622 & -1.36500 \end{bmatrix} \begin{bmatrix} 1 \\ -864 \\ 55 \end{bmatrix},$$

or

$$\mathbf{P} \qquad \mathbf{b} = \qquad\qquad \mathbf{G} \qquad\qquad \mathbf{v},$$

and the solutions can be obtained as

$$b \quad = \quad P^{-1} \quad\quad\quad Gv$$

$$= \begin{bmatrix} 0.000346 & 0.000668 \\ 0.000668 & 0.128839 \end{bmatrix} \begin{bmatrix} 3928.86 \\ -137.01 \end{bmatrix}$$

$$= \begin{bmatrix} 1.269 \\ -15.028 \end{bmatrix}$$

Thus, $b_1 = 1.269$ and $b_2 = -15.028$.

In the above example, covariances were calculated from correlations and heritabilities published in the literature. While this is a common practice, it should be noted that rounding errors can be quite large when covariances are obtained in this way. Because of this, it is always advisable to use actual estimates of covariances when setting up normal equations, rather than values derived from correlations and heritabilities.

Another point that should be noted is that, irrespective of how the covariances are obtained, a check should be made to ensure that all the covariances are compatible, in the sense that they correspond to a set of heritabilities and correlations, including partial correlations, that are 'possible', i.e. none have absolute values greater than 1.0. A useful method for checking the compatibility of a set of covariances is described by Hayes and Hill (1980), and a technique for overcoming incompatibilities is given by Hayes and Hill (1981).

17
Breed structure

17.1 Introduction

Not all animals in a breed contribute to genetic improvement within that breed; in most cases, breeds are structured so that only a minority of animals are given that opportunity.

The aim of this chapter is to describe the structure of breeds and to discuss the ways in which the structure can be modified so as to increase the overall genetic merit of the breed.

17.2 The traditional pyramid

The structure of most breeds consists of a series of subgroups called tiers. In many situations there are three tiers, namely nucleus, multiplier and commercial. Occasionally there are only two (nucleus and commercial), and sometimes there are more than three (more than one level of multiplier). Irrespective of the number of tiers, the basic structure is much the same, being commonly represented in the shape of a pyramid, as shown in Fig. 17.1.

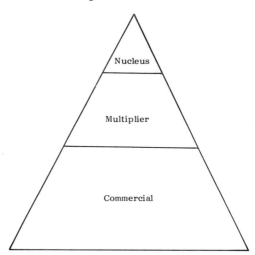

Fig. 17.1. A typical breed structure consisting of three tiers in the shape of a pyramid. The area of the pyramid devoted to each tier reflects the relative numbers of animals in each tier.

In a traditional pyramid, the *nucleus* tier consists of herds or flocks that breed their own male and female replacements. In some cases, they may occasionally import a sire or dam from another nucleus herd or flock. *Multipliers* take males and sometimes females from usually just one nucleus herd or flock, with the aim of producing sufficient breeding stock (males, and sometimes females) to satisfy the demands from herds or flocks in the commercial tier.

17.3 Closed nucleus breeding schemes

In the traditional pyramid, as described above, there is a one-way flow of genes within the pyramid, *downwards* from top to bottom. This means that the only source of cumulative genetic progress in commercial populations is improvement that occurs at the top of the pyramid in the nucleus populations. If the nucleus tier makes no genetic improvement, then neither will the rest of the breed. However, even if the nucleus tier does make progress, this improvement is not seen immediately at lower levels of the pyramid; it takes time for genetic progress in one tier to be transmitted to the next tier. The resultant difference in performance between any two adjacent tiers is called the *improvement lag*, which is usually expressed in terms of the number of years of genetic improvement that are represented by the difference in performance between adjacent tiers.

There are two factors that affect the size of the improvement lag. They are the age structure in the lower tiers, and the source and merit of sires and dams used in the lower tiers.

For example, the lag can be reduced by keeping sires and dams in the lower tiers for shorter periods of time before replacing them with younger stock. It can also be reduced by transferring females as well as males downwards between tiers. Even greater reductions result if some nucleus parents can be transferred directly to the commercial tier.

The effect of altering some of these factors is illustrated for the case of pigs in Fig. 17.2. The lag in a traditional pig structure, in which only sires are passed down, and only between adjacent tiers, is around 4 years. In contrast, if all sires come from the nucleus tier and if in addition, dams are transferred downwards between adjacent tiers, then the lag can be reduced to around $1\frac{1}{2}$ years, which is a substantial reduction.

In all the breeding schemes discussed above, there is a downward-only flow of genes from the nucleus to the lower tiers. Because no genes flow into the nucleus tier, these schemes are called *closed nucleus breeding schemes*. This type of structure is the one most commonly encountered in practice; most of the traditional breeds of livestock and most of the modern pig and poultry breeding programmes have a closed nucleus structure.

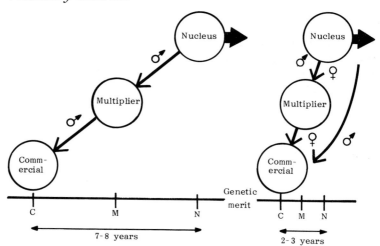

Fig. 17.2. The reduction in improvement lag brought about by altering the source of sires and dams used in the lower tiers of a pig pyramid. Thick arrows indicate the direction of genetic improvement in the nucleus tier, and thin arrows represent the direction of gene flow between tiers. In this example, the lag values are those expected when unselected animals are passed down. If selected animals are passed down, the lag values will be smaller. After Bichard (1978), based on the results of calculations by Bichard (1971).

17.4 Open nucleus breeding schemes

Even if there is a substantial improvement lag between tiers, there is usually sufficient variation within the lower tiers to result in some animals that are born in those tiers having a higher performance than some animals in a higher tier. To the extent that higher performance is evidence of a superior breeding value, it would make good sense to transfer superior animals from a lower tier into a higher tier, and even (if they are sufficiently superior) into the nucleus tier. This amounts to opening the nucleus. Since traditionally the nucleus tier has consisted of studs of 'pedigree' animals registered with the relevant breed society, the above proposal raises the possibility of introducing commercial (non-'pedigree') animals into 'pedigree' herds or flocks, which of course is not exactly in keeping with traditional stud breeding philosophy. And yet there are definite advantages to be gained by opening the nucleus: the annual response to selection is increased, and the rate of inbreeding in the nucleus is substantially reduced.

Some idea of the magnitude of these two effects can be gained from the results of calculations by James (1977), who showed that for a well-designed sheep or beef cattle open nucleus scheme, response to selection could be increased by 10 per cent to 15 per cent, and rate of inbreeding could be halved, in comparison with a closed nucleus scheme of the same overall size.

The most popular form of open nucleus breeding scheme involves a group of breeders agreeing to cooperate in the formation and subsequent running of an open nucleus, in return for a regular supply of breeding stock (mostly males) from the nucleus, for use in their own herd or flocks. There are quite a number of such schemes in practice now in sheep and to a lesser extent in cattle. They are often called group breeding schemes or cooperative breeding schemes. As an illustration of the scale of operations that can be involved, Fig. 17.3 shows the open nucleus structure of a group breeding scheme being conducted with Merino sheep in Western Australia.

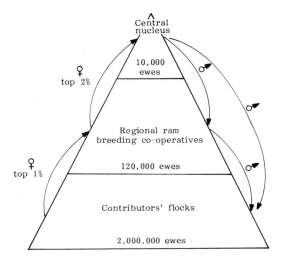

Fig. 17.3. A group breeding scheme in Merino sheep. Note the two-way flow of genes, with sires being transferred down the pyramid, and with the top 1% or 2% of dams (in this case) being transferred up the pyramid. Note also that the improvement lag is reduced by using nucleus males in both of the lower tiers, through the extensive use of artificial insemination. (After Shepherd 1976).

One of the biggest practical difficulties with group breeding schemes is that very close and continuing cooperation is required between the cooperating breeders. The successful establishment of a group breeding scheme is therefore as much a matter of sociology as of genetics. However, a sufficient number of group breeding schemes have been running successfully for a sufficient time now to indicate that they can, given the right circumstances, be very beneficial to the cooperating breeders and to the industry as a whole.

If the advantages of open nucleus schemes are as great as described above, why do pig and poultry breeders still persist with closed nucleus schemes? The main reason is that in pig and poultry breeding, nucleus herds and flocks are usually maintained under strict quarantine, with the aim of excluding as many diseases as possible. Opening these nucleus populations to regular importations from other herds or flocks would involve an unacceptably high risk of introducing disease. This is an important point. It illustrates clearly that practical breeding programmes cannot be designed solely on the basis of genetic requirements. Quite often there are non-genetic constraints, such as requirements in relation to disease, that have a major influence on the final design of a breeding programme.

17.5 Summary

The usual structure of a breed or strain consists of a pyramid with tiers corresponding to nucleus, multiplier, and commercial herds or flocks. In this structure the aim is to maximize response to selection in the nucleus tier, and to minimize the improvement lag, which is the time taken for genetic progress to be passed down the pyramid from one tier to the next. In the traditional pyramid, which is still very common today, there is only a one-way flow of genes, downwards from the closed nucleus.

An alternative idea involves opening the nucleus to the best animals occurring in the lower tiers, giving rise to a two-way flow of genes in what is called an open nucleus breeding scheme. Efficiently-designed open nucleus schemes can lead to a 10 per cent to 15 per cent increase in annual response to selection and to a substantial reduction in rate of inbreeding in the nucleus, when compared with closed nucleus schemes. A popular form of open nucleus scheme is the so-called group breeding scheme, in which several breeders form a cooperative which assembles and runs a nucleus herd or flock.

17.6 Further reading

Barton, R. A. and Smith, W. C. (Eds) (1982). *Proceedings of the World Congress on Sheep and Beef Cattle Breeding*. Dunsmore Press, Palmerston North, New Zealand. (Vol. II of this Proceedings contains a section devoted to group breeding schemes.)

Bichard, M. (1971). Dissemination of genetic improvement through a livestock industry. *Animal Production* 13, 401-11. (A discussion of the concept of improvement lag, and an investigation into various ways of minimizing lag.)

Guy, D. R. and Smith, C. (1981). Derivation of improvement lags in a livestock industry. *Animal Production* 32, 333-6. (A simple method for calculating the improvement lag.)

James, J. W. (1977). Open nucleus breeding schemes. *Animal Production* 24, 287-305. (The basic theory of open nucleus breeding schemes, including recommendations for optimum design as well as predictions of response to selection and of rates of inbreeding.)

Robertson, A. (1953). A numerical description of breed structure. *Journal of Agricultural Science* 43, 334–6. (This is the paper in which the methods of breed structure analysis were formally derived.)

Robertson, A. and Asker, A. A. (1951). The genetic history and breed structure of British Friesian cattle. *Empire Journal of Experimental Agriculture* 19, 113–30. (This is the original study of breed structure. It represents the first description of the traditional pyramid, and includes the first discussion of the concept of improvement lag.)

Tomes, G. J., Robertson, D. E., and Lightfoot, R. J. (Eds) (1976). *Sheep breeding.* Western Australian Institute of Technology, Perth. (This book is the proceedings of a conference on sheep breeding. It contains several papers on cooperative breeding schemes, written both from the theoretical and the practical points of view.)

18

Crossing

18.1 Introduction

Crossing is the mating of individuals from different populations which can be strains, breeds, or species. It is the second major method of exploiting genetic variation, selection being the first.

Animals that result from crossing are called *crossbreds*, as distinct from those animals that result from matings within a population, which are called *straightbreds*. We shall use this latter term in preference to the more common term, purebreds, because the term purebreds is strongly associated in the minds of many people with those animals officially registered as members of the so-called pure or pedigreed breeds of animals. The topics in this chapter certainly apply to such animals, but they apply equally to many other animals as well. In fact they apply, as the first sentence of this introduction says, to any member of a population, which, as defined in Section 5.2, is a group of intermating individuals who share a common pool of genes.

Many people think of crossing in terms of crosses between inbred lines, partly because of the considerable success achieved by this breeding method in plants, especially in maize. But as described in Section 13.6, the development of inbred lines in animals is very wasteful and costly, and hence is hardly ever used now, except by one or two poultry breeding organizations. Instead, crossing in animals is generally carried out between populations that have not been deliberately inbred, but which have been kept isolated from each other for varying lengths of time.

The aim of this chapter is to describe the main methods of crossing that are used in practice, and to comment on the implications of each method.

18.2 Regular crossing

Regular or systematic crossing occurs when the same cross is made on a regular basis, with the aim of producing a particular type of progeny. Regular crossing may be advantageous from two points of view.

Firstly, the crossbred progeny usually show *heterosis* or *hybrid vigour* for certain characters. Heterosis occurs when the average performance of crossbred progeny is superior to the average performance of the two parents. Various types of heterosis are recognized, including *parental heterosis* (either *maternal* or *paternal*), referring to the performance of animals as parents, and *individual heterosis*, referring to non-parental performance. The amount of heterosis for a particular character can vary considerably, depending on the environment and on the populations being crossed. However, in general

> *(1) heterosis is greatest in characters most closely associated with the ability to survive and reproduce, and (2) the greater the genetic diversity between two populations of domesticated animals, the greater the heterosis in crosses between them.*

A detailed theoretical explanation for these common observations is beyond the scope of this book; readers seeking further information should consult Falconer (1981, Chapter 14). Briefly, heterosis occurs only if there is non-additive gene action (dominance and/or epistasis) for the character concerned; and heterosis increases with increases in the difference in gene frequencies between the two populations being crossed together. Recalling from Section 13.6 that inbreeding depression occurs only if there is non-additive gene action, it is not surprising to find in general that the same characters that show heterosis also show inbreeding depression.

The second advantage of regular crossing is that *complementarity* may occur. This refers to the additional profitability obtained from crossing two populations, resulting not from heterosis but from the manner in which two or more characters complement each other (Cartwright 1970).

For example, consider two populations of pigs, one (A) with a very high food conversion ratio but low litter size, and the other (B) with poor food conversion ratio but high litter size. Suppose that boars from A are regularly mated with sows from B to produce a crossbred pig that is raised solely for meat production. Even if there is no heterosis for food conversion ratio, i.e. even if the average performance of the crossbred pigs is exactly intermediate between that of the two parental populations, and if the litter size of B sows is the same when mated to either A or B boars, then the overall profitability of this particular operation is likely to be much greater than the profitability of running either A or B alone, because the same total number of pigs will be produced as from the most prolific population, B, but the food conversion ratio of each offspring will be higher than that of offspring from B X B matings.

Certain crosses show much more complementarity than others,

depending on the extent to which the populations differ in reproductive performance and in production characters, and also depending on the 'direction' of the cross. The latter point can be illustrated by noting that there will be far greater complementarity when the most prolific population is used as a source of dams rather than of sires.

Bearing in mind the above discussion of heterosis and complementarity, it is evident that crossing exploits genetic variation in two ways. Given that those characters most likely to show heterosis are those with considerable non-additive genetic variation, it can be seen that by enabling heterosis to be expressed, crossing exploits the non-additive portion of genetic variation, namely V_D and V_I. However, we saw above that in relation to complementarity, crossing exploits differences in average performance between populations, i.e. differences in additive effects between populations. We can therefore conclude that:

> *Crossing exploits non-additive gene effects (through heterosis) and additive gene effects (through complementarity).*

There are two basic methods of regular crossing, and we shall now discuss each one in turn.

18.2.1 Specific crossing

The simplest form of specific crossing involves the *two-way cross*, in which animals from one population, A, are regularly mated to animals of a second population, B.

In symbolic terms,

$$A \times B$$
$$\downarrow$$
$$(AB),$$

in which A and B are straightbred parents, and (AB) represents the crossbred progeny, which are known as *first-cross* or F1 progeny.

Sometimes the male parent in the cross always comes from one population and the female parent always comes from the other population so as to obtain maximum benefit from complementarity. When this happens, the two parental populations are often called the *sire line* and the *dam line* respectively. Obviously, both males and females must be mated together within a sire line and within a dam line in order to maintain the lines as separate entities; the terms sire line and dam line are simply a means of describing the contribution of each population to the cross.

Examples of two-way crossing include Hereford ♂ × Friesian ♀, Border Leicester ♂ × Merino ♀, and Border Leicester ♂ × Blackface ♀. In all these examples, one of the main reasons for each breed supplying only one sex for the cross is the availability of females; there are always large numbers of Friesian, Merino, and Blackface females available that are no longer required or were never required for mating to males of their own breed. However, there is often some other justification as well. For example, a Hereford X Friesian crossbred raised on its dam will reach slaughter weight more rapidly if its dam is a Friesian rather than a Hereford, because of the superior milk production of the Friesian cow.

In other situations, it makes little difference as to which way the cross is made, and consequently both types of two-way cross are popular. The most popular cross in pig breeding for many years, for example, has been the Large White X Landrace, which produces a similar type of crossbred progeny from matings in either direction.

The two-way specific cross produces offspring that are 100 per cent heterozygous, in the sense that they have one gene from population A and one gene from population B at every locus. As such, they show 100 per cent of individual heterosis. However, two-way crossing does not provide any opportunity for the breeder to benefit from maternal or paternal heterosis, because the parents in a two-way cross are never themselves crossbred; they are always straightbred. In order to exploit maternal and/or paternal heterosis, other types of crossing must be used.

One of these is the *backcross*, which involves mating crossbred individuals from a two-way cross back to one of the two parental breeds. Using the same symbols as with the two-way cross, this can be written as

$$(AB) \times A \quad \text{or} \quad (AB) \times B$$
$$\downarrow \qquad\qquad\qquad \downarrow$$
$$(AB)A \qquad\qquad (AB)B$$

Since maternal heterosis is generally of more importance than paternal heterosis, the crossbred parent in backcrossing is usually the female.

In pigs, for example, the majority of commercial offspring sold in recent years in Britain have been the result of either of the above two backcrosses, in which the two populations, A and B, are Large White and Landrace.

With backcrossing, the crossbred parent is 100 per cent heterozygous, and hence shows 100 per cent parental heterosis. However, the backcross progeny are the result of the fusion of a gamete containing all, say, A genes (from the parental breed) and a gamete containing an average of one-half A genes and one-half B genes (from the crossbred parent). This means that, at half of their loci on average, backcross progeny have two A genes, i.e. are homozygous for genes from the A population, and at the other half of their loci, they have one A gene and one B gene, i.e. are heterozygous. In other words, backcross progeny on average are 50 per cent less heterozygous than first-cross,

AB, progeny. Consequently, they show on average only 50 per cent of individual heterosis.

A better way to exploit heterosis in the growing, commercial animal as well as in reproductive ability is to use a *three-way cross*, in which a first-cross animal, AB, is mated to an animal of a third population, C. Once again, because it is generally female reproductive ability that needs boosting, a three-way cross is usually of the form

$$(AB)\,♀ \quad × \quad C\,♂$$
$$↓$$
$$(AB)C$$

This cross enables full utilization of maternal heterosis because the dam is 100 per cent heterozygous. It also enables full utilization of individual heterosis because the progeny are also 100 per cent heterozygous, being AC at one half of their loci and BC at the other half of their loci, on average. It should be noted that even though first-cross and three-way cross progeny both show full utilization of individual heterosis, in the sense that they are both 100 per cent heterozygous, they may not show the same amount of heterosis. The reason for this is that different crosses are involved in each case and, as noted in Section 18.2, the amount of heterosis is greater with some crosses than with others. For example, if the crosses AB and AC exhibit the same amount of individual heterosis, but the cross BC exhibits less than the other two, then the amount of individual heterosis will be less in the above three-way cross progeny than in the first-cross progeny on page 492.

Because it is so successful at exploiting heterosis in the female parent and in the growing, commercial animal, and because it also enables exploitation of complementarity, the three-way cross has been used widely throughout the world. Most prime lambs, for example, are produced in this way, where A is a Merino or a hill breed, B is a long-wool breed such as Border Leicester, and C is a down or lowland breed such as Dorset Horn. Similar situations occur in beef cattle, although less extensively than with sheep. Three-way crosses are used extensively in the poultry breeding industry, and their use is gradually increasing in pig breeding. The main reason for the delay in switching from back-crossing to three-way crossing in pigs has been the lack of a suitable third population to use as a sire line for mating to the Large White × Landrace female. Recently, however, suitable populations have started to become available, by means that will be described in Section 18.3.

The final form of specific crossing is the *four-way cross*, in which the crossbred progeny from two separate two-way crosses are mated together to produce the commercial progeny.

In symbols we have

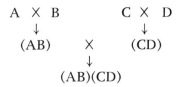

Since both the male and the female parent of the commercial progeny are crossbred, this cross enables full exploitation of both maternal and paternal heterosis, as well as individual heterosis. Once again, it should be noted that the actual amount of individual heterosis in the commercial progeny may not be the same as that observed in, for example, either of the two types of first-cross progeny.

Although used only occasionally in most domestic species, four-way crossing is used regularly by several poultry breeding organizations. The biggest difficulty encountered in practice is finding four different populations that have sufficiently high genetic merit to warrant their inclusion in the cross: there is not much sense in going to a great deal of trouble and expense to obtain maximum utilization of heterosis and complementarity if one or more of the parental lines have such low performance that the resultant crossbreds are still not as profitable as offspring from other populations.

18.2.2 Rotational crossing

In all forms of specific crossing, as described above, both parents of the commercial progeny are either straightbred or are the crossbred progeny of straightbred parents. It follows that a commercial breeder who wishes to produce only the final commercial progeny will have to obtain all breeding replacements, both male and female, from another source. Being obliged always to introduce all breeding replacements from other flocks or herds is not always desirable, from a general management point of view and, especially in the case of pigs, from a disease point of view. It also deprives interested breeders of any possibility of breeding at least some of their own replacements.

Rotational crossing is an alternative form of regular crossing that overcomes some of these difficulties. It generally involves the use of males from two or three different populations, in regular sequence. If rotational crossing is continued in a commercial herd or flock for several years, and if all replacement females are bred in the herd or flock concerned, then all females will soon be crossbred, containing varying proportions of genes from the two or three populations from which males were obtained. Symbolically, we have the situation as illustrated in Table 18.1.

A major disadvantage of rotational crossing is that it does not allow any exploitation of complementarity, because the populations involved in the crossing cannot be restricted in use to a single purpose, as they

Table 18.1. The two most common forms of rotational crossing, together with the proportions of genes from each parental breed, and average heterozygosity, in crossbred progeny in the early generations, and at equilibrium

Two breed

	Proportion of genes from		Ave het.*
	A	B	
A × B →			
[AB] × A	1/2	1/2	1
[(AB)A] × B	3/4	1/4	1/2
[((AB)A)B] × A	3/8	5/8	3/4
. . .			

Three breed

	Proportion of genes from			Ave het.*
	A	B	C	
A × B →				
[AB] × C	1/2	1/2	0	1
[(AB)C] × A	1/4	1/4	1/2	1
[((AB)C)A] × B	5/8	1/8	1/4	3/4
. . .				

Equilibrium at generation:

Two breed:

	A	B	Ave het.*
t	1/3	2/3	2/3
t + 1	2/3	1/3	2/3
t + 2	1/3	2/3	2/3

Three breed:

	A	B	C	Ave het.*
t	1/7	2/7	4/7	6/7
t + 1	2/7	4/7	1/7	6/7
t + 2	4/7	1/7	2/7	6/7

* Average heterozygosity of animals in square brackets, relative to a first-cross individual.

are in specific crossing. Because of this, rotational crossing is a much more attractive proposition where the populations available for crossing show little or no complementarity, but do show some economically useful heterosis.

Although not widely used at present, except in the USA pig industry, rotational crossing could become much more widespread in the future, especially among breeds of pigs that have similar levels of performance, and amongst certain strains of dairy cattle. In the former, it offers considerable health and management advantages, because only males, or semen, need be brought in from outside the herd. And in dairy cattle, where frozen semen is readily available and where artificial insemination is used so widely, it could be used, for example, amongst the various Holstein or Friesian strains that exist around the world.

18.2.3 A comparison of different forms of regular crossing

In order to obtain a valid comparison of different forms of regular crossing, we need to examine data collected in similar circumstances for each form of regular crossing being considered. In relation to complementarity, the comparisons require estimates of average performance of all potential parental populations, for all characters of interest. With respect to heterosis, comparable information is required on the performance of progeny resulting from all types of regular crossing. Despite many crossbreeding experiments over the years, and the collection of considerable data, there are still insufficient data to enable a valid comparison of all methods of regular crossing, in relation to heterosis. Fortunately, in the majority of cases where reliable data have been collected, it has been found that the amount of heterosis observed is highly correlated with average heterozygosity. Consequently, the best way to obtain an overall comparison of types of regular crossing in relation to heterosis is to compare them in terms of average heterozygosity. This is done in Table 18.2.

In order to illustrate the practical implications of the information summarized in Table 18.2, we shall now consider some examples from particular species of domestic animals.

In beef cattle, the results of an extensive set of experiments involving Herefords, Angus, and Shorthorns were reviewed by Gregory and Cundiff (1980). Many of their results can be summarized in terms of 'weight of calf weaned per cow exposed to breeding', which is a major determinant of profitability. In Fig. 18.1, it can be seen that, on average, there was 8.5 per cent individual heterosis and 14.8 per cent maternal heterosis, giving a total increase of 23.3 per cent in weight of calf weaned per cow exposed to breeding, if both individual and maternal heterosis were fully exploited. Thus, there are substantial benefits to be gained from utilizing heterosis. Given these estimates of individual and maternal heterosis, and the expectations listed in

Table 18.2. Fraction of heterosis, as indicated by average heterozygosity, expected in the most common types of regular crossing

Type of regular crossing	Fraction of heterosis		
	Individual	Maternal	Paternal
Straight bred	0	0	0
Two-breed cross			
A♀ × B♂	1	0	0
Backcross			
AB♀ × (A♂ or B♂)	1/2	1	0
(A♀ or B♀) × AB♂	1/2	0	1
Three-breed cross			
AB♀ × C♂	1	1	0
C♀ × AB♂	1	0	1
Four-breed cross			
AB♀ × CD♂	1	1	1
Rotational cross			
2 sire breeds	2/3	2/3	0
3 sire breeds	6/7	6/7	0

Table 18.2, we can predict the increase in weight of calf weaned per cow exposed to breeding for various types of regular crossing. For example, in a backcross of the type AB ♀ × A or B ♂, we expect to see one-half of individual heterosis and all of maternal heterosis, giving a total increase of $(\frac{1}{2} \times 8.5\%) + 14.8\% = 19$ per cent. The observed increase, as reported by Gregory and Cundiff (1980), was 16 per cent, which is very close to the expectation. With two-breed rotational crossing, Table 18.2 shows that we expect $\frac{2}{3}$ of individual and maternal heterosis, giving an increase of $\frac{2}{3} \times (8.5\% + 14.8\%) = 16$ per cent. For three-breed rotational crossing, we expect $\frac{6}{7} \times (8.5\% + 14.8\%) = 20$ per cent increase. The results of these calculations are summarized in Table 18.3.

In sheep, Ch'ang and Evans (1982) reported the results of an extensive set of crossing experiments involving Dorset Horns, Merinos, and Corriedales. The overall results, in terms of weight of lambs weaned per ewe joined, are summarized in Table 18.4. It can be seen that all three types of heterosis (individual, maternal, and paternal) made substantial contributions.

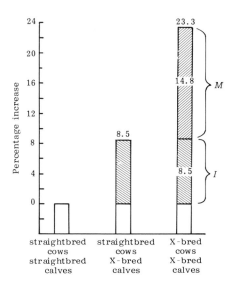

Fig. 18.1. Percentage increase in weight of calf weaned per cow exposed to breeding, as a result of mating straightbred cows to bulls of a different breed (centre), or mating first-cross cows to bulls of a third breed (right). I = individual heterosis, M = maternal heterosis. Results were obtained from an experiment involving all relevant crosses among Hereford, Angus, and Shorthorn cattle. (After Cundiff *et al.* 1982*b*.)

Table 18.3. Performance observed or expected from various types of regular crossing in cattle, based on extensive data obtained from Hereford, Angus, and Shorthorn cattle in the USA

Type of regular crossing	Heterosis utilized*	Weight of calf weaned per cow exposed to breeding, as a percentage of straightbred performance
Straightbred	—	100
Two-breed cross	I	108
Backcross	$\frac{1}{2}I + M$	119
Three-breed cross	$I + M$	123
Two-breed rotation	$\frac{2}{3}I + \frac{2}{3}M$	116
Three-breed rotation	$\frac{6}{7}I + \frac{6}{7}M$	120

* I = individual heterosis; M = maternal heterosis. From data presented by Gregory and Cundiff (1980).

Table 18.4. Performance observed from various types of regular crossing amongst Dorset Horn, Merino, and Corriedale sheep in Australia

Type of regular crossing	Heterosis utilized*	Weight of lamb weaned per ewe joined, as a percentage of straightbred performance
Straightbred	—	100
Two-breed cross	I	107
Three-breed cross		
C♀ × AB♂	$I + P$	137
AB♀ × C♂	$I + M$	152
Other		
AB♀ × AB♂	$\frac{1}{2}I + M + P$	166

    * I = individual heterosis; M = maternal heterosis; P = paternal heterosis. From data presented by Ch'ang and Evans (1982).

    The final example we shall consider comes from pigs. Making use of available data, and using these to make predictions for the cases where useful data were not available, Bichard (1977) summarized the effects of heterosis on pig production, as shown in Table 18.5.

Table 18.5. Performance observed or expected from various forms of regular crossing in pigs

Form of regular crossing	Heterosis utilized*	Pigs per litter	Pigs reared per sow per year, as percentage of straight-bred performance
Straightbred	—	8.3	100
Two-breed cross	I	8.8	106
Backcross	$\frac{1}{2}I + M$	9.3	115
Three-breed cross	$I + M$	9.6	118
Two-breed rotational cross	$\frac{2}{3}I + \frac{2}{3}M$	9.1	112

    * I = individual heterosis; M = maternal heterosis. Adapted from Bichard (1977).

    Taking all of the above results into account, it can be seen that with regular crossing, there are substantial benefits to be gained from heterosis.

    The choice of which type of crossing is best for a particular situation depends on the relative magnitudes of individual, maternal, and paternal heterosis, and on the extent of complementarity and the amount of heterosis between various combinations of available populations. It also depends on the management and logistical factors described earlier. Because there are so many factors involved, it is not possible to draw general conclusions about the choice of a crossing system. In practice, each case must be considered on its merits, taking account of all the factors discussed above.

It must be emphasized that the performance of particular crosses between particular breeds can sometimes be much better or much worse than the average figures given in Tables 18.3 to 18.5. Thus, the data presented in these tables should be taken only as a general guide, rather than as an exact indication of the heterosis to be expected in a particular cross. The final point to emphasize is that the characters discussed in the examples above are closely associated with reproduction and/or viability and hence show substantial heterosis. Other characters less closely associated with reproduction and/or viability will show much less heterosis, and some characters (such as carcase characters) show essentially zero heterosis.

18.3 Crossing to produce a synthetic

An alternative to regular crossing is to perform one or a few crosses between two or more populations in order to produce a single population of animals containing genes from each of the populations involved. A single population that is a mixture of various populations is called a *synthetic* or *composite*. Once a synthetic has been formed, then the main aim is to improve it as rapidly as possible by selection within it, following the guidelines laid down in Chapter 16. In many situations, the end result of selection within a synthetic is a new breed. For example, the Santa Gertrudis, the Jamaican Hope, the Norwegian Red and White, the Luing, the Australian Milking Zebu, the Australian Friesian Sahiwal, and the Belmont Red are all cattle breeds formed comparatively recently by selection within a synthetic, and similar examples can be found in other species of domestic animals. Some synthetics are based almost entirely on just one traditional breed, and were founded with selected animals from a number of different strains within the breed. Others, such as many of the present-day broiler chicken populations maintained by breeding companies, are mixtures of several different breeds, each of which was thought to have at least some desirable characters in relation to the aim of the breeders concerned.

A particularly popular combination in the formation of synthetics in cattle has been that of *Bos taurus* with *Bos indicus*, in order to produce new breeds capable of improved production under tropical and semi-tropical conditions. The aim has usually been to combine the improved production ability of *Bos taurus* breeds with the heat tolerance and tick resistance of *Bos indicus* breeds. The Belmont Red is one such synthetic. A brief history of its formation is illustrated in Fig. 18.2.

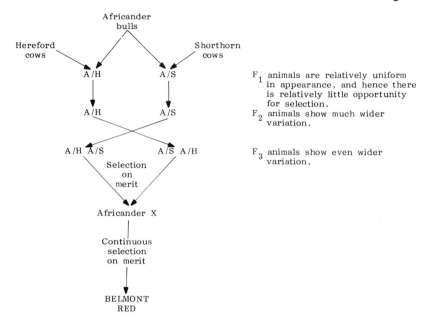

F_1 animals are relatively uniform in appearance, and hence there is relatively little opportunity for selection.

F_2 animals show much wider variation.

F_3 animals show even wider variation.

Fig. 18.2. A brief history of the formation of the Belmont Red. (After Anon. 1970.)

Of course, if we go back far enough in time, then it becomes evident that many of the breeds that we know today started as synthetics of one form or another. In some cases, the crossing phase lasted a long time. But sooner or later, in most cases, the resultant synthetic was closed and a new breed began to emerge. That is not to say that once a synthetic is finally formed it should remain closed for ever. On the contrary, in most cases it is advantageous to introduce some new genetic variation from time to time, and this continues to be done in a number of cases.

The main guidelines to be followed in crossing to produce a synthetic are: (1) Ensure that the animals used in the original crossings have been intensely selected in terms of relevant characters; it is no use starting a synthetic with inferior animals. (2) Maximize variance in breeding values amongst the foundation animals in the synthetic, by using as many unrelated animals as possible from each of the contributing populations, bearing in mind the previous requirement that they be highly selected.

With respect to heterosis, the simplest theoretical expectation is that if there are n populations contributing equally to the synthetic, then $1 - 1/n$ of the heterosis present in the F1 (the first-cross progeny) will be retained in the F2 (the progeny resulting from matings among F1's), and in subsequent generations (Dickerson 1973). However, if the synthetic is not maintained at an adequate size, or if selection is very intense within the synthetic, inbreeding depression may whittle away the remaining heterosis. The main advantage of synthetics is that only one population has to be maintained, rather than the two or more parental populations required for a regular crossing programme.

18.4 Grading-up

Grading-up involves a succession of backcrosses from one population into another population, with the aim of introducing a new gene or specific genes into one of the populations, or with the aim of substituting one population for the other. We shall now consider each of these situations in turn.

18.4.1 Introducing a new gene or new genes

Suppose, for example, that some breeders wish to create a polled strain of a horned breed, and have one mutant individual that is heterozygous for the dominant polled gene, P, at their disposal. The polled individual does not necessarily have to belong to the same breed in which the polled strain is to be produced; its only requirement is to be able to produce fertile offspring when mated to members of that breed. The breeding programme to be followed in introducing the polled gene to a horned breed is illustrated in Table 18.6. The procedure is very straightforward. All that is required is for all horned offspring to be culled each year so that only polled offspring remain to be mated back to the horned breed. A very critical requirement is that a large and representative sample of animals from the horned breed should be used in the backcrossing programme, to ensure that the polled strain has as much genetic variation as the horned breed from which it arose. In most situations, and especially if the above guideline has been followed, then three or at the most four crosses, i.e. the original cross and two or three backcrosses, will be sufficient. By this stage, the average proportion of genes from the horned breed in the new strain will be either 7/8 (two backcrosses) or 15/16 (three backcrosses) and the polled offspring will be indistinguishable from members of the horned breed, especially if there has been selection among polled offspring for characters associated with the horned breed.

The final phase of a grading-up programme for the introduction of a new gene or genes depends on whether the gene is dominant or recessive. If it is recessive, then it is simply a matter of forming the new

Table 18.6. The use of grading-up to create a polled strain of a horned breed

Generation	Mating programme	Average proportion of genes from horned breed in polled offspring
0	Polled × Horned	
1	Selected polled offspring } × Horned	$\frac{1}{2} = \left(\frac{1}{2}\right)^1$
2	Selected polled offspring } × Horned	$\frac{3}{4} = 1 - \left(\frac{1}{2}\right)^2$
3	Selected polled offspring } × Horned	$\frac{7}{8} = 1 - \left(\frac{1}{2}\right)^3$
	. . .	
t		$1 - \left(\frac{1}{2}\right)^t$

strain from all offspring exhibiting the character concerned; they will all breed true for the recessive gene. If, however, the gene is dominant, as is the case with the polled gene, then all offspring showing the character will be heterozygous and will not therefore breed true. The task is then to arrange matings among these individuals, and then in subsequent generations to distinguish carriers from non-carriers, with the aim of identifying and culling all carriers. The general requirements for this were outlined in Chapter 11 and will not be repeated here.

18.4.2 Breed substitution

The late 1960s and early 1970s witnessed an awakening of world-wide interest in long-established continental European breeds of cattle such as Charolais and Simmental. A few animals and many doses of semen from a number of such breeds were exported to many countries in which these breeds were previously unknown, and in each country a race began to establish 'purebred' herds of these 'migrant' breeds as quickly as possible, by means of a grading-up procedure like that illustrated in Table 18.7.

Table 18.7. Grading-up to a migrant breed, M, from any readily available local animals, L

Generation		Designation (grade) of animals in square brackets	Proportion of migrant genes in animals in square brackets		
			Min.	Av.	Max.
0	L × M				
1	[LM] × M	$\frac{1}{2}$ bred	$\frac{1}{2}$	$\frac{1}{2}$	$\frac{1}{2}$
2	[(LM)M] × M	$\frac{3}{4}$ bred	$\frac{1}{2}$	$\frac{3}{4}$	1
3	[((LM)M)M] × M	$\frac{7}{8}$ bred	$\frac{1}{2}$	$\frac{7}{8}$	1
4	[(((LM)M)M)M] × M	$\frac{15}{16}$ bred	$\frac{1}{2}$	$\frac{15}{16}$	1
5	etc.	$\frac{31}{32}$ bred	$\frac{1}{2}$	$\frac{31}{32}$	1

In the early stages of establishing a breed using this type of grading-up procedure, much controversy often arises as to the number of back-crosses that should be required before progeny are eligible for official registration as purebreds. Indeed, one of the continental European breed societies in Australia went so far as to conduct a referendum among its members, asking them to decide between 15/16 and 31/32. This is a completely senseless question to ask, and illustrates a complete lack of understanding of the genetic basis of grading-up.

The most important point to realize about grading-up is that, except for $\frac{1}{2}$ breds:

> *There is variation among the progeny of all other generations, with respect to the proportion of 'local' and 'migrant' genes.*

For example, it is possible that a so-called 3/4 bred could have more than 31/32 migrant genes. Equally likely, it could have just over $\frac{1}{2}$ migrant genes. The reason for this is that segregation during the formation of gametes in $\frac{1}{2}$ breds and in all subsequent crosses gives rise to a spectrum of different types of gametes, ranging theoretically from a gamete containing only local genes, to a gamete consisting solely of migrant genes. Thus, as shown in Table 18.7, the proportion of migrant genes in 3/4 breds and all subsequent crosses can range from $\frac{1}{2}$ to 1. The most likely proportion of migrant genes in any cross is the average, which is the proportion that is used to describe that cross; and extreme deviations from these proportions are less likely than small deviations.

But the fact remains that by the time breeders have reached the second or third backcross, they may have offspring with proportions of migrant genes ranging from nearly $\frac{1}{2}$ to nearly 1. Now, although breeders cannot tell exactly what proportion of migrant genes any particular animal has simply by looking at it, they can be certain that:

> *On average, animals whose appearance corresponds more closely to that of the migrant breed are likely to have a larger than average proportion of migrant genes, within any particular grade.*

And if breeders do any selection in favour of animals showing migrant-like appearance during their grading-up programme, then by the time they have reached the second backcross, i.e. 7/8, their selected progeny will on average have a proportion of migrant genes far higher than 7/8, and probably higher than 31/32.

It is now obvious why the debate about 15/16 versus 31/32 is pointless, even if the only aim of a grading-up programme was to obtain as high as possible a proportion of migrant genes within a particular herd. But there is a more important reason why the debate is pointless. It is that:

> *Breeders would improve economically important characteristics more rapidly if they stopped grading-up after the first or at most the second backcross, and concentrated solely on continual selection.*

In this way, the migrant genes would be allowed to find their own optimum proportion for the particular production system. If, for example, it turns out that approximately 65 per cent of migrant genes is optimum for production characteristics, then continual selection will produce that proportion.

Continual backcrossing simply removes all the local genes, and achieves nothing more than replicas of migrant animals that were alive 15 or 20 years ago. Not much genetic progress there! More importantly, such a programme implies that the local animals have absolutely nothing to offer by way of adaptation to local environments (both natural and managerial) that are often very different from those that exist in the country of origin of the migrant breed. In fact, of course, local animals are usually well-adapted to their local environment, and therefore do have some genes that should make a useful contribution to the new breed being developed.

We can conclude that:

> *Continual backcrossing in a grading-up programme is both unnecessary and undesirable.*

18.4.3 Preservation of rare or unwanted breeds

A particularly important issue arises in situations where a breed that is native to a particular area appears to have no further use in that area or elsewhere, and consequently is in danger of being graded-up into extinction. The general question raised by this situation is whether such a breed should be preserved.

The arguments in favour of preservation are that we do not know what type of animals we will require in the future, and that we should therefore preserve all available genetic variation as an insurance against the unknown future. Against this it is argued that breeders who aim to earn a living from animals cannot afford to look too far into the future; they appreciate the arguments in favour of preservation, but are not able to meet the relatively high cost of preserving populations that they are unlikely to ever want to utilize during their own lifetimes. Not surprisingly, therefore, the financial responsibility for preservation of unwanted breeds often falls on the public sector or on organizations set up specifically for that purpose. At both the international level, e.g. FAO, and at the local level, e.g. the Rare Breeds Trust in Great Britain, concerted efforts are being made to gather relevant data on breeds that seem threatened with extinction, and to act, where possible, to save them.

Interestingly, the two areas that are probably of greatest concern are at either end of the spectrum of animal improvement. At one end we have a large variety of locally adapted native populations in the developing countries that are under threat from the influx of 'improved' breeds and strains bred in developed countries. And at the other end we have an increasing number of poultry selection lines that are discarded when yet another independent poultry breeder is taken over by a larger and often transnational breeding company.

There are no easy solutions to these problems. The only comfort that can be gained from present activity in this area around the world is that more and more people are becoming concerned and are consequently giving more thought to possible solutions.

18.5 Summary

Crossing is the mating of individuals from different populations which can be strains, breeds or species.

Regular crossing occurs when the same cross is made on a regular basis, with the aim of producing a particular type of progeny. The main benefits of regular crossing arise from complementarity (the additional profit resulting from the manner in which two or more characters complement each other), and heterosis (superiority in performance of offspring over the average of parents). Heterosis is greatest in characters most closely associated with the ability to survive and reproduce, and the greater the genetic differences between two populations of domesticated animals, the greater the heterosis in crosses between them.

Regular crossing can be either specific (e.g. two-way cross, three-way cross, backcross) or rotational, in which males from two or more populations are mated on a rotational basis with females obtained from earlier crosses in the series. Rotational crossing has the advantage that all female replacements are obtained from the crossing programme itself and do not therefore have to be specifically bred in a separate population, as they do in the case of specific crossing. On the other hand, rotational crossing produces less benefits from heterosis and allows no exploitation of complementarity.

Crossing can be used to create a synthetic or composite, which is a single population consisting of various proportions of genes from two or more other populations. The aim in producing a synthetic is to maximize genetic variation and then to select within that population as rapidly as possible.

Grading-up involves a succession of backcrosses from one population into another population, with the aim of introducing a new gene or specific genes into one of the populations, or with the aim of substituting one population (migrant) for the other (local). In the second case, there is often controversy as to how many backcrosses to the migrant population should be required before the resultant animals can be registered as 'pure'. Such controversies are pointless from two points of view. Firstly, if any selection in favour of migrant appearance has occurred during the first one or two backcrosses, then the proportion of migrant genes actually present in offspring is likely to be far higher than the average figures of 3/4 and 7/8 expected for all offspring. Secondly, it is very likely that the local population has some useful genes, especially in terms of local adaptation. Consequently, in most cases it is best to forget about grading-up and instead to combine together the local and migrant populations into a synthetic. By selecting in this synthetic for economic performance in the local environment, the optimum proportion of local and migrant genes should automatically be achieved.

However, even the formation of such a synthetic means the loss of the local population as a separate entity, and this almost certainly involves the loss of at least some genes that may have been unique to that local population. While these genes may not have any immediate value, there is increasing belief around the world that in addition to the formation of synthetics, local populations should be preserved as separate entities, as an insurance against the unknown future.

18.6 Further reading

Anon. (1974*a*). *Proceedings of the Working Symposium on Breed Evaluation and Crossing Experiments with Farm Animals*. Research Institute for Animal Husbandry 'Schoonoord', Zeist, The Netherlands. (A large collection of papers dealing with crossing in cattle, pigs, and sheep, with contributions from research workers in New Zealand, Canada, U.S.A., and many European countries. Some of the general reviews are very useful, particularly the one by Dickerson.)

Anon. (1981, 1984). *Animal Genetic Resources*. (FAO Animal Production and Health Papers, 24, 44/1 and 44/2) FAO/UNEP, Rome. (A set of thorough reviews.)

Barker, J. S. F., Mukherjee, T. K., Turner, H. N., and Sivarajasingam, S. (Eds) (1981). *Evaluation of Animal Genetic Resources in Asia and Oceania* (Proceedings of the Second SABRAO Workshop on Animal Genetic Resources, Kuala Lumpur, May 5-6, 1981) Society for the Advancement of Breeding Researches in Asia and Oceania, Kuala Lumpur. (A review of what is known about the performance of many native breeds.)

Barlow, R. (1981). Experimental evidence for interaction between heterosis and environment in animals. *Animal Breeding Abstracts* **49**: 715-37. (A review showing that the amount of heterosis observed for most characters varies with variation in the environment.)

Dickerson, G. E. (1969). Experimental approaches to utilizing breed resources. *Animal Breeding Abstracts* **37**, 191-202. (A very important review.)

Dickerson, G. E. (1973). Inbreeding and heterosis in animals. In *Animal Breeding and Genetics* (Proceedings of a Symposium in Honour of Dr J. L. Lush) (Anon., Ed.). pp. 54-77. American Society of Animal Science, and American Dairy Science Association, Champaign, Illinois. (Another important review.)

Hayman, R. H. (1974). The development of the Australian Milking Zebu. *World Animal Review* (11), 31-5. (An account of the formation of, and subsequent selection within, a synthetic of *Bos taurus* and *Bos indicus*.)

Katpatal, B. G. (1977*a*, *b*). Dairy cattle crossbreeding in India. 1 and 2. *World Animal Review* (22), 27-33, and (23), 2-9. (Two reviews of dairy crossing in India.)

Lauvergne, J. J. (1977). The current status of cattle breeds in Europe. *World Animal Review* (21), 42-7. (The report of a study of endangered cattle breeds in Europe.)

Livestock Production Science **11**(1), Feb., 1984. (This issue contains a set of papers that discuss the conservation of animal genetic resources.)

Madalena, F. E. (1977). Crossbreeding systems for beef production in Latin America. *World Animal Review* (27), 27–33. (A review of the likely role of crossing in Latin American beef breeding programmes.)

Nitter G. (1978). Breed utilization for meat production in sheep. *Animal Breeding Abstracts* 46, 131–43. (A review of crossing in sheep.)

Sellier, P. (1976). The basis of crossbreeding in pigs: a review. *Livestock Production Science* 3, 203–26. (A review of crossing in pigs.)

Sheridan, A. K. (Ed.) (1982). Nature and utilisation of heterosis. *Proceedings of the Second World Congress on Genetics Applied to Livestock Production* 6, 185–262. (A series of papers summarizing current knowledge concerning heterosis in domestic animals.)

19
Selection and regular crossing

19.1 Introduction

Traditionally, many stud breeders have regarded crossing as an insult to them and to their animals, and they have also regarded crossing as a threat to their livelihoods. The aim of this chapter is to explain why these opinions are incorrect, and to illustrate that the most profitable future for both stud breeders and commercial breeders lies in breeding programmes that involve both selection and regular crossing.

19.2 Selection

As we saw in Chapter 16, a well-designed selection programme within a population can lead to a steady improvement in breeding value and hence in profitability. However, the average level of inbreeding will also show a steady increase, even if mating is entirely at random. In addition, the greater the selection applied in any population, the greater will be the increase in average level of inbreeding in that population. Thus, successful selection programmes in closed populations incur inbreeding depression which, as we saw in Chapter 13, has greatest effect on characters associated with viability and reproductive ability. If effective population size (N_e) is not too small (see Section 16.5.1), then the increase in inbreeding depression may be unimportant from a practical point of view. However, if N_e is relatively small, then the steady increase in inbreeding depression may begin to adversely affect profitability.

Of course, this problem can be offset in practice by the occasional introduction of unrelated animals into the otherwise closed population. However, as soon as breeders start buying animals from other sources, then the response in their own population is determined in part by the breeding value of the purchased animals; and if response to selection within their own population has been quite rapid, they may have difficulty finding other sources from which to purchase stock that will in any way match the average breeding value of their own stock. This is becoming an increasingly common dilemma.

19.3   Selection and regular crossing

We saw in Chapter 18 that the characters that show the greatest inbreeding depression are those that also show the greatest heterosis when different populations are crossed. This observation immediately points the way to a solution of the above dilemma.

> *We can get the best of both worlds by continually selecting within each of several populations and by making regular crosses between them in order to produce the final commercial product.*

In this way we are able to get the most out of available genetic variation; continual selection exploits variation in breeding values, V_A, and regular crossing exploits non-additive genetic variation, V_D and V_I. Results from the practical application of these proposals during more than 30 years of a chicken breeding programme are illustrated in Fig. 19.1.

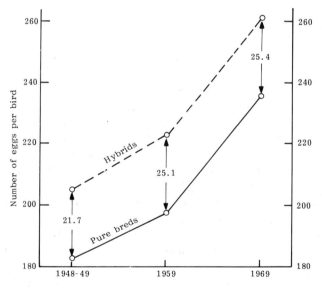

Fig. 19.1. Improvement in egg production to 500 days, per hen completing test, obtained from continual selection within two lines (pure breds), and from crosses between them (hybrids). The solid line labelled purebreds is the average of the two parental lines. The data have been corrected to remove the effect of changes in the environment. Only three sets of points are given, because crosses were produced on only three occasions in this experiment. Notice that the improvement from heterosis is about the same on each occasion that crosses were made. (From Cole and Hutt 1973. Reproduced by permission of Commonwealth Agricultural Bureaux, SL2 3BN, UK.)

The major difficulty with breeding programmes involving both selection and regular crossing is the organizational and management problems associated with maintaining at least two different straightbred populations together with various crossbred populations. In some situations it requires cooperation amongst several breeders, with each breeder maintaining just one population. However, the financial benefits from combining continual selection with regular crossing are sufficient to make this extra effort well worthwhile in most circumstances. An important point to emerge from this is that:

> *Regular crossing can occur only if someone is maintaining and improving the straightbred lines.*

Thus, rather than presenting a threat to stud breeders, regular crossing actually provides them with a substantial incentive to maintain and improve their straightbred populations.

Stud breeders who realize this in time are well placed to capitalize on it. For example, there are several cases in Great Britain where Large White and Landrace stud breeders have successfully joined forces to produce and market first-cross gilts for sale to commercial pig producers. These breeders still enjoy all the traditional activities of stud breeding; they still maintain straightbred herds of Large White or Landrace, the animals in these herds can still be registered with the relevant breed society and, if the breeders are so inclined, their straightbred stock can still be exhibited at shows. But within their herds, they are continually selecting for economically important characters along the lines described in Chapter 16; and most of the gilts sold by them to commercial producers are the offspring of Large White $\times$ Landrace matings conducted on a regular basis. Given the data summarized in Table 18.5, the stud breeders really had no choice but to begin breeding crossbreds; no commercial producer who wished to stay in business would buy a straightbred gilt in preference to a crossbred gilt derived from two straightbred lines.

Although the situation in other species and in other countries is not exactly the same as that described above, the lessons to be learnt from the above example are obvious:

> *Crossing need not be a threat to the livelihood of stud breeders; on the contrary, stud breeders can, if they wish, exploit crossing to their financial benefit.*

In discussing regular crossing in this section, we have concentrated solely on the benefits to be gained from heterosis. However, it was emphasized in Chapter 18 that there is more to crossing than just heterosis. In fact, it was pointed out that even in the absence of heterosis, certain crosses show sufficient complementarity to justify their continued use. Thus, in certain circumstances, regular crossing may still be an economically viable proposition, even if there is negligible inbreeding depression in the relevant closed populations.

Recalling some points made in Chapters 16 and 18, we can conclude that by carrying out both selection within populations and regular crossing between populations, we are able to exploit additive gene effects *within* populations (through selection), additive gene effects *between* populations (through complementarity), and non-additive gene effects (through heterosis).

A final point should be made in relation to synthetics or composites, which were described in Section 18.3. In certain circumstances, there are very good reasons for crossing various populations on a once-only basis to produce a synthetic. Sooner or later, however, a synthetic has to be considered in the same category as any other closed population described in this section. Thus, even if inbreeding depression is not currently a problem in a particular synthetic, it may still be economically advantageous to use that synthetic in a regular crossing programme, if the synthetic exhibits useful complementarity with one or more other populations.

19.4 A practical breeding programme

It is not the aim of this book to provide a survey of practical breeding programmes in each species. Instead, we will consider briefly just one example which illustrates the successful utilization of continual selection and regular crossing, as well as the use of single genes. It is a breeding programme for broiler chickens, the general aspects of which are illustrated in Fig. 19.2.

In order to describe a breeding programme such as this one, we need to think in terms of the pyramid structure described in Chapter 17. At the top of the pyramid are several nucleus populations or lines (A, B, and C), each maintained separately. Animals in the nucleus lines are great-grandparents (GGP) of the final end product. Each GGP line is really a synthetic made up of several different chicken breeds, according to the requirements for each line, as described below. Within each GGP line, continual selection is conducted following the guidelines described in Chapter 16, with the aim of maximizing response per year for a different combination of characters in each line, according to the contribution of each line to the final product, as we shall see below.

The second layer of the pyramid involves matings between grandparents (GP) of the final product. These are straightbred females from line A crossed with straightbred males from line B. The progeny of this cross represent the parents (P) of the final product, which results from females of this cross being mated to males from line C. Notice that each line or cross contributes only one sex in the crossing programme; lines B and C are sire lines while line A and the first-cross AB are dam lines.

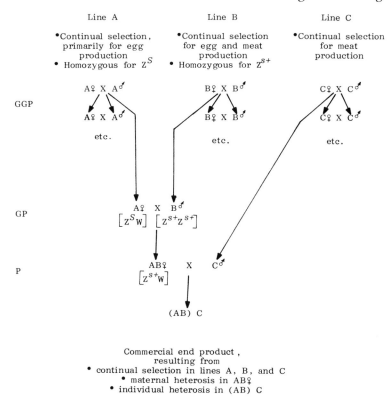

Fig. 19.2. A breeding programme for broiler chickens, illustrating an efficient combination of continual selection within lines, and regular crossing between them. Symbols for Z and W chromosomes have been omitted from line C because they are not relevant there. S = silver allele, which is dominant to the gold, s^+, allele. These alleles are used at the P level, to separate AB ♂ [$Z^S Z^{s+}$ = silver], which are unwanted, from AB ♀ [$Z^{s+}W$ = gold]. If paternal heterosis were to be exploited, a fourth line, D, would be mated to C at the GP level, to produce a hybrid CD male to be mated with AB females.

The sire line C makes its only contribution as sire of the final product, and because only few males are needed relative to females, it does not really matter if viability and reproductive ability are relatively low in this line. Consequently, line C was formed solely from various meat-type breeds, and a selection programme is conducted within line C so as to maximize improvement in efficiency of meat production. This is achieved by selecting on body weight at a certain age, which is the most common selection criterion used in broiler sire lines. Some attention will also be paid to breast shape, but most other characters will be ignored, unless a particular problem such as leg weakness begins to cause trouble.

Although line B is a sire line, it makes its contribution as the sire of the first-cross AB hen. Now, the AB hen is required to have the maximum possible reproductive ability, which of course will be enhanced by heterosis. But it is no good having heterosis for reproductive ability if the average reproductive ability of the two lines is low; thus, although it is a sire line, line B may have incorporated initially some breeds noted for egg laying, and will be selected for egg production as well as efficiency of meat production, with emphasis probably on the latter.

Finally, the dam line A is required to have an especially good reproductive ability, so as to be able to produce sufficient crossbred AB eggs. Thus, the breeds that contribute initially to line A will be mostly layer-type strains, and the selection criteria will be body weight at a certain age and various measures of egg production, with the latter having a high relative importance. The dam line A therefore will be the least meaty of the three original lines, but will lay considerably more eggs than the other two.

Because each line or cross contributes only one sex to the breeding programme, it is obvious that the sexes have to be separated at each stage of the programme. The point at which the numbers of chickens of one sex required will be greatest is amongst day-old AB chickens. At this point, therefore, our breeding programme makes use of Z-linked sexing, in this case at the gold (s^+)/silver (S) locus (see Fig. 12.2). In order to make this system work, line A has to be homozygous for the dominant, S, allele, while line B must be homozygous for s^+.

Another example of the use of single genes is in relation to white feathering. Many of the meat-type breeds that were used in the formation of all three lines have coloured feathers, which is an undesirable character in the final product. This problem can be solved by ensuring that the sire line C is homozygous for dominant white, I, an allele at an autosomal locus.

19.5 The role of commercial producers

Poultry breeding programmes like the one briefly described above are usually controlled entirely by one breeding organization. The role of the commercial producers is simply to buy the day-old chickens resulting from the AB × C final cross, and to raise them to slaughter weight. Such commercial producers do neither selection nor crossing, but instead rely on the breeding organization to do both. This represents one end of the spectrum of possibilities for commercial producers. Moving along this spectrum, the next possibility is for commercial producers to purchase AB females and males, and to perform the final cross themselves, while still relying on other breeders for the initial crossing and for continual selection in the straightbred lines. This is what often happens in pig and fat lamb production, and to a certain extent in beef production.

Further along the spectrum are the commercial producers who buy GP stock, which involves them in doing all the crossing themselves, but still relying on other breeders for continual selection. This is a system being given some encouragement in pig production, because fewer pigs have to be introduced into the commercial herd. It does, however, involve the commercial producer in maintaining separate P lines as well as raising the commercial product.

At the end of the spectrum are the commercial producers who run their own GGP lines as well as GP and P lines, and thus have complete control over both selection and crossing. The big advantage of this system is that the commercial producer does not have to rely on regular supplies of breeding stock from elsewhere. But the managerial and other problems involved in running so many different lines are sufficiently daunting to discourage most commercial producers from going this far. Indeed, any producers who do take on the whole task of selection and crossing very soon find that they become suppliers of breeding stock to other commercial producers. The optimum strategy for any particular producer can be determined only when all the current and local details relevant to that producer have been taken into account. The above discussion does not cover all possibilities, but it does give some idea of the various strategies that can be adopted by commercial producers.

19.6 Summary

The most complete utilization of genetic resources can be achieved by (1) continual selection within each of two or more populations, and (2) regular crossing between these populations in order to produce the final commercial product.

This enables exploitation of additive gene effects *within* populations (through selection), additive gene effects *between* populations (through complementarity), and non-additive gene effects (through heterosis).

Because regular crossing can occur only if someone is maintaining the straightbred lines, crossing need not be a threat to the livelihood of stud breeders; on the contrary, stud breeders can, if they wish, exploit regular crossing to their financial benefit.

All commercial producers should be able to benefit from continual selection and from regular crossing. They can do this by relying on other breeders to do all or part of the job of selection and crossing for them, or by doing both jobs themselves. Because of the complexities involved in conducting several different selection programmes and producing one or more crosses, most commercial producers leave some or all of the selection and crossing to breeders who specialize in these tasks.

19.7   Further reading

Bichard, M. (1977). Economic efficiency of pig breeding schemes: a breeding company view. *Livestock Production Science* 4, 245–54. (A very informative discussion of selection and crossing in pigs, by the person in charge of the breeding programme of a large, pig-breeding organization.)

Cole, R. K. and Hutt, F. B. (1973). Selection and heterosis in Cornell White Leghorns: a review, with special consideration of interstrain hybrids. *Animal Breeding Abstracts* 41, 103–18. (An illustration of the benefits that can be achieved from combining continual selection within lines, with crossing between them.)

Moav, R. (1966*a*, *b*, *c*). Specialised sire and dam lines. *Animal Production* 8, 193–202; 203–11; 365–74. (A set of three papers concerned with the choice of appropriate sire and dam lines using economic criteria.)

Moav, R. and Hill, W. G. (1966). Specialised sire and dam lines. *Animal Production* 8, 375–90. (The final paper in the series, dealing with selection within lines.)

Robertson, A. (1971). Optimum utilization of genetic material, with special reference to cross-breeding in relation to other methods of genetic improvement. *Proceedings of the 10th International Congress of Animal Production*, 57–68. (A review of crossing in relation to selection.)

Smith, C. (1964). The use of specialised sire and dam lines in selection for meat production. *Animal Production* 6, 337–44. (One of the first attempts to examine the question of whether specialized sire and dam lines are economically worthwhile.)

Books on particular species

The following books provide more detailed information on genetics in relation to particular species.

Cat:

Robinson, R. (1977). *Genetics for cat breeders* (2nd edn). Pergamon Press, Oxford.

Dog:

Burns, M. and Fraser, M. N. (1966). *Genetics of the dog* (2nd edn). Oliver and Boyd, Edinburgh.

Hutt, F. B. (1979). *Genetics for dog breeders*. W. H. Freeman, San Francisco.

Little, C. C. (1957). *The inheritance of coat colour in dogs*. Cornell University Press. (Reprinted by Howell Book House, New York.)

Robinson, R. (1982). *Genetics for dog breeders*. Pergamon Press, Oxford.

Willis, M. B. (1976). *The German shepherd dog: its history, development and genetics*. K & R Books, Leicester.

Sheep:

Dolling, C. H. S. (1970). *Breeding merinos*. Rigby, Adelaide.

Land, R. B. and Robinson, D. W. (1984). *The genetics of reproduction in sheep*. Butterworth, London.

Turner, H. N. and Young, S. S. Y. (1969). *Quantitative genetics in sheep breeding*. Macmillan, Melbourne.

Cattle:

Hinks, C. J. M. (1983). *Breeding dairy cattle*. Farming Press, Ipswich.

Schmidt, G. M. and Van Vleck, L. D. (1974). *Principles of dairy science*. W. H. Freeman, San Francisco.

Horse:

Evans, J. W., Borton, A., Hintz, H. F., and Van Vleck, L. D. (1977). *The horse*. W. H. Freeman, San Francisco.

Jones, W. E. (1982). *Genetics and horse breeding*. Lea and Febiger, Philadelphia.
Lasley, J. F. (1970). *Genetic principles in horse breeding*. University of Missouri Press, Columbia.

General:

Hutt, F. B. and Rasmusen, B. A. (1982). *Animal genetics* (2nd edn). John Wiley and Sons, New York.

References

Abraham, E. P. (1981). The beta-lactam antibiotics. *Scientific American* 244(6), 76–86.

Adalsteinsson, S. (1983). Inheritance of colours, fur characteristics and skin quality traits in North European sheep breeds: a review. *Livestock Production Science* 10, 555–67.

Adalsteinsson, S., Sigurjónsson, T., and Jónsson, G. (1979). Variation in colour gene frequencies among Icelandic cats. *Carnivore Genetics Newsletter* 3, 359–72.

Altman, P. L. and Katz, D. D. (1979). *Inbred and genetically defined strains of laboratory animals*. Parts 1 and 2. FASEB, Bethesda, Maryland.

Anderson, R. M. and May, R. M. (1982). *Population biology of infectious diseases*. Springer-Verlag, Berlin.

Anderson, W. F. and Diacumakos, E. G. (1981). Genetic engineering in mammalian cells. *Scientific American* 245(1), 60–93.

Andresen, E., Jensen, P., Jonsson, P., and Staun, H. (1980). Malignant hyperthermia syndrome (MHS) in pigs: relative risks associated with various genotypes. *Zeitschrift für Tierzüchtung und Züchtungsbiologie* 97, 210–16.

Andrews, E. J., Ward, B. C., and Altman, N. H. (Eds) (1979). *Spontaneous animal models of human disease*. Volumes I and II. Academic Press, New York.

Anon. (1969). *Joint Committee on the Use of Antibiotics in Animal Husbandry and Veterinary Medicine* (Swann Report) Cmnd 4190. HMSO, London.

Anon. (1970). Beef cattle for the tropics. *Rural Research in CSIRO* (69), 18–23.

Anon. (1974a). *Proceedings of the Working Symposium on Breed Evaluation and Crossing Experiments with Farm Animals*. Research Institute for Animal Husbandry, 'Schoonoord', Zeist, The Netherlands.

Anon. (1974b). *Germ Plasm Evaluation Program*. Report No. 1. US Meat Animal Research Center, Nebraska.

Anon. (1981, 1984). *Animal genetic resources*. (FAO Animal Production and Health Papers, 24, 44/1 and 44/2.) FAO/UNEP, Rome.

Antaldi, G. G. V. (1980). [Hereditary diseases of the horse.] Parts I and II. *Revista di Zootecnia e Veterinaria* 1, 53–61; 2, 123–32.

Arber, W. (1979). Promotion and limitation of genetic exchange. *Science* 205, 361–5.

Archer, R. K. and Allen, B. S. (1972). True haemophilia in horses. *Veterinary Record* 91, 655–6.

Archibald, A. L., Bradley, J. S., and Spooner, R. L. (1979). Evidence for phenotypic change at the amylase I locus in cattle—an explanation of the apparent heterozygote excess. *Animal Production* 28, 337–45.

Arnall, L. and Keymer, I. F. (1975). *Bird diseases.* Baillière Tindall, London.

Ashton, G. C., Fallon, G. R., and Sutherland, D. N. (1964). Transferrin (β-globulin) type and milk and butterfat production in dairy cows. *Journal of Agricultural Science* 62, 27–34.

Atkins, K. D. and McGuirk, B. J. (1979). Selection of Merino sheep for resistance to fleece-rot and body strike. *Wool Technology and Sheep Breeding* 27(3), 15–19.

Auer, L. and Bell, K. (1981). The AB blood group system of cats. *Animal Blood Groups and Biochemical Genetics* 12, 287–97.

Austin, C. R. and Edwards, R. G (Eds) (1981). *Mechanisms of sex differentiation in animals and man.* Academic Press, London.

Australian Academy of Science (1980). *Recombinant DNA: an Australian perspective.* Australian Academy of Science, Canberra.

Baker, H. J., Mole, J. A., Lindsey, J. R., and Creel, R. M. (1976). Animal models of human ganglioside storage diseases. *Federation Proceedings* 35, 1193-201.

Baker, H. K., Bech Andersen, B., Colleau, J., Langholz, H., Legoshin, G., Minkema, D., and Southgate, J. (1976). Cattle breed comparison and crossbreeding trials in Europe: a survey prepared by a working party of the European Association for Animal Production. *Livestock Production Science* 3, 1–11.

Ballarini, G. (1977). [Hereditary diseases in veterinary practice.] *Revista di Zootecnia e Veterinaria* 2, 177–86.

Balner, H. (1981). The major histocompatibility complex of primates: evolutionary aspects and comparative histogenetics. *Philosophical Transactions of the Royal Society, Series B* 292, 109–19.

Baltimore, D. (1976). Viruses, polymerases and cancer. *Science* 192, 632–6.

Bank, A., Mears, J. G., and Ramirez, F. (1980). Disorders of human hemoglobin. *Science* 207, 486–93.

Barker, J. S. F., Mukherjee, T. K., Turner, H. N., and Sivarajasingam, S. (Eds) (1981). *Evaluation of Animal Genetic Resources in Asia and Oceania* (Proceedings of the Second SABRAO Workshop on Animal Genetic Resources, Kuala Lumpur, May 5–6, 1981). Society for the Advancement of Breeding Researches in Asia and Oceania, Kuala Lumpur.

Barlow, R. (1981). Experimental evidence for interaction between heterosis and environment in animals. *Animal Breeding Abstracts* 49, 715–37.

Barlow, R., Hearnshaw, H., and Thompson, J. M. (1978). Progress in crossbreeding research with cattle on coastal N.S.W. *Wool Technology and Sheep Breeding* 26(11), 5–12.

Barnes, I. C., Kelly, D. F., Pennock, C. A., and Randell, J. A. J. (1981). Hepatic beta galactosidase and feline GM1 gangliosidosis. *Neuropathology and Applied Neurobiology* 7, 463–76.

Barr, M. L. and Bertram, E. G. (1949). A morphological distinction between neurones of the male and female, and the behaviour of the nucleolar satellite during accelerated nucleoprotein synthesis. *Nature* 163, 676–7.

Barton, R. A. and Smith, W. C. (Eds) (1982). *Proceedings of the World Congress on Sheep and Beef Cattle Breeding.* Dunsmore Press, Palmerston North, New Zealand.

Basrur, P. K. (1980). Genetics in veterinary medicine. In *Scientific foundations of veterinary medicine* (Eds Phillipson, A. T., Hall, L. W., and Pritchard, W. R.) pp. 393–413. William Heinemann, London.

Baverstock, P. R., Adams, M., Polkinghorne, R. W., and Gelder, M. (1982). A sex-linked enzyme in birds: Z-chromosome conservation but no dosage compensation. *Nature* **296**, 763-6.

Becker, W. A. (1984). *Manual of quantitative genetics* (4th edn). Academic Enterprises, P.O. Box 666, Pullman, Washington.

Behrens, H. (1979). *Lehrbuch der Schafkrankheiten*. Verlag Paul Parey, Berlin.

Belfield, W. O. (1976). Chronic subclinical scurvy and canine hip dysplasia. *Veterinary Medicine and Small Animal Clinician* **71**, 1399-403.

Bell, K. (1983). The blood groups of domestic mammals. In *Red blood cells of domestic mammals* (Eds Agar, N. S. and Board, P. G.) pp. 133-64. Elsevier, Amsterdam.

Benacerraf, B. (1981). Role of MHC gene products in immune regulation. *Science* **212**, 1229-38.

Benirschke, K. (1981). Hermaphrodites, freemartins, mosaics and chimaeras in animals. In *Mechanisms of sex differentiation in animals and man* (Eds Austin, C. R. and Edwards, R. G.) pp. 421-63. Academic Press, London.

Bennett, B. T., Taylor, Y., and Epstein, R. (1975). Segregation of the clinical course of transmissable venereal tumor with DL-A haplotypes in canine families. *Transplantation Proceedings* **7**, 503-5.

Benson, R. E. and Dodds, W. J. (1977). Autosomal factor VIII deficiency in rabbits: size variation of rabbit factor VIII. *Thrombosis and Haemostasis* **38**, 380.

Berg, P. (1981). Dissections and reconstructions of genes and chromosomes. *Science* **213**, 296-303.

Best, P. (1983). Blood transfusion in veterinary medicine. In *Anaesthesia and Intensive Care: Refresher Course for Veterinarians, Proceedings No. 62* (Ed. Anon.) pp. 281-342. Postgraduate Committee in Veterinary Science, Sydney.

Bichard, M. (1971). Dissemination of genetic improvement through a livestock industry. *Animal Production* **13**, 401-11.

Bichard, M. (1977). Economic efficiency of pig breeding schemes: a breeding company view. *Livestock Production Science* **4**, 245-54.

Bichard, M. (1978). Which breeding programme for you? In *New Developments in Scientific Pig Breeding* (A supplement to *Pig Farming*, Volume **27**(11)) pp. 2-6.

Bijnen, A. B., Obertop, H., Joling, P., and Westbroek, D. L. (1980). Genetics of kidney allograft survival in dogs. III. Relevance of histocompatibility matching in immunosuppressed recipients. *Transplantation* **30**, 191-5.

Bindon, B. M. (1984). Reproductive biology of the Booroola Merino sheep. *Australian Journal of Biological Sciences* **37**, 163-89.

Bindon, B. M. and Piper, L. R. (1976). Assessment of new and traditional techniques of selection for reproduction rate. In *Sheep Breeding* (Eds Tomes, G. J., Robertson, D. E. and Lightfoot, R. J.) pp. 357-71. Western Australian Institute of Technology, Perth.

Biozzi, G., Mouton, D., Heumann, A. M., and Bouthillier, Y. (1982). Genetic regulation of immunoresponsiveness in relation to resistance against infectious diseases. *Proceedings of the Second World Congress on Genetics Applied to Livestock Production* **5**, 150-63.

Bird, P. J. W. N. and Mitchell, G. (1980). The choice of discount rate in animal breeding investment appraisal. *Animal Breeding Abstracts* **48**, 499-505.

Bishop. J. A. (1981). A neo-Darwinian approach to resistance: examples from mammals. In *Genetic consequences of man made change* (Eds Bishop, J. A. and Cook, L. M.) pp. 37–51. Academic Press, London.

Bishop, J. A. and Cook, L. M. (Eds) (1981). *Genetic consequences of man made change*. Academic Press, London.

Blackman, B. L., Takada, H., and Kawakami, K. (1978). Chromosomal rearrangement involved in insecticide resistance of *Myzus persicae*. *Nature* **271**, 450-2.

Blakemore, W. F. (1975). Lysosomal storage diseases. *Veterinary Annual* **15**, 242-5.

Blazak, W. F. and Fechheimer, N. S. (1981). Gonosome–autosome translocations in fowl: the development of chromosomally unbalanced embryos sired by singly and doubly heterozygous cockerels. *Genetical Research* **37**, 161-71.

Blood, D. C., Henderson, J. A., and Radostits, O. M. (1979). *Veterinary medicine* (5th edn). Baillière Tindall, London.

Blue, M. G., Bruère, A. N., and Dewes, H. F. (1978). The significance of the XO syndrome in infertility of the mare. *New Zealand Veterinary Journal* **26**, 137-41.

Bodmer, W. F. (1981). HLA—the major human tissue typing system. In *The biological manipulation of life* (Ed. Messel, H.) pp. 217–44. Pergamon Press, Sydney.

Bodmer, W. F. and Cavalli-Sforza, L. L. (1976). *Genetics, evolution and man*. W. H. Freeman, San Francisco.

Böhme, R. E., Schönfelder, E., and Schlaaf, S. (1978). Prognostische Untersuchungen zur Verbreitung der Hüftgelenksdysplasie beim Deutschem Schäferhund in der D.D.R. *Monatshefte für Veterinärmedizin* **33**, 93-6.

Boothroyd, J. C., Highfield, P. E., Cross, G. A. M., Rowlands, D. J., Low, P. A., Brown, F., and Harris, T. J. R. (1981). Molecular cloning of foot and mouth disease virus genome and nucleotide sequences in the structural protein. *Nature* **290**, 800-2.

Bowman, J. C. (1983). *An introduction to animal breeding* (2nd edn). Edward Arnold, London.

Bowman, J. C. and Falconer, D. S. (1960). Inbreeding depression and heterosis of litter size in mice. *Genetical Research* **1**, 262-74.

Bremermann, H. J. (1980). Sex and polymorphism as strategies in host–pathogen interactions. *Journal of Theoretical Biology* **87**, 671-702.

Breukink, H. J., Hart, H. C., Arkel, C., Veldon, N. A., and Watering, C. C. (1972). Congenital afibrinogenemia in goats. *Zentralblatt für Veterinärmedizin, Reihe A* **19**, 661-76.

Briles, W. E. and Allen, C. P. (1961). The B blood group system of chickens. II. The effect of genotype on livability and egg production in seven commercial inbred lines. *Genetics* **46**, 1273-93.

Briles, W. E. and Briles, R. W. (1982). Identification of haplotypes of the chicken major histocompatibility complex (B). *Immunogenetics* **15**, 449-59.

Briles, W. E., Stone, H. A., and Cole, R. K. (1977). Marek's disease: effects of B histocompatibility alloalleles in resistant and susceptible chicken lines. *Science* **195**, 193-5.

Briles, W. E., Bumstead, N., Ewert, D. L., Gilmour, D. G., Gogusev, J., Hála, K., Koch, C., Longenecker, B. M., Nordskog, A. W., Pink, J. R. L., Shierman, L. W.,

Simonsen, M., Toivanen, A., Toivanen, P., Vainio, O., and Wick, G. (1982). Nomenclature for chicken major histocompatibility (B) complex. *Immunogenetics* 15, 441–7.

Brinkhous, K. M., Davis, P. D., Graham, J. B., and Dodds, W. J. (1973). Expression and linkage of genes for X-linked hemophilias A and B in the dog. *Blood* 41, 577–85.

Britten, R. J. and Kohne, D. E. (1970). Repeated segments of DNA. *Scientific American* 222(4), 24–31.

Brown, A. W. A. (1977). Epilogue: resistance as a factor in pesticide management. In *Proceedings of the XV International Congress of Entomology, Washington DC, 1976* (Ed. White, D.) pp. 816–24.

Bruère, A. N. (1975). Further evidence of normal fertility and the formation of balanced gametes in sheep with one or more different Robertsonian translocations. *Journal of Reproduction and Fertility* 45, 323–31.

Bruère, A. N., Scott, I. S., and Henderson, L. M. (1981). Aneuploid spermatocyte frequency in domestic sheep heterozygous for three Robertsonian translocations. *Journal of Reproduction and Fertility* 63, 61–6.

Buckland, R. A., Fletcher, J. M., and Chandley, A. C. (1976). Characterization of the domestic horse (*Equus caballus*) karyotype using G- and C-banding techniques. *Experientia* 32, 1146–9.

Bulmer, M. G. (1980). *The mathematical theory of quantitative genetics*. Clarendon Press, Oxford.

Bunting, S. and Van Emden, H. F. (1980). Rapid response to selection for increased esterase activity in small populations of an apomictic clone of *Myzus persicae*. *Nature* 285, 502–3.

Burns, M. and Fraser, M. (1966). *Genetics of the dog* (2nd edn). Oliver and Boyd, Edinburgh.

Cartwright, T. C. (1970). Selection criteria for beef cattle for the future. *Journal of Animal Science* 30, 706–11.

Cavalli-Sforza, L. L. and Bodmer, W. F. (1971). *The genetics of human populations*. W. H. Freeman, San Francisco.

Centerwall, W. R. and Benirschke, K. (1975). An animal model for the XXY Klinefelter's syndrome in man: tortoiseshell and calico male cats. *American Journal of Veterinary Research* 36, 1275–80.

Chambon, P. (1981). Split genes. *Scientific American* 244(5), 48–59.

Chandley, A. C. (1981). Does 'affinity' hold the key to fertility in the female mule? *Genetical Research* 37, 105–9.

Ch'ang, T. S. and Evans, R. (1982). Heterotic basis of breeding policy for lamb production. *Proceedings of the Second World Congress on Genetics Applied to Livestock Production* 8, 796–801.

Chapman, A. B. (Ed.) (1985). *General and quantitative genetics*. (Vol. A4, *World Animal Science* series.) Elsevier, Amsterdam.

Chapman, H. M. and Bruère, A. N. (1977). Chromosome morphology during meiosis of normal and Robertsonian translocation-carrying rams (*Ovis aries*). *Canadian Journal of Genetics and Cytology* 19, 93–102.

Cherubin, C. E. (1981). Antibiotic resistance of *Salmonella* in Europe and the United States. *Reviews of Infectious Diseases* 3, 1105–26.

Cho, D. Y. and Leipold, H. W. (1977). Congenital defects of the bovine central nervous system. *Veterinary Bulletin* 47, 489–504.

Chou, T-H., Hill, E. J., Bartle, E., Woolley, K., LeQuire, V., Olson, W., Roelofs, R., and Park, J. H. (1975). Beneficial effects of penicillamine treatment on hereditary avian muscular dystrophy. *Journal of Clinical Investigation* **56**, 842–9.

Clarke, B. (1976). The ecological genetics of host–parasite relationships. In *Genetic aspects of host–parasite relationships* (Eds Taylor, A. E. R. and Muller, R.) pp. 87–103. Blackwell, Oxford.

Clowes, R. C. (1973). The molecule of infectious drug resistance. *Scientific American* **228**(4), 19–27.

Cohen, S. N. and Shapiro, J. A. (1980). Transposable genetic elements. *Scientific American* **242**(2), 36–45.

Cole, R. K. (1968). Studies on genetic resistance to Marek's disease. *Avian Diseases* **12**, 9–28.

Cole, R. K. and Hutt, F. B. (1973). Selection and heterosis in Cornell White Leghorns: a review, with special consideration of interstrain hybrids. *Animal Breeding Abstracts* **41**, 103–18.

Cole, R. K., Kite, J. H., and Witebsky, E. (1968). Hereditary autoimmune thyroiditis in the fowl. *Science* **160**, 1357–8.

Comings, D. E. and Okada, T. A. (1971). Triple chromosome pairing in triploid chickens. *Nature* **231**, 119–21.

Cork, L. C., Munnell, J. F., Lorenz, M. D., Murphy, J. V., Baker, H. J., and Rattazzi, M. C. (1977). GM_2-ganglioside lysosomal storage disease in cats with beta-hexosaminidase deficiency. *Science* **196**, 1014–17.

Cornelius, C. E. (1969). Animal models—a neglected medical resource. *New England Journal of Medicine* **281**, 934–44.

Cotter, S. M., Brenner, R. M., and Dodds, W. J. (1978). Hemophilia A in three unrelated cats. *Journal of the American Veterinary Medical Association* **172**, 166–8.

Crick, F. H. C. (1963). On the genetic code. *Science* **139**, 461–4.

Crick, F. H. C. (1979). Split genes and RNA splicing. *Science* **204**, 264–71.

Cundiff, L. V., Gregory, K. E., and Koch, R. M. (1982a). Selection for increased survival from birth to weaning. *Proceedings of the Second World Congress on Genetics Applied to Livestock Production* **5**, 310–37.

Cundiff, L. V., Gregory, K. E., and Koch, R. M. (1982b). Effects of heterosis in Hereford, Angus and Shorthorn rotational crosses. In *Beef research program progress report No. 1* (ARM-NC-21) (Ed. Anon.) pp. 3–5. Roman L. Hruska US Meat Animal Research Center, Nebraska.

Cunningham, E. P. (1969). *Animal breeding theory.* Landbruksbokhandelen, Universitforlaget Vollebekk, Oslo.

Cunningham, E. P. (1975). Multi-stage index selection. *Theoretical and Applied Genetics* **46**, 55–61.

Curnow, R. N. and Smith, C. (1975). Multifactorial models for familial diseases in man. *Journal of the Royal Statistical Society, A* **138**, 131–69.

Curtis, C. F. (1981). Chromosome manipulations for insect pest control. *Chromosomes Today* **7**, 138–47.

Dalton, D. C. (1976). *Animal breeding: first principles for farmers.* Aster Books, Wellington.

Dalton, D. C. (1980). *An introduction to practical animal breeding.* Granada, London.

Darwin, C. (1859). *The origin of species by means of natural selection* (1st edn). John Murray, London.

Dausset, J. (1981). The Major Histocompatibility Complex in man: past, present and future concepts. *Science* 213, 1469–74.

Davern, C. I. (ed.) (1981). *Genetics: readings from Scientific American*. W. H. Freeman, San Francisco.

Davie, A. M. (1979). The singles method for segregation analysis under incomplete ascertainment. *Annals of Human Genetics* 42, 507–12.

Dennis, S. M. and Leipold, H. W. (1979). Ovine congenital defects. *Veterinary Bulletin* 49, 233–9.

Devonshire, A. L. and Sawacki, R. M. (1979). Insecticide-resistant *Myzus persicae* as an example of evolution by gene duplication. *Nature* 280, 140–1.

Dezco, H. (1974). [*Heritable anatomical abnormalities and diseases of farm animals.*] Akademiai Kiado, Budapest.

Dickerson, G. E. (1962). Implications of genetic–environment interaction in animal breeding. *Animal Production* 4, 47–63.

Dickerson, G. E. (1965). Random sample performance testing of poultry in the USA. *World's Poultry Science Journal* 21, 345–57.

Dickerson, G. E. (1969). Experimental approaches to utilizing breed resources. *Animal Breeding Abstracts* 37, 191–202.

Dickerson, G. E. (1973). Inbreeding and heterosis in animals. In *Animal Breeding and Genetics* (Proceedings of a Symposium in Honor of Dr J. L. Lush). (Ed. Anon.) pp. 54–77. American Society for Animal Science, and American Dairy Science Association, Champaign, Illinois.

Dickerson, G. E. and Hazel, L. N. (1944). Effectiveness of selection on progeny performance as a supplement to earlier culling in livestock. *Journal of Agricultural Research* 69, 459–76.

Dickinson, A. G., Fraser, H., and Outram, G. W. (1975). Scrapie incubation time can exceed natural lifespan. *Nature* 256, 732–3.

Dodds, W. J. (1973). Canine factor X (Stuart-Power factor) deficiency. *Journal of Laboratory Clinical Medicine* 82, 560–6.

Dodds, W. J. (1975). Inherited hemorrhagic disorders. *Journal of the American Animal Hospital Association* 11, 366–73.

Dodds, W. J. (1977). Inherited hemorrhagic defects. In *Current veterinary therapy VI. Small animal practice* (ed. R. W. Kirk) pp. 438–45. Saunders, Philadelphia.

Dodds, W. J. (1978). Inherited bleeding disorders. *Canine Practitioner* 5, 49–58.

Dodds, W. J. (1981). Second international registry of animal models of thrombosis and hemorrhagic diseases. *ILAR News* 24(4), R1–R50.

Dodds, W. J., Moynihan, A. C., Fisher, T. M., and Trauner, D. B. (1981). The frequencies of inherited blood and eye diseases as determined by genetic screening programs. *Journal of the American Animal Hospital Association* 17, 697–704.

Dodgson, J. B., Strommer, J., and Engel, J. D. (1979). Isolation of the chicken β-globin gene and a linked embryonic beta-like globin gene from a chicken DNA recombinant library. *Cell* 17, 879–87.

Dolling, C. H. S. (1970). *Breeding merinos*. Rigby, Adelaide.

Done, J. T. (1981). Hereditary disease in cattle: detection of carrier bulls. *Veterinary Annual* 21, 100–5.

Donelson, J. E. and Turner, M. J. (1985). How the trypanosome changes its coat. *Scientific American* **252**(2), 32–9.

Donnelly, W. J. C. and Sheahan, B. J. (1981). GM_1 gangliosidosis of Friesian calves: a review. *Irish Veterinary Journal* **35**, 45–55.

Doolittle, W. F. and Sapienza, C. (1980). Selfish genes, the phenotype paradigm and genomic evolution. *Nature* **284**, 601–3.

Dougan, G., Dowd, G., and Kehoe, M. (1983). Organisation of K88ac-encoded polypeptides in the *Escherichia coli* cell envelope: use of minicells and outer membrane protein mutants for studying assembly of pili. *Journal of Bacteriology* **153**, 364–70.

Dover, G. and Doolittle, W. F. (1980). Modes of genome evolution. *Nature* **288**, 646–7.

Dunn, H. O., McEntee, K., Hall, C. E., Johnson, R. H. (Jr), and Stone, W. H. (1979). Cytogenetic and reproductive studies of bulls born co-twin with free-martins. *Journal of Reproduction and Fertility* **57**, 21–30.

Dunn, H. O., Johnson, R. H., and Quaas, R. L. (1981). Sample size for detection of Y-chromosomes in lymphocytes of possible freemartins. *Cornell Veterinarian* **71**, 297–304.

Dyrendahl, I. and Gustavsson, I. (1979). Sexual functions, semen characteristics and fertility of bulls carrying the 1/29 chromosome translocation. *Hereditas* **90**, 281–9.

Eldridge, F. E. (1985). *Cytogenetics of livestock*. AVI Publishing Co., Westport, Connecticut.

Eldridge, F. and Blazak, W. F. (1976). Horse, ass and mule chromosomes. *Journal of Heredity* **67**, 361–7.

Eldridge, F. and Blazak, W. F. (1977). Chromosomal analysis of fertile female heterosexual twins in cattle. *Journal of Dairy Science* **60**, 458–63.

Eldridge, F. and Suzuki, Y. (1976). A mare mule–dam or foster mother? *Journal of Heredity* **67**, 353–60.

Elston, R. C., Namboodiri, K. K., and Kaplan, E. B. (1978). Resolution of major loci for quantitative traits. In *Genetic epidemiology* (Eds Morton, N. E. and Chung, C. S.) pp. 223–53. Academic Press, New York.

Emery, A. E. H. (1976). *Methodology in medical genetics*. Churchill Livingstone, Edinburgh.

Emsley, A., Dickerson, G. E., and Kashyap, T. S. (1977). Genetic parameters in progeny-test selection for field performance of strain-cross layers. *Poultry Science* **56**, 121–46.

Entrikin, R. K., Swanson, K. L., Weidoff, P. M., Patterson, G. T., and Wilson, B. W. (1977). Avian muscular dystrophy: functional and biochemical improvement with diphenylhydantoin. *Science* **195**, 873–5.

Epstein, C. J. and Travis, B. (1979). Preimplantation lethality of monosomy for mouse chromosome 19. *Nature* **280**, 144–5.

Epstein, C. J., Tucker, G., Travis, B., and Gropp, A. (1977). Gene dosage for isocitrate dehydrogenase in mouse embryos trisomic for chromosome 1. *Nature* **267**, 615–16.

Erickson, F., Saperstein, G., and Leipold, H. W. (1977). Congenital defects of dogs. *Canine Practice* **4**(4), 54–61; (5), 51–61; (6), 40–53.

Evans, D. A. P. (1980). Genetic factors in adverse reactions to drugs and chemicals. In *Pseudo-allergic reactions. Involvement of drugs and chemicals.*

(Vol. 1) (Eds Dukor, P., Kallow, P., Schlumberger, H. D., and West, G. B.) pp. 1–27. S. Karger, Basel, Switzerland.

Evans, J. W., Borton, A., Hintz, H. F., and Van Vleck, L. D. (1977). *The horse*. W. H. Freeman, San Francisco.

Falconer, D. S. (1965). The inheritance of liability to certain diseases, estimated from the incidence among relatives. *Annals of Human Genetics* 29, 51–76.

Falconer, D. S. (1981). *Introduction to quantitative genetics* (2nd edn). Longman, London.

Falconer, D. S. (1983). *Problems in quantitative genetics*. Longman, London.

Farrell, D. F., Baker, H. J., Herndon, R. M., Lindsey, J. R., and McKhann, G. M. (1973). Feline GM_1-gangliosidosis: biochemical and ultrastructural comparisons with the disease in man. *Journal of Neuropathology and Experimental Neurology* 32, 1–18.

Farrow, B. R. H., Hartley, W. J., Pollard, A. C., Fabbro, D., Grabowski, G. A., and Desnick, R. J. (1982). Gaucher disease in the dog. In *Gaucher disease: a century of delineation and research* (Eds Desnick, R. J. and Gatt, S.) pp. 645–53. Alan R. Liss Inc., New York.

Fechheimer, N. S. (1979). Cytogenetics in animal production. *Journal of Dairy Science* 62, 844–53.

Fechheimer, N. S. (1981). Origins of heteroploidy in chicken embryos. *Poultry Science* 60, 1365–71.

Fenner, F. and Ratcliffe, F. N. (1965). *Myxomatosis*. Cambridge University Press, Cambridge.

Festenstein, H. and Démant, P. (1978). *HLA and H-2: Basic immunogenetics, biology and clinical relevance*. Edward Arnold, London.

Festing, M. F. W. (1979). *Inbred strains in biomedical research*. Macmillan, London.

Fiddes, J. C. (1977). The nucleotide sequence of a viral DNA. *Scientific American* 237(6), 54–67.

Fjolstad, M. and Helle, O. (1974). A hereditary dysplasia of collagen tissues in sheep. *Journal of Pathology* 112, 183–8.

Foley, C. W., Lasley, J. F., and Osweiler, G. D. (1979). *Abnormalities of companion animals: analysis of heritability*. Iowa State University Press, Ames, Iowa.

Ford, C. E., Pollock, D. L., and Gustavsson, I. (1980). Proceedings of the first international conference for the standardization of banded karyotypes of domestic animals. *Hereditas* 92, 145–62.

Foster, G. G. (1980). Genetic control of sheep blowfly (*Lucilia cuprina*) and the logistics of the CSIRO control programme. *Wool Technology and Sheep Breeding* 28(1), 5–10.

Foster, G. G. and Maddern, R. H. (1978). Genetic manipulation of insect pests: a possible control technique for the future. *CSIRO Central Information Service* Sheet No. 1–16.

Foster, G. G., Whitten, M. J., Prout, T., and Gill, R. (1972). Chromosome rearrangement for the control of insect pests. *Science* 176, 875–80.

Foster, T. J. and Kleckner, N. (1980). Properties of drug resistance transposons, with particular reference to Tn10. In *Plasmids and transposons* (Eds Stuttard, C. and Rozee, K. R.) pp. 207–24. Academic Press, New York.

Frankel, O. H. and Soulé, M. E. (1981). *Conservation and evolution*. Cambridge University Press, Cambridge.

Freifelder, D. (Ed.) (1978). *Recombinant DNA: readings from Scientific American.* W. H. Freeman, San Francisco.

Fretz, P. B. and Hare, W. C. D. (1976). A male pseudohermaphrodite horse with 63XO?/64XX/65XXY mixoploidy. *Equine Veterinary Journal* **8**, 130-2.

Freudiger, U., Schärer, V., Buser, J. C., and Mühlebach, R. (1973). Die Resultate der Hüftgelenksdyplasie-Bekämpfung beim D. Schäfer in der Zeit von 1965 bis 1972. *Schweizer Archiv für Tierheilkunde* **115**, 169-73.

Frisch, J. E., Nishimura, H., Cousins, K. J., and Turner, H. G. (1980). The inheritance and effect on production of polledness in four crossbred lines of beef cattle. *Animal Production* **31**, 119-26.

Gahne, B. (1980). Immunogenetics: a review and future prospects. *Livestock Production Science* **7**, 1-12.

Gardner, E. J. and Snustad, D. P. (1984). *Principles of genetics* (7th edn). John Wiley and Sons, New York.

Gardner, E. J., Shupe, J. L., Leone, N. C., and Olson, A. E. (1975). Hereditary multiple exostosis: a comparative genetic evaluation in man and horses. *Journal of Heredity* **66**, 318-22.

Gartler, S. M. and Riggs, A. D. (1983). Mammalian X-chromosome inactivation. *Annual Review of Genetics* **17**, 155-90.

Gavora, J. S., Chesnais, J., and Spencer, J. L. (1983). Estimation of variance components and heritability in populations affected by disease: lymphoid leukosis in chickens. *Theoretical and Applied Genetics* **65**, 317-22.

Gentry, P. A., Crane, S., and Lotz, F. (1975). Factor XI (plasma thromboplastin antecedent) deficiency in cattle. *Canadian Veterinary Journal* **16**, 160-3.

Gershwin, M. E. and Merchant, B. (Eds) (1981). *Immunological defects in laboratory animals.* (Vols 1 and 2). Plenum Press, New York.

Gibbons, R. A., Sellwood, R., Burrows, M., and Hunter, P. A. (1977). Inheritance of resistance to neonatal *E. coli* diarrhoea in the pig: examination of the genetic system. *Theoretical and Applied Genetics* **51**, 65-70.

Gilbert, W. (1981). DNA sequencing and gene structure. *Science* **214**, 1305-12.

Gilbert, W. and Villa-Komaroff, L. (1980). Useful proteins from recombinant bacteria. *Scientific American* **242**(4), 68-82.

Glass, R. E. (1982). *Gene function: E. coli and its heritable elements.* Croom Helm, London.

Gluhovschi, N., Bistriceanu, M., Palicica, R., Codreanu, N., Marschang, F., and Bratu, M. (1972). A contribution to the clinical and cytogenetic study of bovine dwarfism. *Veterinary Medical Review* (2), 107-15.

Gotze, D. (Ed.) (1977). *The major histocompatibility system in man and animals.* Springer-Verlag, Berlin.

Gowe, R. S. and Fairfull, R. W. (1980). Some lessons from selection studies in poultry. In *Proceedings of the World Congress on Sheep and Beef Cattle Breeding* (Vol 1: Technical) (Eds Barton, R. A. and Smith, W. C.) pp. 261-81. Dunmore Press, Palmerston North, New Zealand.

Grant, G. M. (1976). A method to predict the probabilities of homozygous recessives and of heterozygotes. *Journal of Heredity* **67**, 393-6.

Gray, A. P. (1972). *Mammalian hybrids: a check list with bibliography* (2nd edn). Commonwealth Agricultural Bureaux, Farnham Royal, Slough, England.

Greaves, J. H., Redfern, R., Ayres, P. B., and Gill, J. E. (1977). Warfarin resistance: a balanced polymorphism in the Norway rat. *Genetical Research* 30, 257–63.

Green, E. L. (1981). *Genetics and probability in animal breeding experiments.* Oxford University Press, New York.

Greene, H. J., Saperstein, G., Schalles, R., and Leipold, H. W. (1978). Internal hydrocephalus and retinal dysplasia in Shorthorn cattle. *Irish Veterinary Journal* 32, 65–9.

Gregory, I. P. (1982*a*). Genetic studies of South Australian Merino sheep. III. Heritabilities of various wool and body traits. *Australian Journal of Agricultural Research* 33, 355–62.

Gregory, I. P. (1982*b*). Genetic studies of South Australian Merino sheep. IV. Genetic, phenotypic and environmental correlations between various wool and body traits. *Australian Journal of Agricultural Research* 33, 363–73.

Gregory, K. E. and Cundiff, L. V. (1980). Crossbreeding in beef cattle: evaluation of systems. *Journal of Animal Science* 51, 1224–42.

Grobstein, C. (1977). The recombinant DNA debate. *Scientific American* 237(1), 22–33.

Gross, W. G., Siegel, P. B., Hall, R. W., Domeruth, C. H., and DuBoise, R. T. (1980). Production and persistence of antibodies in chickens to sheep erythrocytes. 2. Resistance to infectious diseases. *Poultry Science* 59, 205–10.

Guillaume, J. (1976). The dwarfing gene *dw*: its effects on anatomy, physiology, nutrition and management: its application in the poultry industry. *World's Poultry Science Journal* 32, 285–304.

Gundel, H. and Reetz, I. (1981). Exclusion probabilities obtainable by biochemical polymorphisms in dogs. *Animal Blood Groups and Biochemical Genetics* 12, 123–32.

Gustavsson, I. (1966). Chromosome abnormalities in cattle. *Nature* 211, 865–6.

Gustavsson, I. (1969). Cytogenetics, distribution and phenotypic effect of a translocation in Swedish cattle. *Hereditas* 63, 68–169.

Gustavsson, I. (1979). Distribution and effects of the 1/29 Robertsonian translocation in cattle. *Journal of Dairy Science* 62, 825–35.

Gustavsson, I. (1980*a*). Banding techniques in chromosome analysis of domestic animals. *Advances in Veterinary Science and Comparative Medicine* 24, 245–89.

Gustavsson, I. (1980*b*). Chromosome aberrations and their influence on the reproductive performance of domestic animals—a review. *Zeitschrift für Tierzüchtung und Züchtungsbiologie* 97, 176–95.

Gustavsson, I. and Hageltorn, M. (1976). Staining technique for definite identification of individual cattle chromosomes in routine analysis. *Journal of Heredity* 67, 175–8.

Gustavsson, I. and Rockborn, G. (1964). Chromosome abnormality in three cases of lymphatic leukaemia in cattle. *Nature* 203, 990.

Gustavsson, I., Hageltorn, M., and Zech, L. (1976). Recognition of the cattle chromosomes by the Q- and G-banding techniques. *Hereditas* 82, 157–66.

Guy, D. R. and Smith, C. (1981). Derivation of improvement lags in a livestock industry. *Animal Production* 32, 333–6.

Guyre, P. M., Zsigray, R. M., Collins, W. M., and Dunlop, W. R. (1982). Major histocompatibility complex (B): effect on the response of chickens to a second challenge with Rous sarcoma virus. *Poultry Science* 61, 829–34.

Hageltorn, M. and Gustavsson, I. (1981). XXY-trisomy identified by banding techniques in a male tortoiseshell cat. *Journal of Heredity* 72, 132–4.

Hageltorn, M., Gustavsson, I., and Zech, L. (1973). The Q- and G-banding patterns of a t(11p+; 15q−) in the domestic pig. *Hereditas* 75, 147–51.

Hala, K., Boyd, R., and Wick, G. (1981). Chicken major histocompatibility complex and disease. *Scandinavian Journal of Immunology* 14, 607–16.

Hammond, K. (1973). Population size, selection response and variation in quantitative inheritance. Unpublished Ph.D. thesis. University of Sydney, Sydney.

Hancock, J. L. (1959). Polyspermy of pig ova. *Animal Production* 1, 103–6.

Hansen, K. M. (1973a). Q-band karyotype of the goat (*Capra hircus*) and the relation between goat and bovine Q-bands. *Hereditas* 75, 119–30.

Hansen, K. M. (1973b). The karyotype of the domestic sheep (*Ovis aries*) identified by quinacrine mustard staining and fluorescence microscopy. *Hereditas* 75, 233–40.

Hanset, R. and Ansay, M. (1967). Dermatosparaxie (peau dechiree) chez le veau un degaut general du tissu conjonctef, de nature hereditare. *Annales de Médecine Vétérinaire* 111, 451–70.

Hansmann, I., Zmarsly, R., Probeck, H. D., Schäfer, J., and Jenderny, J. (1979). Aneuploidy in mouse fetuses after paternal exposure to X rays. *Nature* 280, 228–9.

Hare, W. C. D. and Singh, E. L. (1979). *Cytogenetics in animal reproduction.* Commonwealth Agricultural Bureaux, Farnham Royal, Slough, England.

Hare, W. C. D., Singh, E. L., Betteridge, K. J., Eaglesome, M. D., Randall, G. C. B., Mitchell, D., Bilton, R. J., and Trounson, A. O. (1980). Chromosomal analysis of 159 bovine embryos collected 12 to 18 days after estrus. *Canadian Journal of Genetics and Cytology* 22, 615–26.

Harris, H. (1980). *The principles of human biochemical genetics* (3rd edn). North Holland, Amsterdam.

Hartl, D. L. (1980). *Principles of population genetics.* Sinauer, Sunderland, Massachusetts.

Hartl, D. L. (1981). *A primer of population genetics.* Sinauer, Sunderland, Massachusetts.

Hartley, W. J., Barker, J. S. F., Wanner, R. A., and Farrow, B. R. H. (1978). Inherited cerebellar degeneration in the rough coated Collie. *Australian Veterinary Practitioner* 8, 79–85.

Hartmann, W. (1974). Random sample poultry tests; underlying principles, achievements and future prospects. *World Animal Review* (11), 44–9.

Haskins, M. E., Jezyk, P. F., and Patterson, D. F. (1979a). Mucopolysaccharide storage disease in three families of cats with arylsulfatase B deficiency: leukocyte studies and carrier identification. *Pediatric Research* 13, 1203–10.

Haskins, M. E., Jezyk, P. F., Desnick, R. J., McDonough, S. K., and Patterson, D. F. (1979b). Alpha-L-iduronidase deficiency in a cat: a model of mucopolysaccharidosis I. *Pediatric Research* 13, 1294–7.

Hayes, J. F. and Hill, W. G. (1980). A reparameterisation of a genetic selection index to locate its sampling properties. *Biometrics* 36, 237–48.

Hayes, J. F. and Hill, W. G. (1981). Modifications of estimates of parameters in the construction of genetic selection indices ('Bending'). *Biometrics* 37, 483–93.

Hayman, R. H. (1974). The development of the Australian Milking Zebu. *World Animal Review* (11), 31–5.

Hazel, L. N. (1943). The genetic basis for constructing selection indices. *Genetics* 28, 476–90.

Healy, P. J., Seaman, J. T., Gardner, I. A., and Sewell, C. A. (1981). β-mannosidase deficiency in Anglo-Nubian goats. *Australian Veterinary Journal* 57, 504–7.

Healy, P. J., Farrow, B. R. H., Nicholas, F. W., Hedberg, K., and Ratcliffe, R. (1984). Canine fucosidosis: a biochemical and genetic investigation. *Research in Veterinary Science* 36, 354–9.

Hedhammer, A., Olsson, S. E., Andersson, S. A., Persson, L., Pettersson, L., Olausson, A., and Sundgren, P. E. (1979). Canine hip displasia: study of heritability in 401 litters of German Shepherd dogs. *Journal of the American Veterinary Medical Association* 174, 1012–16.

Hegreberg, G. A., Padgett, G. A., Gorham, J. R., and Henson, J. B. (1969). A connective tissue disease of dogs and mink resembling the Ehlers–Danols syndrome of man. II. Mode of inheritance. *Journal of Heredity* 60, 249–54.

Henderson, C. R. (1949). Estimation of changes in herd environment. *Journal of Dairy Science* 32, 706.

Henderson, C. R. (1973). Sire evaluation and genetic trends. In *Animal Breeding and Genetics*. (Proceedings of a Symposium in Honor of Dr J. L. Lush.) (Ed. Anon.) pp. 10–41. American Society of Animal Science, and American Dairy Science Association, Champaign, Illinois.

Henderson, C. R. (1974). General flexibility of linear model techniques for sire evaluation. *Journal of Dairy Science* 57, 963–72.

Henderson, C. R. (1976). A simple method for computing the inverse of a numerator matrix used in prediction of breeding values. *Biometrics* 32, 69–84.

Henricson, B. and Bäckström, L. (1964). Translocation heterozygosity in a boar. *Hereditas* 52, 166–70.

Henson, J. B. and Noel, J. C. (1979). Immunology and pathogenesis of African animal trypanosomiasis. *Advances in Veterinary Science and Comparative Medicine* 23, 161–82.

Hers, H. G. (1965). Inborn lysosomal diseases. *Gastroenterology* 48, 625–33.

Herschler, M. S. and Fechheimer, N. S. (1968). The role of sex chromosome chimerism in altering sexual development of mammals. *Cytogenetics* 6, 204–12.

Heuertz, S. and Hors-Cayla, M. C. (1981). Cattle gene mapping by somatic cell hybridization study of 17 enzyme markers. *Cytogenetics and Cell Genetics* 30, 137–45.

Hill, W. G. (1971). Investment appraisal for national breeding programmes. *Animal Production* 13, 37–50.

Hill, W. G. (1974). Size of experiments for breed or strain comparisons. In *Proceedings of the Working Symposium on Breed Evaluation and Crossing Experiments with Farm Animals* (Ed. Anon.) pp. 43–54. Research Institute for Animal Husbandry, 'Schoonoord', Zeist, The Netherlands.

Hill, W. G. (1978). How reliable is CPE? *Pig Farming* 26(2), 40–3.

Hill, W. G. (1982). Genetic improvement of reproductive performance in pigs. *Pig News and Information* 3, 137–41.

Hines, H. C., Zikakis, J. P., Haenlein, G. F. W., Kiddy, C. A., and Trowbridge, C. L. (1981). Linkage relationships among loci of polymorphisms in blood and milk of cattle. *Journal of Dairy Science* 64, 71–6.

Hinks, C. J. M. (1983). *Breeding dairy cattle*. Farming Press, Ipswich.

Hintz, R. L. (1980). Genetics of performance in the horse. *Journal of Animal Science* 51, 582–94.

Hoare, M., Davies, D. C., and Pattison, I. H. (1977). Experimental production of scrapie-resistant Swaledale sheep. *Veterinary Record* 101, 482–4.

Hopwood, D. A. (1981). The genetic programming of industrial microorganisms. *Scientific American* 245(3), 66–78.

Howell, J. McC., Dorling, P. R., Cook, R. D., Robinson, W. F., Bradley, S., and Gawthorne, J. M. (1981). Infantile and late onset form of generalised glycogenesis type II in cattle. *Journal of Pathology* 134, 266–77.

Hoy, M. A. and McKelvey, J. J. (Jr) (Eds) (1979). *Genetics in relation to insect management*. The Rockefeller Foundation, New York.

Hsu, T. C. and Benirschke, K. (1967 *et seq.*). *An atlas of mammalian chromosomes*. (Volume I *et seq.*) Springer-Verlag, New York.

Huston, R., Saperstein, G., and Leipold, H. W. (1977). Congenital defects in foals. *Journal of Equine Medicine and Surgery* 1, 146–62.

Huston, R., Saperstein, G., Schoneweis, D., and Leipold, H. W. (1978). Congenital defects in pigs. *Veterinary Bulletin* 48, 645–75.

Hutt, F. B. (1979). *Genetics for dog breeders*. W. H. Freeman, San Francisco.

Hutt, F. B. and Rasmusen, B. A. (1982). *Animal genetics*. John Wiley and Sons, New York.

Inouye, E. and Nishimura, H. (Eds) (1977). *Gene-environment interactions in common diseases*. University Park Press, Baltimore.

Irvin, A. D. (1976). Techniques and applications of cell fusion and hybridisation: an introduction. *Veterinary Record* 98, 351–6.

Jacobs, P. A. (1972). Chromosome mutations: frequency at birth in humans. *Humangenetik* 16, 137–40.

Jain, H. K. (1980). Incidental DNA. *Nature* 288, 647–8.

James, J. W. (1977). Open nucleus breeding schemes. *Animal Production* 24, 287–305.

James, J. W. (1979). Selection theory with overlapping generations. *Livestock Production Science* 6, 215–22.

James, J. W. (1982). Construction, uses, and problems of multitrait selection indices. *Proceedings of the Second World Congress on Genetics Applied to Livestock Production* 5, 130–9.

Jasiorowski, H., Reklewski, Z., and Stolzman, M. (1983). Testing of different strains of Friesian cattle in Poland. I. Milk performance of F1 paternal Friesian strain crosses under intensive feeding conditions. *Livestock Production Science* 10, 109–22.

Johansson, I. and Rendel, J. (1968). *Genetics and animal breeding*. W. H. Freeman, San Francisco.

John, M. E. and John, M. (1977). A new hemoglobin β chain variant in sheep. *Animal Blood Groups and Biochemical Genetics* 8, 183–90.

Johnson, J. L., Leipold, H. W., Snider, G. W., and Baker, R. D. (1980). Progeny testing for bovine syndactyly. *Journal of the American Veterinary Medical Association* 176, 549–50.

Johnson, K. H. (1970). Globoid leukodystrophy in the cat. *Journal of the American Veterinary Medical Association* 157, 2057–64.

Jolly, R. D. (1975). Mannosidosis of Angus cattle: a prototype control program for some genetic diseases. *Advances in Veterinary Science and Comparative Medicine* 19, 1–21.

Jolly, R. D. (1977*a*). Lysosomal storage disease of animals. *Records of the Adelaide Childrens Hospital* 1, 346–53.

Jolly, R. D. (1977*b*). The founder effect and genetic disease of cattle. *New Zealand Veterinary Journal* 25, 109–10.

Jolly, R. D. (1978). Mannosidosis and its control in Angus and Murray Grey cattle. *New Zealand Veterinary Journal* 26, 194–8.

Jolly, R. D. and Blakemore, W. F. (1973). Inherited lysosomal storage diseases: an essay in comparative medicine. *Veterinary Record* 92, 391–400.

Jolly, R. D. and Hartley, W. J. (1977). Storage diseases of domestic animals. *Australian Veterinary Journal* 53, 1–8.

Jolly, R. D. and Townsley, R. J. (1980). Genetic screening programmes: an analysis of benefits and costs using the bovine mannosidosis scheme as a model. *New Zealand Veterinary Journal* 28, 3–6.

Jolly, R. D., Thompson, K. G., and Tse, C. A. (1974). Evaluation of a screening programme for identification of mannosidosis heterozygotes in Angus cattle. *New Zealand Veterinary Journal* 22, 185–90.

Jolly, R. D., Dodds, W. J., Ruth, G. R., and Trauner, D. B. (1981). Screening for genetic diseases: principles and practice. *Advances in Veterinary Science and Comparative Medicine* 25, 245–76.

Jolly, R. D., Thompson, K. G., Murphy, C. E., Manktelow, B. W., Bruère, A. N., and Winchester, B. G. (1976). Enzyme replacement therapy—an experiment of nature in a chimeric mannosidosis calf. *Pediatric Research* 10, 219–24.

Jones, G. W. and Rutter, J. M. (1972). Role of the K88 antigen in the pathogenesis of neonatal diarrhoea caused by *Escherichia coli* in piglets. *Infection and Immunity* 6, 918–27.

Jones, M. Z. and Dawson, G. (1981). Caprine β-mannosidosis: inherited deficiency of β-D-mannosidase. *Journal of Biological Chemistry* 256, 5185–8.

Jones, T. C. (1978). Hereditary disease. In *Pathology of laboratory animals*. (Vol. II) (Eds Benirschke, K., Garner, F. M., and Jones, T. C.) pp. 1981–2064. Springer-Verlag, Berlin.

Jones, T. C., Capen, C. C., Hackel, D. B., and Migaki, G.M. (Eds) (1972 *et seq.*). *Handbook: animal models of human disease*. Registry of Comparative Pathology, Armed Forces Institute of Pathology, Washington, DC.

Jones, W. E. (1982). *Genetics and horse breeding*. Lea and Febiger, Philadelphia.

Jonker, M. and Balner, H. (1980). A review: current knowledge of the D/DR region of the major histocompatibility complex of rhesus monkeys and chimpanzees. *Human Immunology* 1, 305–16.

Kammermann, B., Gmür, J., and Stünzi, H. (1971). [Afibrinogenaemia in the dog.] *Zentralblatt für Veterinärmedizin, Reihe A* 18, 192–205.

Kästli, F. and Hall, J. G. (1978). Cattle twins and freemartin diagnosis. *Veterinary Record* 102, 80–3.

Katpatal, B. G. (1977*a*). Dairy cattle crossbreeding in India. 1. Growth and development of crossbreeding. *World Animal Review* (22), 27–33.

Katpatal, B. G. (1977*b*). Dairy cattle crossbreeding in India. 2. The results of the All India Coordinate Research Project on cattle. *World Animal Review* (23), 2–9.

Kehoe, M., Sellwood, R., Shipley, R., and Dougan, G. (1981). Genetic analysis of K88-mediated adhesion of enterotoxigenic *Escherichia coli*. *Nature* 291, 122–6.

Kidwell, J. F. (1970). On the use of *a priori* probability in progeny tests for recessive alleles. *Journal of Heredity* 61, 55–8.

Kidwell, J. F. and Hagy, G. W. (1973). On the optimum structure of progeny tests for recessive alleles. *Theoretical and Applied Genetics* 43, 35–8.

Kier, A. B., Bresnahan, J. F., White, F.J., and Wagner, J. E. (1980). The inheritance pattern of factor XII (Hageman) deficiency in domestic cats. *Canadian Journal of Comparative Medicine* 44, 309–14.

Kimberlin, R. H. (1979a). Aetiology and genetic control of natural scrapie. *Nature* 278, 303–4.

Kimberlin, R. H. (1979b). An assessment of genetical methods in the control of scrapie. *Livestock Production Science* 6, 233–42.

Kimberlin, R. H. (1982). Reflections on the nature of scrapie agent. *Trends in Biochemical Sciences* 7, 392–4.

King, J. W. B., Curran, M. K., Standal, N., Power, P., Heaney, I. H., Kallweit, E., Schröder, J., Maijala, K., Kangasniemi, R., and Walstra, P. (1975). An international comparison of pig breeds using a common control stock. *Livestock Production Science* 2, 367–79.

King, R. C. (Ed.) (1975). *Handbook of genetics. Vol. 4. Vertebrates of genetic interest*. Plenum, New York.

King, W. A., Linares, T., and Gustavsson, I. (1981). Cytogenetics of preimplantation embryos sired by bulls heterozygous for the 1/29 translocation. *Hereditas* 94, 219–24.

Kleid, D. G., Yansura, D., Small, B., Dowbenko, D., Moore, D. M., Grubman, M. J., McKercher, P. D., Morgan, D. O., Robertson, B. H., and Bachrach, H. L. (1981). Cloned viral protein vaccine for foot and mouth disease: responses in cattle and swine. *Science* 214, 1125–9.

Koch, R. M., Gregory, K. E., and Cundiff, L. V. (1982). Critical analysis of selection methods and experiments in beef cattle and consequences upon selection programs applied. *Proceedings of the Second World Congress on Genetics Applied to Livestock Production* 5, 514–26.

Kopp, E., Mayr, B., Czaker, R., and Schleger, W. (1981). Nucleolus organizer regions in the chromosomes of the domestic horse. *Journal of Heredity* 72, 357–8.

Küpper, H., Keller, W., Kurz, C., Forss, S., Schaller, H., Franze, R., Strohmaier, K., Marguardt, O., Zaslavsky, G., and Hofschneider, P. H. (1981). Cloning of cDNA of major antigen of foot and mouth disease virus and expression in *E. coli*. *Nature* 289, 555–9.

La Chance, L. E. (1979). Genetic strategies affecting the success and economy of the sterile insect release method. In *Genetics in relation to insect management* (Eds Hoy, M. A. and McKelvey, J. J. (Jr)) pp. 8–18. The Rockefeller Institute, New York.

Lalouel, J. M. (1978). Recurrence risks as an outcome of segregation analysis. In *Genetic epidemiology* (Eds Morton, N. E. and Chung, C. S.) pp. 255–84. Academic Press, New York.

Lamb, R. C., Arave, C. W., and Shupe, J. L. (1976). Inheritance of limber legs in Jersey cattle. *Journal of Heredity* 67, 241–4.

Lamberson, W. R. and Thomas, D. L. (1984). Effects of inbreeding in sheep: a review. *Animal Breeding Abstracts* 52, 287–97.

Land, R. B. and Hill, W. G. (1975). The possible use of superovulation and embryo transfer in cattle to increase response to selection. *Animal Production* 21, 1–12.

Land, R. B. and Robinson, D. W. (Eds) (1984). *The genetics of reproduction in sheep*. Butterworths, London.

Landauer, W., Clark, E. M., and Larner, M. M. (1976). Cholinomimetic teratogens. IV. Effects of the genotype for muscular dystrophy in chickens. *Teratology* 14, 281-6.

Landsteiner, K. (1900). Zür Kenntnis der antifermentativen, lytischen und agglutinierenden Wirkungen des Blutserums und der Lymphe. *Zetralblatt für Bakteriologie, Parasitenkunde, Infektionskrankbeiten und Hygiene* 27, 357-62.

Lasley, J. F. (1970). *Genetic principles in horse breeding*. University of Missouri Press, Columbia.

Lasley, J. F. (1978). *Genetics of livestock improvement* (3rd edn). Prentice-Hall, Englewood Cliffs, New Jersey.

Lauvergne, J. J. (1968). Catalogue des anomalies hereditaires des bovins. *Bulletin Technique du Département de Génétique Animale, Number 1*, INRA, France.

Lauvergne, J. J. (1977). The current status of cattle breeds in Europe. *World Animal Review* (21), 42-7.

Lauvergne, J. J. (1978). [Hereditary and non-hereditary congenital abnormalities in Normandy cattle.] *Annales de Génétique et de Sélection animale* 10, 131-4.

Leder, P. (1982). The genetics of antibody diversity. *Scientific American* 246(5), 72-83.

Leipold, H. W. (1977). Nature and causes of congenital defects of dogs. *Veterinary Clinics of North America* 8, 47-77.

Leipold, H. W. and Peeples, J. G. (1981). Progeny testing for bovine syndactyly. *Journal of the American Veterinary Medical Association* 179, 69-70.

Leipold, H. W., Dennis, S. M., and Huston, K. (1972). Congenital defects of cattle: nature, cause and effect. *Advances in Veterinary Science and Comparative Medicine* 16, 103-50.

Leipold, H. W., Dennis, S. M., and Huston, K. (1973). Syndactyly in cattle. *Veterinary Bulletin* 43, 399-403.

Leipold, H. W., Huston, K., and Dennis, S. M. (1983). Bovine congenital defects. *Advances in Veterinary Science and Comparative Medicine* 27, 197-271.

Le Jambre, L. F. (1978). Host genetic factors in helminth control. In *The epidemiology and control of gastrointestinal parasites of sheep in Australia* (Eds Donald, A. D., Southcott, W. H., and Dineen, J. K.) pp. 137-41. C.S.I.R.O., Melbourne.

Le Jambre, L. F. and Royal, W. M. (1980). Meiotic abnormalities in backcross lines of hybrid *Haemonchus*. *International Journal for Parasitology* 10, 281-6.

Le Jambre, L. F., Southcott, W. H., and Dash, K. M. (1976). Resistance of selected lines of *Haemonchus contortus* to thiabendazole, morantel tartrate, and levamisole. *International Journal for Parasitology* 6, 217-22.

Lerner, I. (1958). *The genetic basis of selection*. John Wiley and Sons, New York.

Lerner, I. and Donald, H. P. (1966). *Modern developments in animal breeding*. Academic Press, London.

Levin, E.Y. (1974). Comparative aspects of porphyria in man and animals. *Annals of the New York Academy of Science* 241, 347-59.

Lewin, B. (1980). *Gene expression Vol. 2. Eukaryotic chromosomes* (2nd edn). John Wiley and Sons, New York.

Lewis, J. W. (1981). On the coevolution of pathogen and host. I. General theory of discrete time coevolution. *Journal of Theoretical Biology* 93, 927–51.

Lillie, F. R. (1916). The theory of the free-martin. *Science* 43, 611–13.

Little, C. C. (1957). *The inheritance of coat colour in dogs.* Cornell University Press (reprinted by Howell Book House), Ithaca.

Ločnišcar, F., Gustavsson, I., Hageltorn, M., and Zech, L. (1976). Cytological origin and points of exchange of a reciprocal chromosome translocation (1p—; 6q+) in the domestic pig. *Hereditas* 83, 272–5.

Long, C. R. and Gregory, K. E. (1978). Inheritance of the horned, scurred, and polled condition in cattle. *Journal of Heredity* 69, 395–400.

Long, S. E. (1977). Cytogenetic examination of pre-implantation blastocysts of ewes mated to rams heterozygous for the Massey I (t_1) translocation. *Cytogenetics and Cell Genetics* 18, 82–9.

Long, S. E. (1985). Centric fusion translocations in cattle: a review. *Veterinary Record* 116, 516–18.

Long, S. E. and David, J. S. E. (1981). Testicular feminisation in an Ayrshire cow. *Veterinary Record* 109, 116–18.

Long, S. E. and Williams, C. V. (1980). Frequency of chromosomal abnormalities in early embryos of the domestic sheep (*Ovis aries*). *Journal of Reproduction and Fertility* 58, 197–201.

Long, S. E., Gruffydd-Jones, T., and David, M. (1981). Male tortoiseshell cats: an examination of testicular histology and chromosome complement. *Research in Veterinary Science* 30, 274–80.

Longenecker, B. M., and Mosmann, T. R. (1981). Structure and properties of the Major Histocompatibility Complex of the chicken. *Immunogenetics* 13, 1–24.

Loughman, W. D. and Frye, F. L. (1974). XY/XYY bone marrow karyotype in a male Siamese-crossbred cat. *Veterinary Medicine and Small Animal Clinician* 69, 1007–11.

Luft, B. (1972). Deletion of one arm of an acrocentric autosome of two ewes and two rams with brachygnathia superior. *Proceedings of the Seventh International Congress of Animal Reproduction and A.I., Munich* 2, 1123–7.

Lush, J. L. (1945). *Animal breeding plans* (3rd edn). Iowa State University Press, Ames, Iowa.

Lust, G., Farrell, P. W., Sheffy, B. E., and Van Vleck, L. D. (1978). An improved procedure for genetic selection against hip dysplasia in dogs. *Cornell Veterinarian* 68, 41–7.

Lyon, M. F. (1961). Gene action in the X-chromosome of the mouse (*Mus musculus*). *Nature* 190, 372–3.

McFeely, R. A. (1967). Chromosome abnormalities in early embryos of the pig. *Journal of Reproduction and Fertility* 13, 579–81.

McKay, W. M. (1975). The use of antibiotics in animal feeds in the United Kingdom; the impact and importance of legislative controls. *World's Poultry Science Journal* 31, 116–28.

McPhee, C. P., Brennan, P. J., and Duncalfe, F. (1979). Genetic and phenotypic parameters of Australian Large White and Landrace boars performance-tested when offered food *ad libitum*. *Animal Production* 28, 79–85.

Macera, M. J. and Bloom, S. E. (1981). Ultrastructural studies of the nucleoli in diploid and trisomic chickens. *Journal of Heredity* 72, 249–52.

Maciejowski, J. and Zieba, J. (1983). *Genetics and animal breeding*. Elsevier, Amsterdam.

Madalena, F. E. (1977). Crossbreeding systems for beef production in Latin America. *World Animal Review* (27), 27–33.

Maijala, K. and Hanna, M. (1974). Reliable phenotypic and genetic parameters in dairy cattle. *Proceedings of the First World Congress on Genetics Applied to Livestock Production* 1, 541–63.

Malik, V. S. (1981). Recombinant DNA technology. *Advances in Applied Microbiology* 27, 1–84.

Manktelow, B. W. and Hartley, W. J. (1975). Generalised glycogen storage disease in sheep. *Journal of Comparative Pathology* 85, 139–45.

Marcum, J. B. (1974). The freemartin syndrome. *Animal Breeding Abstracts* 42, 227–42.

Marangu, J. P. and Nordskog, A. W. (1974). The allograft reaction as an index of genetic diversity in inbred chickens. *Theoretical and Applied Genetics* 45, 215–21.

Matthey, R. (1945). L'évolution de la formule chromosomiale chez les vertébrés. *Experientia* 1, 50–6, 78–86.

Mayr, B., Krutzler, H., Auer, H., and Schleger, W. (1983). Reciprocal translocation 60,XY, t(8; 15)(21; 24) in cattle. *Journal of Reproduction and Fertility* 69, 629–30.

Maxam, A. M. and Gilbert, W. (1977). A new method for sequencing DNA. *Proceedings of the National Academy of Sciences* 74, 560–4.

Messel, H. (Ed.) (1981). *The biological manipulation of life*. Pergamon, Sydney.

Mikami, H. and Fredeen, H. T. (1979). A genetic study of cryptorchidism and scrotal hernia in pigs. *Canadian Journal of Genetics and Cytology* 21, 9-19.

Miller, R. H. and Pearson, R. E. (1979). Economic aspects of selection. *Animal Breeding Abstracts* 47, 281–90.

Mitchell, D. (1977). Sexing of embryos. In *Embryo transfer in farm animals*. (Monograph 16) (Ed. Betteridge, K. J.) pp. 26–7. Canada Department of Agriculture, Ottowa.

Mitchell, G., Smith, C., Makower, M., and Bird, P. J. W. N. (1982). An economic appraisal of pig improvement in Great Britain. I. Genetic and production aspects. *Animal Production* 35, 215–24.

Moav, R. (1966a). Specialised sire and dam lines. I. Economic evaluation of crossbreds. *Animal Production* 8, 193–202.

Moav, R. (1966b). Specialised sire and dam lines. II. The choice of the most profitable parental combinations when component traits are genetically additive. *Animal Production* 8, 203–11.

Moav, R. (1966c). Specialised sire and dam lines. III. The choice of the most profitable parental combination when component traits are genetically non-additive. *Animal Production* 8, 365–74.

Moav, R. and Hill, W. G. (1966). Specialised sire and dam lines. IV. Selection within lines. *Animal Production* 8, 375–90.

Moore, N. W., Halnan, C. R. E., McDowell, G. H., and Martin, I. C. A. (1980). Hybridization between a Barbary ram (*Ammotragus lervia*) and goat does (*Capra hircus*). *Proceedings of the Twelfth Conference of the Australian Society for Reproductive Biology*, 44.

Morton, N. E. and Chung, C. S. (Eds) (1978). *Genetic epidemiology.* Academic Press, New York.

Mostafa, I. E. (1970). A case of glycogenic cardiomegaly in a dog. *Acta Veterinaria Scandinavica* 11, 197–208.

Motulsky, A. G. (1957). Drug reactions, enzymes and biochemical genetics. *Journal of the American Medical Association* 165, 835–7.

Mulvihill, J. J. (1972). Congenital and genetic disease in domestic animals. *Science* 176, 132–7.

Murray, M., Morrison, W. I., Murray, P. K., Clifford, D. J., and Trail, J. C. M. (1979). Trypanotolerance: a review. *World Animal Review* (31), 2–12.

Nathans, D. (1979). Restriction endonucleases, Simian virus 40, and the new genetics. *Science* 206, 903–9.

Newcombe, H. B. (1964). Untitled remarks in panel discussion. In *Papers and Discussions of the Second International Conference on Congenital Malformations* (Ed. Fishbein, M.) pp. 345–9. International Medical Congress, New York.

Newman, M. J. and Antczak, D. F. (1983). Histocompatibility polymorphisms of domestic animals. *Advances in Veterinary Science and Comparative Medicine* 27, 2–76.

Nicholas, F. W. (1975). Hip dysplasia: the results of an analysis of New South Wales Labradors. *Royal Agricultural Society Kennel Control Journal* 18(11), 15–18.

Nicholas, F. W. (1980). Size of population required for artificial selection. *Genetical Research* 35, 85–105.

Nicholas, F. W. (1984). Simple segregation analysis: a review of the methodology. *Animal Breeding Abstracts* 52, 555–62.

Nicholas, F. W. and Smith, C. (1983). Increased rates of genetic change in dairy cattle by embryo transfer and splitting. *Animal Production* 36, 341–53.

Nicholas, F. W., Muir, P., and Toll, G. L. (1980). An XXY male Burmese cat. *Journal of Heredity* 71, 52–4.

Nitter, G. (1978). Breed utilization for meat production in sheep. *Animal Breeding Abstracts* 46, 131–43.

Nordskog, A. W. (1983). Immunogenetics as an aid to selection for disease resistance in the fowl. *World's Poultry Science Journal* 39, 199–209.

Nordskog, A. W., Rishell, W. A., and Briggs, D. M. (1973). Influence of B locus blood groups on adult mortality and egg production in White Leghorn chickens. *Genetics* 75, 181–9.

Nordskog, A. W., Pevzner, I. Y., Trowbridge, W. L., and Benedict, A. A. (1977). Immune response and adult mortality associated with the B locus in chickens. In *Avian Immunology* (Ed. Benedict, A. A.) pp. 245–56. Plenum Press, New York.

O'Brien, S. J. (Ed.) (1984). *Genetic Maps* (Vol. 3). Cold Spring Harbor Laboratory, New York.

O'Brien, S. J. and Nash, W. G. (1982). Genetic mapping in mammals: chromosome map of domestic cat. *Science* 216, 257–65.

Ohno, S. (1967). *Sex chromosomes and sex-linked genes.* Springer-Verlag, Berlin.

Ohno, S. (1968). Veterinary medical cytogenetics. *Advances in Veterinary Science and Comparative Medicine* 12, 1–31.

Ohno, S. (1973). Ancient linkage groups and frozen accidents. *Nature* 244, 259–62.

Ohno, S. (1979). *Major sex-determining genes*. Springer-Verlag, Berlin.

Ohno, S. and Gropp, A. (1965). Embryological basis for germ cell chimerism in mammals. *Cytogenetics* 4, 251–61.

Ohno, S., Christian, L. D., Wachtel, S. S., and Koo, G. C. (1976). Hormone-like role of H-Y antigen in bovine freemartin gonad. *Nature* 261, 597–99.

Old, R. W. and Primrose, S. B. (1985). *Principles of gene manipulation* (3rd edn). Blackwell Scientific Publications, Oxford.

Ollivier, L. (1974). Optimum replacement rates in animal breeding. *Animal Production* 19, 257–71.

Ollivier, L. (1981). *Éléments de génétique quantitative*. Masson, Paris.

Olney, P. J. S. (Ed.) (1977). *1977 international zoo yearbook* (Vol. 17). Zoological Society of London, London.

Orgel, L. E. and Crick, F. H. C. (1980). Selfish DNA: the ultimate parasite. *Nature* 284, 604–7.

Orgel, L. E., Crick, F. H. C., and Sapienza, C. (1980). Selfish DNA. *Nature* 288, 645–6.

Paatsama, S. (1979). Hip dysplasia in dogs: a controlled restrictive breeding programme in Finland. *Twenty-first World Veterinary Congress Summaries, Moscow* 4, 42–3.

Padgett, G. A. (1968). The Chediak–Higashi syndrome. *Advances in Veterinary Science* 12, 239–84.

Padgett, G. A. (1979). Chediak–Higashi syndrome. In *Spontaneous animal models of human disease* (Vol. 1) (Eds Andrews, E. J., Ward, B. C., and Altman, N. H.) pp. 256–9. Academic Press, New York.

Parry, H. B. (1979). Elimination of natural scrapie in sheep by sire genotype selection. *Nature* 277, 127–9.

Partridge, G. G. (1979). Relative fitness of genotypes in a population of *Rattus norvegicus* polymorphic for warfarin resistance. *Heredity* 43, 239–246.

Patterson, D. F. (1974). Pathological and genetic studies of congenital heart disease in the dog. *Advances in Cardiology* 13, 210–49.

Patterson, D. F. (1975). Diseases due to single mutant genes. *Journal of the American Animal Hospital Association* 11, 327–41.

Patterson, D. F. (1976). Congenital defects of the cardiovascular system of dogs: studies in comparative cardiology. *Advances in Veterinary Science and Comparative Medicine* 20, 1–37.

Patterson, D. F. (1977). A catalogue of genetic disorders of the dog. In *Current veterinary therapy. VI. Small animal practice* (Ed. Kirk, R. W.) pp. 73–88. Saunders, Philadelphia.

Patterson, D. F. (1979). [Genetics in small animal medicine. New findings on hereditary diseases of metabolism and congenital abnormalities in dogs and cats.] *Praktische Tierarzt* 60, 1061–82.

Patterson, D. F. and Minor, R. R. (1977). Hereditary fragility and hyperextensibility of the skin of cats. A defect in collagen fibrilogenesis. *Laboratory Investigation* 37, 170–9.

Patterson, D. F. and Pyle, R. L. (1971). In *Newer knowledge about dogs. Proceedings of the twenty-first Gaines veterinary symposium, Ames, Iowa* pp. 20–8. Gaines Dog Research Center, White Plains, New York.

542 *Veterinary Genetics*

Patterson, D. F., Pyle, R. L., and Buchanan, J. W. (1972). Hereditary cardiovascular malformation of the dog. *Birth Defects: Original Article series, Part XV. The Cardiovascular System* 8(5), 160–74.

Patterson, D. F., Pyle, R. L., Van Mierop, L., Melbin, J., and Olson, M. (1974). Hereditary defects of the conotruncal septum in Keeshond dogs: pathologic and genetic studies. *American Journal of Cardiology* 34, 187–205.

Pech, M., Streeck, R. E., and Zachau, H. G. (1979). Patchwork structure of a bovine satellite DNA. *Cell* 18, 883–93.

Perryman, L. E. (1979). Primary and secondary immune deficiencies of domestic animals. *Advances in Veterinary Science and Comparative Medicine* 23, 23–52.

Peters, J. A. (Ed.) (1959). *Classic papers in genetics.* Prentice-Hall, Englewood Cliffs, New Jersey.

Pevzner, I. Y., Kujdych, I., and Nordskog, A. W. (1981). Immune response and disease resistance in chickens. II. Marek's disease and immune response to GAT. *Poultry Science* 60, 927–32.

Pevzner, I. Y., Trowbridge, C. L., and Nordskog, A. W. (1978). Recombination between genes coding for immune response and the serologically determined antigens in the chicken B system. *Immunogenetics* 7, 25–33.

Pidduck, H. (1985). Is this disease inherited? A discussion paper with some guidelines for canine conditions. *Journal of Small Animal Practice* 26, 279–91.

Piper, L. R., Bindon, B. M., and Davis, G. W. (1984). The single gene inheritance of the prolificacy of the Booroola Merino. In *The genetics of reproduction in sheep* (Eds Land, R. B. and Robinson, D. W.) pp. 115–25. Butterworths, London.

Pirchner, F. (1983). *Population genetics in animal breeding* (2nd edn). Plenum, New York.

Pollock, D. L., Fitzsimons, J., Deas, W. D., and Fraser, J. A. (1979). Pregnancy termination in the control of the tibial hemimelia syndrome in Galloway cattle. *Veterinary Record* 104, 258–60.

Popescu-Vifor, St., Sarbu, I., Ciupercescu, D. D., and Grosu, M. (1980). *Genetica si eredopatologie. [Genetics and inherited pathological conditions].* Editura Didactica si Pedagogica, Bucharest, Romania.

Porter, D. D., Larsen, A. E., and Porter, H. G. (1980). Aleutian disease of mink. *Advances in Immunology* 29, 261–86.

Prasse, K. W. (1977). Pyruvate kinase deficiency anemia. In *Current veterinary therapy. VI. Small animal practice* (Ed. Kirk, R. W.) pp. 434–5. Saunders, Philadelphia.

Predojevic, M. R. (1973). [Hereditary diseases in cattle.] *Veterinarski Glasnik* 27, 49–53.

Prichard, R. K., Hall, C. A., Kelly, J. D., Martin, I. C. A., and Donald, A. D. (1980). The problem of anthelmintic resistance in nematodes. *Australian Veterinary Journal* 56, 239–51.

Propping, P. (1978). Pharmacogenetics. *Reviews of Physiology, Biochemistry and Pharmacology* 83, 124–73.

Prota, G. and Searle, A. G. (1978). Biochemical sites of gene action for melanogenesis in mammals. *Annales de Génétique et de Sélection animale* 10, 1–8.

Prusiner, S. B. (1984). Prions. *Scientific American* 251(4), 48–57.

Queinnec, G. (1975). [Genetic anomalies of swine.] *Revue de Médecine Vétérinaire* 126, 983–94.

Queinnec, G. (1977). [Diseases of genetic origin in calves.] In *Le veau* (Eds Mornet, P. and Espinasse, J.) pp. 419–65. Maloine, Paris.

Ralls, K., Brugger, K., and Ballou, J. (1979). Inbreeding and juvenile mortality in small populations of ungulates. *Science* 206, 1101–3.

Rapaport, F. T. and Bachvaroff, R. J. (1978). Experimental transplantation and histocompatibility systems in the canine species. *Advances in Veterinary Science and Comparative Medicine* 22, 195–219.

Rasmusen, B. A. (1975). Blood-group alleles of domesticated animals. In *Handbook of genetics, Vol. 4, Vertebrates of genetic interest* (Ed. King, R. C.) pp. 447–57. Plenum, New York.

Rathie, K. A. and Nicholas, F. W. (1980). Artificial selection with differing population structures. *Genetical Research* 36, 117–31.

Rattazzi, M. C. and Cohen, M. M. (1972). Further proof of genetic inactivation of the X chromosome in the female mule. *Nature* 237, 393–6.

Rendel, J. M. (1971). Myxomatosis in the Australian rabbit population. *Search* 2, 89–94.

Rendel, J. M. and Robertson, A. (1950). Estimation of genetic gain in milk yield by selection in a closed herd of dairy cattle. *Journal of Genetics* 50, 1–8.

Richardson, R. H., Ellison, J. R., and Averhoff, W. W. (1982). Autocidal control of screwworms in North America. *Science* 215, 361–70.

Riddle, C. (1975). Pathology of developmental and metabolic disorders of the skeleton of domestic chickens and turkeys. I. Abnormalities of genetic or unknown origin. *Veterinary Bulletin* 45, 629–40.

Rieger, R., Michaelis, A., and Green, M. M. (1976). *A glossary of genetics and cytogenetics* (4th edn). Springer-Verlag, New York.

Riser, W. H. (1973). Hip dysplasia in military dogs. *Journal of the American Veterinary Medical Association* 162, 664 only.

Rittman, L. S., Tennant, L. L., and O'Brien, J. S. (1980). Dog G_{M1} gangliosidosis: characterization of the residual liver acid β-galactosidase. *American Journal of Human Genetics* 32, 880–9.

Roberts, R. J. (1983). Restriction and modification enzymes and their recognition sequences. *Nucleic Acids Research* 11, r135–r167.

Robertson, A. (1953). A numerical description of breed structure. *Journal of Agricultural Science* 43, 334–6.

Robertson, A. (1965). The interpretation of genotypic ratios in domestic animal populations. *Animal Production* 7, 319–24.

Robertson, A. (1971). Optimum utilization of genetic material, with special reference to cross-breeding in relation to other methods of genetic improvement. *Proceedings of the Tenth International Congress of Animal Production*, 57–68.

Robertson, A. and Asker, A. A. (1951). The genetic history and breed structure of British Friesian cattle. *Empire Journal of Experimental Agriculture* 19, 113–30.

Robertson, A. and Rendel, J. M. (1950). The use of progeny testing with artificial insemination in dairy cattle. *Journal of Genetics* 50, 21–31.

Robertson, W. R. B. (1916). Chromosome studies. I. Taxonomic relationships shown in the chromosomes of Tettigradae and Acrididae V-shaped chromosomes and their significance in Acrididae, Locustidae and Gryllida: chromosomes and variation. *Journal of Morphology* 27, 179–331.

Robinson, R. (1977). *Genetics for cat breeders* (2nd edn). Pergamon, Oxford.

Robinson, R. (1982). *Genetics for dog breeders*. Pergamon, Oxford.

Roitt, I. M. (1984). *Essential immunology* (5th edn). Blackwell Scientific Publications, Oxford.

Rose, F. C. and Behan, P. O. (Eds) (1980). *Animal models of neurological disease*. Pitman Medical Company, Tunbridge Wells, England.

Ross, J. (1982). Myxomatosis: the natural evolution of the disease. In *Animal disease in relation to animal conservation* (ed. M. A. Edwards and U. McDonnell) pp. 77–95. Academic Press, London.

Ruddle, F. H. (1981). A new era in mammalian gene mapping: somatic cell genetics and recombinant DNA methodologies. *Nature* 294, 115–20.

Ruddle, F. H. and Kucherlapati, R. S. (1974). Hybrid cells and human genes. *Scientific American* 231(1), 36–44.

Ruth, G. R., Schwartz, S., and Stephenson, B. (1977). Bovine protoporphyria: the first nonhuman model of this hereditary photosensitizing disease. *Science* 198, 199–201.

Rutter, J. M. (1981). Gene manipulation and biotechnology. *Veterinary Record* 109, 192–4.

Ryder, D. A., Epel, N. C., and Benirschke, K. (1978). Chromosome banding studies of the Equidae. *Cytogenetics and Cell Genetics* 20, 323–50.

Ryder, M. L. (1980). Fleece colour in sheep and its inheritance. *Animal Breeding Abstracts* 48, 305–24.

Sandstrom, B., Westman, J., and Ockerman, P. A. (1969). Glycogenesis of the central nervous system in the cat. *Acta Neuropathologica* 14, 194–200.

Sanger, F. (1981). Determination of nucleotide sequences in DNA. *Science* 214, 1205–10.

Sanger, F. and Coulson, A. R. (1975). A rapid method for determining sequences in DNA by primed synthesis with DNA polymerase. *Journal of Molecular Biology* 94, 441–8.

Sanger, F., Nicklen, S., and Cohen, A. R. (1977). DNA sequencing with chain-terminating inhibitors. *Proceedings of the National Academy of Sciences* 74, 5463–7.

Saperstein, G., Harris, S., and Leipold, H. W. (1976). Congenital defects in domestic cats. *Feline Practice* 6(4), 18–27.

Saperstein, G., Leipold, H. W., and Dennis, S. M. (1975). Congenital defects of sheep. *Journal of the American Veterinary Medical Association* 167, 314–22.

Saperstein, G., Leipold, H. W., Kruckenberg, S. M., and Mickenhirn, N. A. (1977). Congenital defects of wild and zoo mammals. *ILAR News* 20(4), A1–A23.

Saunders, J. R. (1981). Human impact on microbial evolution. In *Genetic consequences of man made change* (Eds Bishop, J. A. and Cook, L. M.) pp. 249–94. Academic Press, London.

Saxton, A. M. (1982). A note on a computer program for independent culling. *Animal Production* 35, 295–7.

Schimke, R. T. (1980). Gene amplification and drug resistance. *Scientific American* 243(5), 50–9.

Schmidt, G. M. and Van Vleck, L. D. (1974). *Principles of dairy science*. W. H. Freeman, San Francisco.

Schulz-Schaeffer, J. (1980). *Cytogenetics*. Springer-Verlag, New York.

Scott, A. M. and Jeffcott, L. B. (1978). Haemolytic disease of the newborn foal. *Veterinary Record* 103, 71–4.

Searle, A. G. (1968). *Comparative genetics of coat colour in mammals*. Logos, London.

Seifert, G. W. (1971). Variations between and within breeds of cattle in resistance to field infestations of the cattle tick (*Boophilus microplus*). *Australian Journal of Agricultural Research* 22, 159–68.

Selden, J. R., Moorhead, P. S., Dehlert, M. L., and Patterson, D. F. (1975). The Giemsa banding pattern of the canine karyotype. *Cytogenetics and Cell Genetics* 15, 380–7.

Sellier, P. (1976). The basis of crossbreeding in pigs: a review. *Livestock Production Science* 3, 203–26.

Shalev, A., Short, R. V., and Hamerton, J. L. (1980). Immunogenetics of sex determination in the polled goat. *Cytogenetics and Cell Genetics* 28, 195–202.

Shanahan, G. J. and Roxburgh, N. A. (1974). The sequential development of insecticide resistance problems in *Lucilia cuprina* Wied in Australia. *Pest Articles and News Summaries* 20, 190–202.

Shepherd, J. H. (1976). The Australian Merino Society nucleus breeding scheme. In *Sheep breeding* (Eds Tomes, G. J., Robertson, D. E., and Lightfoot, R. J.) pp. 188–99. Western Australian Institute of Technology, Perth.

Sheridan, A. K. (Ed.) (1982). Nature and utilisation of heterosis. *Proceedings of the Second World Congress on Genetics Applied to Livestock Production* 6, 185–262.

Shoffner, R. N. (1981). Marker chromosomes and G-banding for location of genes in the chicken. *Poultry Science* 60, 1372–5.

Shull, R. B., Munger, R. J., Spellacy, E., Hall, C. W., Constantopoulos, G., and Neufeld, E. F. (1982). Mucopolysaccharidosis I. *American Journal of Pathology* 109, 244–8.

Siegel, P. B. and Gross, W. B. (1980). Production and persistence of antibodies in chickens to sheep erythrocytes. I. Directional selection. *Poultry Science* 59, 1–5.

Silverudd, M. (1978). Genetic basis of sexing automation in the fowl. *Acta Agriculturae Scandinavica* 28, 169–95.

Sinsheimer, R. L. (1977). Recombinant DNA. *Annual Review of Biochemistry* 46, 415–38.

Smith, C. (1964). The use of specialised sire and dam lines in selection for meat production. *Animal Production* 6, 337–44.

Smith, C. (1976). A note on the efficiency of different forms of comparison among foreign breeds. *Animal Production* 23, 413–16.

Smith, C. (1978). The effect of inflation and form of investment on the estimated value of genetic improvement in farm livestock. *Animal Production* 26, 101–10.

Smith, C. and Bampton, P. (1977). Inheritance of reaction to halothane anaesthesia in pigs. *Genetical Research* 29, 287–92.

Smith, C. and Webb, A. J. (1981). Effects of major genes on animal breeding strategies. *Zeitschrift für Tierzüchtung und Züchtungsbiologie* 98, 161–9.

Smith, G. S., Van Camp, S. D., and Basrur, P. K. (1977). A fertile female co-twin to a male calf. *Canadian Veterinary Journal* 18, 287–9.

Smith, H. O. (1979). Nucleotide sequence specificity of restriction endonucleases. *Science* 205, 455–62.

Smith, H. W. (1970). The transfer of antibiotic resistance between strains of enterobacteria in chickens, calves and pigs. *Journal of Medical Microbiology* 3, 165–80.

Smith, H. W. (1975). Antibiotic-resistant bacteria in animals: the dangers to human health. *World's Poultry Science Journal* 31, 104–15.

Snell, G. D. (1981). Studies in histocompatibility. *Science* 213, 172–8.

Snodgrass, D. R., Chandler, D. S., and Makin, T. J. (1981). Inheritance of *Escherichia coli* K88 adhesion in pigs: identification of nonadhesive phenotypes in a commercial herd. *Veterinary Record* 109, 461–3.

Sobey, W. R. (1969). Selection for resistance to myxomatosis in domestic rabbits (*Oryctolagus cuniculus*). *Journal of Hygiene* 67, 743–54.

Sobey, W. R., Conolly, D., Haycock, P., and Edmonds, J. W. (1970). Myxomatosis. The effect of age upon survival of wild and domestic rabbits (*Oryctolagus cuniculus*) with a degree of genetic resistance and unselected domestic rabbits infected with myxoma virus. *Journal of Hygiene* 68, 137–49.

Solomon, K. R. (1983). Acaracide resistance in ticks. *Advances in Veterinary Science and Comparative Medicine* 27, 273–96.

Somes, R. G. (Jr) (1982). Linked loci of the chicken. *Genetic Maps* 2, 310–18.

Soulé, M. E. and Wilcox, B. (Eds) (1980). *Conservation biology*. Sinauer Associates, Sunderland, Massachusetts.

Southern, E. M. (1975). Detection of specific sequences among DNA fragments separated by gel electrophoresis. *Journal of Molecular Biology* 98, 503–17.

Splitter, G. A., Perryman, L. E., Magnuson, N. S., and McGuire, T. C. (1980). Combined immunodeficiency of horses: a review. *Developmental and Comparative Immunology* 4, 21–32.

Spurling, N. W. (1980). Hereditary disorders of haemostasis in dogs: a critical review of the literature. *Veterinary Bulletin* 50, 151–73.

Steinberg, A. G. (1959). Methodology in human genetics. *Journal of Medical Education* 34, 315–34.

Stevenson, A. C. and Davison, B. C. C. (1976). *Genetic counselling* (2nd edn). William Heinemann Medical Books, London.

Stolzman, M., Jasiorowski, H., Reklewski, Z., Zarnecki, A., and Kalinowska, G. (1981). Friesian cattle in Poland: preliminary results of testing different strains. *World Animal Review* (38), 9–15.

Stone, H. A., Briles, W. E., and McGibbon, W. H. (1977). The influence of the major histocompatibility locus on Marek's disease in the chicken. In *Avian immunology* (Ed. Benedict, A. A.) pp. 299–307. Plenum, New York.

Stormont, C. (1975). Neonatal isoerythrolysis in domestic animals: a comparative review. *Advances in Veterinary Science and Comparative Medicine* 19, 23–45.

Stranzinger, G., Dolf, G., Fries, R., and Stocker, H. (1981). Some rare cases of chimerism in twin cattle and their proposed use in determining germinal cell migration. *Journal of Heredity* 72, 360–2.

Streeck, R. E. (1981). Inserted sequences in bovine satellite DNA's. *Science* 213, 443–5.

Sutherland, R. A., Webb, A. J. and King, J. W. B. (1985). A survey of world pig breeds and comparisons. *Animal Breeding Abstracts* **53**, 1–22.

Suzuki, D. T., Griffiths, A. J. F., and Lewontin, R. C. (1981). *An introduction to genetic analysis* (2nd edn). W. H. Freeman, San Francisco.

Suzuki, Y., Miyatake, T., Fletcher, T. F., and Suzuki, K. (1974). Glycosphingolipid β-galactosidases. III. Canine form of globoid cell leukodystrophy: comparison with the human disease. *Journal of Biological Chemistry* **249**, 2109–12.

Swanson, C. P., Merz, T., and Young, W. J. (1981). *Cytogenetics* (2nd edn). Prentice-Hall, Englewood Cliffs, New Jersey.

Symons, R. H. (1981). Structure and replication of DNA. In *The biological manipulation of life* (Ed. Messel, H.) pp. 13–30. Pergamon, Sydney.

Temin, H. M. (1976). The DNA provirus hypothesis. *Science* **192**, 1075–80.

Threlfall, E. J., Ward, L. R., Ashley, A. S., and Rowe, B. (1980). Plasmid-encoded trimethoprim resistance in multiresistant epidemic *Salmonella typhimurium* types 204 and 193 in Britain. *British Medical Journal* **280**, 1210–11.

Tizard, I. R. (1982). *An introduction to veterinary immunology* (2nd edn). Saunders, Philadelphia.

Todd, N. B. (1977). Cats and commerce. *Scientific American* **237**(5), 100–7.

Tomes, G. J., Robertson, D. E., and Lightfoot, R. J. (Eds) (1976). *Sheep breeding*. Western Australian Institute of Technology, Perth.

Tonegawa, S. (1983). Somatic generation of antibody diversity. *Nature* **302**, 575–81.

Trommershausen-Smith, A. (1980). Aspects of genetics and disease in the horse. *Journal of Animal Science* **51**, 1087–95.

Trommershausen-Smith, A., Suzuki, Y., and Stormont, C. (1976). Use of blood typing to confirm principles of coat-colour genetics in horses. *Journal of Heredity* **67**, 6–10.

Trueman, J. W. H. (1978). A programme to reduce the incidence of combined immuno-deficiency. *Theriogenology* **10**, 365–70.

Tudor, D. C. (1979). Congenital defects in poultry. *World's Poultry Science Journal* **35**, 20–6.

Turner, H. G. (1975). The tropical adaptation of beef cattle: an Australian study. *World Animal Review* (13), 16–21.

Turner, H. N. and Young, S. S. Y. (1969). *Quantitative genetics in sheep breeding*. Macmillan, Melbourne.

Utech, K. B. W. and Wharton, R. H. (1982). Breeding for resistance to *Boophilus microplus* in Australian Illawarra Shorthorn and Brahman X Australian Illawarra Shorthorn cattle. *Australian Veterinary Journal* **58**, 41–6.

van Dam, R. H. (1981). Definition and biological significance of the Major Histocompatibility System (MHS) in man and animals. *Veterinary Immunology and Immunopathology* **2**, 517–39.

Van der Velden, N. A. (1979). [Heritable defects in dogs.] *Tijdschrift voor Diergeneeskunde* **114**, 424–30.

Van Vleck, L. D. (1967). Effect of artificial insemination on frequency of undesirable recessive genes. *Journal of Dairy Science* **50**, 201–4.

Van Vleck, L. D. (1977). Genetics of the horse. In *The horse* (by Evans, J. W., Borton, A., Hintz, H. F., and Van Vleck, L. D.) pp. 427–552. W. H. Freeman, San Francisco.

Van Vleck, L. D. (1979). *Notes on the theory and application of selection principles for the genetic improvement of animals*. Department of Animal Science, Cornell University, Ithaca, New York.

Van Vleck, L. D. and Hintz, R. L. (1976). Prediction of genetic value of stallions. *Proceedings of the International Symposium on Genetics and Horse-Breeding, Royal Dublin Society, Dublin*, 19–23.

Vogel, F. and Motulsky, A. G. (1979). *Human genetics: problems and approaches*. Springer-Verlag, Berlin.

Wachtel, S. S. and Koo, G. C. (1981). H-Y antigen in gonadal differentiation. In *Mechanisms of sex differentiation in animals and man* (Eds Austin, C. R. and Edwards, R. G.) pp. 255–99. Academic Press, London.

Wagner, T. E., Hoppe, P. C., Jollick, J. D., Scholl, D. R., Hodinka, R. L., and Gault, J. B. (1981). Microinjection of a rabbit β-globin gene into zygotes and its subsequent expression in adult mice and their offspring. *Proceedings of the National Academy of Sciences* 78, 6376–80.

Wakelin, D. (1978). Genetic control of susceptibility and resistance to parasitic infections. *Advances in Parasitology* 16, 219–308.

Watson, J. D. (1963). Involvement of RNA in the synthesis of proteins. *Science* 140, 17–26.

Watson, J. D. (1976). *Molecular biology of the gene* (3rd edn). Benjamin, New York.

Watson, J. D. and Crick, F. H. C. (1953). Molecular structure of nucleic acids. A structure for deoxyribose nucleic acid. *Nature* 171, 737–8.

Watson, J. D., Tooze, J., and Kurtz, D. T. (1983). *Recombinant DNA: a short course*. Scientific American Books, New York.

Weaver, A. D. (1975). Dwarfism in cattle. *Veterinary Annual* 15, 7–9.

Webb, A. J. (1981). Age effect on expression of halothane gene in pigs. *Annual Report of the Animal Breeding Research Organisation, Edinburgh, Scotland*, 43–4.

Webster, W. P., Mandel, S. R., Strike, L. E., Penick, G. D., Griggs, T. R., and Brinkhous, K. M. (1976). Factor VIII synthesis: hepatic and renal allografts in swine with von Willebrand's disease. *American Journal of Physiology* 230, 1342–8.

Wegner, W. (1976). Defekte und Dispositionen. *Tierärztliche Umschau* 31, 494–502, 556–62.

Wegner, W. (1977). Defekte und Dispositionen. *Tierärztliche Umschau* 32, 30–6, 80–6, 138–45, 210–16, 260–6, 326–31, 386–90, 429–34, 484–90, 546–53, 605–13, 687–90.

Wegner, W. (1978). Defekte und Dispositionen. *Tierärztliche Umschau* 33, 53–9, 112–16, 166–72, 234–8, 275–82, 340–4, 389–93, 441–5, 502–10, 614–22, 693–7.

Wegner, W. (1979). Defekte und Dispositionen. *Tierärztliche Umschau* 34, 55–60, 120–4, 193–201, 262–72.

Weiden, P. L., Storb, R., Graham, T. C., and Schroeder, M. L. (1976). Severe hereditary haemolytic anaemia in dogs treated by marrow transplantation. *British Journal of Haematology* 33, 357–62.

Wenger, D. A., Sattler, M., Kudoh, T., Snyder, S. P., and Kingston, R. S. (1980). Niemann–Pick disease: a genetic model in Siamese cats. *Science* 208, 1471–3.

Wharton, R. H. (1976). Tick-borne livestock diseases and their vectors. 5. Acaricide resistance and alternative methods of tick control. *World Animal Review* (20), 8–15.

White, J. M., Vinson, W. E., and Pearson, R. E. (1981). Dairy cattle improvement and genetics. *Journal of Dairy Science* 64, 1305–17.

White, W. T. and Ibsen, H. L. (1936). Horn inheritance in Galloway–Holstein cattle crosses. *Journal of Genetics* 32, 33–49.

Whitten, M. J. (1979). The use of genetically selected strains for pest replacement or suppression. In *Genetics in relation to insect management* (Eds Hoy, M. A. and McKelvey J. J. (Jr)) pp. 31–40. The Rockefeller Foundation, New York.

WHO (1973–1982) *Bibliography on congenital defects in animals.* Veterinary Public Health Unit, Division of Communicable Diseases, World Health Organisation, Geneva, Switzerland.

Wick, G., Boyd, R., Hála, K., Thunold, S., and Kofler, H. (1982). Pathogenesis of spontaneous autoimmune thyroiditis in Obese strain (OS) chickens. *Clinical and Experimental Immunology* 47, 1–18.

Wickham, G. A. (1978). Development of breeds for carpet wool production in New Zealand. *World Review of Animal Production* 14, 33–40.

Wiesner, E. and Willer, S. (1974). *Veterinärmedizinische Pathogenetik [Veterinary Pathogenetics]*. VEB Gustav Fisher Verlag, Jena, East Germany.

Wijeratne, W. V. S. and Curnow, R. N. (1978). Inheritance of ocular coloboma in Charolais. *Veterinary Record* 102, 513 only.

Wilkins, M. H. F. (1963). Molecular configuration of nucleic acids. *Science* 140, 941–50.

Williams, D. L., Gartman, S. C., and Hourrigan, J. L. (1977). Screwworm eradication in Puerto Rico and the Virgin Islands. *World Animal Review* (21), 31–5.

Willis, M. B. (1976). *The German Shepherd dog: its history, development and genetics*. K. & R. Books, Leicester.

Windon, R. G. and Dineen, J. K. (1981). The effect of selection of both sire and dam on the response of F1 generation lambs to vaccination with irradiated *Trichostrongylus colubriformis* larvae. *International Journal of Parasitology* 11, 11–18.

Winter, H. and Pfeffer, A. (1977). Pathogenic classification of intersex. *Veterinary Record* 100, 307–10.

Wood, R. J. (1981). Insecticide resistance: genes and mechanisms. In *Genetic consequences of man made change* (Eds Bishop, J. A. and Cook, L. M.) pp. 53–96. Academic Press, London.

Wooster, W. E., Fechheimer, N. S., and Jaap, R. G. (1977). Structural rearrangements of chromosomes in the domestic chicken: experimental production by X-irradiation of spermatozoa. *Canadian Journal of Genetics and Cytology* 19, 437–46.

Wright, S. (1922). Coefficients of inbreeding and relationship. *American Naturalist* 63, 274–9.

Wright, S. (1931). Evolution in Mendelian populations. *Genetics* 16, 97–159.

Wurster, D. H. and Benirschke, K. (1968). Chromosome studies in the superfamily Bovoidea. *Chromosoma* 25, 152–71.

Young, F. E. and Mayer, L. (1979). Genetic determinants of microbial resistance to antibiotics. *Reviews of Infectious Diseases* 1, 55–62. University of Chicago Press, Chicago.

Young, G. B. (1967). Hereditary diseases in livestock. *Veterinary Record* 81, 606–17.

Zimmerman, T. S., Ruggeri, Z. M., and Fulcher, C. A. (1983). Factor VIII/von Willebrand factor. *Progress in Hematology* 13, 279–309.

Author index

This index includes all authors of all papers cited in the text. If a name does not appear on the page indicated, then that person is a co-author of a paper cited on that page.

Subject index